WOODHEAD PUBLISHING IN MATERIALS

影印版

镁合金的腐蚀防护

Corrosion prevention of magnesium alloys

Edited by Guang-Ling Song

哈爾濱工業大學出版社
HARBIN INSTITUTE OF TECHNOLOGY PRESS

黑版贸审字08-2017-080号

Corrosion prevention of magnesium alloys
Guang-Ling Song
ISBN: 978-0-85709-437-7

Elsevier (Singapore) Pte Ltd.
3 Killiney Road
#08-01 Winsland House I
Singapore 239519
Tel: (65) 6349-0200
Fax: (65) 6733-1817

First Published 2017

图书在版编目（CIP）数据

镁合金的腐蚀防护=Corrosion prevention of magnesium alloys：英文 / 宋光铃(Guang-Ling Song)主编.—影印本.—哈尔滨：哈尔滨工业大学出版社，2017.10

ISBN 978-7-5603-6362-2

Ⅰ.①镁… Ⅱ.①宋… Ⅲ.①镁合金-腐蚀-英文 ②镁合金-防腐-英文 Ⅳ.①TG146.2

中国版本图书馆CIP数据核字（2016）第310906号

责任编辑 杨　桦　张秀华　许雅莹
出版发行 哈尔滨工业大学出版社
社　　址 哈尔滨市南岗区复华四道街10号 邮编 150006
传　　真 0451-86414749
网　　址 http://hitpress.hit.edu.cn
印　　刷 哈尔滨市石桥印务有限公司
开　　本 660mm × 980mm 1/16 印张 36.5 插页 1
版　　次 2017 年 10 月第 1 版 2017 年 10 月第 1 次印刷
书　　号 ISBN 978-7-5603-6362-2
定　　价 360.00元

（如因印刷质量问题影响阅读，我社负责调换）

影印版说明

本书综述了镁合金表面处理的方法，介绍了提高镁合金耐蚀性的多种表面技术，包括激光表面处理技术、化学转移膜技术、阳极氧化技术、冷喷涂技术、溶胶–凝胶技术和电镀技术等。

第一部分（1~3 章）介绍了腐蚀和耐腐蚀的原理，探讨了缓蚀和防腐蚀的措施。

第二部分（4~10 章）介绍了表面的处理和转化，从概述表面清洁和预处理开始，进而讨论了表面处理、合金化、激光处理、化学转化等对镁合金耐蚀性能的作用。

第三部分（11~17 章）讲述了不同的电镀技术，介绍了冷喷涂涂层、溶胶–凝胶涂层和化学电泳涂层等提高镁合金耐蚀性的原理和技术。

第四部分（18、19 章）通过精选的案例介绍了防腐技术在汽车和医药行业的应用。

本书主要适合从事镁合金表面处理和应用的科研人员、技术人员使用，也可供高等院校相关专业的师生参考。

Guang-Ling Song　美国 ORNL（Oak Ridge National Laboratory）科学家，澳大利亚昆士兰大学兼职教授，国际腐蚀与防护领域知名专家，曾在通用汽车设计中心工作 5 年。

材料科学与工程图书工作室

联系电话 0451–86412421

0451–86414559

邮　　箱 yh_bj@aliyun.com

xuyaying81823@gmail.com

zhxh6414559@aliyun.com

Corrosion prevention of magnesium alloys

Edited by
Guang-Ling Song

Oxford Cambridge Philadelphia New Delhi

Contents

Contributor contact details

(* = main contact)

Editor and chapters 1, 3, 5, 9, 13 and 16

Guang-Ling Song
GM Global Research and Development
General Motors Corporation
30500 Mound Road
Warren
MI48090
USA

Guang-Ling Song has moved from GM to ORNL National Lab.
Current correspondence:

Materials Science and Technology Division
Oak Ridge National Lab
One Bethel Valley Road
P.O. Box 2008, MS-6156
Oak Ridge,
TN 37831-6156,
USA

E-mail: guangling.song@gmail.com; songg@ornl.gov

Chapter 2

Jian Meng*, Wei Sun, Zheng Tian, Xin Qiu and Deping Zhang
State Key laboratory of Rare Earth Resource Utilization
Changchun Institute of Applied Chemistry
Chinese Academy of Sciences
Changchun 130022
China

E-mail: jmeng@ciac.jl.cn

Chapter 3

Xingpeng Guo
School of Chemistry and Chemical Engineering
Huazhong University of Science and Technology
1037 Luoyu Road
Wuhan 430074
China

E-mail: guoxp@mail.hust.edu.cn

Chapter 4

Daniel Höche
Department of Corrosion and Surface Technology
Institute of Materials Research
Helmholtz-Zentrum Geesthacht
Zentrum für Material- und Küstenforschung GmbH
Max-Planck-Straße 1

21502 Geesthacht
Germany

E-mail: daniel.hoeche@hzg.de

Chapter 6

Jyotsna Dutta Majumdar and Indranil Manna
Department of metallurgical and materials engineering
Indian Institute of Technology Kharagpur
Kharagpur
W. B. – 721302
India

E-mail: jyotsna@metal.iitkgp.ernet.in

Chapter 7

Professor Bai-ling Jiang* and Yan-feng GE
School of Material Science and Engineering
Xi'an University of Technology
No. 5 South Jinhua Road
Xi'an 710048
China

E-mail: jiangbail@vip.163.com; geyanf@gmail.com

Chapter 8

S. A. Salman* and M. Okido
Department of Materials Science and Engineering
School of Engineering
Nagoya University
Furo-cho
Chikusa-ku Nagoya 464-8603
Japan

E-mail: sa.salman@yahoo.com

and

S. A. Salman
Department of Mining, Metallurgy and Petroleum Engineering
School of Engineering
Al-Azhar University
Nasr City
Cairo 11371
Egypt

Chapters 10 and 11

Xiaobo Chen
Monash University
Wellington Road
Clayton
VIC 3800
Australia

E-mail: xiaobo.chen@monash.edu

Chapter 12

Kazuhisa Azumi
Graduate School of Engineering
Hokkaido University
N13W8, Kitaku
Sapporo 060-8628
Japan

E-mail: azumi@eng.hokudai.ac.jp

Chapter 13

Professor Zhoucheng Wang
College of Chemistry & Chemical Engineering
Xiamen University
422 Siming South Road
Siming
Xiamen 361005
Fujian
China

E-mail: zcwang@xmu.edu.cn

Chapter 14

Wen-Ta Tsai
Department of Materials Science and Engineering
National Cheng Kung University
1, Ta-Hsueh Road
Tainan 70101
Taiwan

E-mail: wttsai@mail.ncku.edu.tw

I-Wen Sun
Department of Chemistry
National Cheng Kung University
Taiwan

E-mail: iwsun@mail.ncku.edu.tw

Chapter 15

Victor K. Champagne and Brian Gabriel
US Army Research Laboratory
Aberdeen Proving Ground
MD
USA

E-mail: victor.k.champagne.civ@mail.mil

Julio Villafuerte*
CenterLine (Windsor) Ltd
Windsor
ON
Canada

E-mail: julio.villafuerte@cntrline.com

Chapter 17

Qing Li
School of Chemistry and Chemical Engineering
Southwest University
No.2 Tiansheng Road
BeiBei District
Chongqing 400715
China

E-mail: liqingswu@yeah.net

Chapter 18

Gerald S. Cole
Light Weight Strategies LLC
25800 Romany Way
Franklin
MI 48025
USA

E-mail: geraldscole@att.net

Chapter 19

Ke Yang* and Lili Tan
Institute of Metal Research
Chinese Academy of Sciences
72 Wenhua Road
Shenyang
Liaoning 110016
China

E-mail: kyang@imr.ac.cn; lltan@imr.ac.cn

Preface

Magnesium (Mg) alloys, as the lightest engineering structural materials, have found more and more potential applications in the automotive, aerospace, electronic and defense industries. Because of its unique physical and chemical properties, Mg is also an attractive functional material in the energy, medical and corrosion prevention fields. However, due to its highly active nature, the practical use of Mg and its alloys in these areas critically depends upon the successful control of their corrosion in actual service environments. For example, corrosion has been termed as a 'show stopper' in the auto industry, because corrosion problems severely limit many exciting applications of Mg alloys in vehicles. These problems make mitigation of corrosion damage of Mg alloys presently an extremely important topic.

Generally speaking, fundamental corrosion research is primarily aimed at a comprehensive scientific understanding of corrosion behavior or phenomena. This fundamental research provides the foundation for developing solutions to current and future corrosion problems. Prevention, protection or corrosion resistance enhancement is ultimately an important goal of corrosion studies. Today, Mg alloys are particularly needed in industries due to the ever-growing pressures of energy and environment conservation. It is more urgent than ever to develop techniques that can effectively prevent Mg alloy components from corrosion attack. In the past decade, the development of prevention/protection methods for Mg alloys has dramatically intensified based on the continuously broadened and deepened fundamental research into Mg alloy corrosion. It is essential now to summarize systematically and consolidate the newly developed protection and prevention knowledge and expertise which may be utilized to address practical corrosion issues with Mg alloy components in industry.

This book is a continuation of the previously published volume 'Corrosion of magnesium alloys' (Song ed., WP, 2011). Following the fundamental understanding regarding corrosion mechanisms and behavior of Mg and its alloys summarized in that volume, this book focuses on recently developed corrosion prevention technologies for Mg alloys. It provides a compilation of comprehensive overviews and some technical details of important developments in corrosion prevention theories and practices. Four linked parts compose the main body of this book: Part I-Alloying and inhibition, Part II-Surface treatment and conversion, Part III-Coatings and Part IV-Case studies.

Part I begins with a review of corrosion behavior of Mg and its alloys, which is a natural continuation of the previous book 'Corrosion of magnesium alloys'. Based on the understanding regarding the dependence of corrosion performance on metallurgical and environmental factors, corrosion issues and protection strategies associated with corresponding corrosion characteristics are discussed. Following corrosion resistance enhancement by rare earth element alloying, inhibition techniques are proposed for Mg alloys. Alloying and inhibition are two basic methodologies tackling corrosion problems from the material and environment points of view. They originate from the nature of corrosion – a reaction between metal and its environment.

In Part II, this book delves into the core of corrosion prevention engineering. Corrosion is a surface process and thus the most effective solution to corrosion problems must be a surface technique. This part of the book reviews how Mg alloy surfaces are modified or changed in state, composition and microstructure by means of cleaning, machining, alloying, beam (electron, ion or laser) treatment, anodizing and conversion. After processing, a new surface state or layer is generated. The original substrate material actively participates in the physical and chemical reactions during the new surface formation and becomes an important and dominating component in the new surface layer. The treated or converted surface is much more stable and corrosion resistant than the original surface.

The more popular and effective surface engineering processes are presented in Part III. They include metallic and organic coatings on Mg alloys, such as normal electro- and electroless plating, innovative ionic liquid plating, recently developed cold spraying, traditional organic painting, new electroless E-coating and novel sol-gel layer formation. Unlike the processes demonstrated in Part II, the coatings in Part III are externally applied, and their interaction with the substrate Mg alloys is confined to the coating/substrate interface only. The coating is a completely different material separated by a distinct boundary from the substrate Mg alloy. Such a coating can be tailored to have some special physical and chemical properties, particularly a much enhanced corrosion resistance.

Part IV is the concluding section of this book giving two examples to illustrate corrosion control designs and corrosion resistance enhancement skills for Mg alloys in the automotive and medical industries. They convey the idea that in dealing with practical corrosion problems, there is no universal or common method, but a practical solution tailored to particular circumstances.

With only 19 chapters it is impossible to cover all the impressive achievements in corrosion prevention of Mg alloys. However, I have tried to include representative results in all the important aspects of the subject. It is my intention to make the contents of this book useful to the researchers and engineers who are interested in Mg alloys. It would be my greatest honor if this book could serve as an informative resource for professors, scientists, students, engineers, designer, manufacturers and suppliers.

This book together with its sister volume 'Corrosion of magnesium alloys' is an important symbolic milestone in my research. These books also reflect my many years of research in the automotive industry. It is therefore my hope that automotive companies keen to use Mg alloys in their vehicles will benefit from the knowledge documented in this book.

I am confident that the industrial use of Mg alloys will continue to increase. Although a few companies stepped back when they encountered some stubborn corrosion problems, the strong faith that success is just one step away is inspiring more scientists and engineers to pursue corrosion resistant Mg alloys and components. I am greatly excited by the fact that more and more researchers are now committed to the development of Mg corrosion protection techniques. I believe that further significant breakthroughs will occur soon and more practical applications will be realized. I look forward to some techniques documented in this book being used on the products of those far-sighted companies that have supported research in this important subject.

Having stated the possible impact of this book, we should not forget the chapter authors' contributions in this book. I would like to take this opportunity to thank all the authors for sharing their interesting and impressive work with us. Sincere appreciation also goes to Francis Dodds and Emily Cole of Woodhead Publishing for their assistance in coordinating the author team and editing the book. Their hard work has significantly sped up this book's publication.

Lastly, I would also like to take the liberty to express my gratitude to my wife for her support in book edition. Her continuous encouragement helped me to overcome frustrations, brave difficulties and rise to more challenges in my research career. Without the accompaniment of her support, this book may not have been possible. Dedicating this book to my beloved wife has been an unchanged incentive for me to continue working diligently.

Guang-Ling Song

Part I

Alloying and inhibition

1 Corrosion behavior and prevention strategies for magnesium (Mg) alloys

G.-L. SONG,
General Motors Corporation, USA

DOI: 10.1533/9780857098962.1.3

Abstract: Magnesium (Mg) alloys have low corrosion resistance and exhibit unusual corrosion behavior in aqueous environments. Because of this unique corrosion performance, some special corrosion prevention techniques have to be employed for Mg alloys in their applications. This chapter briefly summarizes the corrosion characteristics of Mg alloys, and also presents a strategy and methodologies to mitigate the corrosion damage of Mg alloys in applications.

Key words: magnesium, corrosion, protection.

Note: This chapter is a revised and expanded version of Chapter 1 'Corrosion behaviour of magnesium alloys and protection techniques' by G.-L. Song, originally published in *Surface engineering of light alloys: aluminium, magnesium and titanium alloys*, ed. Hanshan Dong, Woodhead Publishing Limited, 2010, ISBN: 978-1-84569-537-8.

1.1 Introduction

Magnesium and its alloys have a high strength/density ratio and have found many successful applications, particularly in the automotive and aerospace industries (Makar and Kruger, 1993; Polmear, 1996; Aghion and Bronfin, 2000; Bettles *et al.*, 2003; Song, 2005b, 2006). However, the poor corrosion resistance of existing magnesium alloys in some service environments has limited the further expansion of their application. Existing investigations have clearly suggested that the corrosion behavior of Mg alloys is very different from that of a convention metal, such as steel (Song, 2004a, 2005b, 2006, 2009a, 2009c; Winzer *et al.*, 2005, 2007, 2008; Wan *et al.*, 2006; Song and Atrens, 2007; Wang *et al.*, 2007). For more successful applications of Mg alloys, it is important to understand their characteristic corrosion phenomena and have a practical corrosion prevention strategy.

1.2 Corrosion characteristics and implications in protection

The corrosion characteristics of Mg alloys critically concern their applications. Only if the corrosion behavior is comprehensively understood, can a Mg alloy be correctly used in suitable applications, and also the most effective protection be selected for this alloy.

1.2.1 Electrochemical corrosion mechanism

Mg alloys follow a corrosion mechanism different from other engineering metallic materials. On pure Mg or a Mg alloy, the overall corrosion reaction can be written as (Makar and Kruger, 1993; Song and Atrens, 1999; Song, 2005b, 2006):

$$Mg + 2H^+ \rightarrow Mg^{2+} + H_2 \quad \text{(in an acidic solution) [1.1]}$$

or

$$Mg + 2H_2O \rightarrow Mg^{2+} + 2OH^- + H_2 \quad \text{(in a neutral or basic solution) [1.2]}$$

This overall corrosion can be decomposed into anodic and cathodic reactions. The cathodic process is (Song, 2005b, 2006):

$$2H^+ + 2e \rightarrow H_2 \quad \text{(in an acidic solution) [1.3]}$$

or

$$2H_2O + 2e \rightarrow H_2 + 2OH^- \quad \text{(in a neutral or alkaline solution) [1.4]}$$

and the anodic process (Song, 2005b, 2006):

$$Mg + \left[\frac{1}{(1+y)}\right]H^+ \rightarrow Mg^{2+} + \left[\frac{1}{(2+2y)}\right]H_2 + \left[\frac{(1+2y)}{(1+y)}\right]e \quad \text{(in an acidic solution) [1.5]}$$

or

$$Mg + \left[\frac{1}{(1+y)}\right]H_2O \rightarrow Mg^{2+} + \left[\frac{1}{(1+y)}\right]OH^- + \left[\frac{1}{(2+2y)}\right]H_2 + \left[\frac{(1+2y)}{(1+y)}\right]e \quad \text{(in a neutral or basic solution) [1.6]}$$

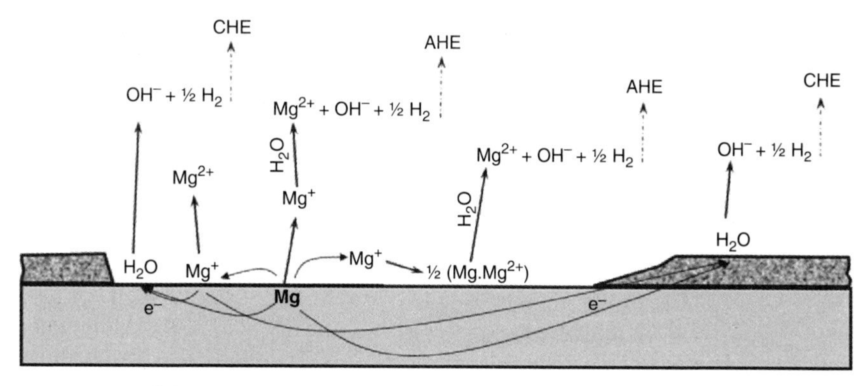

1.1 Schematic illustration of anodic and cathodic reactions involved in the self-corrosion of Mg.

y is the ratio of further anodic reaction over hydrogen production reactions. The detailed anodic and cathodic reactions under a steady corrosion condition have been illustrated previously (Song *et al.*, 1997a, 1997b; Song and Atrens, 1999, 2003; Song, 2005b). In general the anodic dissolution occurs mainly in a film-free area, while in a film-covered area the anodic dissolution is negligible. The cathodic reaction, which is mainly a hydrogen evolution process, can take place in both film-free and film-covered areas, and the reaction rate in a film-free area is much faster than the rate on a film-covered surface, particularly if impurity particles are present there.

Both the anodic and cathodic reactions occur simultaneously on Mg. In fact, more detailed reaction steps can be involved in the anodic and cathodic processes, which have been demonstrated by Song (2011) and are schematically illustrated in Fig. 1.1.

Based on this corrosion model, the subsequent corrosion phenomena and behavior can be predicted. Correspondingly, what these corrosion characteristics imply can also be understood.

1.2.2 Hydrogen evolution

The overall corrosion reactions [1.1] and [1.2] suggest that the dissolution of Mg is always accompanied by hydrogen evolution. This corrosion-related hydrogen evolution is also applicable to Mg alloys in aqueous solutions (Song *et al.*, 1998, 2001, 2005a; Song and Atrens, 1998, 2003; Song and St John, 2002), including engine coolants (Song and St John, 2004, 2005) and simulated body fluids (SBF) (Song and Song, 2006, 2007; Song, 2007b). In a severe corrosion process, Mg particle undermining may take place (Mg particle falling into solution due to the surrounding material becoming completely corroded). However, this process has been shown to have no influence upon either reaction [1.1] or [1.2] (Song *et al.*, 1997b). Therefore, for Mg and Mg alloys in an aqueous solution, the hydrogen evolution phenomenon is one of the most important features and includes cathodic hydrogen evolution (CHE) and anodic hydrogen evolution (AHE).

Having realized that hydrogen evolution is closely associated with the corrosion of Mg and its alloys, a simple hydrogen evolution measurement technique was first employed by Song *et al.* (1997b) to estimate the corrosion rate of Mg. Following further illustration of the theory and analysis of possible errors inherent to this method (Song *et al.*, 2001; Song, 2005b, 2006), it has also been widely used on many Mg alloys (Krishnamurthy *et al.*, 1988; Hallopeau *et al.*, 1999; Bonora *et al.*, 2000; Eliezer *et al.*, 2000; Mathieu *et al.*, 2000; Song and St John, 2002; Song, 2005a; Pu *et al.*, 2012; Song and Xu, 2012).

According to reaction [1.1] or [1.2], dissolution of one Mg atom always corresponds to the generation of one hydrogen gas molecule. In other words, if there is one mole of hydrogen evolved, then there must be one mole of magnesium dissolved (Song *et al.*, 2001; Song, 2005b, 2006). Measuring the volume of hydrogen evolved is equivalent to measuring the weight-loss of a corroding Mg alloy, and the measured hydrogen evolution rate is equal to the weight-loss rate if they are both converted into the same unit (e.g. mole per minute).

The experimental set-up for measuring hydrogen evolution is straightforward (Song *et al.*, 2001) and can simply consist of a burette, a funnel and a beaker (see Fig. 1.2a). This set-up can also be combined into an electrolytic cell for electrochemical measurements (see Fig. 1.2b) (Song and St John, 2002). The overall theoretical error of this technique is less than 10% (Song *et al.*, 2001). In practice, the corrosion rates of various Mg alloy specimens measured by the hydrogen evolution method have been found to be in very good agreement with those measured by weight-loss (Song *et al.*, 2001). Furthermore, the hydrogen evolution measurement has several advantages over the traditional weight-loss measurement (Song, 2005b, 2006):

1. smaller theoretical and experimental errors;
2. easy to set-up and operate;
3. suitable for corrosion monitoring of magnesium and its alloys; and
4. no need to remove corrosion products.

While hydrogen evolution is an important corrosion characteristic and can even be employed to reliably monitor the corrosion of Mg and its alloys, its detrimental effect on the corrosion prevention/protection performance should not be overlooked. First, the hydrogen evolution from a corroding area will disturb the solution, which will to a great degree eliminate concentration difference and diffusion effect in the adjacent solution. Without a limited diffusion current density controlling the process, the corrosion rate can continuously increase to a very high level. Second, owing to the

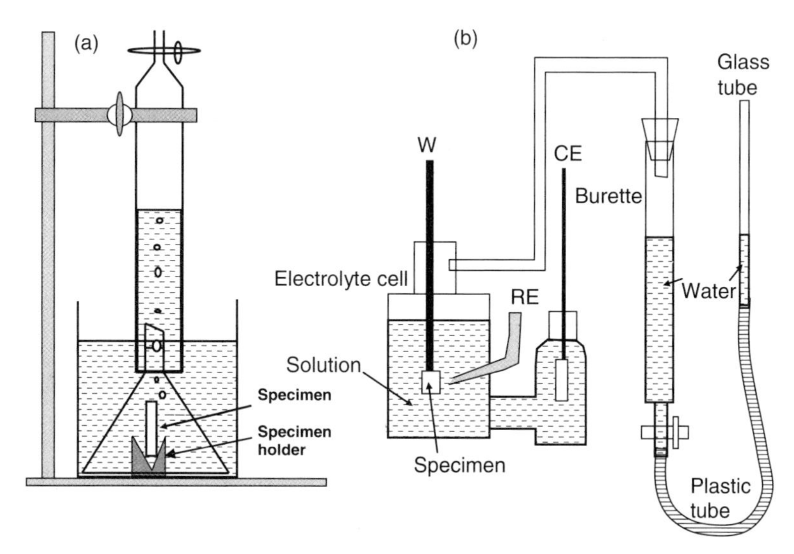

1.2 Schematic illustration of the set-up used to measure the volume of hydrogen evolved. (a) Simple set-up for short-term corrosion measurements (Song *et al.*, 1997b). (b) Combination of hydrogen evolution and electrochemical measurements (according to Song and St John, 2002).

hydrogen evolution, corrosion products are unlikely to precipitate in a corroded area and the corrosion product film which forms is very loose and porous. Since the naturally formed surface film on Mg is not protective, the self-limitation effect caused by the products during corrosion is weak. Third, the hydrogen evolution resulting from corrosion can cause damage to an externally applied coating/plating and explains why blistering phenomena are especially common for a Mg alloy coated in an organic coating.

1.2.3 Alkalization

A relationship between the dissolution of Mg and the generation of OH^- or consumption of H^+ during the corrosion of Mg alloys is also given by either reaction [1.1] or [1.2]. Similar to hydrogen evolution in an acidic solution, the production of OH^- or consumption of H^+ always accompanies the corrosion of Mg and its alloys in a neutral or alkaline environment, which can result in an increased pH value for the solution. However, the pH increase levels out at ~10.5 even though the corrosion will continue at this pH value. This is a result of the deposition of $Mg(OH)_2$ (Song, 2005b, 2006): the additional hydroxyls generated at this pH level are consumed by dissolved Mg^{2+} through deposition of $Mg(OH)_2$, which stabilizes the pH value of the solution at approximately ~10.5.

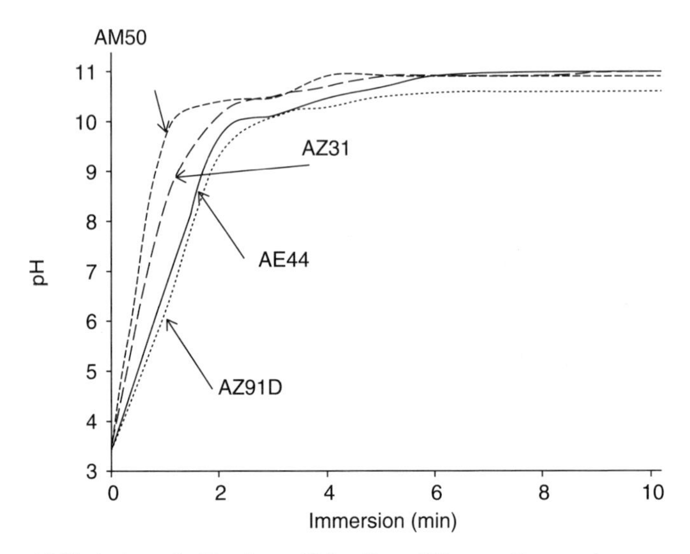

1.3 Variation of pH value of Mg alloys (10 cm × 7 cm × 1 mm coupons) immersed in a diluted HCl solution (100 mL; initial pH = 3.5) with immersion time (Song and Liu, 2011).

The surface of a corroding Mg alloy always experiences an alkalization process because either hydroxyls are generated or protons are consumed directly in the corroding area. It has been estimated that the local pH value of the solution adjacent to a Mg surface can be around 10.5 even if the bulk solution is acidic (Nazarrov and Mikhailovskii, 1990; Song, 2005b, 2006). The alkalization effect varies slightly with Mg alloy. Figure 1.3 shows the variation of solution pH value with time after different Mg alloys are immersed in the solution (Song and Liu, 2011).

The alkalization effect can be affected by the ratio of the surface area of a Mg alloy to the volume of the solution it is exposed to. A more significant alkalization effect is likely with a larger Mg alloy specimen in a smaller volume of solution. For example, under an atmospheric corrosion condition, only a thin aqueous film stays on the surface of a Mg alloy. This is a case of a large specimen with a small volume of solution which can easily result in a strong alkalization effect. Hence, the corrosion of a Mg alloy in an atmospheric environment is normally less severe than under immersion conditions (Song *et al.*, 2004b, 2006b; Wan *et al.*, 2006).

The alkalization effect has also been utilized to monitor the corrosion rate of a Mg alloy. For example, Hur *et al.* (1996) and Weiss *et al.* (1997) tried to measure the change in pH value or concentration of H^+ of the solution in order to monitor the corrosion process of a Mg alloy. The pH method may be convenient in monitoring the corrosion of Mg alloys in a neutral solution in the initial stage. However, it is not as reliable as the hydrogen evolution technique. Firstly, the alkalization is dependent on the volume of the testing solution which can vary in different experiments; it is difficult to compare corrosion rates measured at different ratios of surface/volume. Secondly, in the later stages of corrosion, as considerable Mg has been dissolved and the solution has become $Mg(OH)_2$ saturated, the pH value of the testing solution will be stable at 10.5 and will not increase further, even though the corrosion rate may still be high. Thirdly, the dissolution of atmospheric carbon dioxide could influence the pH value of the testing solution and result in a faulty corrosion indication. Lastly, in a strongly acidic or basic solution the variation of the pH value is not sufficiently sensitive to reflect the change of proton or hydroxyl concentration.

As some organic polymers dislike alkalinity and degrade quickly in an alkaline environment, the Mg surface alkalization effect hinders the application of some organic coatings on Mg alloy surfaces. However, for those organics that are by nature resistant to alkalinity, the surface alkalization effect can be utilized to enhance the coating formation and performance. Electroless E-coating is an example of such an organic coating resulting from the Mg surface alkalization effect (Song, 2009c, 2009d, 2009e, 2010a, 2011a; Song and Liu, 2011).

1.2.4 Macro-galvanic corrosion

Mg has the highest chemical activity of the engineering metals. Because the standard equilibrium potential of Mg/Mg^{2+} is as negative as –2.4 V~NHE (Perrault, 1978; Song, 2005b), Mg and its alloys normally have a very negative open-circuit (or corrosion) potential (around –1.5 V~NHE) in a neutral aqueous environment. This means that Mg alloys are always the anode when in contact with other engineering metals.

Macro-galvanic corrosion is an anodic dissolution process of an anode accelerated by the cathodic reactions on a cathode that is in electrical contact with the anode. In theory, the galvanic corrosion rate (i_g) is determined by (Song *et al.*, 2004b):

$$i_g = \frac{(E_c - E_a)}{(R_a + R_c + R_s + R_m)} \qquad [1.7]$$

where E_c and E_a are the open-circuit (corrosion) potentials of cathode and anode respectively in a given electrolyte; R_c and R_a are the cathode and anode polarization resistances; and R_s and R_m are the solution and electronic resistances between the anode and cathode, respectively.

The corrosion potential difference (E_c–E_a) between the anode and cathode is the first critical variable determining the galvanic corrosion rate i_g. The corrosion potential of Mg in a NaCl solution is the least noble among those engineering metals and in fact, is over 600 mV more negative than zinc which is second in the galvanic series. Based on corrosion potentials, engineering metals exposed to seawater can be ranked in a galvanic series from active (negative) to passive (noble) (Hack, 1995):

> Magnesium→ magnesium alloys→ zinc→ galvanized steel→ aluminum 1100→ aluminum 6053→ Al clad→ cadmium→ aluminum 2024→ mild steel→ wrought iron→ cast iron→13% Cr stainless steel, type 410 (active)→ 18-8 stainless steel, type 304 (active)→ 18-12-3 stainless steel, type 316 (active)→ lead-tin solders→ lead→ tin→ muntz metal→ manganese bronze→ naval brass→ nickel (active)→ 76Ni-6Cr-7Fe alloy (active)→ 60Ni-30Mo-6Fe-1Mn→yellow brass→ admiralty brass→ aluminum brass→ red brass→ copper→ silicon bronze→70:30 cupro nickel→ G-Bronze→ Silver solder→ Nickel (passive)→ 76Ni-16Cr-7Fe alloy (passive)→ 13% Cr stainless steel, type 410 (passive)→ titanium→18-8 stainless steel, type 304 (passive)→ 18-12-3 stainless steel, type 316 (passive)→ silver→ graphite→ gold→ platinum.

Mg and its alloys always act as anodes if in contact with other engineering metals. Because of the large difference between their corrosion potentials (E_c–E_a), the galvanic corrosion tendency is always significant between a Mg alloy and another engineering metal. Equation [1.7] also suggests that the galvanic corrosion is determined by the anodic and cathodic polarization resistances (R_a and R_c). Mg and its alloys normally have a low anodic polarization resistance (Song *et al.*, 1997a, 1997b, 2004a; Song, 2004b, 2009b) while the cathodic polarization resistance of other metals in a neutral solution resulting from diffusion controlled oxygen reduction is relatively large. For a specific Mg alloy in a given solution whose corrosion potential and anodic resistance are fixed, the corrosion potential and the cathodic polarization resistance of a cathode metal will have a decisive influence on the galvanic corrosion of the Mg alloy (Song *et al.*, 2004b). A metal with a noble potential and relatively low cathodic polarization resistance will severely accelerate the corrosion of a Mg alloy. It is well known that Fe, Co, Ni, Cu, W, Ag and Au metals are much nobler than Mg alloys in an aqueous solution (Table 1.1) and have relatively low hydrogen evolution over-potentials (lower than 500 mV, i.e. low cathodic polarization resistance). Thus, these metals in contact with a Mg alloy can form galvanic corrosion couples and result in serious galvanic corrosion damage to Mg alloys. This type of galvanic couple should be avoided. Unfortunately, aluminum, steel and galvanized steel are widely used engineering materials and are quite often used together with Mg alloys. In this case, steel is the most detrimental and Al the least harmful to Mg alloys (Avedesian and Baker, 1999; Song *et al.*, 2004b).

In Equation [1.7], R_m can also significantly affect the galvanic corrosion current. If it is large enough, for example, where the anode and cathode are separated by an insulator, then the galvanic corrosion will stop. Unfortunately, the anode and cathode are normally electrically connected

Table 1.1 Standard equilibrium potentials of typical metals in aqueous solution

Reactions	Standard equilibrium potential (V~NHE)
$Au^+ + e = Au$	1.68
$Pt^{2+} + 2e = Pt$	1.2
$Pd^{2+} + 2e = Pd$	0.83
$Ag^+ + e = Ag$	0.7996
$PtCl_4^{2-} + 2e = Pt + 4Cl^-$	0.73
$Cu^+ + e = Cu$	0.522
$Ag_2O + H_2O + 2e = 2\,Ag + 2OH^-$	0.342
$Cu^{2+} + 2e = Cu$	0.3402
$Pb^{2+} + 2e = Pb$	−0.1263
$Sn^{2+} + 2e = Sn$	−0.1364
$Ni^{2+} + 2e = Ni$	−0.23
$Co^{2+} + 2e = Co$	−0.28
$PbSO_4 + 2e = Pb + SO_4^{2-}$	−0.356
$Cd^{2+} + 2e = Cd$	−0.4026
$Fe^{2+} + 2e = Fe$	−0.409
$Cr^{2+} + 2e = Cr$	−0.557
$Ni(OH)_2 + 2e = Ni + 2OH^-$	−0.66
$Zn^{2+} + 2e = Zn$	−0.7628
$Mn^{2+} + 2e = Mn$	−1.029
$ZnO_2^{2-} + 2H_2O + 2e = Zn + 4OH^-$	−1.216
$Al^{3+} + 3e = Al$ (0.1 M NaOH)	−1.706
$Mg + 2e = Mg^{2+}$	−2.4

Source: Bard and Faulkner, 1980.

and thus $R_m = 0$ in practice. Therefore, R_m is not seriously considered in many galvanic corrosion studies.

The parameters involved in Equation [1.7] can determine the overall galvanic corrosion current, i_g, which represents the overall galvanic corrosion of an anode (Song, 2010). As long as all the parameters of Equation [1.7] are measured, the overall galvanic corrosion damage can be estimated. Many applied studies have tried to estimate the compatibility of various materials (including fasteners) with Mg alloy parts using this equation (Hawke, 1987; Skar, 1999; Gao *et al.*, 2000; Senf *et al.*, 2000; Starostin *et al.*, 2000; Boese, *et al.*, 2001).

However, in many practical applications, the distribution of galvanic current density, i_g, over a Mg alloy component surface is of greater concern as it cannot simply be predicted by Equation [1.7]. The distribution of galvanic current density is closely associated with parameter R_s which is dependent on the geometric configuration of the solution path for the galvanic current between the anode and cathode. For a simple one-dimensional galvanic couple consisting of a Mg alloy and another metal, an analytical prediction of the galvanic current density distribution is possible (Waber and Rosenbluth, 1955; Kennard and Waber, 1970;

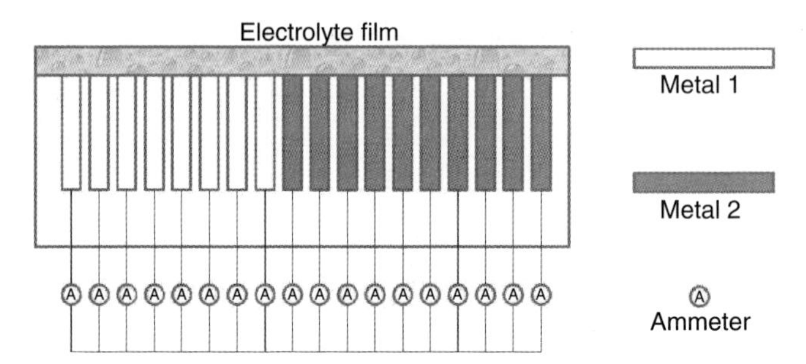

1.4 Configuration of a 'sandwich' like probe used to measure the galvanic current (Song *et al.*, 2006b).

Gal-Or *et al.*, 1973; McCafferty, 1976, 1977; Melville, 1979, 1980). It has been reported that the galvanic current density has an exponential distribution (Song *et al.*, 2004b; Song, 2010). This can be confirmed experimentally by direct measurement of the distribution of galvanic current density of a Mg alloy in contact with another metal under a standard salt spray condition (Song *et al.*, 2004b).

A specially designed 'sandwich'-like galvanic probe (see Fig. 1.4) (Song *et al.*, 2004b; Song, 2010) can be used for these measurements. The measured galvanic current density exponentially decreases with the thickness of an insulating spacer which implies that the galvanic corrosion can still be significant even when the thickness of the insulating spacer between a Mg alloy and steel is as large as 9 cm under a salt spray condition (Song *et al.*, 2004b; Song, 2005b, 2006). It seems that insertion of an insulating spacer may reduce but cannot eliminate galvanic corrosion. The reason is that the insulating spacer does not completely block the ionic current path.

When the geometry of a galvanic couple becomes complicated, experimental measurement of the galvanic corrosion rate using such a 'sandwich' galvanic probe will be difficult. In this case, computer modeling is an option (Klingert *et al.*, 1964; Doig and Flewitt, 1979; Sautebin *et al.*, 1980; Helle *et al.*, 1981; Fu, 1982; Munn, 1982; Kasper and April, 1983; Munn and Devereux, 1991a, 1991b; Miyaska *et al.*, 1990; Aoki and Kishimoto, 1991; Hack, 1997). There are already a few successful studies of the galvanic corrosion of Mg alloys by computer modeling (Jia *et al.*, 2004, 2005a, 2005b, 2006, 2007). Nevertheless, experimental measurement of the polarization curves of the coupling metals is still important in numerical analysis and computer modeling. Polarization curves are required as boundary conditions in computer modeling. Due to the negative difference effect, alkalization effect, 'poisoning' effect, 'short-circuit' effect, etc. (Song *et al.*, 2004b; Song, 2005b, 2006), current computer modeling techniques cannot at this time comprehensively simulate a practical macro-galvanic process and quite often there are significant deviations between computer-modeled data and measured galvanic corrosion results.

The most important conclusion regarding macro-galvanic corrosion is that galvanic effect is responsible for most corrosion problems with practical Mg components, as the contact with other metals cannot be avoided in industry. An isolation spacer in theory cannot completely eliminate the galvanic effect, but can significantly reduce the corrosion intensity. The most cost-effective way of mitigating galvanic corrosion is to disconnect the electronic circuit between the anode and cathode in design, but this is impractical. In industry, an insulating coating layer separating the anode or cathode or both from contacting the electrolyte may be a practical approach. An insulating spacer between the anode and cathode is also usually adopted under atmospheric conditions.

1.2.5 Micro-galvanic effect

The surface of Mg or a Mg alloy cannot be perfectly uniform and as such it is impossible for anodic and cathodic reactions to occur uniformly throughout the entire surface. Due to the non-uniformity in terms of the composition, microstructure and even crystal orientation, galvanic couples can be formed within a Mg alloy (see Fig. 1.5). These micro-galvanic cells dominate the corrosion of the alloy. Within the Mg alloy, the Mg matrix always acts as a micro-anode and is preferentially corroded (Song, 2005b, 2006, 2007a).

Micro-cathodes include:

- grains with different orientations and areas with significantly different solid solution concentrations of alloying elements within the matrix phase;
- secondary phases along grain boundaries; and/or
- impurity containing particles.

An important implication of the non-uniformity is that Mg and Mg alloys will be corroded unevenly. Localized corrosion will be the dominating damage. Localized corrosion will be the main threat to Mg alloy components that we need to battle against in practice. In some cases, the non-uniformity may diminish as corrosion proceeds. For example, the difference between the matrix and secondary phases and that of different alloying element segregation zones can become smaller after the easily corroded zones or phases have been dissolved. In some other cases, the factors causing the non-uniformity remain unchanged with time in corrosion. For example, when the galvanic corrosion is caused by different grain orientations, before the grains are corroded completely, this kind of difference remains.

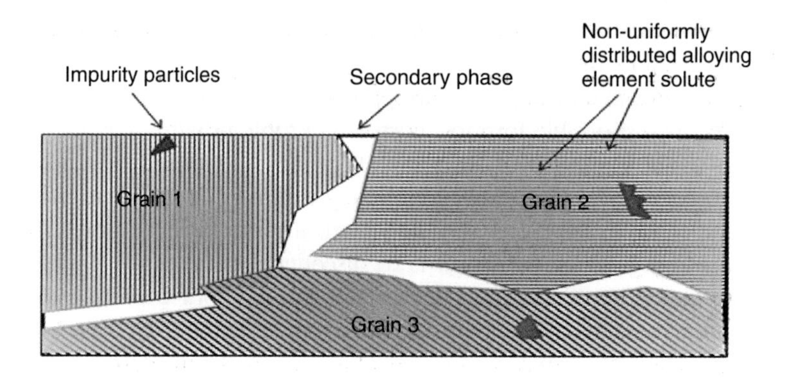

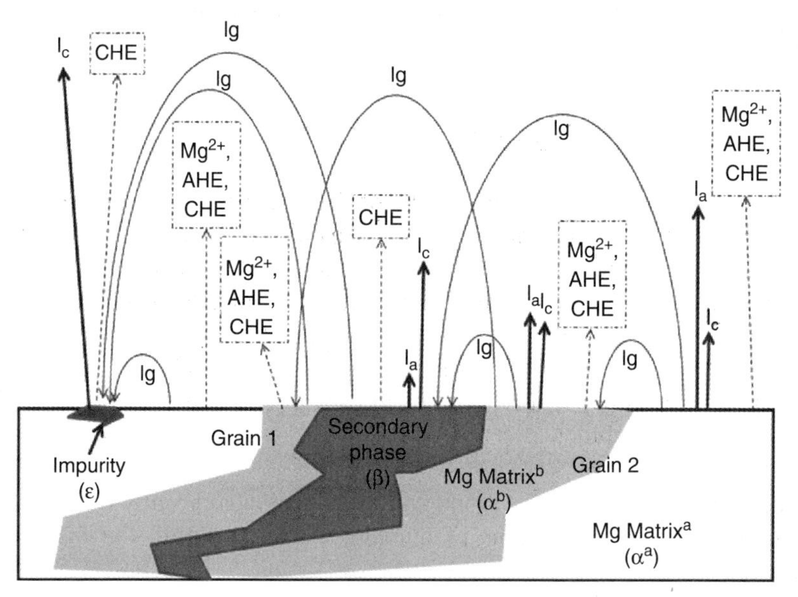

1.5 Schematic illustration of the non-uniformity within a Mg alloy that can result in various micro-galvanic cells. AEH and CHE refer to anodic hydrogen evolution and cathodic hydrogen evolution, respectively; i_g is galvanic current caused by the differences in the Mg alloy.

Because of the non-uniformity, the practical corrosion of Mg and its alloys becomes unpredictable or can significantly deviate from a theoretical model. Thus, computer modeling of a corrosion process of Mg and its alloys poses a great challenge in corrosion science. The unpredictable localized corrosion attack also complicates the selection of the corrosion prevention/protection techniques. More importantly, the non-uniformity can interfere with a film or coating formation process. If surface treatments or coatings are employed to prevent the corrosion of a Mg alloy component, the heredity of the non-uniformity from the substrate to the surface layer/coating will lead to a premature failure. In the following sections, the non-uniformity resulting from the microstructure of the substrate will be discussed further.

1.2.6 Corrosion of matrix phase

In a magnesium alloy, the matrix phase is a major constituent and its corrosion performance is crucial to the alloy. The corrosion of the matrix phase is responsible for the corrosion behavior of the alloy. Therefore, the corrosion of a magnesium alloy is actually a problem of the matrix phase. In the matrix, which is a solid solution, its solute has an important influence on its corrosion behavior. For an Al-containing Mg alloy, the matrix phase is a Mg-Al single phase which becomes less active as the Al content increases (Song *et al.*, 1998, 1999, 2004a; Song, 2005a). More importantly, the difference in solid solution concentration can result in differences in corrosion potential and anodic/cathodic activity (Song *et al.*, 1998, 2004a). These differences can lead to formation of a micro-galvanic cell within a grain because the Al content is higher along the grain boundary than the grain interior for a cast Mg-Al alloy. The Al content in the solid solution can vary from 1.5 wt.% in the grain centre to about 12 wt.% along the grain boundary (Dargusch *et al.*, 1998). Therefore, the difference in electrochemical activity and corrosion resistance can be significant even within the same grain of a Mg-Al alloy. Experimental evidence for the differences between the grain center and boundary comes from the corrosion morphology of an AZ91E alloy (Song, 2005a). The corrosion occurred in the interior areas of α grains and the grain boundary areas were much less corroded.

In the group of non-Al containing alloys, the matrix typically contains Zr as a grain refiner. The role of Zr in corrosion is remarkable and is as important as Al is for the Al-containing alloys. For example, in the Zr-containing alloy MEZ, the central areas of many grains remain uncorroded while the grain boundaries had been severely corroded which is converse to the corrosion morphology of Al-containing Mg alloys. It was found (Song and St John, 2002; Song, 2005a) that the distribution of Zr in the grain is not uniform. The grain centre is rich in Zr which may stabilize Mg through forming a more protective Zr-containing surface film.

In high purity Mg (in which there is no micro-galvanic effect caused by impurities or different solid solution concentrations), different exposed crystal planes, that have different crystal orientations, can still produce different electrochemical activities. Different Mg grains have different depths of corrosion after immersion in an acidic solution (Liu *et al.*, 2008). The corrosion depths of different crystal planes were compared and it was found that the corrosion rates on the basal plane were more stable than the prismatic planes (Liu *et al.*, 2008).

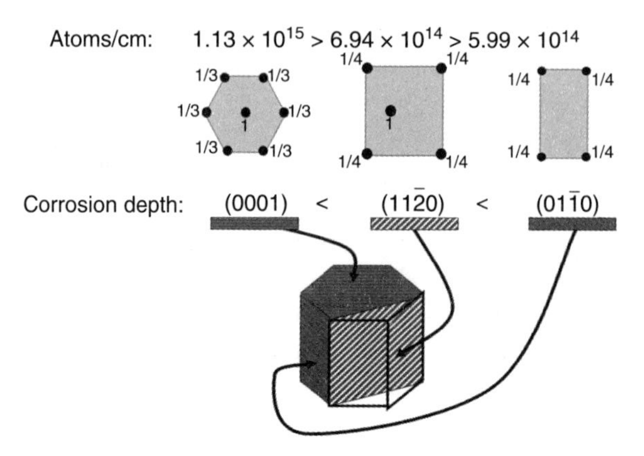

1.6 Corrosion resistance of the basal and prismatic planes (according to Liu *et al.*, 2008).

The different corrosion rates of individual grains can be ascribed to their unique orientation or the unique crystal plane exposed to the corrosive solution. The corrosion rate of a metal can to some extent be correlated to its surface energy (Buck and Henry, 1957; Weininger and Breiter, 1963; Ashton and Hepworth, 1968; Abayarathna *et al.*, 1991; Konig and Davepon, 2001) which in turn, is associated with its atomic density. It is well known that pure Mg has a hexagonal crystallographic cell with axes: a = b = 0.3202 nm, c = 0.5200 nm and angles: $\alpha = \beta = 90°$, $\gamma = 120°$. The atomic densities of the three lowest index planes can be calculated (Konig and Davepon, 2001) to be 1.13×10^{15}, 6.94×10^{14} and 5.99×10^{14}, respectively (see Fig. 1.6). Different crystal planes have different electrochemical activities when they are exposed to an aqueous solution. The atoms in the lower surface energy planes are more difficult to dissolve or corrode away. Therefore, the lowest index plane (0001) has the lowest energy and should be dissolved more slowly than the other surfaces. This has been evidenced by the dependence of corrosion resistance of a crystal plane on its atomic density (Liu *et al.*, 2008).

Recently, the grain orientation-dependent anisotropic corrosion behavior has been further verified with Mg alloy AZ31B (Song and Xu, 2010, 2012; Song *et al.*, 2010). The basal dominated rolling surface of this alloy is much more corrosion resistant than the cross-section surface which is mainly composed of the prismatic planes. In fact, the grain orientation influences the formation and protection performance of the surface corrosion product film, and thus affects the corrosion behavior of Mg (Song and Xu, 2012a). The important lesson from the effect of the matrix on corrosion suggests that successful passivation of the matrix phase is a critical prerequisite for gaining high corrosion resistance of a Mg alloy. Alloying, surface conversion and anodization that can enhance the matrix passivity will lead to significantly improved corrosion performance of Mg alloys.

1.2.7 Influence of secondary phase

Almost all the Mg intermetallic phases are nobler than the magnesium matrix itself and many of them exist in commercial magnesium alloys as secondary phases (Nisancioglu *et al.*, 1990; Song *et al.*, 1998, 1999; Song and St John, 2002). These phases are cathodic to the matrix phase and can act as micro-galvanic cathodes to accelerate the corrosion of the matrix. For example, the β phase not only has a corrosion potential about 400 mV more positive but also a cathodic current density much larger than the α matrix phase, suggesting that the β phase is an active cathode to the matrix phase. Therefore, in a magnesium alloy the matrix can be subjected to micro-galvanic corrosion attack caused by the β phase.

It is well known that the galvanic corrosion current density of a galvanic couple increases with cathode electrode surface area. Due to the galvanic effect of the secondary phase, a Mg alloy is expected to be subjected to increasingly severe corrosion with an increasing amount of the secondary phase in the alloy. This has been verified by some experimental results. It was found (Song, 2005b, 2006; Shi *et al.*, 2005) that the corrosion rate of a Mg-Al alloy first increased and then decreased as the amount of the secondary phase β increased. The corrosion rate increasing with the amount of the β phase before the maximum corrosion rate suggests that the micro-galvanic effect was dominating the corrosion of the alloy. However, it has also been widely reported (Nisancioglu *et al.*, 1990; Lunder *et al.*, 1993, 1995, 1989; Ambat *et al.*, 2000; Uzan *et al.*, 2000; Yim and Shin, 2001) that the corrosion rate of a Mg alloy decreases as the volume fraction of the β phase increases. Even within the same alloy sample, some areas with a large amount of the secondary phase can display higher corrosion resistance than other areas having only a small amount of the secondary phase. For example, a cold-chamber die-cast AZ91D has different microstructures in its interior and surface layer. The surface layer has a larger amount of the finely and continuously distributed β-phase than the interior, and the corrosion rate of the surface layer is about 10 times lower than the interior under the same salt solution (Song and Atrens, 1998; Song *et al.*, 1999; Song, 2005b). In this case, the galvanic effect of the secondary phase does not appear to be responsible for the corrosion behavior.

In fact, the secondary phase can also play another important role within a Mg alloy as a barrier against corrosion. It has been found that the secondary phase in Mg alloys is much more stable than the magnesium matrix (Song *et al.*, 1998, 2007; Song and Atrens, 2003). In AZ alloys, the β-phase is normally not corroded if exposed to an NaCl solution (Song *et al.*, 1998; Zhao *et al.*, 2008a, 2008c) and in the corroded areas where the matrix phase has been severely corroded, the β-phase is still intact. Similarly, the secondary phase in a Mg-RE alloy is also superior to the matrix in terms of corrosion resistance (Song and St John, 2002).

The high corrosion resistance of the secondary phase in a Mg alloy can retard the corrosion of the alloy though a barrier mechanism. A piece of direct evidence is that the corrosion within the matrix phase stops upon reaching the β phase in an AZ Mg alloy (Song, 2005a). The same barrier effect from the secondary phase of a non-aluminum containing alloy has also been found in a Mg-RE-Zr alloy (Song and St John, 2002). The barrier effect and the micro-galvanic effect are two different aspects of the role that the secondary phase plays in corrosion. In other words, the secondary phase has a dual-role in the corrosion (Song and Atrens, 1998, 2003; Song *et al.*, 1999, 2004a; Song, 2005). Whether the barrier effect or the galvanic effect of the secondary phase dominates the corrosion of a Mg alloy depends on the continuity and amount of the secondary phase. If the amount of the secondary phase is small and the distribution is discontinuous, then the galvanic effect will govern the corrosion of an alloy. On the contrary, the secondary phase will effectively separate grains and thus act as a barrier to retard the corrosion development from grain to grain (Song and Atrens, 1999; Song *et al.*, 1999; Song and St John, 2000, 2002; Song, 2005b). Figure 1.7 schematically summarizes the dual-role of the secondary phase in corrosion of a Mg alloy.

The dual-role model can explain many corrosion phenomena of Mg alloys. For example, the higher corrosion rate of Mg-5%Al can be attributed to the galvanic effect of the β phase in a discontinuous distribution; whereas for Mg-10%Al, there is sufficient β phase continuously distributed

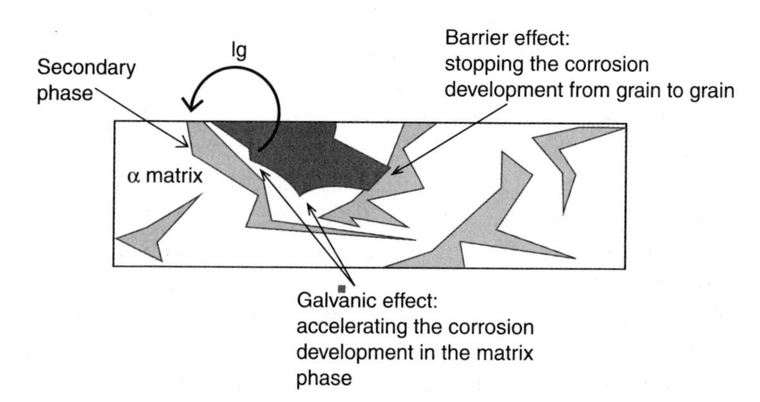

1.7 Schematic illustration of the dual-role (galvanic effect and barrier effect) of the secondary phase in corrosion of a Mg alloy.

along the grain boundaries to stop corrosion from progressing across grain boundaries, and hence its corrosion rate is lower than that of Mg-5%Al.

Based on the dual-role model, the corrosion resistance of a Mg alloy can be enhanced through decreasing the galvanic effect and increasing the barrier effect. Increasing the alloying level is a straightforward approach of producing more continuous network of the secondary phase to retard the corrosion development. In practice, alloying the surface with a passivating element or selective dissolution of the matrix phase in the surface layer is a cost-effective way.

1.2.8 Detrimental effect of impurities

A very small amount of Fe, Ni, Co or Cu addition can dramatically increase the corrosion rate of Mg through a micro-galvanic effect (Hanawalt *et al.*, 1942; Nisancioglu *et al.*, 1990; Lunder *et al.*, 1995). Currently, Fe, Cu and Ni, which are likely to be contained in Mg alloys, are defined as Mg alloy impurities. It is well known that the over-voltages for hydrogen evolution on Fe, Co, Ni, Cu, etc., are less than 0.7 V. This means that Fe, C, Ni and Cu are very active under cathodic polarization and that hydrogen evolution from them is easy and fast. Moreover, they are cathodic to Mg in most aqueous solutions. If these impurities are present in Mg alloys as separate phases, they will be effective micro-galvanic cathodes and can significantly accelerate the corrosion of the matrix phase.

A very small amount of impurity present in Mg or its alloys can result in strikingly deteriorated corrosion performance. For each impurity in Mg and its alloys, there is a critical concentration, also known as the impurity tolerance limit, above which the corrosion rate increases significantly. Corrosion rates are considerably higher and can be accelerated by a factor of 10 to 100 when the concentrations of the impurities Fe, Ni and Cu are increased above their tolerance limits (Hillis and Murray, 1987). If the impurity levels are lower than their tolerance limits, the corrosion rates are low and the impurities have an insignificant influence on the corrosion rates. A higher level of the impurity can lead to lower corrosion resistance for Mg and most Mg alloys (Emley, 1966; Hillis, 1983; Busk, 1987b; Avedesian and Baker, 1999). Studies (Aune, 1983; Hillis, 1983; Frey and Albright, 1984) have shown that dramatic improvement of corrosion resistance can be realized through controlling impurity levels in a Mg alloy. For this reason, Dow has recommended that the following specific impurity tolerance limits should be used to ensure optimum salt water corrosion performance for AZ91: Fe < 50 ppm, Ni < 5 ppm, Cu < 300 ppm (Hillis, 1983).

It seems that the impurity tolerance limit is associated with its solubility in Mg alloys. When these impurities are below their tolerance limits, they are present in the form of solutes in the Mg solid solution. No micro-galvanic cells between the impurities and the Mg matrix are formed. Only after the content of the impurities reach their solubility in a Mg alloy, can they precipitate as separate phases in the Mg alloy and act as galvanic cathodes to accelerate the corrosion. It is calculated by Liu *et al.* (2009) that the Fe tolerant limit in Mg corresponds well to the solubility of Fe in Mg. The Mg-Fe system has a eutectic temperature of ~650°C and a eutectic Fe concentration of 180 ppm. For a content of 180 ppm Fe in the Mg system, the liquid Mg undergoes eutectic solidification at 650°C and forms α-Mg containing about 10 ppm iron in solid solution. However, the region of (Mg + α-Mg) is extremely narrow, such that both the pre-eutectic and eutectic reactions would be suppressed in practical ingot or casting production. Thus, commercial Mg containing less than 180 ppm Fe would solidify to a single α-Mg phase with a super-saturation of Fe in Mg solid solution. Consequently, Mg can have up to 180 ppm of Fe dissolved in the Mg solid solution. At a concentration higher than 180 ppm, Fe will precipitate out via eutectic reaction. The value of 180 ppm Fe corresponds well to the Fe tolerance level of 170 ppm (Song *et al.*, 2004a; Song and Shi, 2006) or 150 ppm (Song and St John, 2004) noted for pure Mg.

Other alloying elements present in Mg can shift the eutectic point and thereby alter the impurity tolerant limits (Emley, 1966). For example, when a few percent of Al is added to Mg, the tolerance limit of iron decreases from 170 ppm to a few ppm. It is calculated (Liu *et al.*, 2008) that the eutectic point is shifted to a lower Fe content after Al is added and thus the Fe tolerance limit decreases rapidly with increasing Al content. With a higher addition of Al in Mg, Fe and Al can form a Fe-Al phase (i.e. $FeAl_3$) particles and precipitate out in Mg-Al alloys, potentially acting as galvanic cathodes (Loose, 1946b; Hawke, 1975; Linder *et al.*, 1989). This is why a Mg alloy with 7% Al can tolerate about 5 ppm Fe; while the tolerance limit becomes too low to be determined when Al concentration is increased to 10 wt.% (Loose, 1946). Therefore, it is also understandable that different Mg alloys have different impurity tolerance limits (Albright, 1988; Song, 2007a).

The solubility-related critical concentration can be extended to some other elements. There is a rough correspondence between the critical concentrations and the solubility of some elements in Mg alloys (Roberts, 1960). Only after the element concentration levels exceed their solubility in the Mg matrix (which is a solid solution) do they start to form separate phases in a Mg alloy, and the subsequent corrosion of the alloy is dramatically accelerated. Moreover, the various cathodic activities of these elements are responsible for the different levels of their corrosion acceleration after their concentrations exceed their solubility.

It should be noted that Mn and Zr are two special elements in terms of their beneficial influence on the impurity tolerance limits. They cannot improve the corrosion resistance of a Mg alloy by themselves and, in the case that their additions exceed their solubility in a Mg alloy, they may even deteriorate the corrosion performance of a high purity Mg alloy due to the galvanic effect of their precipitates (Song, 2005b, 2006). However, they can effectively reduce the detrimental effect of impurities in low purity Mg alloys. It is now well known that the iron tolerance limit in a Mg-Al alloy depends on the Mn concentration. A small addition (0.2%) of Mn can reduce the detrimental effect of impurities when their tolerance limits are exceeded (Hillis, 1983), resulting in increased corrosion resistance of the Mg alloy (Polmear, 1992; Makar and Kruger, 1993). Mn increases the iron tolerance limit to 20 ppm for Mg-Al alloys (Emley, 1966). It can also increase the Ni tolerance limit (Makar and Kruger, 1993). Similarly, Zr additions can also lead to a higher purity and hence a more corrosion resistant Mg alloy.

Since the iron tolerance limit is dependent upon the Mn content in a Mg alloy, it is understandable that the Fe/Mn ratio is a critical factor determining the impurity tolerance limits (Zamin, 1981; Hillis and Shook, 1989). A nearly direct proportionality has been observed between the Fe/Mn ratio and the corrosion rate (Nisancioglu *et al.*, 1990). It is assumed that Mn reduces the corrosion rate by the following two mechanisms. First, Mn combines with Fe in the molten Mg alloy during melting and forms intermetallic compounds which settle to the melt bottom thereby lowering the iron content of the alloy. Second, Mn encapsulates the iron particles that remain in the metal during solidification thereby making them less active as micro-galvanic cathodes. The beneficial effect of Zr is also associated with the reaction of Zr with impurities to form heavier intermetallic particles which quickly settle out due to their high density in the molten Mg. Therefore, the presence of Zr can purify Mg alloys or increase impurity tolerance limits.

The concept of impurity tolerance limit has brought many high purity Mg alloys into the commercial market, which is one of the greatest breakthroughs in Mg alloy research and development. Its implication is profound today in corrosion prevention of Mg alloys: impurity is not only an issue in the alloying process, it is also a big concern in casting, forming and even heat-treatment, which may change the amount and distribution of impurities in the alloy. Although purification can lead to significantly improved corrosion resistance of a Mg alloy, it is not very helpful in dealing with galvanic corrosion attack, the most common practical corrosion problem with Mg alloy components in industrial applications.

1.3 Corrosion mitigation strategy

The purpose of comprehensively understanding corrosion is to effectively mitigate the corrosion damage of metals. A successful corrosion mitigation strategy includes various effective approaches. In each approach, the key is selection and development of corrosion protection/prevention techniques. There are no fundamental differences in corrosion protection principles for Mg alloys versus other conventional metals. However, due to the unique corrosion characteristics and behavior of Mg alloys, the detailed corrosion prevention techniques will have some unusual requirements. Numerous corrosion prevention/protection techniques have been, and are being, developed for Mg alloys in various applications. In this section, the advantages and disadvantages of a few typical approaches will be briefly compared.

1.3.1 Reasonable design

An ideal design can ensure that a Mg alloy in a given service environment has the lowest thermodynamic possibility and the highest corrosion kinetic resistance. A successful corrosion protection design is one of the most cost-effective corrosion mitigation approaches. A simple modification in design may lead to huge savings in subsequent corrosion protection. For example, under atmospheric corrosion conditions, addition of a simple Al alloy washer between a Mg alloy component and a steel bolt may significantly reduce the galvanic corrosion attack to the Mg component. For a Mg alloy part in industry, a small change in geometry may eliminate the potential for moisture to accumulate and thus avoid corrosion damage to the Mg part.

There are many factors to consider in design although, due to a large number of limitations in practical service, sometimes a perfect design is unlikely to be achieved. Nevertheless, a reasonable design should be as simple, low-cost and effective as possible based upon careful selection of all the existing corrosion protection/prevention techniques for the Mg alloy and consideration of the corrosivity of the service environment to this Mg alloy after comprehensively understanding the alloy corrosion characteristics in the environment.

1.3.2 Cathodic protection

Cathodic protection of a metal normally requires the applied cathodic protection potential to be more negative than the equilibrium potential of the metal to be protected. Mg has a very negative equilibrium potential (~-2.7 V). An existing engineering metal cannot act as a sacrificial anode to offer cathodic protection for a Mg alloy, as it is unlikely to have a potential more negative than this value. Nevertheless, cathodic protection can theoretically still be provided by an imposed cathodic current. However, this requires a very high cathodic current density which can lead to intensive hydrogen evolution and low current efficiency, and may not be acceptable in practical applications. Moreover, the cathodic protection may charge hydrogen into a Mg alloy (Mg can store a certain amount of hydrogen at room temperature), which could lead to an embrittlement problem. Therefore, traditional cathodic protection is not a practical approach for Mg alloys. It is at least not recommended for use currently.

However, Mg dissolution can become nearly zero while the hydrogen evolution rate reaches the minimum at a potential between its corrosion and equilibrium potentials (Song *et al.*, 1997b). This means that there is a possibility that effective cathodic protection of Mg alloys can be achieved at a less negative potential than that theoretically required. This would make cathodic protection possible for Mg alloys in practice. Certainly, to make such an 'insufficient' cathodic protection technique feasible and practical, several issues should firstly be addressed. For example, is there always such a minimum hydrogen evolution point for Mg alloys in any corrosive environment? Is this point stable in practice? How effective is it in a long-period? It should be noted that cathodic protection is only applicable under immersion conditions. Therefore, the 'insufficient' cathodic protection cannot be used to mitigate the atmospheric corrosion of a Mg component in practice.

1.3.3 Development of corrosion resistant alloys

Improving the corrosion resistance of Mg alloys through a change in their chemical composition, phase constituent and distribution and/or other micro-structural characteristics is the most reliable and trouble-free corrosion mitigation approach. However, it is difficult and remains a long-term goal. The critical part of a Mg alloy is its matrix, which is a weak region preferentially suffering from corrosion attack according to the analysis in previous sections. Therefore, the key to this approach is to make the matrix phase inert or passive.

From a thermodynamic point of view, there may be a theoretical possibility of reducing the tendency of oxidation or dissolution of Mg through alloying with inert elements. However, due to the high chemical activity of Mg, large amounts of the alloying elements need to be added, which will significantly change the physical properties of Mg. To date, it is known that elements lighter than Mg are usually more active than Mg, thus making it impossible for Mg to become inert through alloying with these light elements. Even though an inert Mg alloy is achievable in theory by alloying with heavy noble metals, the alloy will be quite heavy and have lost the advantage of low density.

From a kinetic point of view, corrosion resistant Mg alloys may be achievable through alloying with highly passive elements. This idea of significantly improving the passivity of Mg alloys appears to be more practical than the proposal of an inert Mg alloy, but it is far from easy. All the known highly passive elements, such as Ti, Al, Cr and Ni, etc., have a limited solid solubility in the Mg matrix phase. None of them, even with an addition up to its solid solubility limit, can really make Mg passive in a corrosive solution (e.g. 5 wt.% NaCl). Nevertheless, there is still a possibility that some highly passive alloying elements form a super-saturated solid solution with Mg in the matrix, thereby making the matrix passive. Certainly, such a super-saturated passive alloy is difficult to obtain through traditional alloying production methods. Some innovative techniques should be considered, including new casting processes (e.g. semi-solid casting, rapid solidification, vapor deposition (both chemical (CVD) and physical (PVD)), sputtering, cold spray, etc.) (Song and Haddad, 2010; Haddad *et al.*, 2011; Xu *et al.*, 2011).

1.3.4 Alloy modification

Alloy modification is different from the approach of developing corrosion resistant alloys. It modifies an existing alloy only slightly, without significantly changing its basic composition and phase constituent. Generally speaking, the improvement of corrosion resistance by this approach is not dramatic, but is attractive for its compatibility with Mg alloy production and processing. For example, heat-treatment is a common practice for some Mg alloys and it has been found that the corrosion resistance of some Mg alloys can be improved through carefully designed heat-treatments (Song *et al.*, 2004a; Zhao *et al.*, 2008a, 2008b, 2008c). A disadvantage of this approach is that some mechanical properties of Mg alloys may also be changed after alloy modification, which could be undesirable in some cases.

1.3.5 Surface modification and treatment

Sometimes it is difficult to improve the corrosion resistance without sacrificing the mechanical performance of an alloy. In this case, surface modification or treatment is a favorable option, because surface modification or treatment only occurs on the surface layer of a Mg alloy and the bulk alloy is left unchanged. Surface modification and treatment can lead to enhancement of anodic and/or cathodic polarization resistance; isolation of anodic or cathodic phase; and elimination or reduction of electrochemical differences between anodic and cathodic phases. These effects can be achieved by various methods, such as: surface machining, surface cleaning, ion implantation, laser treatment, hot diffusion, surface conversion, anodizing, etc. It should be stressed that surface machining can be selectively applied to a region of interest while chemical cleaning is much more practical in industry

(Song and Xu, 2010; Pu *et al.*, 2011, 2012); hot diffusion is relatively inexpensive and suitable for a large surface area (Zhu *et al.*, 2005; Zhu and Song, 2006) compared with ion implantation or laser treatment; and anodizing is more corrosion resistant than surface conversion (Shi *et al.*, 2001, 2005, 2006a, 2006b, 2006c; Blawert *et al.*, 2006).

1.3.6 Coating

A protective coating is an independent layer applied on a Mg alloy surface. It is different from the surface modification or treatment, as the Mg alloy surface does not normally participate in a reaction or experience a significant change underneath the coating. The coating can (1) effectively separate the Mg alloy substrate from the corrosive environment and/or (2) considerably increase the polarization resistance of the alloy substrate and hence dramatically retard its corrosion.

Coatings can be obtained by many methods for Mg alloys and they can be metallic, oxide/hydroxide, organic, etc. Deposition on a Mg surface in a normal aqueous plating bath is difficult due to the intensive hydrogen evolution and the rapid oxide or hydroxide formation. In addition, a modern electro- or electroless-plating technique must be cost-effective, non-toxic and environmentally friendly (Wang *et al.*, 2012a, 2012b). This makes the development of a new plating technique for Mg alloys more difficult. Pinholes are common defects in a plating layer and hence in many cases a plating layer with pinholes may even worsen the corrosion performance of a Mg alloy due to the micro-galvanic corrosion in the pinholes. The most attractive feature of a plated coating for Mg alloys is its metallic characteristics that are important in some practical applications. CVD and PVD techniques can also deposit metallic coatings on Mg alloys. Recently, hot-spray, cold spray and magnetic sputtering techniques have been tried in order to form a corrosion resistant coating for Mg alloys (Tao *et al.*, 2010). They are all capable of producing a metallic coating on Mg alloys, but at a relatively high cost.

Organic coating can be a cost-effective method to separate a Mg alloy from its corrosive environment and hence protect it from corrosion attack. Organic coating is a mature corrosion protection technique and there are many organic coating brands available for Mg alloys (Hu *et al.*, 2012b). As long as Mg alloys are properly surface-treated, these coatings can be applied in the same manner as they are used for conventional metals. Many organic coatings can be applied to Mg alloys, such as normal paints, E-coatings, sol-gel coats, electro-polymerized coats, etc. (Sen *et al.*, 2012a, 2012b). However, due to the high surface alkalinity of Mg, which is not favorable to many organics, Mg alloys normally need to be surface-treated prior to the application of an organic coating. However, by utilizing the surface alkalization effect, an electroless E-coating was recently developed, which is a very quick and easy, low-cost dipping process selectively for Mg alloys, and provides adequate protection performance (Song, 2009a, 2009c, 2009d, 2009e, 2010a, 2011a; Song and Liu, 2011, 2012).

1.3.7 Environmental modification

Environmental parameters are as important to corrosion as the composition and microstructure of an alloy. A change in environment can significantly alter the corrosion rate of a Mg alloy. Hence, modification of the environment can be a strategic method for corrosion prevention of a Mg alloy. For example, under atmospheric conditions, keeping the air dry is a very effective method for controlling the corrosion of Mg alloys (Song *et al.*, 2006b). If environmental modification is achievable, it is relatively cost-effective compared with other approaches. However, environmental impact and health hazards are major concerns.

It is difficult to modify an open environment. If a Mg alloy is in a closed service environment, modification of the environment is possible. It is also possible to modify a local environment after an open system is partially isolated. Basically, modification of the environment includes increasing the pH value, reducing the concentration of aggressive species and adding passivators or inhibitors (Song, 2007a; Song and Song, 2007). Recently, some organic and inorganic inhibitors as well as combinations thereof have been developed for Mg alloys (Hu *et al.*, 2011, 2012a; Huang *et al.*, 2011, 2012). Due to the high electrochemical activity and the presence of the corrosion product film, identification of a highly efficient inhibitor for Mg is currently still difficult. It should be noted that an inhibitor effective for Mg may be detrimental to other metals, such as Al, in corrosion. Therefore, a practical inhibitor for Mg component should be very carefully selected.

1.4 Selection of corrosion protection techniques

Mitigating the corrosion of a Mg alloy is not a matter of simply reducing the corrosion rate. Many issues other than corrosion performance need to be considered as well, as there may be conflicts between improving corrosion resistance and other key properties. The following basic principles should be followed in selecting a corrosion protection approach.

1.4.1 Enhancement of corrosion resistance

Any approach or technique being considered should firstly be able to reduce the corrosion damage of Mg alloys. Mg alloy corrosion mechanisms differ from those of conventional metals and thus it is quite possible that some corrosion prevention measures suitable for conventional metals may not be so effective for Mg alloys. An inappropriate corrosion protection technique may even accelerate corrosion damage. It is dangerous to estimate the effectiveness of existing corrosion mitigation techniques, which were originally developed for conventional metals, on Mg alloys simply based on previous experience.

1.4.2 Compatibility with other properties

Corrosion resistance is only one of the important properties of Mg alloys. It is quite common to find designers, engineers or users are more concerned about the mechanical performance of a Mg alloy rather than corrosion performance. Usually, only after the mechanical properties of a Mg alloy have met application requirements, is the alloy tested and improved for its corrosion performance. In this case, the improvement of corrosion resistance should not deteriorate the mechanical or other properties that have met the design requirements. Moreover, low density is one of the most attractive features of Mg alloys for which they have found many applications. Improving the corrosion performance of a Mg alloy should not compromise this advantage. There are still other unique properties of Mg alloys which make them irreplaceable in some applications. These should be preserved when corrosion protection measures are taken.

1.4.3 Combination of various corrosion mitigation approaches

In applications, there are usually many practical requirements for corrosion protection. Meeting all these requirements is difficult with one corrosion mitigation technique alone. However, it is possible that some of the requirements are met by one corrosion mitigation method and others by a different corrosion protection technique. Hence, all the requirements may be met through a combination of several corrosion prevention techniques. Each corrosion mitigation technique has its own advantages and disadvantages. When various corrosion mitigation methods are combined, they may strengthen the advantages and offset the drawbacks each other.

1.4.4 Cost, toxicity and environmental impact

A direct economic cost should include immediate investment and ongoing expenses and is always a critical issue in selecting corrosion prevention techniques. Mg alloys are in a position to compete against Al alloys in many cases. If the overall economic cost of a corrosion mitigation scheme for a

Mg alloy is higher than the benefits gained from its mass savings over an Al alloy, the corrosion mitigation method is unlikely to be adopted, unless the Mg alloy has some unique properties that Al alloys or other materials in the application cannot meet. Apart from the direct economic cost, there are significant indirect social costs, such as health and environmental impacts. It is a pity that some very effective corrosion prevention techniques are toxic to humans or hazardous to the environment.

1.5 Future trends

For their unique properties, Mg alloys have a variety of potential applications, particularly in the automotive and aerospace industries. There is no doubt that the corrosion problem of Mg alloys is becoming a significant issue, especially as the demand for lighter engineering metals increases and the cost of corrosion prevention for Mg alloys becomes relatively low compared to the increasing energy cost for using other materials. Therefore, some topics will become 'hot' in the field of Mg corrosion and protection:

1.5.1 Innovative surface conversion/treatment and coating techniques

Surface conversion/treatment and coating techniques are the most cost-effective approaches of immediately improving the corrosion performance of Mg alloys in some applications. Currently, it is already a 'hot' topic in corrosion and protection of Mg alloys. In theory, all the surface engineering approaches used for conventional metals may be applicable to Mg alloys. However, due to the unique surface properties of Mg alloys, further developments of innovative technical details are still required. In fact, new surface conversion/treatment and coating techniques are urgently needed for Mg alloys in many applications and the relevant research should be prioritized.

1.5.2 Corrosion database for Mg alloys in typical environments

As a relatively new material, there is a lack of understanding of the corrosion behavior of Mg alloys in given service environments. The application of Mg alloys requires this information, particularly the long-term corrosion behavior. Unfortunately, Mg alloys have only recently been increasingly considered for various applications. It is impossible to estimate or predict the long-term corrosion behavior of a Mg alloy based on the existing experience of other conventional metals or some short-term lab corrosion results of the Mg alloy. Therefore, a collection of long-term corrosion exposure results will be an essential task in the field of Mg alloy corrosion and protection. On top of this, establishment of a database for Mg alloys with some typical protection systems in typical environments will be an even longer term task. Corrosion resistant designs for Mg alloy components in industry need such valuable information.

1.5.3 Evaluation and prediction of corrosion damage for Mg alloys

It is impossible to measure the long-term corrosion performance of Mg alloys in all possible service environments. Therefore, evaluation or prediction of the corrosion of Mg alloys in various corrosion environments based on their existing corrosion results evaluated or measured under a few typical corrosion environments is important and essential. This work includes establishment of lab and field corrosion testing methods, planning of *in-situ* long-term corrosion exposure of specimens, analysis of corroded components from real service environments, development of the relationship between experimental results and real component corrosion damage, etc. This is a challenging area in the corrosion and protection of Mg. Only after the corrosion behavior of a Mg alloy in its service environment can be reasonable predicted, can a long-term protection strategy be proposed.

1.5.4 Understanding of corrosion protection principles

To reasonably predict and prevent the corrosion of Mg alloys, the corrosion mechanisms and prevention principles of Mg alloys should be understood. Using a corrosion mechanism to predict the corrosion performance of a Mg alloy which is actually governed by a different mechanism can generate a misleading result. Also, a misunderstanding of the prevention principle can also result in either under-protection or over protection. Furthermore, with new Mg alloys and new applications being developed, associated new corrosion mechanisms should also be investigated. Correspondingly, new protection techniques need to be developed. Therefore, the fundamental study of the corrosion mechanism and protection technology will be an important area.

1.6 References

Abayarathna D., Hale E.B., O'Keefe T.J., Wang Y.M. and Radovic D. (1991), *Corrosion Science*, **32**:755.

Aghion E. and Bronfin B. (2000), 'Magnesium Alloys Development towards the 21st Century', *Materials Science Forum*, **350–351**:19–28.

Albright D.L. (1988), 'Relationship of Microstructure and Corrosion Behavior in Magnesium Alloy Ingots and Castings', Edited by Henry G. Paris and W.H. Hunt, 'Advances in Magnesium Alloys and Composites', International Magnesium Association and the Non-Ferrous Metals Committee, The Minerals, Metals, and Materials Society, Phoenix, Arizona, January 26, pp 57–75.

Ambat R., Aung N.N. and Zhou W. (2000), 'Evaluation of microstructural effects on corrosion behaviour of AZ91D magnesium alloy', *Corrosion Science*, **42**, 1433–1455.

Aoki S. and Kishimoto K. (1991), *Mathematical and Computer Modelling*, **15**:11.

Ashton R.F. and Hepworth M.T. (1968), *Corrosion* **24**:50.

Aune T.K. (1983), 'Minimizing Base Metal Corrosion on Magnesium Products. The Effect of Element Distribution (Structure) on Corrosion Behavior', Proceedings of the 40th world magnesium conference, Toronto, June.

Avedesian M.M. and Baker H. (1999), 'ASM Specialty Handbook, Magnesium and Magnesium Alloys', Chapter 'Cleaning and Finishing' pp 138–162, Chapter 'Corrosion Behaviour' pp 194–210, Chapter 'Stress Corrosion' pp 210–215, Chapter 'Fatigue and Fracture Resistance' pp 173–176, ASM.

Bard A.J. and Faulkner L.R. (1980), *'Electrochemical Methods, Fundamentals and Applications'*, John Wiley and Sons, Inc.

Bettles C.J., Forwood C.T. and Griffiths J.R., *et al.* (2003), 'AMC-SC1: new magnesium alloy suitable for powertrain applications', SAE Technical paper# 2003-01-1365, SAE World Congress, Detroit, Mach.

Bettles C.J., Forwood C.T. and St John D., *et al.* (2003), 'AMC-SC1: An elevated temperature magnesium alloy suitable for precision sand casting of powertrain components', in H Kaplan (ed), *Magnesium Technology*, TMS, San Diego, pp 223–226.

Blawert C., Dietzel W., Ghali E. and Song G. (2006), 'Anodizing Treatments for Magnesium Alloys and Their Effect on Corrosion Resistance in Various Environments – A Critical Review', *Advanced Engineering Materials*, **8**(7):511–533.

Boese E., Gollner J., Heyn A. and Strunz J. (2001), *Materials and Corrosion*, **52**:247.

Bonora P L, Andrei M., Eliezer A., Gutman E. (2000), 'Corrosion Behaviour of Stressed Magnesium Alloys', E. Aghion and D. Eliezer (eds), Magnesium 2000, Proceedings of the Second Israeli International Conference on Magnesium Science and Technology, (Dead Sea, Israel, 2000), pp 410–416.

Buck I.W.R. and Henry L. (1957), *Journal of the Electrochemical Society* **104**:474.

Busk R.S. (1987b), Magnesium Products Design, International Magnesium Association.

Dargusch M.S., Dunlop G.L. and Pettersen K. (1998), in *Magnesium Alloys and Their Applications* (Wolfsburg, Germany: Werkstoff-Informations GmbH, 1998). pp 277–282.

Doig P. and Flewitt P.E.J. (1979), *Journal of the Electrochemical Society*, **126**:2057.

Eliezer A., et al. (2000), 'Dynamic and Static Corrosion Fatigue of Magnesium Alloys in Electrolytic Environment', E. Aghion and D. Eliezer, ed., Magnesium 2000, Proceedings of the Second Israeli International Conference on Magnesium Science and Technology, (Dead Sea, Israel, 2000), 356–362.

Emley E.F. (1966), 'Principles of Magnesium Technology', Chapter XX, Pergamon Press.

Frey D. and Albright L.L. (1984), 'Development of a Magnesium Alloy Structural Truck Component', Proceedings of the 41st world magnesium conference, London, June.

Froats A., Aune T.Kr. and Hawke D. 'Corrosion of Magnesium and Magnesium Alloys', A.S.M, 'Metals Handbook', 740–754.

Fu J.W. (1982), *Corrosion*, **38**:295.

Gal-Or L., Raz Y. and Yahalon J. (1973), *Journal of the Electrochemical Society*, **120**:598.

Gao G., Cole G., Richetts M., Balzer J. and Frantzeskakis P. (2000), 'Effects of fastener surface on galvanic corrosion of automotive magnesium components', in: E. Aghion and D. Eliezer (eds), Magnesium 2000, Proc. of the second Israeli international conference on magnesium science and technology, Dead Sea, Israel, 2000, pp 321–338.

Hack H.P. (1995), Chapter 20 'Galvanic', in R. Baboian (ed.), 'Corrosion tests and standards: application and interpretation' ASTM, pp 186–196.

Hack H.P. (1997), *Corrosion Review*, **15**:195.

Haddad D., Song G.-L. and Cheng Y.T. (2011), 'Structure and mechanical properties of magnesium-Titanium solid solution thin film alloys prepared by magnetron-sputter deposition', Magnesium Technology 2011, Wim H. Sillekens, Sean R. Agnew, Neale R. Neelameggham, and Suveen N. Mathaudhu (Eds), TMS 2011, San Diego, March, 2011, pp 617–621.

Hallopeau X., Beldjoudi T., Fiaud C. and Robbiola L. (1999), 'Electrochemical Behaviour and Surface Characterisation of Mg and AZ91E Alloy in Aqueous Electrolyte Solutions Containing XOyn- Inhibiting Ions', in G.W. Lorimer (ed.), Proceedings of the Third International Magnesium Conference, The Institute of Materials, London, 1997, 713–724.

Hanawalt J.D., Nelson C.E. and Peloubet J.A. (1942), 'Corrosion Studies of Magnesium and its Alloys', *Transactions of the American Institute of Mining and Metallurgical Engineers*, **147**:273.

Hawke D. (1975), 'Corrosion and Wear Resistance of Magnesium Die Castings', SYCE 8th International Die Casting Exposition and Congress, Detroit, Paper No. G-T75–114.

Hawke D.L. (1987), Galvanic corrosion of magnesium, SDCE 14th International Die Casting Congress and Exposition, Toronto, Canada, May 1987, Paper No. G-T87-004.

Helle H.P.E. and Beck G.H.M., Ligtelijin (1981), *Corrosion*, **37**, 522.

Hillis J.E. (1983), 'The Effects of Heavy Metal Contamination on Magnesium Corrosion Performance', SAE Technical Paper # 830523, Detroit.

Hillis J.E. and Murray R.W. (1987), 'Finishing Alternatives for High Purity Magnesium Alloys', SDCE 14th International Die Casting Congress and Exposition, Toronto, Paper No. G-T87-003.

Hillis J.E. and Shook S.O. (1989), 'Composition and Performance of an Improved Magnesium AS41 Alloy', SAE Technical Paper Series # 890205, Detroit.

Hu J., Huang D., Song G.-L. and Guo X. (2011), 'The Synergistic Inhibition Effect of Organic Silicate and Inorganic Zn Salt on Corrosion of Mg-10Gd-3Y Magnesium Alloy', *Corrosion Science*, **52**:4093–4101

Hu J., Huang D., Zhang G., Song G.-L. and Guo X. (2012a) 'A Study on Tetraphenylporphyrin as a Corrosion Inhibitor for Pure Magnesium', *Electrochemical Solid-State Letters*, **15**(6):C13–C15.

Hu R., Zhang Su., Bu J., Lin C. and Song G.-L. (2012b), 'Recent progress in corrosion protection of magnesium alloys by organic coatings', *Progress in Organic Coatings*, (accepted for publication on 15 Oct, 2011).

Huang D., Hu J., Song G.-L. and Guo X. (2011), 'Inhibition effect of inorganic and organic inhibitors on the corrosion of Mg-10Gd-3Y-0.5Zr alloy in an ethylene glycol solution at ambient and elevated temperatures', *Electrochimica Acta*, **56**:10166–10178

Huang D., Hu J., Song G.-L. and Guo X. (2012) 'Galvanic Corrosion and Inhibition of GW103 and AZ91D Mg Alloys Coupled to an Al Alloy in an Ethylene Glycol Solution at Ambient and Elevated Temperatures', *Corrosion* (accepted on August 6, 2011).

Hur B.-Y., Kin K.-W., Ahn H.-J. and Kim K.-H. (1996), 'A new method for evaluation of pitting corrosion resistance of Mg alloys', ir G.W. Lorimer (ed), 'Proceedings of the Third International Magnesium Conference', Manchester, 1996, pp 557–564.

Jia J., Song G., Atrens A., St John D., Baynham J. and Chandler G. (2004), 'Evolution of the BEASY Program using linear and piecewise linear approaches for the boundary conditions', *Materials and Corrosion*, **55**:845–852.

Jia J., Song G. and Atrens A. (2005a), 'Boundary element method predictions of the influence of the electrolyte on the galvanic corrosion of AZ91D coupled to steel', *Materials and Corrosion*, **56**(4):259–270.

Jia J., Atrens A. and Song G., *et al.* (2005b), 'Simulation of galvanic corrosion of magnesium coupled to a steel fastener in NaCl', *Materials and Corrosion*, **56**(7):468–474.

Jia J., Song G. and Atrens A. (2006), 'The influence of the geometry on the galvanic corrosion of AZ91D coupled to steel', *Corrosion Science*, **48**:2133–2153.

Jia J., Song G. and Atrens A. (2007), 'Experimental measurement and computer simulation of galvanic corrosion of magnesium coupled to steel', *Advanced Engineering Materials*, **9**(1–2):65–74.

Kasper R.G. and April M.G. (1983), *Corrosion*, **39**:181.

Kennard E. and Waber J. (1970), *Journal of the Electrochemical Society*, **117**:880.

Klingert J.A. and Lynne S., Tobias C.W. (1964), *Electrochimica Acta*, **9**:297.

Konig U. and Davepon B. (2001), *Electrochimica Acta*, **47**:149.

Krishnamurthy S., Robertson E. and Froes F.H. (1988), 'Rapidly Solidified Magnesium Alloys Containing Rare Earth Additions', in H.G. Paris and W.H. Hunt (eds), Advances in Magnesium Alloys and Composites, (International Magnesium Association and the Non-Ferrous Metals Committee, The Minerals, Metals, and Materials Society, Phoenix, Arizona, January, 1988, **26**:77–89.

Linder O., Lein J.E. and Aune T.Kr., *et al.* (1989), *Corrosion*, **45**(9):741–747.

Liu M., Qiu D., Zhao M., Song G. and Atrens A. (2008), 'The effect of crystallographic orientation on the active corrosion of pure magnesium', *Scripta Materialia*, **58**:421–442.

Liu M., Uggowitzer P.J., Nagasekhar A.V., Schmutz P., Easton M., Song G. and Atrens A. (2009), 'Calculated phase diagrams and the corrosion of die-cast Mg-Al alloys', *Corrosion Science*, **51**:602–619.

Loose W.S. (1946), 'Corrosion and Protection of Magnesium', in L.M. Pidgeon, J.C. Mathes, N.E. Woldman, J.V. Winkler and W.S. Loos (eds), 'Magnesium', A series of five educational lectures on magnesium presented to members of the A.S.M. during the twenty-seventh national metal congress and exposition, cleveland, February, 4-8, American Society for Metals, pp 173–260.

Lunder O., Lein J.E., Aune T.Kr. and Nisancioglu K. (1989), 'The role of Mg17Al12 phase in the corrosion of Mg alloy AZ91', *Corrosion*, **45**(9):741–748.

Lunder O., Nisancioglu K. and Hansen R.S. (1993), 'Corrosion of Die cast Magnesium-aluminium Alloys, SAE Technical Paper Series #930755, Detroit.

Lunder O., Videm M. and Nisancioglu K. (1995), 'Corrosion Resistant Magnesium Alloys', SAE 1995 Transactions, *Journal of Materials and Manufacturing*, Section 5-Volume **104**:352–357, paper#950428.

Makar G.L., Kruger J. (1993), 'Corrosion of Magnesium', *International Materials Reviews*, **38**(3):138–153.

Mathieu S., Hazan J., Rapin C. and Steinmetz P. (2000), 'Corrosion Behaviour of Die cast and Thixocast AZ91D Alloys – The Use of Sodium Linear Carboxylates as Corrosion Inhibitors', E. Aghion and D. Eliezer, ed., Magnesium 2000, Proceedings of the Second Israeli International Conference on Magnesium Science and Technology, (Dead Sea, Israel, 2000), pp 339–346.

McCafferty E. (1976), *Corrosion Science*, **16**:183.

McCafferty E. (1977), *Journal of the Electrochemical Society*, **124**:1869.

Melville P.H. (1979), *Journal of the Electrochemical Society*, **126**:2081.

Melville P.H. (1980), *Journal of the Electrochemical Society*, **127**:864.

Miyaska M., Hashimoto K., Kishimoto K., Aoki S. (1990), *Corrosion Science*, **30**:299.

Munn R.S. (1982), *Materials Performance*, August, 29.

Munn R.S. and Devereux O.F. (1991a), *Corrosion*, **47**:612.

Munn R.S. and Devereux O.F. (1991b), *Corrosion*, **47**:618.

Nazarrov A.P. and Mikhailovskii Y.N. (1990), *Protection of Metals*, **26**(1):9–14.

Nisancioglu K., Lunder O. and Aune T.Kr. (1990), 'Corrosion Mechanism of AZ91 magnesium alloy', Proceedings of 47th World Magnesium Association, Mcleen, Virginia, pp 43–50.

Nisancioglu K., Lunder O. and Aune T. (1990), presented at 47th World Magnesium Conference, Cannes, France.

Perrault G.G. (1978), Chapter VIII-4 'Magnesium', in A.J. Bard (ed), '*Encyclopedia of Electrochemistry of the Elements*', Marcel Dekker, New York, pp 263–319.

Polmear I.J. (1992), 'Physical Metallurgy of Magnesium Alloys', DGM Informationsgesellschaft, Oberursel, Germany, p 201.

Polmear I.J. (1996), 'Metallurgy of Light Alloys', Halsted Press, NY, p. 362.

Pu Z., Song G.-L., Yang S., Outeiro J.C., Dillon O.W. Jr., Puleo D.A. and Jawahir I.S. (2012), 'Grain refined and basal textured surface produced by burnishing for improved corrosion performance of AZ31B Mg alloy', *Corrosion Science*, **57**:192–201

Pu Z., Yang S., Song G.-L., Dillon O.W., Jr., Puleo D.A. and Jawahir I.S., (2011) 'Ultrafine-grained surface layer on AZ31B Mg alloy produced by cryogenic burnishing for enhanced corrosion resistance', *Scriptal Materialia*, **65**:520–523

Roberts C.S. (1960), 'Magnesium alloy systems' in '*Magnesium and Its Alloys*', John Wiley & Sons, pp 42–80.

Sautebin R., Froidewaux H. and Landolt D. (1980), *Journal of the Electrochemical Society*, **127**:1096.

Senf J., Broszeit E., Gugau M. and Berger C. (2000), 'Corrosion and Galvanic corrosion of die casted magnesium alloys', in H.I. Kaplan, J. Hryn, and B. Clow (eds), Magnesium Technology 2000, TMS Nashville, March 2000, pp 137–142.

Shi Z., Song G. and St John D. (2001), 'A Method for Evaluating the corrosion resistance of anodised magnesium alloys', Corrosion and prevention 2001, ACA, Newcastle, paper # 058.

Shi Z., Song G. and Atrens A. (2005), 'Influence of the β phase on the corrosion performance of anodised coatings on magnesium-aluminum alloys', *Corrosion Science*, **47**:2760–2777.

Shi Z., Song G. and Atrens A. (2006a), 'Corrosion resistance of anodized single-phase Mg alloy', *Surface & Coatings Technology*, **201**:492–503.

Shi Z., Song G. and Atrens A. (2006b), 'Influence of anodizing current on the corrosion resistance of anodized AZ91D magnesium alloy', *Corrosion Science*, **48**:1939–1959.

Shi Z., Song G. and Atrens A. (2006c), 'The corrosion performance of anodized magnesium alloys', *Corrosion Science*, **48**:3531–3546.

Skar J.I. (1999), *Materials and Corrosion*, **50**:2.

Song G. (2004a) (CAST Research Center), 'Environment-friendly, Non-toxic and Corrosion Resistant Magnesium Anodisation' 2004904949 (Australia).

Song G. (2004b), 'Investigation on Corrosion of magnesium and its alloys', *The Journal of Corrosion Science and Engineering (JCSE)*, **6**:C104.

Song G. (2005a), 'Corrosion and Protection of Magnesium alloys: An overview of research undertaken by CAST', *Materials Science Forum*, **488–489**:649–652.

Song G. (2005b), 'Recent progress in corrosion and protection of Mg alloys', *Advanced Engineering Materials*, **7**(7):563–586.

Song G. (2006), '*Corrosion and Protection of Magnesium Alloys*', Chemical Industry Press, (in Chinese).

Song G. (2007a), 'Control of biodegradation of biocompatible magnesium alloys', *Corrosion Science*, **49**:1696–1701.

Song G. (2007b), 'Control of degradation of biocompatible magnesium in a pseudo-physiological environment by a ceramic like anodized coating', *Advanced Materials Research*, **9**(4):1–5.

Song G. (2009a), 'An irreversible dipping sealing technique for anodized ZE41 Mg alloy', *Surface & Coatings Technology*, **203**:3618–3625.

Song G. (2009b), 'Effect of tin modification on corrosion of AM70 magnesium alloy', *Corrosion Science*, **51**:2063–2070.

Song G. (2009c), 'Electroless E-coating: An innovative surface treatment for Magnesium alloys', *Electrochemical and Solid-state Letters*, **12**:D77–D79.

Song G. (2009d), 'An irreversible dipping sealing technique for anodized ZE41 Mg alloy', *Surface & Coatings Technology*, **203**:3618–3625.

Song G. (2009e), 'An electroless E-coating bath sealing technique for a phosphated magnesium alloy', *Materials Science Forum*, **618–619**:269–271.

Song G. (2010), 'Potential and current distributions of one-dimensional galvanic corrosion systems', *Corrosion Science*, **52**: 455–480.

Song G.-L. (2010a), '"Electroless" deposition of pre-film of electrophoresis coating and its corrosion resistance on a Mg alloy', *Electrochimica Acta*, **55**:2258–2268.

Song G.-L. (2011), Chapter 1 'Electrochemistry of Mg and its Alloys' in G.-L. Song (ed.), '*Corrosion of Magnesium Alloys*', Woodhead Publishing Limited, UK.

Song G.-L. (2011a), 'A dipping E-coating for Mg alloys', *Progress in Organic Coatings*, **70**:252–258

Song G. and Atrens A. (1998), 'Corrosion Behaviour of Skin Layer and Interior of Diecast AZ91D', International Conference on Magnesium alloys and their applications, pp 415–419 (Wolfsburg).

Song G. and Atrens A. (1999), 'Corrosion Mechanisms of Magnesium Alloys', *Advanced Engineering Materials*, **1**(1):11–33.

Song G. and Atrens A. (2003), 'Understanding Magnesium Corrosion, a Framework for Improved Alloy Performance', *Advanced Engineering Materials*, **5**(12):837–857.

Song G. and Atrens A. (2007), 'Recent insights into the mechanism of magnesium corrosion and research suggestions', *Advanced Engineering Materials*, **9**(3):177–183.

Song G.-L. and Haddad D. (2010), 'The Grains and Topographies of Magnetron Sputter-deposited Mg-Ti Alloy Thin films', *Materials Chemistry and Physics*, **125**:548–552.

Song G.-L. and Liu M. (2011), 'The Effect of Mg Alloy Substrate on 'Electroless' E-Coating Performance', *Corrosion Science*, **53**:3500–3508.

Song G.-L. and Liu M. (2012), 'The Effect of Surface Pretreatment on the Corrosion Performance of Electroless E-Coating Coated AZ31', *Corrosion Science*, (accepted for publication, 4/21/2012).

Song G. and Shi Z. (2006), 'Characterisation of anodized coatings on a magnesium alloy', M.O. Pekguleryuz and L.W.F. Mackenzie (eds), Magnesium Technology in the Global Age, 45th Annual Conference of Metallurgists of CIM, Montreal, Quebec, October 2006, pp 385–395.

Song G. and Song S. (2006), 'Corrosion Behaviour of Pure Magnesium in a Simulated Body Fluid', *Acta Physico Chimica Sinica*, **22**(10):1222–1226.

Song G. and Song S. (2007), 'A possible biodegradable magnesium implant material', *Advanced Engineering Materials*, **9**(4):298–302.

Song G. and St John D., (2000), 'Corrosion Performance of Magnesium Alloys MEZ and AZ91', *International Journal of Cast Metals Research*, **12**:327–334.

Song G. and St John D. (2002), 'The effect of zirconium grain refinement on the corrosion behaviour of magnesium-rare earth alloy MEZ', *Journal of Light Metals*, **2**:1–16.

Song G. and St John D. (2004), 'Corrosion behaviour of magnesium in ethylene glycol', *Corrosion Science*, **46**:1381–1399.

Song G. and St John D. (2005), 'Corrosion of Magnesium alloys in commercial engine coolant', *Materials and Corrosion*, **56**(1):15–23.

Song G.-L. and Xu Z. (2010), 'The surface, microstructure and corrosion of magnesium alloy AZ31 sheet', *Electrochimica Acta*, **55**:4148–4161.

Song G.-L. and Xu Z. (2012), 'Effect of Microstructure Evolution on Corrosion of Different Crystal Surfaces of AZ31 Mg alloy in a Chloride Containing Solution', *Corrosion Science*, **54**:97–105.

Song G.-L. and Xu Z. (2012a), 'Crystal orientation and electrochemical corrosion of polycrystalline Mg', *Corrosion Science* (submitted).

Song G., Atrens A., St John D.H, Wu X. and Nairn J. (1997a), 'Anodic dissolution of magnesium in chloride and sulphate solutions', *Corrosion Science*, **39**(10–11):1981.

Song G., Atrens A., St John S., Nairn J. and Li Y. (1997b), 'The Electrochemical Corrosion of Pure Magnesium in 1N NaCl', *Corrosion Science*, **39**(5):855–857.

Song G., Atrens A., Wu X. and Zhang B. (1998), 'Corrosion Behaviour of AZ21, AZ501 and AZ91 in Sodium Chloride', *Corrosion Science*, **40**(10):1769–1791.

Song G., Atrens A. and Dargusch M. (1999), 'Influence of Microstructure on the corrosion of diecast AZ91D', *Corrosion Science*, **41**:249–273.

Song G., Atrens A. and St John D. (2001), 'An Hydrogen Evolution Method for the Estimation of the Corrosion Rate of Magnesium Alloys', in J. Hryn (ed.), Magnesium Technology 2001, TMS, New Orleans, 2001, pp. 255–262.

Song G., Bowles A.L. and St John D.H. (2004a), 'Corrosion resistance of aged die cast magnesium alloy AZ91D', *Materials Science and Engineering A***366**:74–86.

Song G., Hapugoda S. and St John D. (2006b), 'Degradation of the surface appearance of magnesium and its alloys in simulated atmospheric environments', *Corrosion Science*, **49**:1245–1265.

Song G., Jonhannesson B., Hapugoda S. and St John D.H. (2004b), 'Galvanic corrosion of Magnesium alloy AZ91D in contact with an aluminum-alloy, steel and zinc', *Corrosion Science*, **46**:955–977.

Song G.-L., Mishra R. and Xu Z. (2010), 'Crystallographic Orientation and Electrochemical Activity of AZ31 Mg Alloy', *Electrochemistry Communications*, **12**:1009–1012

Song G., Shi Z., Hinton B., McAdam G., Talevski J. and Gerrard D. (2006a), 'Electrochemical evaluation of the corrosion performance of anodized magnesium alloys', 14th Asian-Pacific Corrosion Control Conference, Oct. Shanghai 2006, paper# Keynote-11.

Song G., Song S. and Li Z. (2007), 'Corrosion Control of Magnesium as an Implant Biomaterial in Simulated Body Fluid', Ultralight 2007–2nd International Symposium on Ultralight Materials and Structures, Beijing, September 2007.

Song G., St John D. and Abbott T. (2005a), 'Corrosion behaviour of a pressure die cast Magnesium alloy', *International Journal of Cast Metals Research*, **18**(3):174–180

Song G., St John D., Bettles C. and Dunlop G. (2005b), 'Corrosion performance of magnesium alloy AM-SC1 in automotive engine block applications', *Journal of the Minerals, Metals and Materials Society*, **57**(5):54–56.

Song S., Shen W., Liu M. and Song G.L. (2012b), 'Corrosion Study of New Surface Treatment/Coating for AZ31B Magnesium Alloy', *Surface Engineering*, DOI: http://dx.doi.org/10.1179/1743294411Y.0000000056 (accepted for publication, June 14, 2011).

Song S., Song G.-L., Shen W. and Liu M. (2012a), 'Corrosion and Electrochemical Evaluation of E-coated and Powder Coated Magnesium Alloys', *Corrosion*, **68**:015005 (accepted for publication, July 26, 2011).

Starostin M., Smorodin A., Cohen Y., Gal-Or L. and Tamir S. (2000), 'Galvanic corrosion of Magnesium alloys', in E. Aghion and D. Eliezer (eds), Magnesium 2000, Proceed. Of the second Israeli international conference on magnesium science and technology, Dead Sea, Israel, 2000, pp 363–370.

Tao Y., Xiong T., Sun C., Kong L., Cui X., Li T. and Song G.-L. (2010), 'Microstructure and corrosion performance of a cold sprayed aluminium coating on AZ91D magnesium alloy', *Corrosion Science*, **52**(10):3191–3197

Uzan P., Frumin N., Eliezar D. and Aghion E. (2000), 'The Role of Composition and Second Phases on the Corrosion Behaviour of AZ Alloys', in E. Aghion and D. Eliezer (eds), Magnesium 2000, Proceedings Of the 2nd Israeli International Conference on Magnesium Science and Technology, Dead Sea, pp185–191.

Waber J.T. and Rosenbluth M. (1955), *Journal of the Electrochemical Society*, **102**:344.

Wan Y., Tan J., Song G. and Yan C. (2006), 'Corrosion Morphology of AZ91D exposed in a simulated atmospheric environment', *Metallurgical and Materials Transactions*, **37**A(7):2313–2316.

Wang H., Estrin Y., Fu H.M., Song G., and Zúberová Z. (2007), 'The effect of pre-processing and grain structure on the bio-corrosion and fatigue resistance of magnesium alloy AZ31', *Advanced Engineering Materials*, **9**(11):967–972.

Wang Z., Jia F., Yu L., Qi Z.B., Tang Y. and Song G.-L. (2012a), 'Direct Electroless Nickel-Boron Plating on AZ91D Magnesium Alloy', *Surface & Coatings Technology*, **206**(17):3676–3685.

Wang Z., Yu L., Jia F. and Song G.-L. (2012b), 'Effect of additives and heat treatment on the formation and performance of electroless nickel-boron plating on AZ91D Mg alloy', *Journal of the Electrochemical Society* (accepted for Publication, March 25, 2012)

Weininger J.L. and Breiter M.W. (1963), *Journal of the Electrochemical Society*, **110**:484.

Weiss D., Bronfin B., Golub G. and Aghion E. (1997), 'Corrosion resistance evaluation of Magnesium and magnesium alloys by an ion selective electrode', in E. Aghion and D. Eliezer (eds) "Magnesium 97', proceedings of the first Israeli international conference on magnesium science and technology', Dead Sea, 1997, pp 208–213.

Winzer N., Atrens A., Song G., Ghali E., Dietzel W., Kainer K.U., Hort N and Blawert C. (2005), 'A Critical review of the stress corrosion cracking (SCC) of magnesium alloys', *Advanced Engineering Materials*, **7**(8):659–693.

Winzer N., Atrens A., Dietzel W., Song G. and Kainer K.U. (2007), 'Stress corrosion cracking of magnesium alloys', *Journal of the Minerals, Metals and Materials Society*, **August**:49.

Winzer N., Atrens A., Dietzel W., Song G. and Kainer K. (2008), 'Stress corrosion cracking (SCC) in Mg-Al alloys studied using compact specimens', *Advanced Engineering Materials*, **10**(5):453–458.

Xu Z., Song G.-L. and Haddad D. (2011), 'Corrosion performance of Mg-Ti alloys synthesized by magnetron sputtering', Magnesium Technology 2011, in W.H. Sillekens, S.R. Agnew, N.R. Neelameggham, and S.N. Mathaudhu (Eds.), TMS 2011, San Diego, March, 2011, pp 611–615.

Yim C. and Shin E.K.K. (2001), Effect of Heat treatment on Corrosion Behaviour of an AZ91HP Magnesium Alloy, in Proceedings of the 12th Asia Pacific Corrosion Control Conference 2001, Seoul, **2**:1306–1306–7.

Zamin M. (1981), *Corrosion*, **37**(11):627.

Zhao M., Liu M., Song G. and Atrens A. (2008a), 'Influence of Homogenization Annealing of AZ91 on Mechanical Properties and Corrosion Behavior', *Advanced Engineering Materials*, **10**(1–2)93–103.

Zhao M., Liu M., Song G. and Atrens A. (2008b), 'Influence of pH and Chloride ion concentration on the corrosion of Mg alloy ZE41', *Corrosion Science, Advanced Engineering Materials* **10**(1–2):104–111.

Zhao M., Liu M., Song G. and Atrens A. (2008c), 'Influence of the beta phase Morphology on the corrosion of the Mg alloy AZ91', *Corrosion Science*, **50**:1939–1953.

Zhu L., Liu H., Li W. and Song G. (2005), 'Zin alloyed coating on AZ91D Magnesium alloys', *Journal of Beijing University of Aeronautics and Astronautics*, **31**(1):8–12.

Zhu L. and Song G. (2006), 'Improved Corrosion resistance of AZ91D magnesium alloy by an aluminum-alloy coating', *Surface and Coatings Technology*, **200**(8):2834–2840.

2

Corrosion performance of magnesium (Mg) alloys containing rare-earth (RE) elements

J. MENG, W. SUN, Z. TIAN, X. QIU and D. ZHANG,
Chinese Academy of Sciences, China

DOI: 10.1533/9780857098962.1.38

Abstract: In magnesium alloys, the rare-earth (RE) elements first react with the impurities in the alloy, then with alloying elements, and finally form an intermetallic compound with magnesium. Therefore, RE elements play the key role in removing impurity and purifying the matrix in Mg alloys so as to enhance the corrosion resistance. The RE elements have a lower electrode potential resulting in a decrease in the electrode potential of the intermetallic compound in which they participate. The reaction results lead to reduced electrode potential difference between the matrix and second phase. It will play an important role in reducing galvanic corrosion.

Key words: Mg-RE alloys, galvanic corrosion, corrosion potential, cathode current density.

2.1 Introduction

Magnesium (Mg), a fairly reactive metal with a low standard electrode potential, is prone to electrochemical corrosion. A limited number of alloying elements are suitable for magnesium alloys because of potential electrochemical corrosion reactions between Mg and the alloying elements. Common alloying elements that can form stable alloys with Mg include Al, Zn, Mn and interestingly rare-earth (RE) elements. Even though RE elements are very reactive themselves, they exist as RE compounds in magnesium alloys. The electrode potentials of these compounds are very low, hence weak corrosion reactions with the magnesium matrix. Several advantages of the Mg-RE alloys render them an important class in the magnesium alloy family. For instance, the Mg-RE phase has a high melting point and can be alloyed by solid solution strengthening. However, the anticorrosion capability of the Mg-RE alloys varies substantially depending on the alloying components and their structures. The goal of this review is to aid the development of new Mg-RE alloys by examining the general corrosion mechanism of this type of alloys.

Scandium, yttrium, lanthanum, etc., 17 elements in all, are known as RE elements. These elements are divided into two groups: cerium groups and yttrium groups. The cerium group has light RE elements, including lanthanum (La), cerium (Ce), praseodymium (Pr), neodymium (Nd), promethium (Po), samarium (Sm) and europium Eu. Elements in the yttrium group including gadolinium (Gd), terbium (Tb), erbium (Er), thulium (Tm), ytterbium (Yb), lutetium (Lu) and yttrium (Y) are relatively heavy. As chemical properties of the RE elements are very similar, they often coexist in nature and are difficult to separate. In industrial applications mixed RE elements are typically used, not individual pure elements. RE elements in magnesium alloys can effectively influence alloying, degassing, cleaning and purification processes, resulting in high melting point intermetallic compounds, and lead to improved mechanical properties in high-temperature and corrosion resistance.

In general, RE elements have a +3 valence. The electronegativity of RE elements is very low; they can easily become positive ions. They react easily with oxygen, hydrogen, halogen and other harmful substances in the alloys to form stable compounds. At room temperature, RE elements react with air to form stable oxides. However, because RE oxide film is not compact, there is no protective function. The degree of activity of RE elements increases from scandium to yttrium and lanthanum, with lanthanum, cerium and europium most active. Samarium, yttrium, gadolinium and the other RE metals can be dissolved in any concentration of sulfuric acid, hydrochloric acid and nitric acid. However, REs are all stable in alkali. The physical and chemical properties of Mg and RE are listed in Table 2.1.

RE metals are very effective deoxidizing and dehydrogenation agents. RE elements react with oxygen very quickly. As long as a steel contains 10^{-4}% of lanthanum, the oxygen mass fraction can be reduced to 10^{-4}%, so the efficiency of deoxidizing is very high. Hydrogen solubility in RE elements is also high. For example, at 800°C the solubility of hydrogen in cerium is 5700 times that in iron. Therefore, RE elements can absorb the hydrogen in magnesium alloys effectively. The solubility of hydrogen in RE elements increases as temperature drops.

2.2 Factors affecting the Mg-RE alloy corrosion reaction

The primary driving force for the Mg-RE alloy corrosion reaction is the varied electrode potentials on the microscale alloy surface. Hydrogen-evolution corrosion takes place between the spots with high potentials, which act as the cathodes, and the low-potential spots acting as the anodes. The varied potentials can be attributed to the multi-phase alloy system as well as the crystal defects. It is known that the potential at the boundary of two phases is lower than that in the crystal matrix. A deep and wide phase boundary

Table 2.1 Physical and chemical properties of magnesium and RE elements

Elements	Atomic radius (Á)	Maximum solubility in magnesium alloys wt.%	Electronegativity	E^0m^{3+}/mv	Mg-RE compounds
Mg	–	–	1.31	–2.36	–
Sc	2.09	25.91	1.36	–2.03	MgSc
Y	2.27	12.41	1.22	–2.22	$Mg_{24}Y_5$
La	2.74	0.79	1.10	–2.38	$Mg_{12}La$
Ce	2.71	1.62	1.12	–2.34	$Mg_{12}Ce$
Pr	2.67	1.71	1.13	–2.35	$Mg_{12}Pr$
Nd	2.64	3.60	1.14	–2.32	$Mg_{12}Nd$
Pm	2.62	≈2.90	–	–2.42	–
Sm	2.59	5.84	1.17	–2.32	$Mg_{41}Sm_5$
Eu	2.56	≈0.00	1.18	–1.99	$Mg_{17}Eu_2$
Gd	2.54	23.51	1.20	–2.28	Mg_5Gd
Tb	2.51	24.00	1.21	–2.31	$Mg_{24}Tb_5$
Dy	2.49	25.80	1.22	–2.29	$Mg_{24}Dy_5$
Ho	2.47	28.00	1.23	–2.23	$Mg_{24}Ho_5$
Er	2.45	32.70	1.24	–2.30	$Mg_{24}Er_5$
Tm	2.42	31.81	1.25	–2.32	$Mg_{24}Tm_5$
Yb	2.40	3.32	1.26	–2.27	Mg_2Yb
Lu	2.25	41.01	1.27	–2.31	$Mg_{24}Lu_5$

makes intercrystalline corrosion liable to occur. Based on the above analysis, the corrosion studies of the Mg-RE alloys should focus on the microstructures of the alloy crystals. The following discussion exemplifies the structure-based analysis of Mg-RE alloys involving common RE elements, that is Ce, La, Y, Nd and Gd.

2.3 Structural analysis of the Mg-RE alloys containing Ce or La

Corrosion factor dependent on microstructure in the Mg alloys containing Ce or La.

2.3.1 Mg-Ce (La) binary alloys

Ce and La exhibit similar properties in magnesium alloys. Both Ce and La have low maximum solid solubility in Mg with Ce at 1.6 wt.% and La at 0.79 wt.%. In the binary alloy system, La forms the $Mg_{12}La$ phase while Ce forms the $Mg_{12}Ce$ phase. They share the same crystal structure and similar lattice parameters. The La phase and the Ce phase both have high melting points at 613°C and 595°C, respectively.[1] Since magnesium alloy starts to solidify

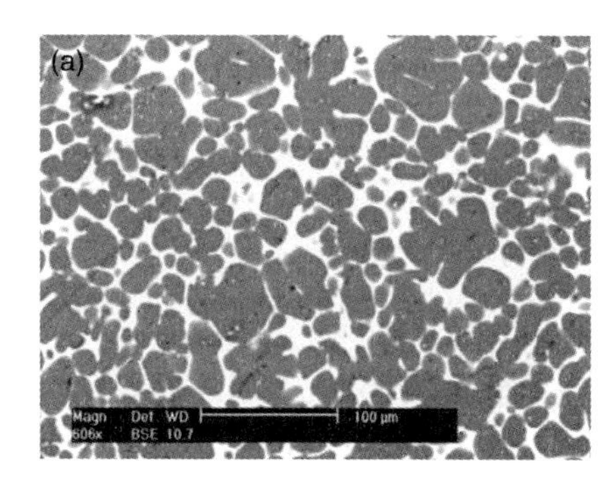

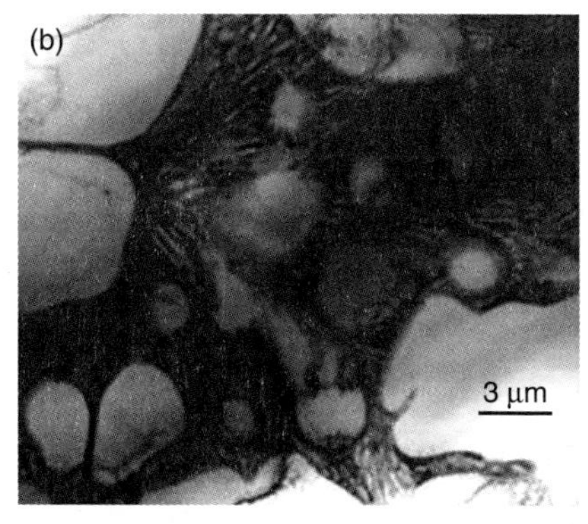

2.1 (a) Scanning electron micrographs (backscattered electron mode) of Mg-5.07La alloys; (b) TEM bright field image of intermetallic phase in Mg-5.07La alloys.[2]

at 650°C, the added Ce or La leads to a narrower solidification temperature range, which is well suited for die-casting processes. The alloy formed after the die-casting process has a eutectic structure at the grain boundary. The corrosion behavior of binary Mg-RE alloys has been thoroughly discussed Birbilis et al.[2] As shown in Fig. 2.1a, the $Mg_{12}La$ phase appears white in the micrograph. The more magnified view (Fig. 2.1b) suggests a eutectic structure at the boundary. The individual intermetallic compounds were observed by using the microcapillary electrochemical cell method. The corrosion potentials of the $Mg_{12}La$ and $Mg_{12}Ce$ phase are –1.6 and –1.5 V corresponding to a difference of +45 and 145 mV, respectively, compared to the magnesium electrode potential. The polarization curve shows a positive correlation between the Ce/La content and the corrosion potential/current. Apparently, micro-galvanic corrosion takes place between the electrode pair composed of the magnesium matrix phase and the only cathode phase, that is the $Mg_{12}La$ or $Mg_{12}Ce$, in the binary alloy. As the amount of RE element increases, the effective surface area of the cathode phase increases leading to higher corrosion rates.

2.3.2 Mg-Al-Ce (La) alloys

When added to the Mg-Al alloy system, the RE elements (in this case La, Ce) preferentially combine with Al in the solidification process to form the needle-shaped AlRE phase (γ phase), which has a high melting point.[3] It can be seen in Fig. 2.2a that the needle-shaped γ phase exists in both α matrix phase and β phase, which suggests that the γ phase solidified before both the α and β phases. The polarization curve shows a decrease of the corrosion potential after the addition of the RE alloying elements. This behavior indicates that the newly-formed γ phase, acting as a weak cathode phase in the

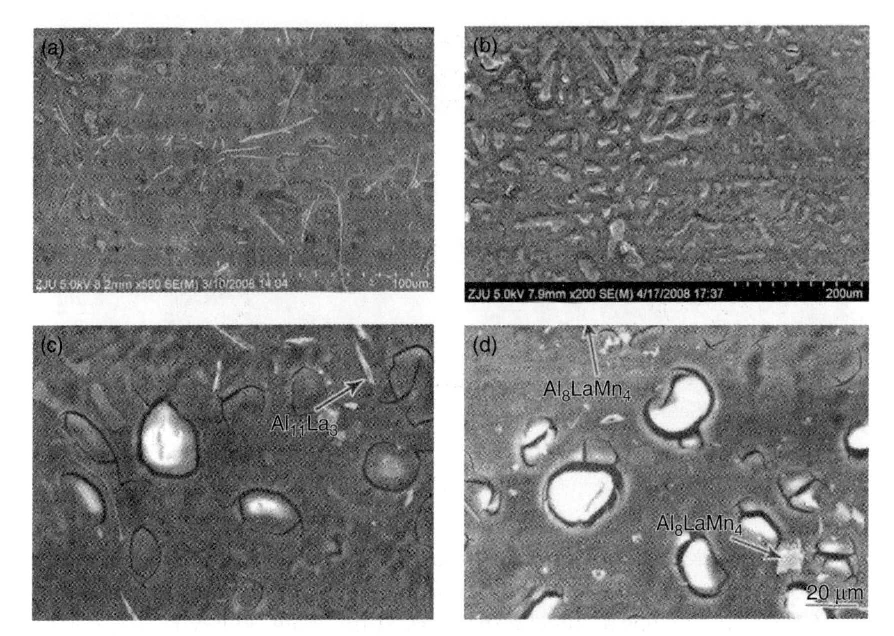

2.2 (a) Microstructure of AM60 magnesium alloys with 1 wt.% Ce addition; (b) the corrosion products formed on AM60 + Ce;[3] (c) SEM morphology of alloy Mg-9Al-1.0La after immersion in 5%NaCl solution for 48 and 72 h (d).[5]

micro-galvanic corrosion process, has a lower corrosion potential than the β phase. Liu's group discovered that the potentials of α, β and γ phases in AZ91 magnesium alloy with RE addition are –1.66, –1.20, –1.44 V (*vs.* SCE). The potential difference between α and β phases is 460 mV and between α and γ phases is merely 220 mV (less than half of the previous value).[4] Therefore, the corrosion process starts in the α matrix phase (Fig. 2.2c and 2.2d). The SEM micrographs (Fig. 2.2b) show branch-shaped concaves on the surface confirming the roles of the γ phase as a weak cathode phase and the β phase as the primary cathode phase.

The immersion test of AZ91 + xCe alloy in 3.5% NaCl solution, which removes all of corrosion products, shows pitting corrosion (Table 2.2).[6] Addition of Ce significantly reduces the corrosion. Ce and Al, Mn, Fe and other elements form intermetallic compounds, reducing the content of impurities. The potential difference between the new phase and the matrix decreases, reducing the driving force for galvanic corrosion. The addition of Ce to AZ91 can improve resistance to pitting corrosion: 0.5% Ce content of AZ91 alloy shows the best performance for corrosion resistance, resulting from grain refinement after addition of Ce.

Usually die-casting is applied with Mg-Al alloys having a high RE content, that is AE44. As Fig. 2.3 shows, the second phase in the alloy produced by die-casting is the eutectic structure formed by the γ phase and the α-Mg. The

Table 2.2 The corrosion rate of AZ91 alloys with different Ce content[6]

	Corrosion rate of weight loss (g/m^2*h)	Annual corrosion depth (mm/a)	Maximum depth of pitting corrosion (mm)
AZ91	2.14	10.3	0.406
AZ91 + 0.5Ce	0.16	0.77	0.241
AZ91 + 1.0Ce	0.26	1.25	0.282
AZ91 + 1.5Ce	0.57	2.74	0.332

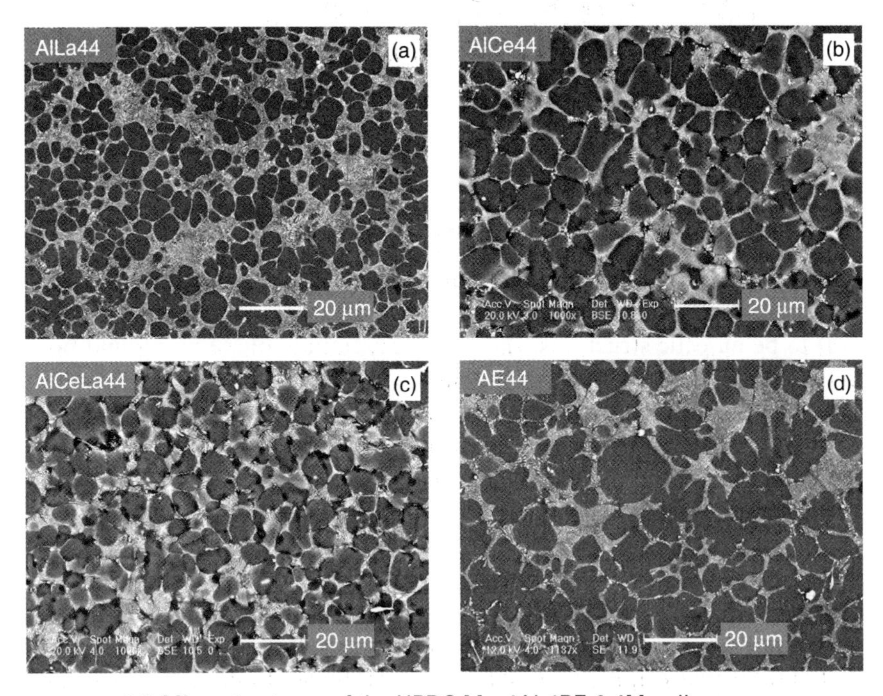

2.3 Microstructures of the HPDC Mg-4Al-4RE-0.4Mn alloys. (a) Mg-4%Al-4%La; (b) Mg-4%Al-4%Ce; (c) Mg-4%Al-2%Ce-2%La; (d) Mg-4%Al-4% mixed RE.[7]

micro-galvanic corrosion effect in AE44 is weaker than that of the AZ91 alloy because of the lower potential of the γ phase compared to the β phase. Furthermore, the lamellar geometry of the eutectic structure can resist corrosion, hence the excellent anticorrosion capability of thc AE44 alloy.

2.3.3 Mg-Zn-Ce (La) alloys

A SEM micrograph (Fig. 2.4) shows the addition of the Ce (La) to the Mg-Zn alloy results in granular crystals with a broad grain boundary. This is

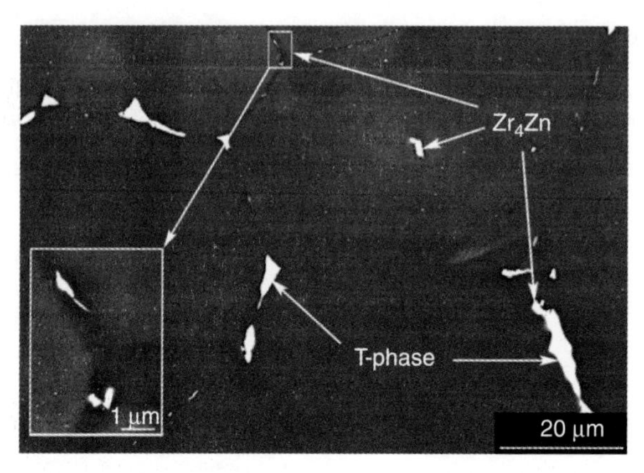

2.4 Backscattered scanning electron micrograph of the as-cast ZE41 alloy.

because of the low solidification points of both the Mg-Zn and Mg-Zn-Ce phases. In other words, both phases only precipitate out as eutectic structures after the α matrix. The SEM observation by Coy *et al.* on the ZA41 alloy[8] showed the second phase T-phase (Mg_7Zn_3RE, where RE = La, Ce, Gd) to be eutectic structures. The Zr_4Zn precipitate phase was found both in the matrix and at the grain boundary.

Neil *et al.* analyzed the corrosion process of the ZE41.[9] The sequence of corrosion initiation and propagation involves pitting initially adjacent to the T-phase, followed by the deep attack at the Zr-rich regions, and pitting within the α-Mg phase (Fig. 2.5).

Coy *et al.*[8] proposed an explanation for the ZE41 alloy corrosion based on the electrode potential data of the matrix and second phase determined by scanning Kelvin probe force microscopy (SKPFM). Compared to the matrix potential, the potential at the grain boundary is 80 mV lower and that of the T-phase at the boundary is 100 mV higher (Table 2.3). Therefore, corrosion occurs first at the boundary due to the steep potential difference. Meanwhile, the Zr-Zn-rich phase, which has a high potential, interacts with the α phase in the micro-galvanic corrosion. The T-phase morphology in the Mg-Zn-Zr alloy can be altered by heat distortion to reduce the effective surface area of the cathode phase and thus improve the anticorrosion capability of the alloy.

2.4 Structural analysis of the Mg-RE alloys containing Y

Corrosion factor dependent on microstructure in the Mg alloys containing Y.

Table 2.3 Volta potential differences of the micro-constituents determined by SKPFM

Micro-constituent	ΔV (mV)
Grain boundary	–80 ± 5
Mg_7Zn_3RE phase	+100 ± 5
Zr-Zn-rich	+180 ± 10

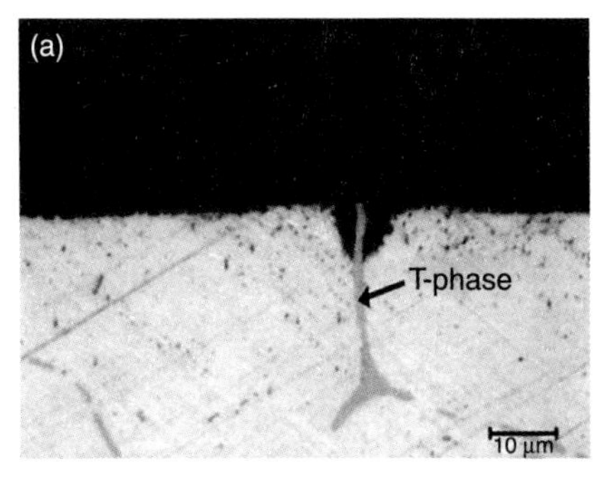

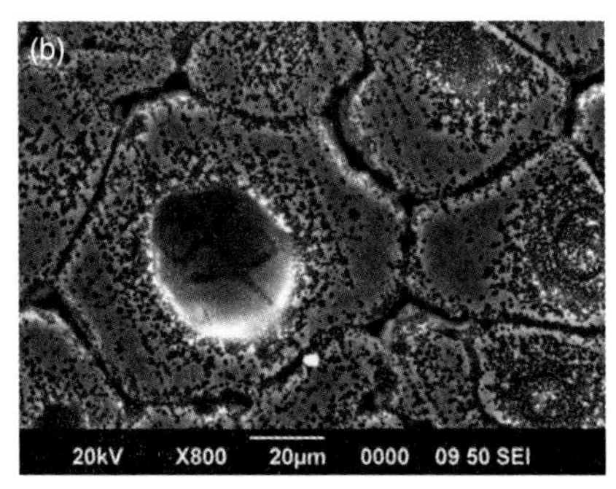

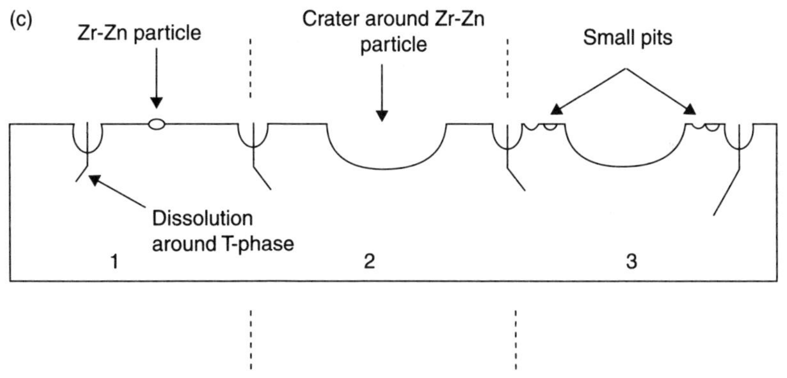

2.5 (a) Optical micrograph showing a cross-section through a corrosion specimen after 18 h immersion in 0.001 M NaCl. (b) SEM–SEI images of ZE41 after 18 h immersion in 0.001 M NaCl with the corrosion product removed. (c) Schematic diagram of sequence of corrosion in ZE41.

2.4.1 Mg-Y binary alloys

The RE element Y is highly soluble in Mg to form the $Mg_{24}Y_5$ phase, which has a high melting point of 567.4°C.[1] Therefore, this phase solidifies ahead of the boundary, leading to a blurred grain boundary. According to Sudholz *et al.*[10] and Liu *et al.*,[11] the corrosion potential of the $Mg_{24}Y_5$ phase is –1.61 V (*vs.* SCE), a difference of +30 mV from the matrix potential. As the Y content in the binary alloy increases, the $Mg_{24}Y_5$ phase volume ratio increases. It is therefore easy to form Y-rich granules with a potential difference from the matrix of between +50 and ~90 mV, depending on the Y content. This leads to an elevated corrosion potential of the alloy. Since the $Mg_{24}Y_5$ phase and Y-rich granules act as the cathode, the anticorrosion properties of the alloy deteriorate with increasing Y content.

2.4.2 Mg-Al-Y alloys

Addition of Y to the Mg-Al alloy results in the formation of the intermetallic compound Al_2Y with a high melting point of 980°C. This compound becomes solid particles during the alloy melting process which are mostly removed by the refining agent as precipitates. Therefore, the addition of Y in the Mg-Al alloy will not only lead to low Y content but also consume Al.

2.4.3 Mg-Zn-Y alloys

Two Mg-Zn-Y phases exist in the Mg-Zn-Y alloys, that is W-phase ($Mg_3Zn_3Y_2$) and I-phase (Mg_3Zn_6Y). It was reported that the melting point temperatures of the W-phase and I-phase are 520°C and 445°C, respectively.[12,13] In the heat extrusion process, the secondary phases did not dissolve solute completely in the Mg matrix. On the contrary, the secondary phases are transformed into nanoparticles in the Mg matrix by the severe plastic deformation. One study[14] suggests that the W-phase has a high melting point and therefore the plasticity of the alloy has been reduced in the heat distortion process. On the other hand, the I-phase, which has a low melting point, functions as the toughening reinforcement during the distortion. However, when Y content reaches 1.54 wt%, there is no I-phase in the Mg matrix. Currently, there are only a few studies of the electrochemical properties of the W-phase and I-phase. In the polarization curve (Fig. 2.6),[13] the Y contents are 0.36, 0.82 and 1.54 wt.% for alloys I, II and III, respectively. As the Y content increases in the alloy, the corrosion potential first decreases and then increases. In this process, the I-phase in the alloy gradually decreases and W-phase gradually increases. This probably indicates that the W-phase, having a lower electrode potential than the I-phase, acts as a relatively weak cathode phase.

Initially, Mg-Zn-Y alloy corrosion occurs around the second phase and gradually expands to the grain center (Fig. 2.7). This is very different from the Mg-Al alloy in which the corrosion occurs in the center of the grain first, resulting, if corrosion is severe, in deep pit corrosion. This is because the Al content around the β phase is higher in the matrix and the Al content away from the center of the β phase is lower. This results in the minimum potential in the grain center, making the alloy vulnerable to corrosion. Because

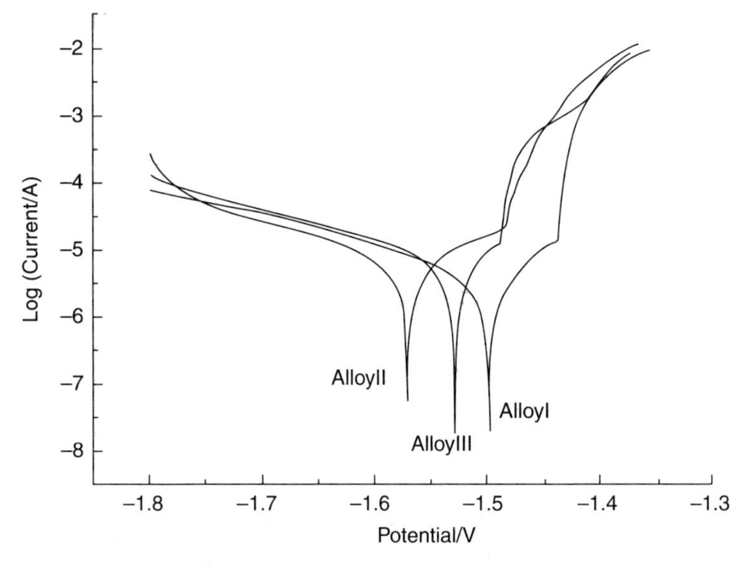

2.6 Tafel curves of the extruded alloys.

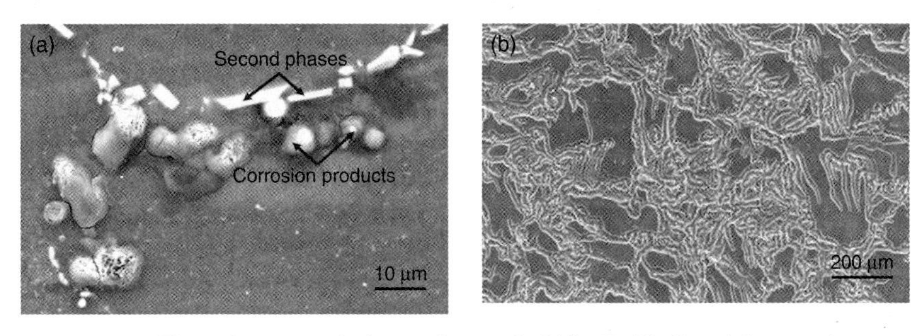

2.7 The micro-morphology of corroded Mg-Zn-Y alloy. (a) corrosion occurs around the second phase at first; (b) corrosion expands by the alloy surface.[12]

the grain boundary is clear between the I-phase and W-phase surrounding the matrix, with no composition transition zone, grain center, and given the interfacial location of the potential, electrochemical corrosion occurs in the interface first. In the Mg-RE alloy containing Zn the grain boundaries are relatively clear so corrosion can expand the boundaries, increasing the depth of corrosion.

2.4.4 Mg-Y-Nd alloys

The Mg-Y-Nd phase has several desirable properties. First, the Mg-Y-Nd phase has a melting point of 536°C,[1] which is lower than the Mg-Y and

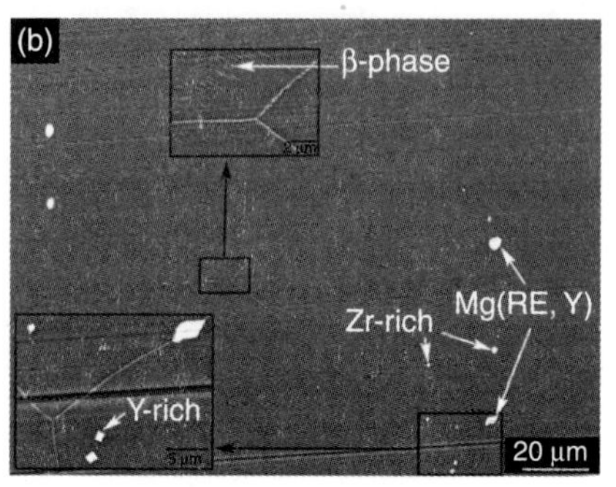

2.8 (a) Microstructure of WE54 after solution treatment at 525°C/8 h.[15] (b) Backscattered scanning electron micrograph of the wrought WE54-T6 alloy.[8]

Mg-Nd phases. Almost all of the Mg-Y-Nd phase can be solubilized in the matrix when the WE43 alloy solid solubilizes at 525°C (Fig. 2.8). A large amount of fine β′ phases form in the aging process, which effectively reduces the cathode surface area and improves the high-temperature strength of the alloy. Second, the potential of the Mg-Y-Nd phase, lower than those of the Mg-Y and Mg-Nd phases, differs from the matrix potential by +25 mV.[8] The β′ phases precipitated after the solid solution treatment have even lower potential difference from the matrix (+15 mV). The WE series of alloys have tightly packed crystals and blurred grain boundaries. Therefore the corrosion at the boundaries is reduced ensuring good anti-corrosion properties of the WE alloys. However, there are unavoidably some Y-rich and Zr-rich particles in the alloy, which act as strong cathode phases and thus weaken anticorrosion capability.

2.5 Structural analysis of the Mg-RE alloys containing Nd

Corrosion factor dependent on microstructure in the Mg alloys containing Nd.

2.5.1 Mg-Nd alloys

In Mg-Nd alloys, Nd forms the $Mg_{12}Nd$ phase, which is similar to the $Mg_{24}Y_5$ and Mg-Y-Nd phases. It is a high-temperature reinforced phase since the melting point of the $Mg_{12}Nd$ is between those of the $Mg_{24}Y_5$ and Mg-Y-Nd phases at 545°C. The $Mg_{12}Nd$ solidifies after the α-Mg matrix and therefore distributes at the crystal boundaries. After solid solution treatment, almost all of them can be solubilized in the matrix. $Mg_{12}Nd$ is a weak cathode phase as its potential is close to that of the $Mg_{24}Y_5$.[2,10] The maximum solid solubility of Nd in Mg is 3.6 wt.%. The Nd solubility decreases with temperature to circa 0.8~ 1.0 wt.% at room temperature.[16] Therefore, the optimum Nd content in the alloy is generally between 3 and 4 wt.%. It is known that the alloy crystals become finer with the addition of Zr. Castability can be improved by the addition of small amounts of Zn to form Mg-4Nd-0.3Zn-0.4Zr alloy. After heat distortion treatment, the $Mg_{12}Nd$ phase in the alloy mainly exists as the precipitated phase in the crystal matrix and the boundary.[17] Furthermore, it has relatively low corrosion potential hence good mechanical behavior at high temperatures and excellent anticorrosion capability. Ding *et al.*[18] compared the anticorrosion behavior of Mg-3Nd-0.2Zn-0.4Zr and AZ91D and found the Mg-Nd alloy had better anticorrosion properties than the AZ91D alloy. The corrosion rate of the former in 5% NaCl solution was only half of that of the latter. It was suggested the AZ91D had a higher cathode/anode surface ratio and higher potential difference, and therefore higher degree of local corrosion than the Mg-Nd alloy.

2.5.2 Mg-Al-Nd alloys

Neodymium in a Mg-Al alloy exists in the form of $Al_{11}Nd_3$, which has a high melting point of 632°C. Its properties are similar to those of the AlCe(La) phase discussed above. Because its potential is lower than the β phase, the addition of Nd lowers the corrosion potential of the alloy thus limiting the electrochemical corrosion tendency.[19] However, the matrix mechanical properties are adversely affected since the $Al_{11}Nd_3$ particles are needle-shaped. Die-casting is therefore preferred.

If the Nd content in the Mg-Al alloy is too high, the Al atoms will be consumed by Nd atoms, and impact $Mg_{17}Al_{12}$ phase formation, leading to the quantity and size reduction of the β phase. The continuous β-phase will be replaced by a discrete Al-Nd phase. The β phase loses the ability to protect the matrix, and the corrosion resistance decreases. When the Nd content is up to 1.5% by mass, the corrosion rate increases significantly (Table 2.4).[20] In practice, more Nd is used in Mg-Zn alloys, than in Mg-Al alloys.

2.5.3 Mg-Zn-Nd (Gd) alloys

In contrast to Mg-Al alloys, corrosion potential increases with the addition of RE elements in Mg-Zn alloys (Table 2.5).[21,22] This suggests that the Mg-Zn-RE phase has a higher potential than the Mg-Zn phase and acts as the strong cathode phase. The grain boundary of the Mg-Zn is clear, which leads to lower potential at the boundary. As a result, electrochemical corrosion is liable to take place between the Mg-Zn-RE phase and α-Mg in the

Table 2.4 The corrosion rate of AZ91 alloys with different Nd content[20]

	Corrosion rate of weight loss (g/m^2*h)	Annual corrosion depth (mm/a)	Maximum depth of pitting corrosion (mm)
AZ91	2.13	10.2	0.406
AZ91 + 0.5Nd	0.17	0.82	0.327
AZ91 + 1.0Nd	0.12	0.57	0.290
AZ91 + 1.5Nd	1.2	5.7	0.340

Table 2.5 Corrosion potentials of Mg-Zn-xNd(Gd)

Alloy	Corrosion potential V
ZK50	−1.6
ZK50 + 1Nd	−1.57
ZK50 + 2Nd	−1.55
ZK60	−1.52
ZK60 + 1.2Gd	−1.5
ZK60 + 1.6Gd	−1.45
ZK60 + 2.0Gd	−1.4

grain boundary, which exhibits as intercrystalline corrosion when viewed macroscopically.

2.5.4 Mg-Nd-Zr alloys

Zirconium forms insoluble precipitates with impurities (especially iron and nickel), so the enrichment of zirconium in the grain centre leads to a higher corrosion resistance in this zone (shown in Fig. 2.9). On the other hand, RE-containing magnesium alloys can also result in a higher purity and corrosion resistance. In different heat treatment conditions, corrosion resistance is: T4> T6> F (shown in Table 2.6). The reason for this is that the galvanic corrosion rate relates strongly to the morphology of the cathode phase $Mg_{12}Nd$ like a Mg-Gd alloy.

2.6 Structural analysis of Mg-RE alloys containing Gd

The maximum solid solubility of Gd in Mg is 23.5 wt.%. When added to Mg, Gd forms a Mg_5Gd phase with a high melting point of 548°C. Figure 2.10[24] shows the as-cast microstructure (a) and the solution-treated microstructure (b) of the Mg-14Gd binary alloy. Apparently, all of the Mg_5Gd

Table 2.6 Corrosion rates of Nd-Zn-Zr, Nd-Zn alloys after immersion in the 5% NaCl solution for 3 days[23]

		Corrosion rate of weight loss (g/m^2*h)	Annual corrosion depth (mm/a)
Mg-3Nd-0.2Zn-0.4Zr	F	0.085	0.35
	T4	0.046	0.19
	T6	0.061	0.25
Mg-3Nd-0.2Zn	F	0.56	2.4
	T4	0.11	0.47
	T6	0.13	0.56

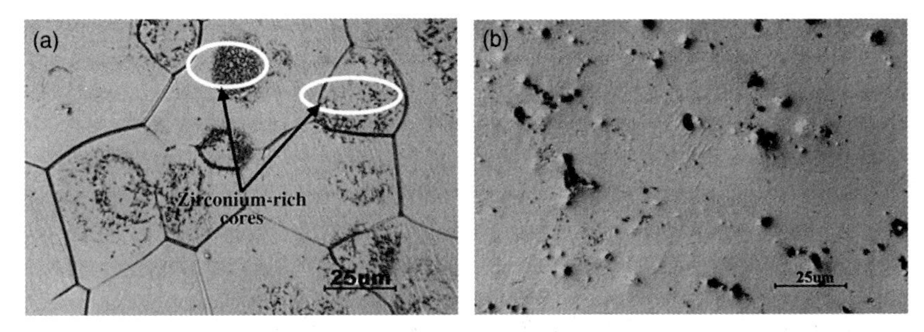

2.9 Optical micrographs of Nd-Zn-Zr (a) and Nd-Zn (b) alloys (T4).[23]

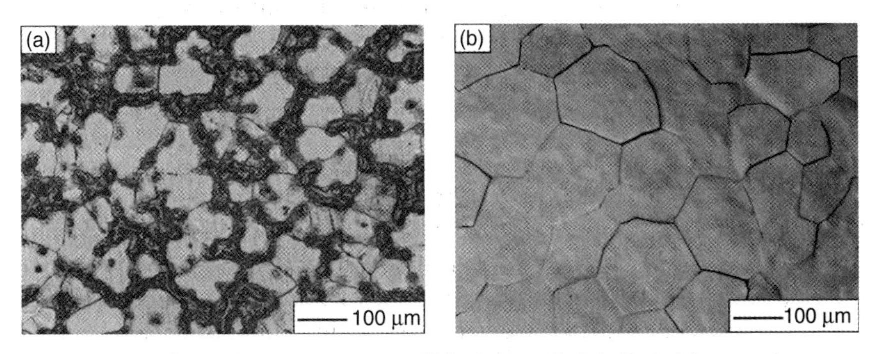

2.10 Optical microstructure of Mg-2.49 at.% Gd alloy: (a) as-cast; (b) solution treated.

produced in the casting process is dissolved in the matrix. In the polarization curve of the Mg-RE alloys, Mg-Gd has a similar corrosion potential to Mg-Y and Mg-Nd, which means that Mg_5Gd is also a low-potential weak cathode phase.

Adding Y to the Mg-Gd binary system can effectively reduce the Gd content and make the strengthening effect improve. However, it results

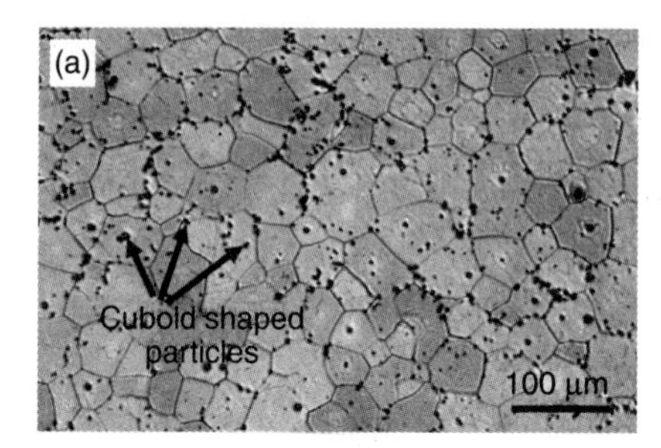

2.11 (a) Typical optical micrograph of 10Gd-3Y-0.4Zr alloy, (b) FE-SEM micrograph of the cuboid-shaped particles in (a).[25]

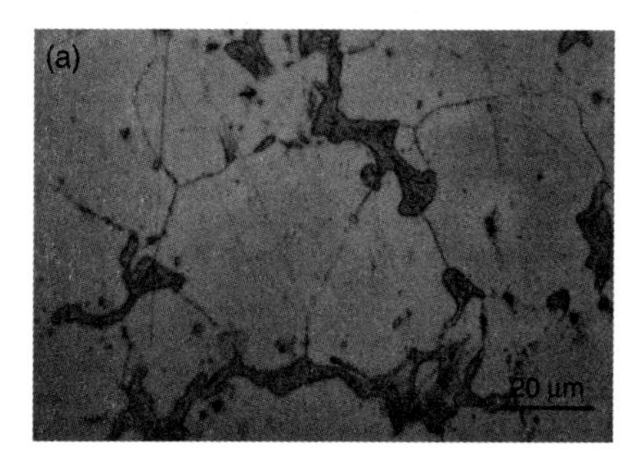

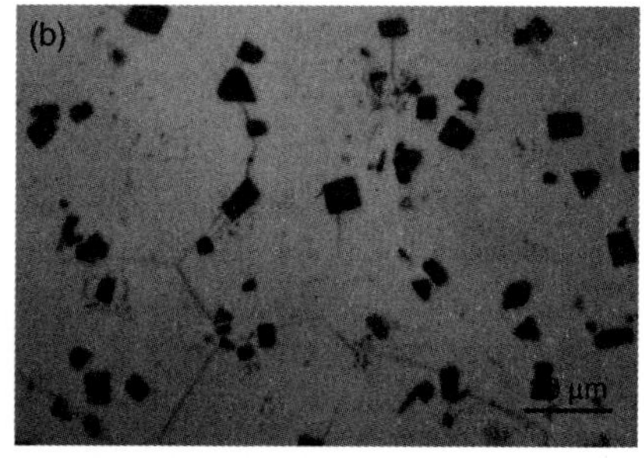

2.12 Optical images of microstructures of Mg-10Gd-4.8Y-0.6Zr cast alloy with different holding time at (a) 520°C/2 h; (b) 520°C/12 h.[26]

in a higher square phase in the matrix. When Y content is increased by 3 wt.%, Gd(Y)-rich particles can be observed after solution treatment, when Y content is 5% by mass. In the alloy matrix there is a large amount of Gd(Y)-rich particles as shown in Figs. 2.11 and 2.12. A large amount of Gd(Y)-rich particles generated after solid solution process cannot be entirely dissolved in the matrix. The higher the content of RE elements in this RE-rich substance, the higher the corrosion potential: they act as a strong cathode phase.

Figure 2.13a and 2.13b are micrographs of the Mg-Gd-Y-Zr alloy after an extrusion process followed by solid solution and aging. The lamellar precipitates create a RE-poor Mg_5Gd or Mg_5Y phase, which generally has a low-potential. A large amount of Gd(Y)-rich particles, which show as white particles in the micrographs, appear after aging or extrusion processes. The addition of Y produces a large quantity of stable Gd(Y)-rich particles, which

Table 2.7 Immersion in 5% NaCl solution for 3 days[25]

	Corrosion rate of weight loss (g/m^2*h)	Annual corrosion depth (mm/a)
Mg-6Gd-3Y-0.4Zr	5.5	21.9
Mg-8Gd-3Y-0.4Zr	8.2	30.8
Mg-10Gd-3Y-0.4Zr	17.1	61.1
Mg-12Gd-3Y-0.4Zr	12.8	43.5

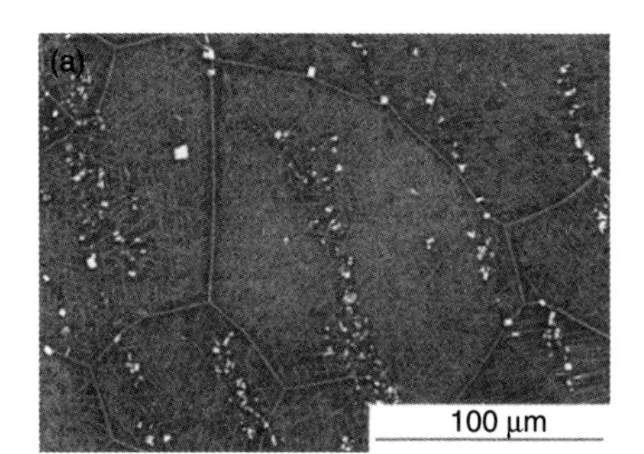

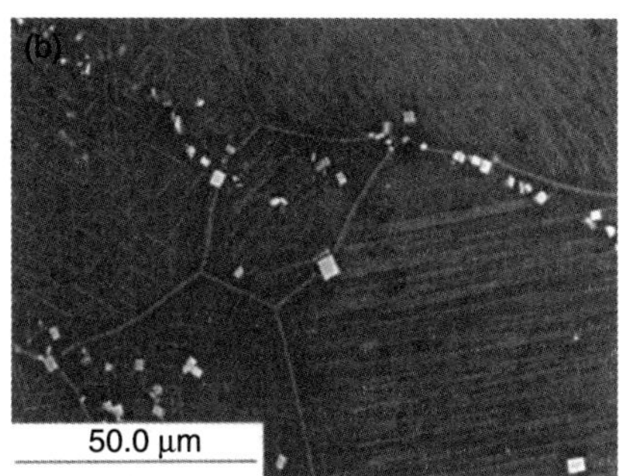

2.13 Mg-13Gd-4Y-0.4Zr alloy (a) microstructure of aged at 300°C for 24 h; (b) morphology of lamellar precipitate phase along special orientations (300°C for 20 h).[27]

act as both a reinforcement of high-temperature mechanical properties and a strong cathode phase in facilitating corrosion.

With the increase of Gd content, the amount of β′ precipitates also increases. When the fraction of β′ phase is low and the distribution is not continuous, the presence of β′ phase will accelerate corrosion. The higher Mg-10Gd-3Y corrosion rate, compared to Mg-12Gd-3Y, can be attributed to the higher fraction and more continuous distribution of β′ precipitates in Mg-12Gd-3Y, β′ precipitates in Mg-12Gd-3Y can act as a corrosion barrier to some degree (shown in Table 2.7).

Heat treatment has significant impact on the mechanical properties and corrosion resistance of Mg-Gd alloys. In the as-cast form, electrochemical corrosion occurs between the second phase at the grain boundary and the matrix. In an as-solid solution, most of the second phase disappears, and the remaining cube Gd(Y)-rich particles have minimum electrochemical corrosion. With the increase of aging time, the cathode phase increases gradually and leads to increased corrosion (Table 2.8).

Table 2.8 Corrosion rates of Nd-Gd alloys after immersion in 5% NaCl solution for 3 days[28]

Mg-10Gd-3Y-0.4Zr	Corrosion rate of weight loss (g/m^2*h)	Annual corrosion depth (mm/a)
As-cast	14.45	51.7
T4	1.86	6.7
T6-0.5h	3.75	13.4
T6-16h	9.43	33.7
T6-193h	10.18	36.4

2.7 The effect of the cathode area and current density on corrosion

2.7.1 The effect of the cathode area on corrosion

The cathode area plays an important role in determining the corrosion rate. All of the added RE elements form RE compounds except the solid-solubilized portion in the magnesium matrix. The solid-solubilized RE elements have little impact on the anticorrosion properties. Although RE compounds are the major factor in determining the corrosion rate, the area ratio of the RE to the matrix (cathode area ratio) has little correlation to the RE content. First, the RE content is not directly correlated with the number of resulting RE phases. Second, heat processing can alter the number and morphology of the RE phases. Heat treatment can also change the RE morphology in the magnesium alloy. For example, bulky RE phases can break into jagged smaller RE phases in the process. Therefore, much interference can be excluded by directly studying the second phase area ratio on alloy corrosion.

2.7.2 The effect of cathode current density on corrosion

Corrosion of magnesium alloys leads to reduced mass (macroscopically) and corrosion current (microscopically). Cathode current determines the corrosion current and in effect controls the corrosion rate.

Calculation of the cathode current density

Besides the α-Mg matrix, binary alloys, that is Mg-Ce, Mg-La, Mg-Nd and Mg-Y, contain only the $Mg_{12}Ce$, $Mg_{12}La$, Mg_3Nd and $Mg_{12}Y_5$ phases. Figure 2.14 shows the relationship between the intermetallic volume percentage and the corrosion current.[2,10] The corrosion current increases with the volume percentage of intermetallic compounds at different rates for different phases. The cathode current density on the surface

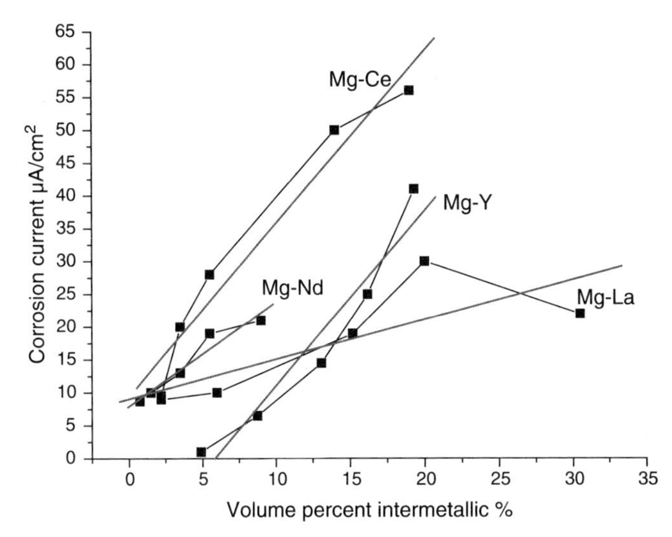

2.14 Volume percent intermetallic *vs* corrosion rate, for HPDC Mg-RE alloys, where RE = Ce/La/Nd/Y.

of the second phase can be estimated by extrapolating the fitted linear curve to 100% intermetallic by volume. Consequently, the polarization resistance can be calculated by dividing the potential difference between the second phase and the matrix by the current density of the second phase (Table 2.9).

The current density on the surface of the cathode phase is an indicator for hydrogen-evolution efficiency, which is positively related with the corrosion sensitivity of the magnesium alloy. The polarization resistance, a barometer for the extent of cathode polarization, is a parameter for the catalytic activity in the hydrogen-evolution reaction on the cathode surface. Zero polarization resistance means non-polarized while ∞ resistance represents complete polarization.[29] Lower cathode current density means higher polarization resistance, which is more beneficial to the anticorrosion properties.

Factors affecting the cathode current density

In the galvanic corrosion process, the current density on the cathode surface is not constant. It depends on the potential difference between the cathode phase and the matrix. The matrix potential also changes with the matrix content. If the surrounding matrix potential decreases, the current density on the surface of the cathode phase increases. For instance, the matrix potential decreases abruptly adjacent to the grain boundary. If RE particles happen to be there, the cathode current density would be higher

Table 2.9 Electrochemical parameters of the Mg-RE cathode phase

Compound	Corrosion potential (V)	Potential difference *vs* Matrix (mV)	Current density (μA/cm²)	Polarization resistance (Ωcm²)
Mg	−1.645	0		
$Mg_{12}La$	−1.6	45	69.7	646
$Mg_{12}Ce$	−1.5	145	273	531
Mg_3Nd	−1.57	75	166	452
$Mg_{12}Y5$	−1.61	35	251	139

than those in the matrix. Consequently, intercrystalline corrosion takes place more easily.

Different RE phases have a different polarization resistance resulting in different current density. It can be seen from Table 2.9 that the $Mg_{12}Y_5$ phase, which has the lowest polarization resistance, will generate a strong corrosion current even in the absence of a high potential difference between the RE phase and the matrix. Therefore, in order to improve the anticorrosion capability of the magnesium alloy, it is advisable to reduce the Y-rich cathode phases in the matrix.

The polarization resistance of the cathode phase is not a constant value either. The cathode potential starts out in a linear relationship with the current over the equilibrium value. Then it becomes linearly related to the log value of the current as shown in the Tafel curve. Figure 2.15 shows the relationship of the cathode current and overpotential (both experimental and calculated).[29] The corrosion current is overestimated when the potential difference is small but underestimated when the potential difference is large.

2.8 Conclusions

The purpose of research into Mg-RE alloy corrosion mechanisms is not to accurately calculate the corrosion rates of each alloy, but to generate a general approach in determining the corrosion properties of Mg-RE alloys and understanding the causes for the deterioration of anticorrosion properties. The high reactivity of RE and Mg causes RE to preferentially form compounds with the noble elements in the alloy such as Fe, Cu, Ni and subsequently Al, Mn, Zn, etc. This is in agreement with other studies claiming the RE elements to be 'cleansing elements'. Moreover, RE elements and their compounds have low electrode potentials, which help lower the overall potentials of the cathode phase and consequently the corrosion current when combined with high potential elements. The AlRE phase has a lower potential than the AlMg phase (β phase), therefore the addition of RE elements can improve the anticorrosion capability of the

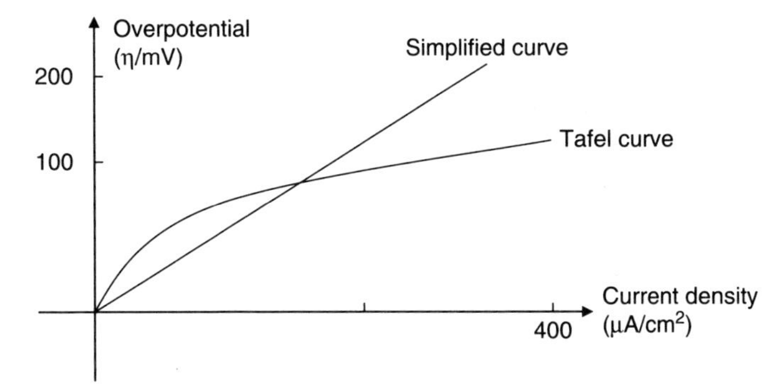

2.15 Experimental and simplified curves of overpotential *vs* cathode current.

AZ series alloys. However, it should be noted that the high melting point AlRE phases precipitate first in the common casting process, leading to the existence of AlRE in both the α and β phases. Mechanical properties may be adversely affected by the fragmentation effect of the AlRE phases on the matrix.

Conversely, the die-casting process results in the formation of eutectic structures between the AlRE phase and β phase at the grain boundary, which improves anticorrosion capability but without undermining the mechanical properties. This is the reason why mostly die-casting was used in the AE series. Alloys based on Mg-Zn are usually used as malleable magnesium alloys. The Mg-Zn phase mostly appears at the grain boundary due to its low solidification point. The Mg-Zn-RE phase formed upon the addition of RE mostly distributed at the grain boundary with a portion existing in the bulk crystal. This type of phases has many varieties due to varied contents and they have higher melting points than Mg-Zn. In the process of high-temperature distortion, they will break into smaller fragments or transform their morphology dramatically, which leads to reinforcement effect. However, the Mg-Zn-RE phase has a higher potential than the Mg-Zn phase, therefore excessive addition of the RE affects the anticorrosion properties adversely.

It has been demonstrated that the Mg-RE binary alloys tend to have high melting points (>530°C) and moderate phase potentials. Nevertheless, their ability to provide reinforcement is not ideal due to the weak solid solution precipitation effect. Many advantages have been discovered for the addition of multiple RE elements in magnesium alloys, which has become the focus of recent research. Studies show that multiple RE phases (MgRE-RE phase, where RE = Y, Nd, Gd) have lower melting points than the Mg-RE phase, which is beneficial to the solid solubility. This feature also reduces the time and temperature required for solid solution process, increases the number of the precipitation phases and at the same time improves the mechanical strength at high temperatures. Electrochemically, the MgRE-RE phase still maintains fairly low potentials. After the solid solution precipitation, the sizes of the phases become smaller leading to even lower potentials. WE43 is by far the most successful MgRE-RE species. It has both excellent mechanical and anticorrosion properties. However, one drawback of the MgRE-RE alloys is the presence of RE-rich particles which have relatively high potentials and high melting points. Even though the high-temperature strength can be improved, the anticorrosion capability is significantly weakened. This behavior is exemplified by the Mg-Gd-Y-Zr alloy.

2.9 References

1. Liu, C., X . Zhu, and H. Zhou, *Magnesium Alloy Phase Atlas*. (2006): Central South University Press, Changsha.
2. Birbilis, N., M.A. Easton, and A.D. Sudholz, On the corrosion of binary magnesium-rare earth alloys. *Corrosion Science*, (2009). **51**(3): 683–689.
3. Wenjuan, L., C. Fahe, and C. Linrong, Effect of rare earth element Ce and La on corrosion behavior of AM60 magnesium alloy. *Corrosion Science*, (2009). **51**(6): 1334–1343.
4. Liu, W., F. Cao, A. Chen, L. Chang, J. Zhang, and C. Cao, Corrosion behaviour of AM60 magnesium alloys containing Ce or La under thin electrolyte layers. Part 1: Microstructural characterization and electrochemical behaviour Original Research Article. *Corrosion Science*, (2010). **52**(2): 627–638.
5. Yang, D., J. Chen, Q. Han, and K. Liu, Effects of lanthanum addition on corrosion resistance of hot-dipped galvalume coating. *Journal of Rare Earths*, (2009). **27**(1): 114–118.
6. Jie, Y., Y.I. Dan-qing, and D. Shu-hao, Effect of trace Ce on the microstructure and corrosion resistance of AZ91 magnesium alloy. *Journal of Chinese Society for Corrosion and Protection*, (2008). **28**(4): 205–209.
7. Zhang, J., and Meng, J., *Study on microstructure and properties of the heat-resistant HPDC Mg-Al-RE alloys*. Ph.D. Thesis (2009), Changchun Institute of Applied Chemistry, Chinese Academy of Sciences, Changchun.
8. Coy, A.E., F. Viejo, P. Skeldon, and G.E. Thompson, Susceptibility of rare-earth-magnesium alloys to micro-galvanic corrosion. *Corrosion Science*, (2010). **52**(12): 3896–3906.
9. Neil, W.C., M. Forsyth, P.C. Howlett, C.R. Hutchinson, and B.R.W. Hinton, Corrosion of magnesium alloy ZE41 – The role of microstructural features. *Corrosion Science*, (2009). **51**(2): 387–394.
10. Sudholz, A.D., K. Gusieva, X.B. Chen, B.C. Muddle, M.A. Gibson, and N. Birbilis, Electrochemical behaviour and corrosion of Mg–Y alloys. *Corrosion Science*, (2011). **53**(6): 2277–2282.
11. Liu, M., P. Schmutz, P.J. Uggowitzer, G. Song, and A. Atrens, The influence of yttrium (Y) on the corrosion of Mg–Y binary alloys. *Corrosion Science*, (2010). **52**(11): 3687–3701.

12. Song, Y., D. Shan, R. Chen, and E.-H. Han, Effect of second phases on the corrosion behaviour of wrought Mg–Zn–Y–Zr alloy. *Corrosion Science*, (2010). **52**(5): 1830–1837.
13. Zhang, E., W. He, H. Du, and K. Yang, Microstructure, mechanical properties and corrosion properties of Mg–Zn–Y alloys with low Zn content. *Materials Science and Engineering*, (2008). **488**(1–2): 102–111.
14. Bae, D.H., S.H. Kim, D.H. Kim, and W.T. Kim, Deformation behavior of Mg–Zn–Y alloys reinforced by icosahedral quasicrystalline particles. *Acta Mater*, (2002). **50**: 2343–2356.
15. Rzychoń, T., J. Michalska, and A. Kiełbus, Effect of heat treatment on corrosion resistance of WE54 alloy. *Journal of Achievements in Materials and Manufacturing Engineering*, (2007). **20**(1–2): 191–194.
16. Xiao-fang, W. S. Yang-shan, W. Qi, X. Feng, B. Jing, X. Shan, and T. Wei-jian, Microstructures and mechanical properties of Mg-2Nd alloy. *The Chinese Journal of Nonferrous Metals*, (2007). **17**(6): 927–933.
17. Zhiliang, N., C. Fuyang, L. Honghui, S. Jianfei, and D. Jianfeng, Microstructure analysis of Mg-2.54Nd-0.26Zn-0.32Zr alloy. *Rare Metal Materials and Engineering*, (2009). **38**(11): 1997–2000.
18. Wen-jiang, D., X. Yazhen, C. Jian-wei, P. Ying-hong, Corrosion and electrochemical behaviour of Mg-Al alloys and Mg-RE alloys in NaCl solution. *The Chinese Journal of Nonferrous Metals*, (2009). **19**(10): 1713–1719.
19. Liu, N., J. Wang, L. Wang, Y. Wu, and L. Wang, Electrochemical corrosion behavior of Mg–5Al–0.4Mn–xNd in NaCl solution. *Corrosion Science*, (2009). **51**(6): 1328–1333.
20. Jie, Y., Y. Danqing, D. Shuhao, W. Bin, and L. Gongqi, Effect of trace Nd on microstructure and corrosion resistance of AZ91 magnesium alloy. *Journal of Materials Science & Engineering*, (2008). **26**(2): 251.
21. Chang, J.-W., J. Duo, Y.-Z. Xiang, H.-Y. Yang, W.-J. Ding, and Y.-H. Peng, Influence of Nd and Y additions on the corrosion behaviour of extruded Mg-Zn-Zr alloys. *International Journal of Minerals, Metallurgy and Materials*, (2011). **18**(2): 203.
22. Ping, W., L. Jianping, and M. Qun, Effects of gadolinium on the microstructure and corrosion resistance properties of ZK60 magnesium alloy. *Rare Metal Materials and Engineering*, (2008). **37**(6).
23. Chang, J.-W., P.-H. Fu, X.-W. Guo, L.-M. Peng, and W.-J. Ding, The effects of heat treatment and zirconium on the corrosion behaviour of Mg-3Nd-0.2Zn-0.4Zr (wt.%) alloy. *Corrosion Science*, (2007). **49**: 2612–2627.
24. Gao, L., R.S. Chen, and E.H. Han, Effects of rare-earth elements Gd and Y on the solid solution strengthening of Mg alloys. *Journal of Alloys and Compounds*, (2009). **481**(1–2): 379–384.
25. Chang, J., X. Guo, S. He, P. Fu, L. Peng, and W. Ding, Investigation of the corrosion for Mg–xGd–3Y–0.4Zr (x = 6, 8, 10, 12 wt%) alloys in a peak-aged condition. *Corrosion Science*, (2008). **50**(1): 166–177.
26. Hui-zhong, L., G. Fei-fei, L. Chu-ming, L. Hong-ting, W. Hai-jun, L. Xiao-peng, and Z. Jia-yan, Behavior of Mg-10Gd-4.8Y-0.6Zr alloy during solution heat treatment at 520°C. *Materials Science and Engineering of Powder Metallurgy*, (2011). **16**(3): 326.

27. Quanying, G., L. Zheng, M. Pingli, and S. Jing, High temperature aged precipitate mechanism of Mg-13Gd-4Y-0.4Zr alloys as squeeze. *Chinese Journal of Raremetals*, (2010). **34**(2): 199–201.
28. Peng, L.-M., J.-W. Chang, X.-W. Guo, A. Atrens, W.-J. Ding, and Y.-H. Peng, Influence of heat treatment and microstructure on the corrosion of magnesium alloy Mg-10Gd-3Y-0.4Zr. *Journal of Applied Electrochemistry*, (2009). **39**: 913.
29. Chu-Nan, C., *Principles of Electrochemistry of Corrosion*. (2008). Chemical Industry Press, Beijing.

3 Corrosion inhibition of magnesium (Mg) alloys

X.-P. GUO, Huazhong University of Science and Technology, China, G.-L. SONG, General Motors Corporation, USA and J.-Y. HU and D.-B. HUANG, Huazhong University of Science and Technology, China

DOI: 10.1533/9780857098962.1.61

Abstract: Addition of inhibitors in service environments is a practical corrosion protection approach for engineering materials in industry. This chapter provides an overview of corrosion inhibition of magnesium (Mg) alloys. After a summary of inhibitor inhibition mechanisms on Mg, a few case studies on inhibitor selection and inhibition efficiency evaluation for Mg and its alloys in various media are presented. Based on promising results, future work in this area is proposed.

Key words: magnesium alloy, corrosion, inhibitor.

3.1 Introduction

An inhibitor is a small amount of chemical additive that can effectively inhibit the corrosion of a metal in a corrosive environment. Application of inhibitors is one of the most practical methods of protecting metals from corrosion attack.[1,2] Inhibitors can be divided into passivation, precipitation and adsorption groups according to their inhibition mechanisms. Passivating inhibitors have a better protective effect on iron family metals by forming a passive film on their surfaces. The commonly used passivating inhibitors include chromate, nitrite and phosphate. However, these groups of inhibitors have little effect on non-ferrous metals, such as zinc and magnesium.[3–5]

In the case of adsorptive inhibitors, their stability and inhibition efficiency are closely related to the interaction of their inhibitor molecules at the metal/solution interfaces.[6] According to the quantum chemistry frontier orbital theory,[7] the unoccupied *d* orbit of transition metals, like Fe, Zn, etc., can accept π electrons from inhibitor molecules and lone-pair electrons from N, S and O atoms. Therefore, it is possible for inhibitor molecules to be adsorbed on the metal surfaces and to form an adsorptive film to retard the dissolution of the metals. Precipitating inhibitors refer to inhibitors that can react with some ions in solution to form insoluble compounds deposited on a metal surface, and thus slow down the metal corrosion. At present, precipitating inhibitors mainly include zinc sulfate, calcium bicarbonate, sodium tripolyphosphate, etc., which can form a precipitation membrane on metals.[8–10]

Compared with other anti-corrosion techniques, an inhibition approach possesses some obvious advantages, such as no requirement for special equipment, simple control and operation as well as low cost. Currently, however, the application of inhibitors in protection of Mg alloys is not very popular. This is because there is a lack of effective inhibitors for lightweight alloys that are much more chemically active than conventional metals.[11–13] Nevertheless, inhibitors must be considered in some applications of Mg alloys.[14] For example, if an engine block is made of an Mg alloy, Mg corrosion inhibitors should be used in the engine cooling systems. Moreover, inhibitors may also be mixed in a water tank made of Mg alloys. They may even be added to a coating or sealant to retard the corrosion of the substrate Mg alloy. In all these cases, selection of suitable and efficient inhibitors for Mg alloys is essential and important.

3.2 Inhibitors for magnesium

Due to the special electrochemistry of Mg, highly efficient adsorptive inhibitors for conventional metals may not be effective on an Mg alloy. This section discusses those inhibitors that may be suitable.

3.2.1 Adsorption-type inhibitors for Mg alloys

In theory, since the energy of the *3d* orbital of Mg is much higher than those of Fe and Zn transition metals, the potential for Mg to accept electrons is lower. Thus, the inhibition efficiency of adsorptive inhibitors cannot be very high on Mg alloys. However, due to the high chemical activity of Mg and the presence of surface film, the experimental inhibition results of Mg and its alloys are usually different from theoretical expectations.

Today, tested adsorptive inhibitors for Mg alloys mainly include sodium dodecylbenzenesulfonate (SDBS), hexamethylenetetramine (HMTA),[15,16] derivatives of lactobionic-acid (LTA), and 6-ring organic compounds containing N-heteroatom.[17] The inhibition effect of HMTA on AZ31 Mg alloy in simulated cooling coolant was studied by Lei *et al.*[16] Their results showed that HMTA has a good inhibition effect on AZ31 Mg alloy surface. HMTA was found to be a mixed-type inhibitor that inhibited both the anodic and cathodic reactions on AZ31. The inhibition efficiency of the derivatives of lactobionic-acid compounds and 6-ring organic compounds containing

N-heteroatom on Mg alloy AZ91 in 50 wt.% aqueous ethylene glycol (EG) solutions has been studied by Slavcheva and Schmitt[17] through polarization curve and electrochemical noise measurements. The results revealed that lactobiono-tallowamide acted as a mixed-type inhibitor reducing the rates of both anodic and cathodic partial corrosion reactions. The inhibiting effect of LTA was a result of its adsorption and formation of an adherent protective film on a metal surface. The good inhibition properties combined with the ecological acceptability of this compound allow consideration of LTA as a promising and practical inhibitor for Mg alloys. The corrosion inhibition of Mg-Al-Zn alloy was investigated in naturally aerated chloride free stagnant neutral solutions using amino acids as environment-friendly corrosion inhibitors by Helal and Badawy.[18] Pure amino acids including: aliphatic amino acids, aromatic amino acids and sulfur containing amino acids were studied. Their results showed that the inhibition efficiency depended on the chemical structure of amino acid and its concentration. The presence of aromatic ring and hetero atoms, such as sulfur in the amino acid structure, were responsible for the significant increased inhibition efficiency. The calculated adsorption free energy of phenylalanine indicated that the inhibition resulted from a physical adsorption obeying Langmuir adsorption isotherm.

3.2.2 Passivation-type inhibitors for Mg alloys

The high chemical activity of Mg makes Mg and its alloy difficult to passivate. Hence, except extremely strong passivators, normal passivation-type inhibitors cannot efficiently inhibit the dissolution of an Mg alloy. Although chromate, permanganates, nitrite, H_2O_2, etc., are passivating-type inhibitors on conventional metals, if they are used on Mg alloys, the addition amounts should be large enough to effectively mitigate Mg alloy corrosion. Since these chemicals react with Mg and form a layer of Mg containing products, which are much thicker than a passive film, on Mg alloys, they are widely used in surface conversion treatments for Mg alloys.[19–21] Currently, there is no report on passivating inhibitors that can effectively inhibit the corrosion of Mg alloys with only a small amount of addition.

3.2.3 Precipitation-type inhibitors for Mg alloys

Fluoride, molybdate and tungstate are precipitation-type corrosion inhibitors for Mg alloys.[11,22,23] For example, the corrosion behavior of Mg (purity 99.6%) in EG was investigated by Song and St John.[11] Their results showed that the corrosion of Mg in EG could be effectively inhibited by addition of fluorides (1 wt.% KF), because F^- reacted with Mg and formed a protective fluoride-containing film on the Mg surface. Molybdate and tungstate salts mainly inhibited the cathodic reactions of a Mg alloy through forming a protective layer on the cathodic active sites.[22,23] It has also been demonstrated[24,25] that organic sodium aminopropyltriethoxysilicate (APTS-Na), inorganic zinc nitrate, sodium phosphate and the synthesized 5,10,15,20-tetraphenylporphyrin (TPP) also have a great inhibition effect on Mg alloys. These inhibitors also react with Mg and form precipitated protective films.

In theory, passivation-type and precipitation-type inhibition mechanisms have different phase film formation processes. The former film is converted from the substrate while the latter is deposited from the environment solution. On Mg, due to the Mg surface alkalization effect and the presence of the naturally formed surface film,[26] it is difficult to tell whether a surface film results from surface conversion or environment deposition. Therefore, in this chapter, passivation-type and precipitation-type inhibitors are reviewed as a group of phase film formation inhibitors, and they will be discussed together. Fortunately, this will not affect their inhibition efficiency and mechanism understanding.

3.3 The behavior of different inhibitors on magnesium and its alloys

Corrosion inhibition behavior can vary significantly from material to material, solution to solution, and inhibitor to inhibitor. So far, as the publications on Mg corrosion inhibition are still very limited, they cannot form a base for a comprehensive and systematic review on inhibitor selection, inhibition mechanism or efficiency evaluation according to alloy type or solution system. This section will present a few case studies, which may help establishment of a systematic understanding of the corrosion inhibition of Mg alloys.

3.3.1 ZnO for magnesium in a phosphating solution

Phosphating is one of the important surface conversion techniques and has been widely used on conventional metals. It would be of great significance if the conventional phosphating method could be directly applied to Mg alloys. Mg has a different electrochemistry. Mg alloys are normally dissolved rapidly in an acidic solution. Unfortunately, phosphating bath solutions with phosphoric acid (H_3PO_4) as an essential component are typically acidic. How to slow down the rapid dissolution of Mg in such an acidic phosphating solution is an important topic. The polarization curves of Mg immersed

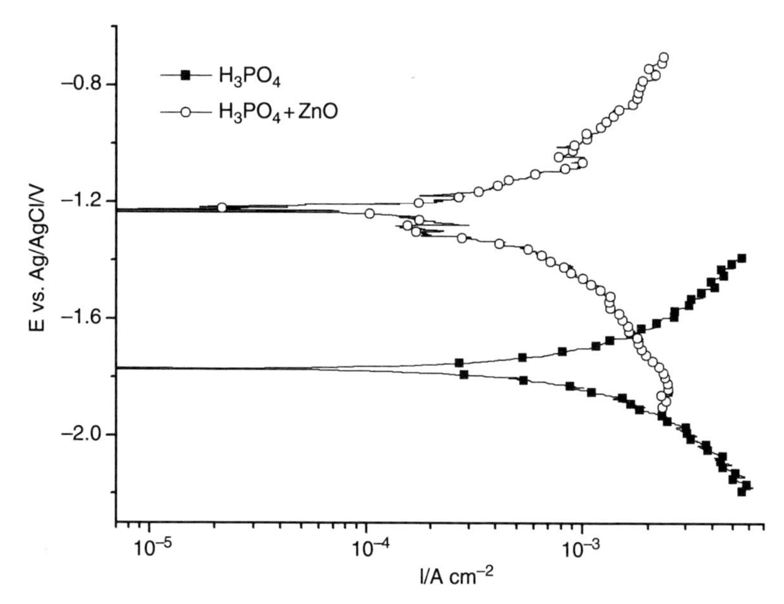

3.1 The polarization curves of Mg in 0.42 M H_3PO_4 (pH = 2.3) and phosphating (0.42 M H_3PO_4 + 0.34 M ZnO, pH = 2.7) solutions.

in 0.42 M H_3PO_4 (pH = 2.3) and phosphating (0.42 M H_3PO_4 + 0.34 M ZnO, pH = 2.7) solutions are shown in Fig. 3.1.

Mg in the phosphating solution shows a much more positive corrosion potential (around −1250 mV) and lower current densities than those in the H_3PO_4 acid. The increase in corrosion potential and decrease in current densities indicate a beneficial effect of ZnO, an essential component in a Zn phosphating bath solution; ZnO has a significant inhibition effect on Mg by promoting the formation of a phosphated layer on Mg. This raises the possibility that the conventional phosphating bath solution may be slightly modified to improve the phosphating performance for Mg alloys by adjusting the ratio of ZnO to phosphoric acid in the bath.

It has been shown that a Mg surface exposed in the ZnO containing phosphating solution consists of $Zn_3(PO_4)_2 \cdot 4H_2O$ (hopeite) and Zn (hexagonal structure). This indicates that Zn can participate in the film formation process in the phosphating solution, which should be responsible for the slower dissolution of Mg in the phosphating solution. In this case, ZnO is inhibitive and its inhibition mechanism is typically a precipitation-type. However, the amount of ZnO needed is relatively large, which means ZnO is not very effective. Moreover, phosphate also has an inhibition effect on Mg. If the pH value of the solution is controlled and unchanged, addition of phosphate can also slow down the dissolution of Mg.

3.3.2 Pyrazine and piperazine for GW103 Mg alloy in ethylene glycol (EG)

As many EG containing commercial coolants cannot offer adequate corrosion protection for Mg alloy components, there is an urgent need for new inhibitors. Mg-10Gd-3Y-0.5Zr (GW103) is a rare-earth containing Mg alloy, which may be used for engine blocks. Selection of effective corrosion inhibitors for this possible engine block alloy in EG is essential and important. The polarization curve results of GW103 alloy in EG solutions mixed with 500 ppm of piperazine and pyrazine, respectively at 25°C are listed in Table 3.1.

The additions of piperazine and pyrazine to the EG blank solution reduce the corrosion rate of the Mg alloy. Similar results can also be obtained at 90°C. (see Table 3.2). Although the corrosion rates of GW103 at 90°C are higher than those at 25°C, the addition of piperazine or pyrazine still improves the corrosion resistance of the Mg alloy at the higher temperature. Pyrazine has a better corrosion inhibition effect than piperazine. The corrosion inhibition effect of the pyrazine organic inhibitors can also be confirmed by weight loss measurements at room and high temperatures (see Table 3.3). Pyrazine has high inhibition efficiencies at both temperatures. Pyrazine and piperazine are both adsorptive inhibitors. Their inhibition behavior should be attributed to the formation of an adsorptive film on GW103.

3.3.3 Tetraphenylporphyrin for magnesium and AZ91D in NaCl solution

Apart from phosphating and EG solutions, most practical corrosive environments normally contain NaCl and are represented by a NaCl solution. It is more important for inhibitors to inhibit the corrosion of Mg and its alloys in such a medium. Many existing inhibitors for conventional metals are organic. However, many organics synthesized in the modern chemical engineering industry cause environmental or health problems. If some species originally from natural creatures can be used as inhibitors, there will be no environmental and health concerns.

Porphyrin is such an example. It is a compound containing a big conjugated ring with pyrrole as its basic component. It has attracted particular attention in chemistry, medicine, materials and information science in recent years because of its great stability and unique optical and electronic properties.[27,29] Chlorophyll, heme and vitamin B12 are important derivatives of pyrrole that can be found widely in nature. These substances have a great chelating ability with many metal ions. It is expected that porphyrin-Mg chelate compounds would be formed on a Mg alloy surface,

Table 3.1 Polarization parameters of GW103 in 50% (vol.%) EG solution mixed with 500 ppm of piperazine and pyrazine at 25°C

	I_{corr} (μA/cm²)	β_c (−mV/dec)	*OCP* (V/SCE)
Blank	27.9	371	−1.391
with 500 ppm piperazine	7.9	608	−1.212
with 500 ppm pyrazine	1.2	868	−1.149

Table 3.2 Polarization parameters of GW103 in 50% (vol.%) glycol solution mixed with 500 ppm of piperazine and pyrazine at 90°C

	I_{corr} (μA cm^{-2})	β_c (−mV dec^{-1})	*OCP* (V/SCE)
Blank	42.5	435	−1.444
With 500 ppm piperazine	15.5	459	−1.347
With 500 ppm pyrazine	7.5	733	−1.218

Table 3.3 Weight loss results of GW103 in 50% (vol.%) EG solution with 500 ppm pyrazine inhibitor after 3 days of immersion

	Weight loss (g)		Weight loss rate (mg/cm².week)		$\eta_{(w)}$%	
	25°C	90°C	25°C	90°C	25°C	90°C
Blank	0.0045	0.0056	2.72	3.39	–	–
With 500 ppm pyrazine	0.0018	0.0016	1.09	0.97	65.9	76.9

and thus there is an inhibiting effect if Mg is exposed to a porphyrin containing solution.

Table 3.4 shows the corrosion rates of pure Mg in a 0.05 wt.% NaCl solution containing tetraphenylporphyrin (TPP) inhibitor at different temperatures and pH values. The measured inhibition efficiency is around 80%. Similarly inhibition results of TPP can also be obtained on Mg alloy AZ91D. Table 3.5 lists the inhibition efficiencies of various concentrations of TPP inhibitor. The inhibition efficiency increases prominently with increasing TPP inhibitor concentration in the low concentration range. After the TPP concentration further increases from 10 to 50 ppm, the increase in inhibition efficiency is not evident.

The inhibition behavior of the TPP can be ascribed to the chelating of TPP with Mg. This postulation has been supported by ultra-violet absorption spectra. For synthesized TPP powder, due to the presence of the conjugate large ring, porphyrin compounds have a very strong absorption belt

Table 3.4 Estimated corrosion rates and inhibition efficiencies (η_e and η_w) based on the polarization curves and weight loss rate ($\bar{v}$) results of pure Mg in basic (0.05 wt.% NaCl), blank (0.05 wt.% NaCl + 5 ppm SDBS + 2200 ppm THF) and blank + 5 ppm TPP solutions at 25°C and 60°C

	b_c (mV/dec)	I_{corr} (A/cm²)	E_{corr} (V/SCE)	η_e%	$\bar{v}$ (mg/cm²/ week)	η(w)%
Basic solution (pH 6.8), 25°C	−216	2.41×10^{-5}	−1.402	–	–	–
Blank solution (pH 7.0), 25°C	−208	2.38×10^{-5}	−1.387	–	3.2671	–
Blank + 5 ppm TPP (pH 7.0), 25°C	−173	2.38×10^{-6}	−1.472	90%	0.5816	83%
Basic solution (pH 6.8), 60°C	−268	5.69×10^{-5}	−1.553	–	–	–
Blank solution (pH 7.0), 60°C	−259	5.48×10^{-5}	−1.524	–	3.9758	–
Blank + 5 ppm TPP (pH 7.0), 60°C	−142	6.02×10^{-6}	−1.575	89%	0.8637	79%

Table 3.5 Calculated parameters for the polarization curves of AZ91D Mg alloy in 0.05 wt.% NaCl solution with and without various concentrations of TPP inhibitor at 60°C

	I_{corr} (A cm⁻²)	η%
Blank solution	2.78×10^{-5}	–
1 ppm	4.17×10^{-6}	85%
5 ppm	3.15×10^{-6}	89%
10 ppm	2.67×10^{-6}	90%
50 ppm	2.27×10^{-6}	91%

at 300–340 nm. This absorption Soret belt (also called B belt) is caused by the $\alpha_{1\mu}(\pi)$–$e_g(\pi^*)$ transaction of porphyrin shown in Fig. 3.2. There are also other four weak absorption Q belt peaks at 517.4, 554.6, 594.2 and 648.7 nm, which are caused by the porphyrin $\alpha_{2\mu}(\pi)$–$e_g(\pi^*)$ transaction. After exposure to TPP-containing solution, the Q belt of the surface film formed on the Mg alloy surface, disappears, which is attributed to the formation of Mg-porphyrin complex. The four N atoms on the porphyrin ring chelate with the central metal Mg ions, which improves the molecular symmetry and the energy levels. Therefore, the Soret belt undergoes red shift (416.0 nm) and the Q belt becomes invisible, suggesting the formation of Mg-porphyrin complex.

XPS spectra of Mg alloy surfaces after 3 days of immersion in 0.05 wt.% NaCl with and without TPP inhibitor also support the chelating inhibition

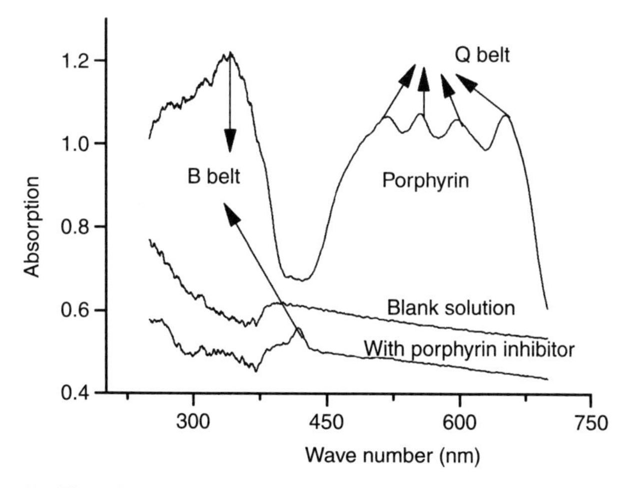

3.2 The ultra-violet absorption spectra of the synthesized TPP powder and the Mg surface films after 7 days of immersion in the blank solution and blank + 5 ppm TPP solution at 60°C.[30]

mechanism. In addition to the usual O1s, C1s, Mg 2p peaks, the peaks of Al 2p and Zn 2p also appear in the blank solution and TPP inhibitor containing solution. Moreover, an N1s peak is also observed in the film formed in the TPP inhibitor containing solution. The peak at 49.79 eV for Mg 2p, which can be attributed to $Mg(OH)_2$, suggests that the Mg alloy surface in the blank solution is mainly covered with a corrosion product film. This film is not compact and can be penetrated by corrosive Cl^- ions. In the solution containing TPP inhibitor, the surface film consists of Mg hydroxide and TPP-Mg complex. The N1s peak is asymmetrical and can be fitted into two peaks corresponding to C=N and C-N, respectively. These indicate that the dissolved Mg^{2+} ions could directly bind with TPP molecules via an N atom and form a protective TPP-Mg film on the Mg alloy surface. These protective complex compounds provide a high inhibition effect on AZ91D Mg alloy.

3.3.4 Sodium dodecylbenzenesulfonate

Sodium dodecylbenzenesulfonate (SDBS) is an organic inhibitor that has been used on steel. It is worthwhile testing its inhibition effect on Mg and Mg alloys. The polarization curve results of Mg, AZ31 and AZ91D in NaCl solution with and without SDBS are listed in Table 3.6. AZ91D is the most corrosion resistant and thus the inhibition efficiency of the tested inhibitor is the lowest compared to Mg and AZ31.

EIS of Mg in NaCl solution with different concentrations of SDBS (25–500 ppm) are shown in Fig. 3.3. The corrosion resistance increases and reaches

Table 3.6 Fitting results of polarization curves of Mg, AZ31 and AZ91D in 1 wt.% NaCl solution with and without 500 ppm SDBS

	I_{corr} (μA/cm²)	E_{corr}(−V)
Blank Mg	89	1.502
Blank Mg + SDBS	47	1.545
Blank AZ31	140	−1.454
Blank AZ31 + SDBS	52	−1.529
Blank AZ91D	17	−1.491
Blank AZ91D + SDBS	14	−1.513

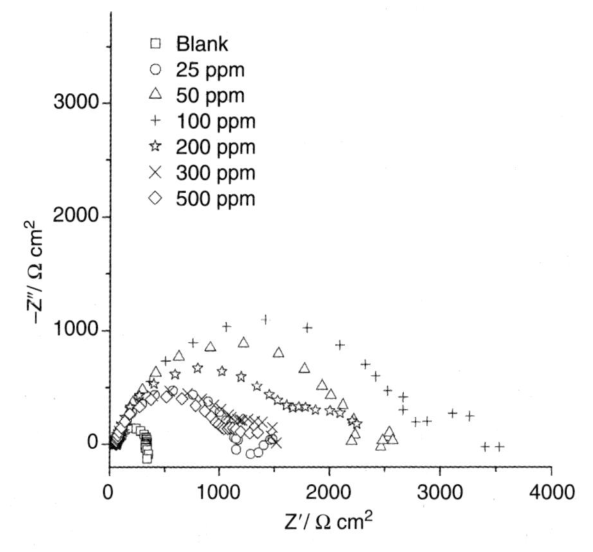

3.3 The EISs of Mg in 1 wt.% NaCl solution with different concentrations of SDBS (25, 50, 100, 200, 300, 500 ppm, pH = 7.3, 7.4, 7.5, 7.7, 7.8, 8.0, respectively).

its maximum value at inhibitor concentration 100 ppm; when the concentration increases further, the corrosion resistance decreases. This indicates that there is a saturation adsorption concentration around 100 ppm for SDBS on the Mg surface. Similarly, SDBS also has a saturation adsorption concentration on AZ31B, but it is around 200 ppm.

It should be noted that the corrosion potential E_{corr} shifts negatively as the concentration of SBDS increases, indicating that the inhibition by SDBS is a cathodic type. SDBS might be preferentially adsorbed on cathodic sites. AZ31 contains many cathodic particles, which allow more SDBS to be adsorbed. Therefore, the saturation concentration of SDBS on AZ31 (200 ppm) is higher than that on Mg (100 ppm).

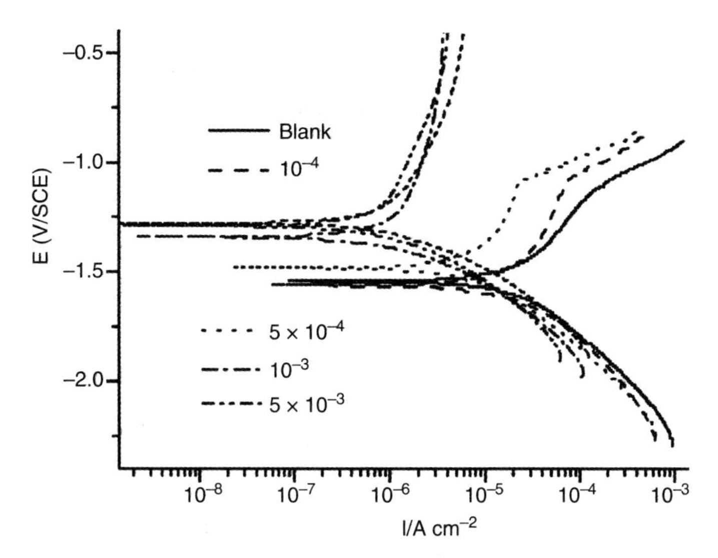

3.4 Potentiodynamic polarization curves of GW103 Mg alloy in ASTM solution with different concentrations of APTS-Na at 25°C.[24]

The Mg surface after immersion in the blank solution is completely covered by a thick, loose and porous corrosion product film, while the specimen surface exposed to the solution containing SDBS has a denser and uniform film. The results suggest that the Mg surface after immersion in the blank NaCl solution is mainly covered by $Mg(OH)_2$, and SDBS is included in the surface film in the SDBS containing solution. The improvement in corrosion resistance by the inhibitor comes through modifying the surface film. Adsorption of SDBS leads to an inhibitor-containing or doped film, which offers a better protection for the substrate.

3.3.5 Inorganic zinc nitrate and organic sodium aminopropyltriethoxysilicate

Inorganic inhibitors and organic inhibitors can be used together as combined inhibitor packages. Figure 3.4 shows the polarization behavior of GW103 Mg alloy in blank ASTM1384-87D solution (148 mg/L Na_2SO_4 + 138 mg/L $NaHCO_3$ + 165 mg/L NaCl, pH = 8.2) and the ASTM solution with different concentrations of sodium aminopropyltriethoxysilicate (APTS-Na) inhibitor at 25°C. The addition of APTS-Na causes a decrease in corrosion rate of the Mg alloy. With increasing APTS-Na concentration, the anodic process is inhibited more prominently than the cathodic process and the corrosion potential shifts positively, which indicates that APTS-Na is primarily an anodic inhibitor. Moreover, as APTS-Na is an alkaline organic salt, its addition can lead to an

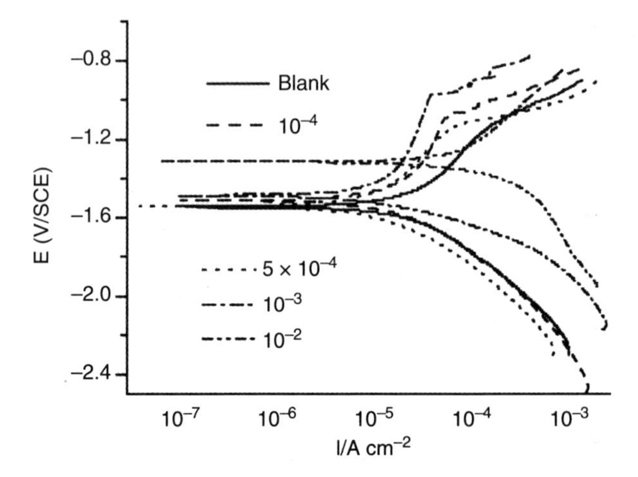

3.5 Potentiodynamic polarization curves of GW103 Mg alloy in ASTM solution containing different concentrations of Zn^{2+} at 25°C.[24]

increase in pH value of the solution. In order to eliminate the effect of solution alkalination on the corrosion of Mg alloy, the concentration of APTS-Na inhibitor was controlled below 5×10^{-3} M.

Figure 3.5 shows the polarization curves of GW103 Mg alloy in ASTM solution with various concentrations of Zn^{2+}. It can be seen that both the anodic and cathodic reactions of the electrode are only slightly inhibited and the corrosion potential does not change remarkably by less than 10^{-3} M Zn^{2+}. However, when 10^{-2} M of Zn^{2+} is added, the polarization current densities, particularly the cathodic one, increase considerably. This can be ascribed to a rapid decrease in pH value of the solution due to the hydrolysis of the large amount of added Zn^{2+}.

If the above two organic and inorganic inhibitors are used together, the inhibition efficiency can be further improved. Figure 3.6 shows the polarization curves of GW103 in ASTM solution containing a total concentration of 6×10^{-4} M of APTS-Na and Zn salt. Both anodic and cathodic reactions are drastically inhibited by the mixed inhibitor package with 10^{-4} M Zn^{2+} and 5×10^{-4} M APTS-Na. On the other hand, the mixed inhibitor package with 10^{-4} M APTS-Na and 5×10^{-4} Zn^{2+} has a lower inhibition effect than the former inhibitor package. The addition of mixed inhibitor packages results in a positive shift of E_{corr} and a remarkable decrease of I_{corr}, suggesting that the anodic reaction of Mg alloy is inhibited more efficiently than the cathodic one by this mixed inhibitor packages. The η of 10^{-4} M Zn^{2+} + 5×10^{-4} M APTS-Na and that of 5×10^{-4} M Zn^{2+} + 10^{-4} M APTS-Na reach 96% and 90%, respectively. Weight loss measurements confirm the polarization curve estimation. Table 3.7 lists the inhibition efficiencies of the single inhibitor and

Table 3.7 Weight loss and inhibition efficiency $\eta_{(w)}$ results of GW103 alloy in the blank solution and the solution containing 5×10^{-4} M APTS-Na + 10^{-4} M Zn^{2+} [24]

	Average weight loss rate (mg/cm^2/week)	$\eta_{(w)}$ (%)
Blank	3.5595	–
5×10^{-4}M APTS-Na	1.2452	65.02%
10^{-4} MZn^{2+}	1.4239	59.99%
5×10^{-4} M APTS-Na + 10^{-4} M Zn^{2+}	0.3002	91.57%

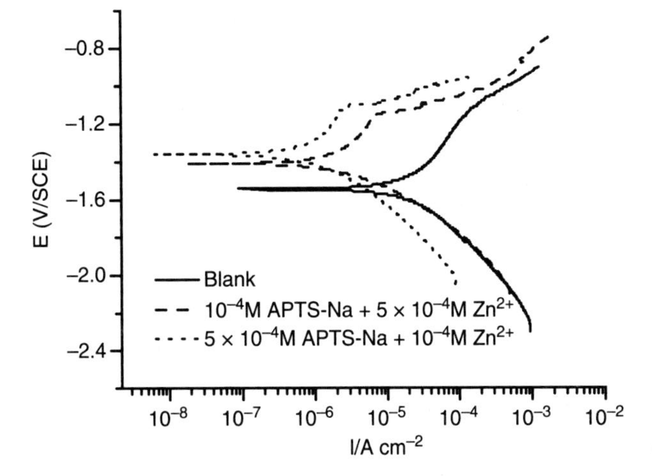

3.6 Potentiodynamic polarization curves of GW103 Mg alloy in ASTM solution containing different concentrations of the mixed inhibitors of APTS-Na and Zn2+ at 25°C.[24]

the mixed inhibitor package (5×10^{-4} M APTS-Na + 10^{-4} M Zn^{2+}) calculated from weight loss measurements.

The synergistic inhibition effect of a mixed inhibitor package can be defined as a parameter s:[31]

$$s = \frac{1 - \eta_A - \eta_B + \eta_A + \eta_A}{1 - \eta_{AB}} \quad [3.1]$$

Where η_A and η_B are individual inhibition efficiencies of inhibitors A and B, respectively; η_{AB} is the inhibition efficiency of the mixed inhibitor A and B. If $s > 1$, the mixed inhibitor package has a synergistic inhibition effect. Otherwise, the inhibition effect is antagonistic. The s values for 5×10^{-4} M APTS-Na + 10^{-4} M Zn^{2+} and 10^{-4} M APTS-Na + 5×10^{-4} M Zn^{2+}

are 3.85 and 2.17, respectively. The former inhibitor package has a much better synergistic effect than the latter.

XPS spectra of the films formed on GW103 alloy in the blank solution and the solution containing 5×10^{-4} M APTS-Na + 10^{-4} M Zn^{2+} suggest that both APTS-Na and Zn^{2+} participate in the film formation. It is likely that the organic silicate and inorganic Zn^{2+} form a complex depositing on the Mg alloy surface to inhibit its corrosion. The deposition of a complex, rather than each of the two inhibitors separately, may account for the synergistic inhibition effect.

3.3.6 Inorganic sodium phosphate and organic sodium dodecylbenzenesulfonate

SDBS can also be used together with phosphate as a combined organic-organic inhibitor package to inhibit the corrosion of a Mg alloy. The inhibition performance of this inhibitor package is evaluated on GW103 in an EG solution at ambient and elevated temperature. The polarization curves of GW103 in the blank EG solution and those with sodium phosphate or SDBS as an inhibitor at 25°C and 90°C are displayed in Fig. 3.7. The addition of sodium phosphate leads to a positive shift of E_{corr} and a retarded anodic process at 25°C, which means that the phosphate inhibitor follows an anodic inhibition mechanism. The addition of SDBS leads to a negatively shifted $E_{corr,}$ suggesting a cathodic inhibition effect which could result from adsorption of SDBS molecules on some cathodic sites on the GW103 surface. The shifts of E_{corr} by addition of phosphate and SDBS at 90°C are much less significant than those at 25°C, indicating that a high temperature is not beneficial for the deposition of phosphate or the adsorption of SDBS on the GW103 surface.

The polarization curves of GW103 alloy at different concentration ratios of the selected inorganic inhibitor to organic inhibitor (sodium phosphate:SDBS) are presented in Fig. 3.8. These curves show that the inhibitor addition does not affect the cathodic polarization curve significantly. Instead, the anodic polarization current densities are evidently reduced. This anodic inhibition behavior is probably due to adsorption and deposition of the organic and inorganic inhibitors on the GW103 surface, which inhibits the anodic dissolution more than the cathodic reaction. A comparison of the polarization curves in Fig. 3.8 suggests that the combination of the organic and inorganic inhibitors at a certain ratio can offer a much better corrosion protection than individual organic or inorganic inhibitors in the EG solution; their inhibition efficiencies are higher at 25°C than those at 90°C in the inhibitor containing solutions.

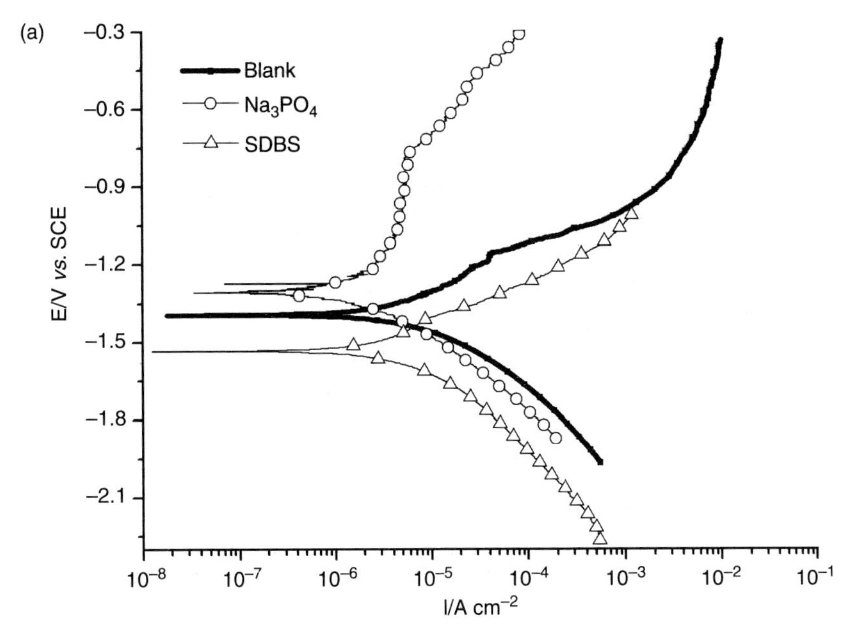

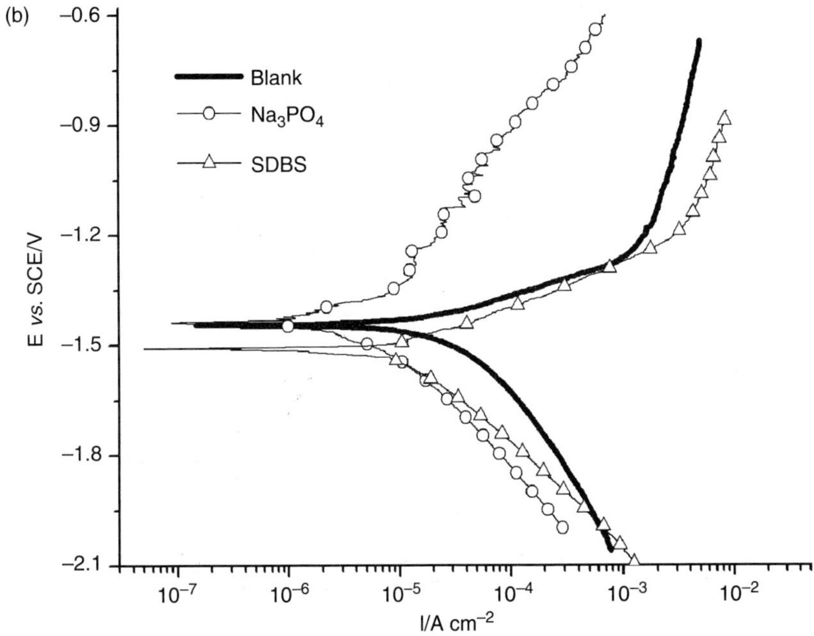

3.7 Polarization curves for GW103 in the 50% (vol.%) EG solution with and without 0.5 g/L Na_3PO_4 or 0.5 g/L SDBS at (a) 25°C; (b) 90°C.

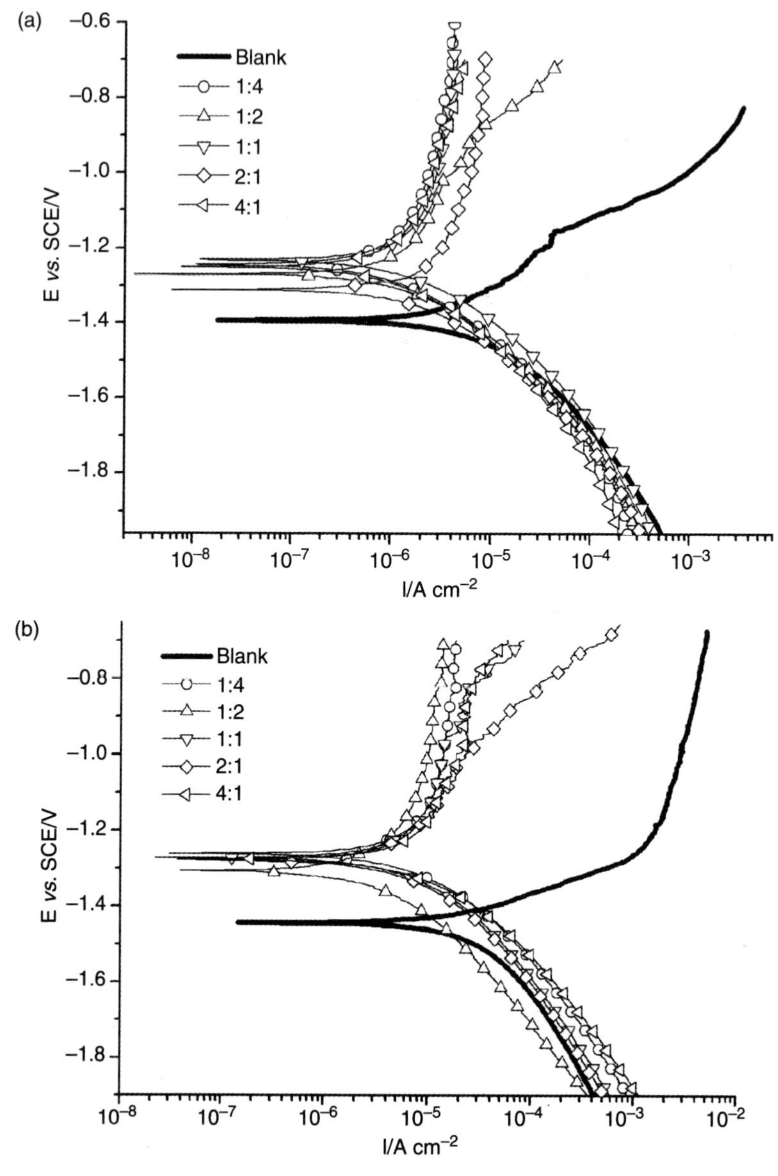

3.8 Polarization curves for GW103 in 50% (vol.%) EG solutions with various ratios of Na_3PO_4 over SDBS (the total inhibitor concentration 1 g/L) at (a) 25°C and (b) 90°C.

It is proposed that sodium phosphate and SDBS act individually as moderate corrosion inhibitors on GW103 Mg alloy in an EG solution, and the combination of sodium phosphate and SDBS as an inorganic-organic inhibitor package has a significantly improved inhibition efficiency. The inhibition efficiency of the phosphate-SDBS inhibitor package increases with increasing total inhibitor concentration, but decreases with increasing temperature. The phosphate-SDBS combination has an anodic inhibition mechanism due to a modified surface film formed on the GW103 surface. The modification of the surface film by the phosphate-SDBS inhibitor package results in a strong synergistic inhibition effect on GW103 in the EG solution at both ambient and elevated temperatures.

3.4 The influence of inhibitors on other materials

In practical applications, Mg alloys are likely to be used together with other metals. It is important to learn if an inhibitor effective for an Mg alloy has any detrimental effect on those metals.

3.4.1 Influence of inhibitors on Al alloy A380

Aluminum alloys are quite often used together with Mg alloy. For example, to mitigate the galvanic corrosion of a Mg alloy component in contact with a steel part, an Al alloy isolator between them is usually recommended. The polarization curves of aluminum alloy A380 in NaCl solutions with KF, Na_3PO_4, SDBS and Na_3PO_4 + SDBS were measured and are shown in Fig. 3.9. All the additives accelerate the corrosion of A380. This is because the pH value of the blank NaCl solution increases with these additives and the Al alloy becomes unstable in the alkalized solutions. The detrimental effect of solution alkalinity on Al can be confirmed by adjusting the pH value of the Na_3PO_4 + SDBS containing solution. Figure 3.10 shows the polarization curves of Al in the blank NaCl solution and the solution containing Na_3PO_4 + SDBS before and after the pH value is adjusted from 9.5 back to 7.0 by HCl.

When Na_3PO_4 + SDBS is added in to the blank solution, the E_{corr} of Al decreases by about 0.75 V and the corrosion rate increases significantly. After the pH value is adjusted back to the original pH 7.0 with HCl, the E_{corr} increases back to the original value and the polarization curve becomes similar to that in the blank NaCl solution. These results suggest that the inhibitors cannot inhibit the corrosion of A380, and in practice its use for Mg alloys should be carefully controlled if Al alloys are included.

3.4.2 Galvanic corrosion and inhibition

The inhibitors that are effective for Mg alloys may alter the galvanic corrosion behavior of the Mg alloys in contact with other metals. The galvanic corrosion of GW103 and AZ91D coupled to an Al alloy (A7005) in a 50% (vol. %) EG solution at ambient (25°C) and elevated temperatures (90°C)

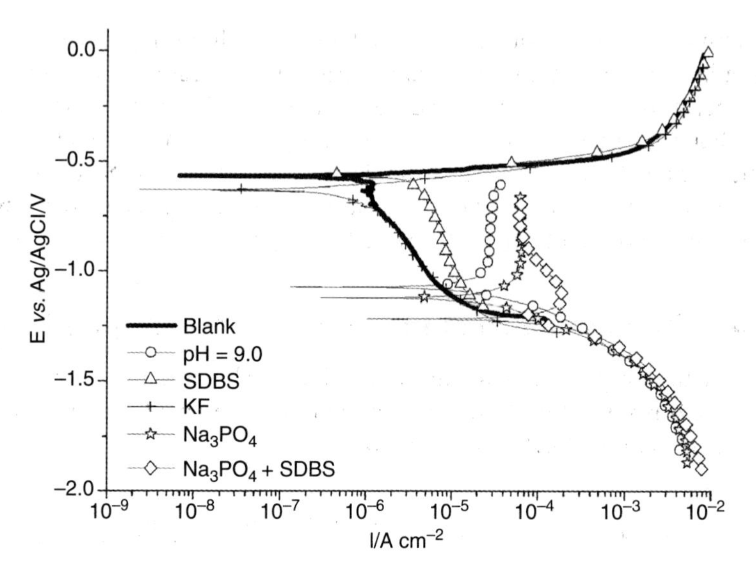

3.9 Polarization curves of A380 in 1 wt.% blank NaCl solution, the solution with 500 ppm KF, 0.4 ppm NaOH (pH = 9.0), 500 ppm Na_3PO_4, 500 ppm SDBS and 250 ppm Na_3PO_4 + 250 ppm SDBS, respectively.

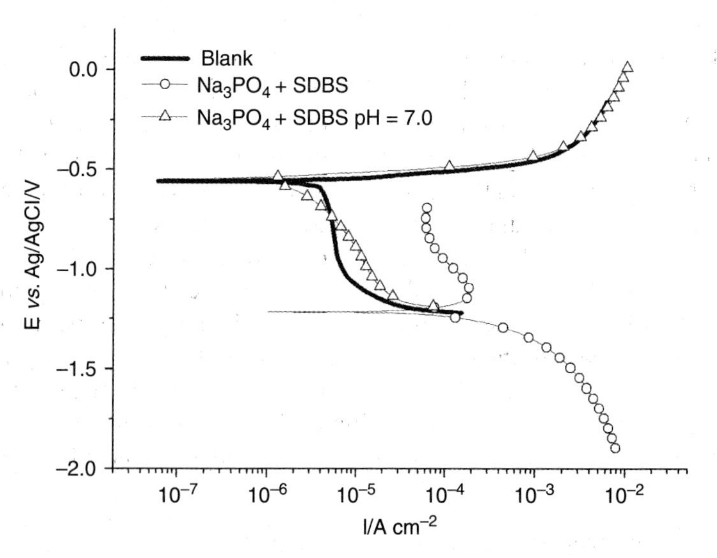

3.10 The polarization curves of A380 in 1 wt.% blank NaCl solution (pH = 7), unadjusted blank + 250 ppm Na_3PO_4 + 250 ppm SDBS solution (pH = 9.5) and adjusted blank + 250 ppm Na_3PO_4 + 250 ppm SDBS solutions (pH = 7.0).

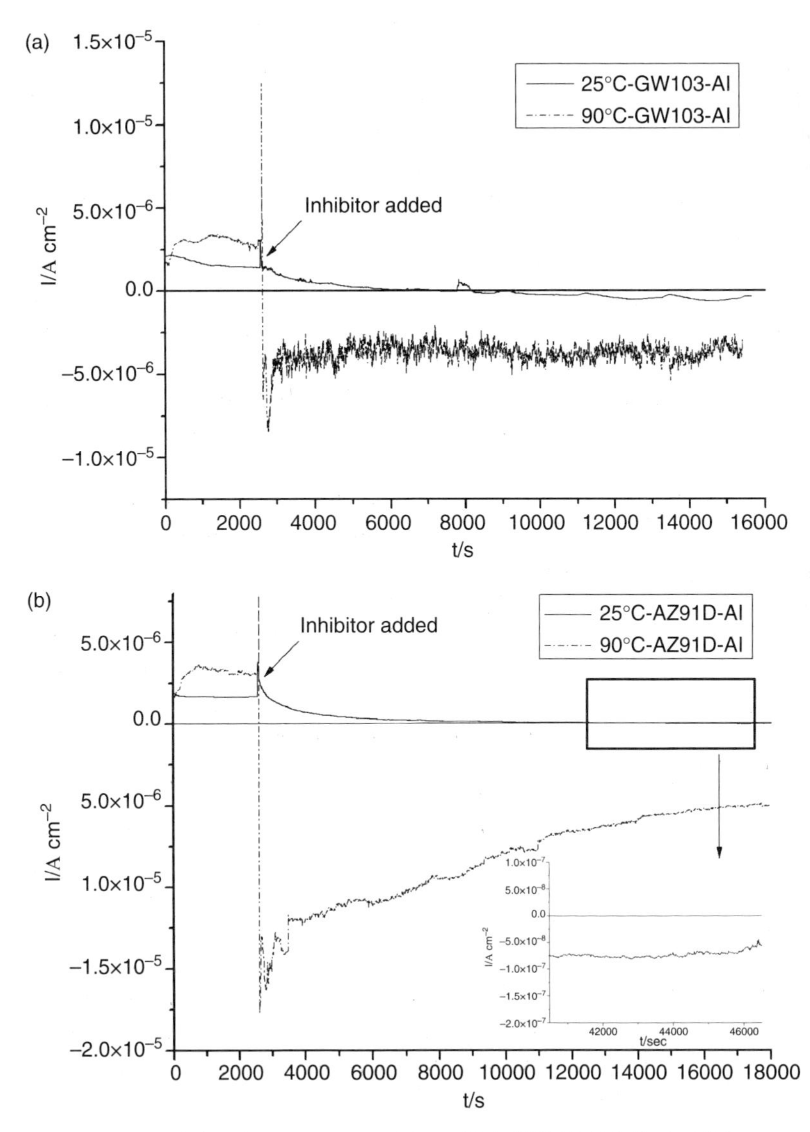

3.11 Galvanic current densities of (a) GW103-A7005 and (b) AZ91D-A7005 couples in 50% (vol.%) blank EG solution with and without 500 ppm Na_3PO_4 + 500 ppm SDBS inhibitors at 25°C and 90°C.

after addition of 500 ppm Na_3PO_4 + 500 ppm SDBS inhibitors is investigated by electrochemical methods. Figure 3.11 shows the time dependence of galvanic current densities of GW103-A7005 and AZ91D-A7005 couples at 25°C and 90°C. At 25°C, the galvanic current density ascends sharply at the moment when 500 ppm Na_3PO_4 + 500 ppm SDBS inhibitors are added into the 50% (vol.%) blank EG solution, then it descends back to the original values rapidly. The sudden increase and decrease in galvanic current density could be due to disturbance of the solution by inhibitor addition. The decrease in galvanic current density is probably related to the fact that the inhibitors gradually reduce the open circuit potential (OCP) differences between the anode and cathode in the galvanic couples. At 90°C the galvanic current density ascends sharply and immediately descends to a negative value on the addition of the inhibitors. The negative values of the galvanic current densities continue all the time after inhibitor addition, which suggests that the anodes and cathodes of the galvanic couples are reversed, that is, A7005 becomes an anode and GW103 or AZ91D becomes a cathode in the galvanic couples after the inhibitor addition.

The galvanic corrosion behavior of GW103 and AZ91D coupled to carbon steel and Cu in 50% (vol.%) EG solution with and without 500 ppm phosphate + 500 ppm SDBS and 500 ppm phosphate + 500 ppm benzoate as inhibitors was investigated by measuring their OCPs and coupling potentials (CPs). Both inhibitors decrease the OCPs of carbon steel and Cu and increase the OCPs of Mg alloys, thus decreasing the OCP differences of the galvanic couples. These mean that addition of sodium phosphate + benzoate could decrease the driving force of galvanic corrosion between Mg alloys and carbon steel or Cu.

The stable galvanic current densities of GW103-carbon steel, AZ91D-carbon steel, GW103-Cu and AZ91D-Cu couples before and after the addition of inhibitors at 25°C are listed in Table 3.8. The addition of Na_3PO_4 + SDBS accelerates the galvanic corrosion of these couples, while Na_3PO_4 + benzoate inhibit the galvanic corrosion of the couples at ambient temperature.

The influence of Na_3PO_4 + benzoate on the galvanic current density of GW103, AZ91D coupled to Cu and carbon steel at 90°C is presented in Table 3.9. The Na_3PO_4 + benzoate inhibitor package cannot protect the Mg alloys from galvanic corrosion at 90°C. Galvanic current densities of Mg-Cu and Mg-carbon steel couples increased after 500 ppm Na_3PO_4 + 500 ppm benzoate is added in the blank EG solution.

Therefore, although both Na_3PO_4 + SDBS and Na_3PO_4 + benzoate may inhibit the corrosion of the Mg alloys, their effects on the galvanic corrosion of these alloys coupled with steel or copper are different. At 25°C, Na_3PO_4 + benzoate has a better inhibition effect on the galvanic corrosion of GW103 or AZ91D coupled to carbon or Cu, whereas Na_3PO_4 + SDBS accelerates the galvanic corrosion. Both inhibitor packages have a detrimental effect on their galvanic corrosion at 90°C.

Table 3.8 Galvanic current densities of GW103-carbon steel, AZ91D-carbon steel, GW103-Cu and AZ91D-Cu couples before and after the addition of different inhibitors at 25°C

	$I_{g\ GW103\text{-carbon steel}}$ (μA/cm²)	$I_{g\ AZ91D\text{-carbon steel}}$ (μA/cm²)	$I_{g\ GW103\text{-Cu}}$ (μA/cm²)	$I_{g\ AZ91D\text{-Cu}}$ (μA/cm²)
Blank	3.4	3.1	4.0	5.5
Na_3PO_4 + SDBS	3.8	13.4	9.8	14.5
Na_3PO_4 + benzoate	1.5	1.1	1.0	3.4

Table 3.9 Galvanic current density of GW103, AZ91D coupled to Cu and carbon steel before and after the addition of 500 ppm Na_3PO_4 + 500 ppm benzoate inhibitor package at 90°C

	I_g (μA/cm²) in blank solution	I_g (μA/cm²) with 500 ppm Na_3PO_4 + 500 ppm benzoate
GW103-Cu	18.7	79.4
AZ91D-Cu	12.4	79.0
GW103-carbon steel	11.9	85.3
AZ91D-carbon steel	9.9	77.9

3.5 Inhibition and other corrosion protection techniques

To further improve the inhibition effect on Mg alloys, inhibitors can sometimes be applied together with other corrosion protection methods, such as coatings. If inhibitors can be embedded or mixed in a coating, then if the coating becomes damaged, the inhibitors will be released from the coating and protect the substrate Mg alloys from corrosion attack in the coating damaged area.

For example, the protection effect of Si sol-gel coatings for AZ91 Mg alloy in 0.5M Na_2SO_4 solution can be enhanced by doping Ce^{3+} or Zr^{4+}.[32] In the inhibition system, Ce^{3+} and Zr^{4+} play an inhibitor role. Galio *et al.*[33] reported that Si sol and Zr sol doped with 8-hydroxyquinoline inhibitor enhanced the corrosion protection of the Mg alloy and did not lead to deterioration of the barrier properties of the sol–gel matrix in 0.005 M NaCl solution. Similar studies include the inhibition effect of 2-methyl piperidine as inhibitor added to Si sol coatings in 0.005%, 0.05% and 0.5% (wt.%) NaCl solutions.[34] Montemor *et al.*[35] investigated the corrosion behavior of AZ31 substrates pre-treated with a water soluble bis-aminosilane modified with multiwall carbon nanotubes (CNTs) in 0.05 M NaCl. The microscopic analysis suggested that the CNTs were homogeneously dispersed in the coating and that the CNTs acted as support for inhibitor storage. The studies listed above suggest that inhibitors can be flexibly used together with other corrosion prevention approaches for Mg alloys.

3.6 Conclusions and future trends

Precipitation inhibitors such as tetraphenylporphyrin, organic sodium APTS-Na and Na_3PO_4 have good inhibition efficiency for Mg and its alloys. These inhibitors could react with Mg^{2+} and form insoluble compounds on the Mg surfaces to slow down the corrosion process. Both $Zn(NO_3)_2$ and APTS-Na have a moderate corrosion inhibition effect on GW103 Mg alloy in the ASTM D1384-87 corrosive water. However, if these two inhibitors are used together, the inhibition efficiency can be above 95%. There is a strong synergistic effect between these two inhibitors. A compact film composed of $Mg(OH)_2$ and zinc silicate deposited on the surface should be responsible for high inhibition efficiency.

The combination of SDBS and Na_3PO_4 as an inorganic-organic inhibitor package also has a significantly improved inhibition effect on GW103 Mg alloy at ambient and high temperatures. The inhibition efficiency of the SDBS + Na_3PO_4 package increases with increasing total inhibitor concentration and decreases with increasing temperature. The Na_3PO_4 + benzoate package has a better inhibition effect on the galvanic corrosion of GW103 and AZ91D coupled to carbon and Cu, whereas the Na_3PO_4 + SDBS package accelerates their galvanic corrosion.

Tetraphenylporphyrin chelates with dissolved Mg^{2+} to form a TPP-Mg compound and precipitates on AZ91D surface. The precipitated TPP-Mg mixes with the original $Mg(OH)_2$ film in the outer layer significantly reducing the porosity of the surface film and thus reduces the corrosion of AZ91D. Pyrazine and piperazine are adsorptive inhibitors. Their inhibition behavior should be attributed to formation of an adsorptive film.

Currently, effective inhibitors for Mg alloys are rare, especially in some severe corrosion conditions. Additionally, the understanding of inhibition mechanisms on Mg alloys is very limited. Therefore, there is no practical application of inhibitors for Mg alloy components in industry. However, the increasing use of the Mg alloys in some applications requires inhibition techniques. An ideal corrosion inhibitor that can be adopted in industry should be highly effective, inexpensive, widely available, non-toxic and environment-friendly. In future development of inhibitors for Mg alloys, apart from understanding the inhibition mechanism, these requirements should also be set as important goals.

3.7 References

1. R. Álvarez-Bustamante, M. Abreu-Quijano and M. Romero-Romo (2009), 'Electrochemical study of 2-mercaptoimidazole as novel corrosion inhibitor for steels', *Electrochim. Acta*, **54**, 5393–5399.

2. F. Matjaz and M. Ingrid (2010), 'Inhibition of copper corrosion by 1, 2, 3-benzotriazole: A review', *Corros. Sci.*, **52**, 2737–2749.
3. C.J. Lan, C.Y. Lee and T.S. Chin (2007), 'Tetra-alkyl ammonium hydroxides as inhibitors of Zn dentrite in Zn-based secondary batteries', *Electrochim. Acta*, **52**, 5407–5416.
4. S. Chaves, S. Marques and M. Amélia Santos (2003), 'Iminodiacetyl-hydroxamate derivatives as metalloproteinase inhibitors: Equilibrium complexation studies with Cu(II), Zn(II) and Ni(II)', *J. Inorg. Biochem.*, **97**, 345–353.
5. M.A. Amin, H.H. Hassan and S.S. Abd El Rehim (2008), 'On the role of NO_2- ions in passivity breakdown of Zn in deaerated neutral sodium nitrite solutions and the effect of some inorganic inhibitors: Potentiodynamic polarization, cyclic voltammetry, SEM and EDX studies', *Electrochim. Acta*, **53**, 2600–2609.
6. K.F. Khaled (2003), 'The inhibition of benzimidazole derivatives on corrosion of iron in 1 M HCl solutions', *Electrochim. Acta*, **48**, 2493–2503.
7. C. Jeyaprabha, S. Sathiyanarayanan, K.L.N. Phani and G. Venkatachari (2005), 'Investigation of the inhibitive effect of poly (diphenylamine) on corrosion of iron in 0.5 M H_2SO_4 solutions', *J. Electroanal. Chem.*, **585**, 250–255.
8. J.M.R. Genin, L. Dhouibi and P. Refait (2002), 'Influence of phosphate on corrosion products of iron in chloride polluted concrete simulating solutions: Ferrihydrite vs green rust', *Corrosion*, 58, 467–478.
9. H. Ashassi-Sorkhabi and E. Asghari (2012), 'Effect of solution hydrodynamics on corrosion inhibition performance of zinc sulfate in neutral solution', *J. Electrochem. Soc.*, **159**, c1–c7.
10. L. Sumari and R.S. Chaudhary (2008), 'Some triphosphate as corrosion inhibitors for mild steel in 3% NaCl solution', *Indian J. Chem. Techn.*, **15**, 364–374.
11. G. Song and D. St John (2005), 'Corrosion of magnesium alloys in commercial engine coolants', *Mater. Corros.*, **56**, 15–23.
12. D.S. Mehta, S.H. Masood and W.Q. Song (2004), 'Investigation of wear properties of magnesium and aluminum alloys for automotive applications', *J. Mater. Process. Technol.*, **155–156**, 1526–1531.
13. G. Song and Z. Xu (2010), 'The surface, microstructure and corrosion of magnesium alloy AZ31 sheet', *Electrochim. Acta*, **55**, 4148–4161.
14. G.L. Song (2010), Chapter 1 "Corrosion Behavior and Mitigation Strategy of Magnesium Alloys", in H. Dong (ed.) *Surface Engineering of Light Alloys – Al, Mg and Ti*, CRC Press, UK, Woodhead publishing Limited.
15. L.J. Li, Z.M. Yao, J.L. Lei, H. Xu, S.T. Zhang and F.S. Pan (2009), 'Adsorption and corrosion inhibition behavior of sodium dodecylbenzenesulfonate on AZ31 magnesium alloy', *Acta Physico-Chimica. Sinica*, **25**, 1332–1336.
16. J.L. Lei, L.J. Li, H.S. Yu, M.L. Chen, S.T. Zhang and F.S. Pan (2008), 'Inhibition of hexamethylenetetramine on corrosion of magnesium alloy in simulated coolant', *Chem. Res. & Appl.*, **20**(4), 461–464.
17. E. Slavcheva and G. Schmitt (2002), 'Screening of new corrosion inhibitors via electrochemical noise analysis', *Mater. Corros.*, **53**(9), 647–655.
18. N.H. Helal and W.A. Badawy (2011), 'Environmentally safe corrosion inhibition of Mg-Al-Zn alloy in chloride free neutral solutions by amino acids', *Electrochim. Acta*, **56**, 6581–6587.
19. M. Mosialek, G. Mordarski and P. Nowak (2011), 'Phosphate-permanganate conversion coatings on the AZ81 magnesium alloy: SEM, EIS and XPS studies', *Surf. Coat. Tech.*, **206**, 51–62.

20. L. Kouisni, M. Azzi, M. Zertoubi and F. Dalard (2004), 'Phosphate coatings on magnesium alloy AM60 part 1: Study of the formation and the growth of zinc phosphate films', *Surf. Coat. Tech.*, **185**, 58–67.
21. S.J. Kim, Y. Zhou. R. Ichino, M. Okido and S. Tanikawa (2003), 'Characterization of the chemical conversion films that form on Mg-Al alloy in colloidal silica solution', *Met. Mater. Int.*, **9**, 207–213.
22. L. Li and Zh. Yao (2008), 'Corrosion inhibition of magnesium alloy in 3.5% NaCl medium by tungstate', *Electrochem.*, **14**(4), 427–430.
23. L. Li and L. Lei (2007), 'Inhibition of molybdate salt on dissolution of magnesium alloys in phosphroic acid', *Surf. Technol.*, **36**(1), 16–18.
24. J.Y. Hu, D.B. Huang, G.L. Song and X.P. Guo (2011), 'The synergistic inhibition effect of organic silicate and inorganic Zn salt on corrosion of Mg-10Gd-3Y magnesium alloy', *Corros. Sci.*, **53**, 4093–4101.
25. D.B. Huang, J.Y. Hu, G.L. Song and X.P. Guo (2011), 'Inhibition effect of inorganic and organic inhibitors on the corrosion of Mg-10Gd-3Y-0.5Zr alloy in an ethylene glycol solution at ambient and eleveated temperatures', *Electrochim. Acta*, **30**, 10166–10178.
26. G.-L. Song (2011), Chapter 1 'Electrochemistry of Mg and its Alloys' in G.-L. Song (ed.), *Corrosion of Magnesium Alloys*, UK, Woodhead publishing Limited.
27. M. Dhifet and M.S. Belkhiria (2010), 'Synthesis, spectroscopic and structural characterization of the hig-spin Fe(II) cyanato-N and thiocyanato-N "picket fence" porphyrin complexes', *Inorg. Chim. Acta*, **363**, 3208–3213.
28. J.T. Lu, L.Z. Wu and J.Z. Jiang (2010), 'Helical nanostructures of an optically active metal-free porphyrin with four optically active binaphthyl moieties: Effect of metal-ligand coordination on the morphology', *Eur. J. Inorg. Chem.*, **25**, 4000–4008.
29. S.K. Ghosh, R. Patra and S.P. Rath (2010), 'Synthesis, structure and photocatalytic activity of a remarkably bent, cofacial ethane-linked diiron (III) μ-oxobisporphyrin', *Inorg. Chim. Acta*. **363**, 2791–2799.
30. J.Y. Hu, D.B. Huang, G.L. Song and X.P. Guo (2012), 'A study on tetraphenylporphyrin as a corrosion inhibitor for pure magnesium', *Electrochem. Solid St.*, **15**(6), c13–c15.
31. H. Tavakoli, T. Shahrabi and M.G. Hosseini (2008), 'Synergistic effect on corrosion inhibition of copper by sodium dodecylbenzenesulphonate (SDBS) and 2-mercaptobenzoxazole', *Mater. Chem. Phys.*, **109**, 281–286.
32. V. Barranco, N. Carmona and J. C. Galvan (2010), 'Electrochemical study of tailored sol-gel thin films as pre-treatment prior to organic coating for AZ91 magnesium alloy', *Prog. Org. Coat.*, **68**(4), 347–355.
33. A.F. Galio, S.V. Lamaka and M.L. Zheludkevich (2010), 'Inhibitor-doped sol-gel coatings for corrosion protection of magnesium alloy AZ31', *Surf. Coat. Tech.*, **204**(9–10), 1479–1486.
34. H. Shi, F. Liu and E. Han (2009), 'Corrosion protection of AZ91D magnesium alloy with sol-gel coating containing 2-methyl piperidine', *Prog. Org. Coat.*, **66**(3), 183–191.
35. M.F. Montemor and M.G.S. Ferreira (2008), 'Analytical characterization and corrosion behavior of bis-aminosilane coatings modified with carbon nanotubes activated with rare-earth salts applied on magnesium alloy', *Surf. Coat. Tech.*, **202**(19), 4766–4774.

Part II

Surface treatment and conversion

4 Surface cleaning and pre-conditioning surface treatments to improve the corrosion resistance of magnesium (Mg) alloys

D. HÖCHE, Helmholtz-Zentrum Geesthacht Centre for Materials and Coastal Research, Germany and
A. NOWAK and T. JOHN-SCHILLINGS, Henkel AG & Co. KGaA, Germany

DOI: 10.1533/9780857098962.2.87

Abstract: This chapter discusses the initial stages involved in the handling of magnesium surfaces and looks at the important issues surrounding the preparation of magnesium alloy surfaces for further treatment, including optimized conditions for finishing processes. It begins with a review of mechanical cleaning procedures such as grinding, chemical cleaning, acid pickling and activation. Later sections go on to explain the method of action of various cleaners and their relationship to common post-processing treatments. The chapter ends with a discussion of current trends in the pre-conditioning of magnesium surfaces in the context of magnesium-based engineering and a brief summary.

Key words: cleaning, pre-treatment, residues, pickling, solvent.

4.1 Introduction

This chapter addresses the topic of physical and chemical surface cleaning and pre-treatments, which has often been overlooked in the research literature. In fact, the topic is particularly important in the case of magnesium: the cleaning of magnesium surfaces and their pre-conditioning for further treatment is the most appropriate method for ensuring top coat adhesion, acceptable surface appearance and improved corrosion performance. The cleaning processes described in this chapter are key in the manufacture of magnesium components, since they allow the production of components with defined surface properties, and support the huge variety of surface treatment technologies, from chemical conversion or anodizing up to painting or plating. Figure 4.1 shows a typical set up of the initial treatment steps applied to magnesium components. This chapter will focus on the first two

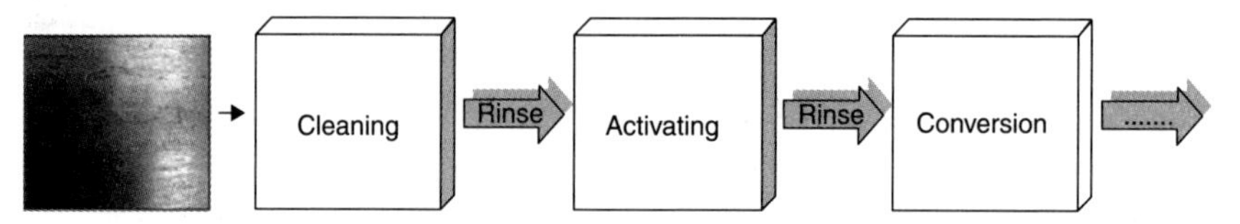

4.1 Standard initial steps of a process chain of handling magnesium surfaces (conversion coating in this example).

steps of the conditioning process: cleaning and preconditioning/activation. The finishing stages, such as conversion treatments and plating, are varied in nature, and their applications are discussed in other chapters.

This chapter will combine industrial experience with scientific knowledge gained from fundamental studies. From a technical point of view, the cleaning of metal surfaces is required because they are exposed to a variety of environmental conditions during their life cycle. During manufacturing or forming processes, it is usually necessary to remove common residues such as:

- oils and fats;
- partially corrosion-induced compounds such as oxides or hydroxides (e.g. humidity);
- adhcring soil particles;
- grinding dust;
- mould release agents (lubricants) and anti backing agents.

All of the above residues interact in different ways as corrosion 'accelerators', since they interfere with subsequent treatments and the quality of the finish.

In principle there are two main reasons for the lack of corrosion resistance in a metal, both of which are affected by pre-treatment. The first of these is internal galvanic corrosion caused by secondary metallic phases or impurities, mainly iron, nickel and copper. These impurities (which can come from the mould, for example) or their compounds act as minute cathodes in a corroding medium, creating micro-cells with an anodic Mg matrix (Bharadwaj *et al.*, 2007), which cause the dissolution of the anodic magnesium. In many manufacturing processes these more noble elements are enriched close to the surface, meaning that these well-known degradation problems occur and counteractive measures become absolutely essential. The second major cause is the fact that the quasi-passive hydroxide film on magnesium alloys in aqueous environments is much less stable than the passive film that forms on metals such as aluminum or stainless steels: at a pH value below 10.5 local breakups can occur in the top layer (Eliezer and Alves, 2002). This quasi-passivity leads to poor pitting resistance in Mg and Mg alloys (Makar and Kruger, 1993).

There are a number of basic principles to take into account in tackling the challenge of corrosion: electrical insulation of magnesium alloy-based parts; avoidance of mistakes in design; and the prevention of surface contamination. The following sections will describe how the technical surfaces of magnesium components (mainly casts or wrought alloys) can be treated and handled to ensure optimized post-processing conditions. Physical/mechanical cleaning processes are discussed, along with possible chemical treatments; guidance is then provided on activating and rinsing magnesium alloy surfaces for particular applications, before some observations on related trends in magnesium surface engineering.

In the context of this chapter, corrosion describes the 'negative' interaction of magnesium surfaces with the environment. Influencing factors include temperature, humidity and air contamination such as salts or pH-value. As a result, many kinds of interaction are possible. This chapter deals with the most common problems arising from conditions such as exposure to water or other solvents at certain temperatures. Magnesium is normally resistant to alkaline solutions but is significantly affected by acids (with an especially high affinity to Cl^{+} and O^{2+} ions). A summary of magnesium's chemical affinity to various compounds and elements can be found in Kurze (2006). In dry air its corrosion performance is quite high and is maintained up to a working temperature of 350°C. The naturally-formed oxide MgO is more or less stable. In wet exposure conditions, however, a strong reaction takes place induced by magnesium's strong reactivity (enthalpy of formation ΔH_f of the hydroxide $Mg(OH)_2$ = −925 kJ/mol) and by galvanic corrosion. The following subsections will discuss the key question of how to tackle this reactive behavior at the beginning of the entire process chain.

4.2 Principles of physical or mechanical cleaning

What is meant by 'mechanical cleaning'? A broad definition would be all processes that induce abrasion or removal of detrimental agents from surface regions affected by various contamination processes (at the micro- or macro-scale) with no chemical reaction. Before each type of treatment, the surface should be cleaned using standard solvents to remove soils, dust or organic residues: this will be discussed in greater depth later in the chapter. Mechanical cleaning offers rapid removal of most kinds of residues, particularly lubricants and dirt, and generally occurs over a broadly macroscopic smoothed surface.

The most common examples of mechanical cleaning methods are grinding and polishing. Areas near the surface of castings (skins), for example, are removed by abrasion in order to reduce the impurity level. This must be carried out with a great deal of care (and with extensive metallurgical experience to ensure correct temperature control, rapid rinsing and adequate drying) as a result of the induced twinning observed in some magnesium alloys. Strength under mechanical deformation is heavily dependent on the type of alloy. HP magnesium, for example is much more prone to twinning than AZ31. The elemental composition of the grinder material (e.g. SiC) does not normally affect the corrosion properties of the alloy. A widely-used method of abrasion is dry abrasive blasting (Avedesian and Baker, 1999), particularly sand blasting, which is most familiar to the public. Song and Xu (2010) carried out a scientific study on AZ31, using glass beads at 60 psi, which led to reduced corrosion resistance due to micro stresses and strain combined with Fe contamination. The abrasive technique should only be applied to the same spot on the surface of the magnesium for a few seconds; negative effects can also be counteracted by the subsequent use of acid pickling. Sandblasting should therefore only constitute the very first step in the surface handling and conditioning process.

A very promising mechanical cleaning method is known as the 'slurry blasting process' which is based on suspended abrasives and can be used as a surface conditioner before electroplating. The main advantages of the process are its efficiency at removing heavy contaminations arising from corrosion reactions and the final surface state that it offers, which is matt and can be treated and finished directly. Other methods used, mainly at the industrial level, are brushing and bowl abrading. One of the main advantages of the brushing method is its in-line capability. The wires used should interact without roughening the surface, thereby removing oxides. Carbon steel brushes should be avoided; rotary files with flutes or a flap brush are preferred, and should be used only on one material in order to minimize secondary element effects. In the bowl abrading method, also known as barrel abrading, casted magnesium components are treated under wet or vibrating conditions (Friedrich and Mordike, 2006), most commonly through the use of ceramic media suspended in a rinse solution. The effective removal of contamination during rinsing should be guaranteed by sufficient flow to avoid recontamination.

In summary, mechanical cleaning should be used for the first rough conditioning of various surface states, mostly in the form of industrial processes like milling, but subsequent chemical processing is normally inevitable. In most cases corrosion behavior is dependent on processes that act on the microscale, arising from impurities or galvanic cells which cannot be removed by 'macro' mechanical cleaning alone. The following subsection on chemical treatments will provide an insight into methods that can act on the results of these microscale processes. It is finally important to note that the handling of magnesium, including any mechanical cleaning processes, has to be carried out in accordance with national and international fire protection standards.

4.3 Principles of chemical cleaning

Chemical cleaning methods are processes such as vapor degreasing, solvent cleaning, emulsion cleaning, alkaline cleaning and acid cleaning (Nwaogu *et al.*, 2009), which are used to remove any kind of surface contamination arising from manufacturing or mechanical cleaning, as well as oxide or hydroxide layers (i.e. former coatings); and to provide optimal surface conditions for post-processing, such as conversion coating and top coats. The treatments are usually performed by dipping, spraying, rinsing or soaking, depending on the requirements for the quality of the end product and on the limitations of the cleaning product. All chemical cleaners are based on chemical reactions at the surface and are characterized by their ability to remove impurities, and to convert and condition the surface. They can principally be divided into two categories, with different subgroups in each. The first category, water-based cleaners, can be subdivided into three groups depending on the pH value: neutral cleaners (pH 5–7) applied during e.g. vapor degreasing; mild (pH 8–10) and strong (pH > 11) alkaline cleaners; and acid cleaners (pH 1–4), which are very strong and are widely used for pickling and sometimes even activating. The second category includes easy to use organic solvent cleaners such as those used in emulsion cleaning (sometimes as an aqueous solution).

For 99% of applications, the performance of a cleaning agent can be benchmarked by its ability to create homogeneous wettable and single-elemental/single-compound metal surfaces. The chemical industry offers a broad range of commercial cleaners, in most cases alkaline cleaners for liquid degreasing; the content of these cleaners is not made publicly available, but they will always contain one or more of the following:

- builders, such as phosphates (STPP), sodium-based carbonates or silicates.
- caustic (soda), in most cases alkaline hydroxides.
- tensides/surfactants.
- complexing agents to bend ions for prevention of oxidizing effects.
- additives, including corrosion protection additives, inhibitors, defoamers and solubilizer.

All commercial cleaners also have a variety of properties that affect the following treatment steps in different ways. The following paragraphs will discuss the varying influences of the different categories and subgroups of chemical cleaners. The problems surrounding the use of chromates will also be discussed, since environmental protection standards in Europe limit the use of chromates to aerospace applications.

4.3.1 Category I: Water-based cleaners

Neutral cleaning

Neutral cleaning is a new innovation and is not yet widely used for magnesium surface treatments. Neutral cleaners are mainly based on non-ionic reagents, and act like a disperser of material such as dirt and soil in the surrounding aqueous environment. In principle, the surfactant consists of both a hydrophilic and a hydrophobic molecule section. The hydrophobic part orients towards the soil, and surrounds, for example, a particle, creating what are known as micelles, which are taken up into the detergent solution that is removed at the end of the process.

Neutral cleaners may additionally contain some additives such as corrosion inhibitors, phosphates and solubilizers to achieve an optimized cleaning effect. To date, the chemical industry has continually improved the properties of the cleaning agents it produces. Neutral cleaners do not significantly attack a magnesium alloy surface and thus allow better stability in the cleaning process itself and consequently a longer component life time. The required chemical attack on the metal can be carried out during subsequent 'harder' treatment steps such as during the acidic activation stage.

Alkaline cleaning

Alkaline cleaning (pre-cleaning) is the most common means of achieving standard surface conditions that meet the requirements of post-processing techniques such as painting or plating, or even of the pickling and activation processes. Care must be taken, however, with regard to the reactivity between aluminum and alkalis. These cleaners are very effective at removing oil, grease and other contaminants, and also have very low abrasive properties. The most frequently used compounds are alkaline salts, normally as a builder, or caustics, silicates and phosphates combined with a balanced amount of surfactants. When sprayed onto the dirty component as an aqueous solution (preferably at high temperatures, ideally 50–70°C), the cleaner generally removes most soils. For industrial applications, soak cleaning stations are generally employed. An electroless pre-plating process permits a different cleaning procedure to be adopted (Sharma *et al.*, 1997), whereby alkaline cleaning is carried out in NaOH, 50 g/L and $Na_3PO_4.12H_2O$ (TSP, trisodium orthophosphate), 10 g/L at 55–65°C for up to 10 min leading to extremely favorable post-processing conditions. Further selected examples of cleaner application methods are shown in Table 4.1: for most processes the level of cleaning does not differ dramatically, and all offer at least a processable (i.e. technically acceptable) surface state.

One very interesting variant is alkaline electrocleaning, which is normally used before electroplating. The process is based on the use of electric

Table 4.1 Some other typical examples of applying alkaline cleaners

Cleaner	Process	Reference
NaOH 45 g/L $Na_3PO_4 \cdot 12H_2O$ 10 g/L	Plating on AZ91d	Zhang *et al.*, 2008
$NaOH/Na_3PO_4/Na_2O$ $\cdot NaSiO_2 \cdot xH_2O$ solution	Chromium free conversion	Zhao *et al.*, 2007
NaOH: 10.0 g/L Na_2CO_3 15.0 g/L Na_2SiO_3 10.0 g/L triethanolamine 5.0 g/L	Conversion coating on AZ31	Liang *et al.*, 2009
Naoh 50 g/L	Ca-p coatings on AZ31, Mg4Y	Singh *et al.*, 2011
Na_3PO_3, Na_2CO_3, Na_2SiO_3 and surfactant (sodium dodecyl benzene sulfonate)	Lanthanum conversion coating on Mg-Li alloy	Song *et al.*, 2011

currents to electrolyze the surrounding liquid. Cathodic cleaning is recommended for the removal of excessive soil, oxides and solid particles (adherents); anodic cleaning leads to pitting and film formation. In the cathodic setup, hydrogen gas is liberated at the surface of the magnesium, and acts as a scrubbing agent (in effect a type of mechanical cleaning). A close eye must be kept on the hydrogen evolution rate, since hydrogen embrittlement might be an issue (Song *et al.*, 2005).

Unlike aluminum, magnesium surfaces are not affected by high pH alkaline cleaners due to their chemical resistance to alkaline substances (Pourbaix diagram) (Avedesian and Baker, 1999). There is only one exception to this: some phosphates with pH <12 cause slight deoxidization of the surface. This behavior leads to the efficient removal of contaminants from cathodic acting surface areas such as beta phases. A particular challenge is posed in the case of the frequently-used magnesium A-alloys. Due to the cathodic acting surface areas of precipitates containing aluminum, for example, the anodic magnesium matrix provides protection against alkaline cleaning of Al containing precipitates. Thus, the combined use of alkaline and acid treatments becomes suitable. In a study carried out by Nwaogu *et al.* (2009), this technique was shown to offer very impressive results with NaOH at a pH value of 13.6.

Acid pickling

It is well-known that surface contamination with common elements such as iron (which causes mould release), nickel and copper induces corrosion when the content is above tolerance limits, due to the formation of micro galvanic cells. Liu *et al.* (2008) provide a useful summary of the elemental limits for several Mg-Al cast alloys. The first goal is therefore to remove these noble elements when they occur close to the surface (Skar *et al.*, 2004). In addition, efficient chemical treatments will also remove natural oxides, residues of lubricants, sands or alkaline cleaners and even chromates.

In this respect, acid pickling can be considered the state of art for magnesium alloys. It prepares surfaces for, among other post-processing procedures, uniform conversion treatments. Problems remain however, particularly with wrought products (such as rollings or forgings) which are heavily contaminated with lubricants; another serious problem is the changing quality of standard alloys, particularly in terms of the varying content of heavy metal contaminants. In the past chromic and hydrofluoric acid and other hazardous mixtures have been used: old handbooks suggest the use of chromic acid combined with ferric nitrate and potassium fluoride, which is effective. A very old commercial example of a typical chromic acid-based treatment is the Dow 17 process. CrO_3 chrome (III) (+HNO_3) is in fact still in use for pickling and plating processes (Jia *et al.*, 2007), but these acids will shortly be replaced, despite their very promising cleaning capabilities, for reasons of occupational and environmental safety. Nitric, phosphoric, organic or even other pickle solutions and mixtures therefore need to be brought up to the same level of performance. To gain an insight into the efficiency of the different acids, some detailed scientific studies on inorganic and organic acids will be described and analyzed below.

One further important point is that the use of surface active substances (surfactants) in acidic cleaners affects the etching rates on magnesium substrates. In this case the surface tension of the acidic etch solution is reduced, which results on the one hand in improved wetting of the magnesium surface, and on the other hand in the fast escape of the generated hydrogen within the interface between the magnesium surface and the acidic liquid. The gas bubbles formed run off faster from the metal surface and the abrasive reaction will be accelerated.

For inorganic acids, one very important and fundamental study is that carried out by Nwaogu *et al.* (2009). These authors tested different concentrations of sulphuric, nitric and phosphoric acids for their cleaning performance, and found that all showed a significant decrease in impurity levels, removing about 5 μm on AZ31 within a reasonable time of 60 s, as shown in Fig. 4.2. Surprisingly, surfaces were not completely clean.

The low impurity level of the bulk was not achieved, but after etching for at least 1 or 2 min the impurity level was lowered, depending on the concentration of the cleaning solution, to below the tolerance limit (Fig. 4.3). Other elements, such as copper or carbon, were not considered, either because their content is too low or because they can easily be removed.

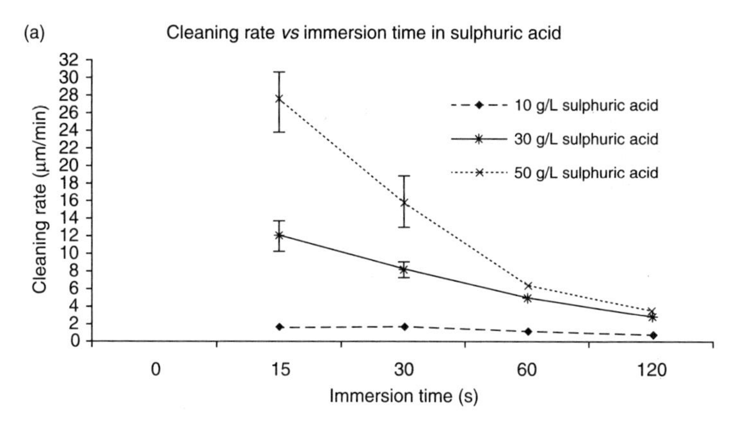

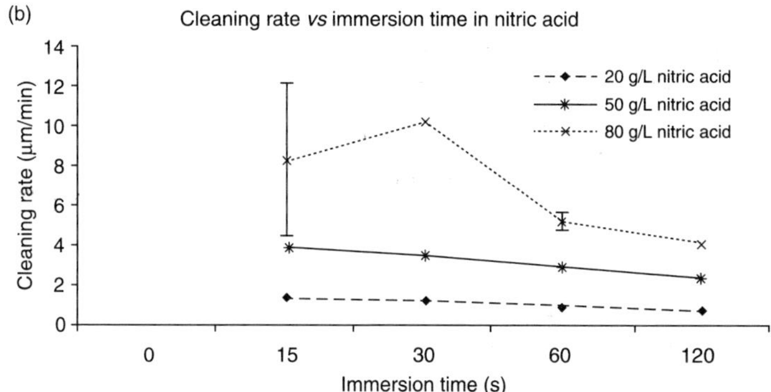

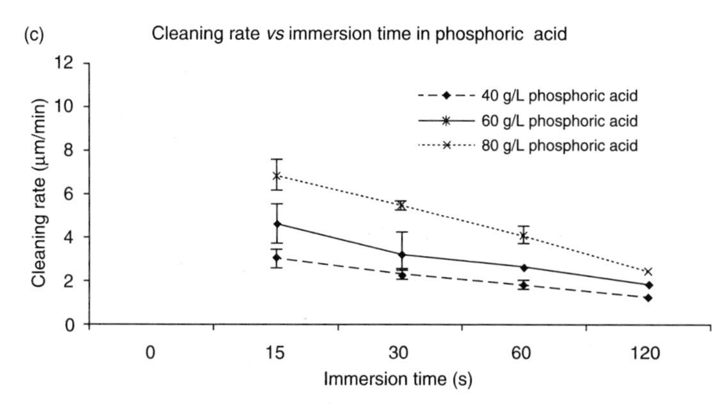

4.2 Variation of cleaning rate with immersion time for (a) sulphuric, (b) nitric and (c) phosphoric acids (Nwaogu *et al.*, 2009).

During the study, some drawbacks were observed. In order to gain a better understanding of these issues, the different aspects of etching must be taken into account. It is not only the removal of contaminants which determines the cleaning performance; a number of parameters can be adjusted to

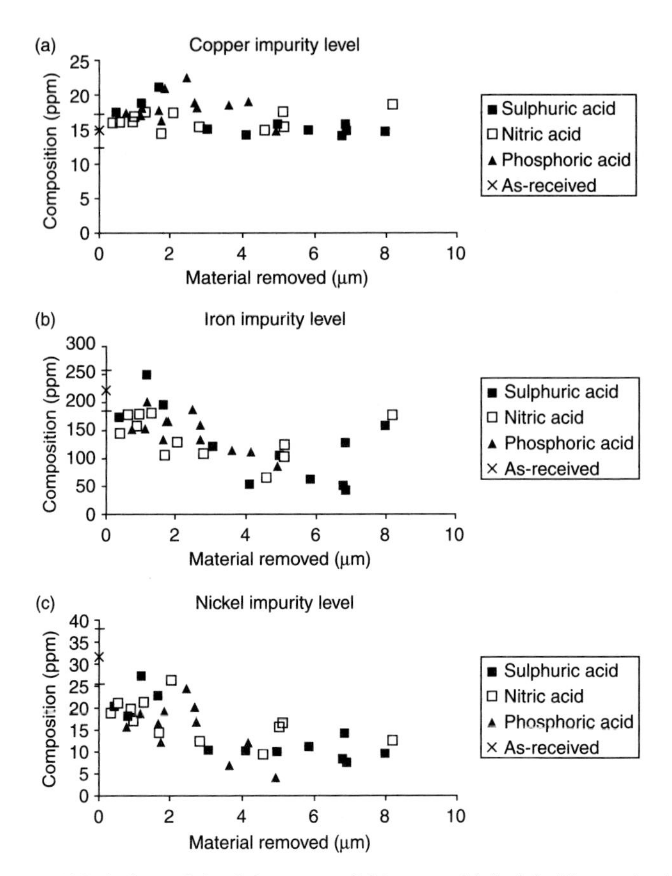

4.3 Variation of the (a) copper, (b) iron and (c) nickel impurity levels with material removed for all the measurements as determined by spark discharge-optical emission spectroscopy (Nwaogu *et al.*, 2009).

improve certain aspects of the cleaning performance, but these can also lead to an overall decrease in efficiency, so a careful balance is required. These factors include:

- An increase of etching rates at the start of the process results in more 'cleaning resistant' surfaces (as is the case with H_2SO_4).
- The formation of a protective coating formation (such as with phosphates) can reduce material removal.
- The formation of gas bubbles around the specimens can reduce or prevent contact with the cleaning solution.
- A spatial gradient on the amount of H^+ ion (pH-value) can arise due to high reactivity (as is the case with H_2SO_4).

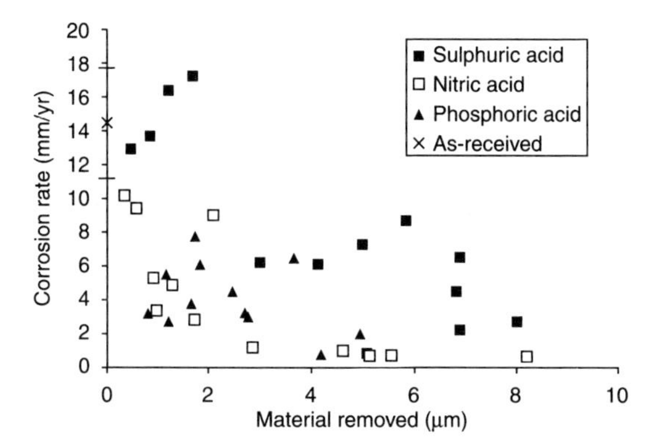

4.4 Variation of corrosion rates of AZ31 in salt spray test with material removed after cleaning in each acid (Nwaogu *et al.*, 2009).

In addition, roughness effects, homogeneity requirements or even just economic considerations can influence the overall result, rendering it impossible to provide a standard strategy for pickling magnesium.

In the case under examination here, nitric and phosphoric acids show a slightly better performance than sulphuric acid. This is because sulphuric acid is less effective at removing intermetallics, and because sulphuric residues have more detrimental effects on corrosion protection than nitrate residues or phosphates. Of HNO_3 and H_3PO_4, the nitric acid showed marginally better corrosion performance, as shown in Fig. 4.4.

The results discussed here provide highly scientific answers to the question of how to treat AZ31, but a general set of 'how to' guidelines is lacking. This is effectively impossible to achieve, but the trends exhibited in this study have been further validated by Supplit *et al.* (2007) for HF, and by Sharma *et al.* (2001) on Mg-Li alloy MLA9 for a combination of chromic acid and ferric nitrate. Other nitrate compounds and mixtures are also in use for the treatment of wrought alloys, including acetic, chromic or glycolic nitrate, and exhibit high performance in removing mill scales. Commercial suppliers offer a wide range of pickling solutions, containing varying amounts of HF and CrO_3 acids, combined with nitric or sulphuric acids; this renders it impossible to provide a qualitative description of these products.

Organic acid cleaning can be used as an alternative to the more commonly used inorganic acid-based processes. The study by Nwaogu *et al.* (2010) demonstrates acceptable cleaning performance (Fig. 4.5). The major mechanism is the same as that of inorganic acids in that they reduce the impurity levels. A more detailed investigation reveals two different phenomena occurring for oxalic, acetic and citric acid, which are quite different to those observed in the previous investigations.

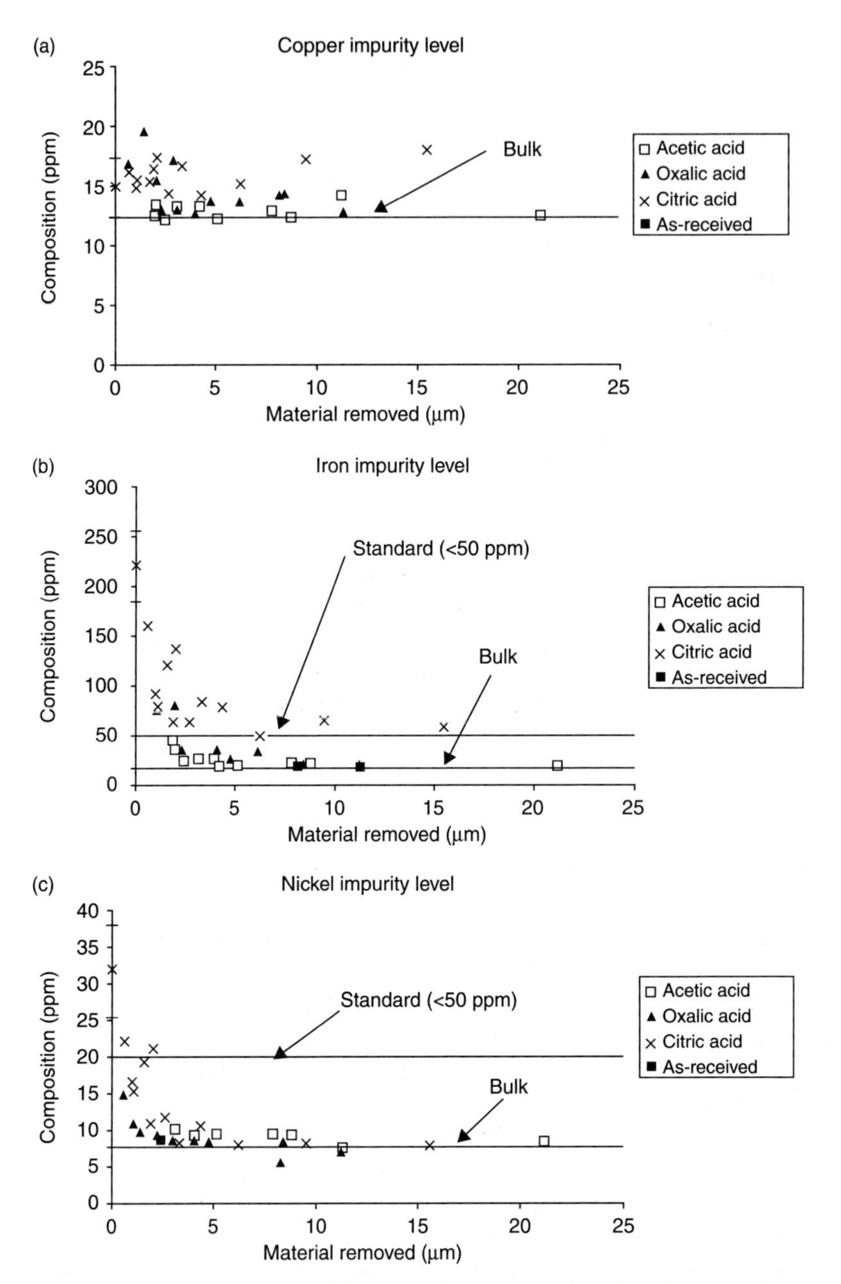

4.5 Variation of the (a) copper, (b) iron and (c) nickel impurity levels with material removed for all the treatments in the three different concentrations of acetic, oxalic and citric acids as determined by spark discharge-optical emission spectroscopy (Nwaogu *et al.*, 2010).

Acetic and citric acids form salts, such as magnesium citrate $Mg_3(C_6H_5O_7)_2$, which are soluble in aqueous solutions, meaning that protective films are not formed. The cleaning of the surface is achieved by the simultaneous dissolution of the matrix including impurities. However, oxalates of Mg such as MgC_2O_4 and of most impurity elements are insoluble, causing conglomerates at the surface with poor adhesion properties. This kind of 'protective film' flakes off during the treatments. Nevertheless, there is an advantage to be gained: impurities are inhibited in the flakes and are removed along with the main oxalates. A summary of the study shows that corrosion rates of less than 1 mm/year can be achieved by adjusting the elemental concentration and immersion time. In the previous AZ31 study on inorganic acids, a removal of 5 μm was demonstrated. One minute etching by organic acids under adequate conditions also meets this target. Figure 4.5 illustrates impurity levels after pickling: around <20 ppm for Fe and <10 ppm for Ni on the surface of AZ31.

The best cleaning performance of any organic acid was demonstrated by acetic acid, as it reached the best rate of efficiency by removing impurities but did not exert a negative influence on the surface. With oxalic acid, the main disadvantage is that it removes the flakes from the surface, despite a good etching performance. Citric acid offers the poorest performance in terms of the removal of impurities, especially iron, from the surface. Table 4.2 summarizes the most common inorganic and organic solutions used for pickling, and their performance in the cleaning and conditioning of a magnesium alloy surface with a technical surface uniformly contaminated with standard impurities.

In most cases the best performance is provided by a mixture of cleaners: the ideal composition of the mixture is strongly dependent on the level of contamination, the elements involved, the oxides, the alloying elements and, of course, on the aqueous concentration. The table above should be used only as a guideline, and the end-user should be aware that alternative solutions may be more suitable for specific problems.

4.3.2 Category II: Organic solvent-based cleaning using degreasers

This type of cleaner, such as acetone, hexane, isopropanol or methyl ethyl ketone will always be used for the removal of oil residues, release agents or other soil material. They are normally divided into three subgroups depending on their permittivity: non-polar, polar aprotic and polar protic solvents. The principles of interaction are the same as those of other standard materials, and a wide variety of commercial products are available for the purpose. Very effective cleaning performance is often achieved by vapor degreasing with trichloroethane TCE, with subsequent sand blasting assistance. It is

Table 4.2 Acid pickling performance of various single solutions on Mg alloys (scale – to +++), the overall performance includes handling and safety factors

Pickle solution	Comment	Cleaning performance	Overall performance
Sulphuric acid	Strong secondary effects	–	+
Phosphoric acid	Protective coating formation can reduce the material removal	+	+
Chromic acid	Hazardous	++	--
Hydrofluoric acid	Hazardous	+++	--
Nitric acid	Handling with respect to nitrous gas development	++	++
Acetic acid	Good handling / smells strong	++	++
Oxalic acid	Flake formation	+	+
Citric acid	Uncritical handling	–	–
Ferric nitrate	Mild interaction	+	+

important to note that a locked system is necessary for this process. The evaporated solvent directly condenses at the surface by greasing the whole component. After short interaction times the contaminated solvent runs off. ASTM D2651 recommends that care is taken in the application of this cleaning process due to the development of a hazardous mixture of oil and particles.

Solvent cleaning in a typical alkane (paraffin) bath can be used for any kind of degreasing and should be carried out before or after mechanical cleaning and whenever surface treatments on magnesium are in progress. Typical organic compounds (e.g. aromatic hydrocarbons such as terpene, citrus D limonene and so on) are normally dissolved and used as emulsion cleaners acting in a very similar way to the neutral aqueous cleaners. In most cases a mild alkaline character provides the best cleaning performance for the removal of soils or detergent residues without affecting the metallic magnesium surface. A thin oil film occurs after emulsion cleaners are applied, which usually provides temporary protection to the surface. This process can be combined with alkaline cleaning to achieve very clean and well conditioned surfaces. Due to the dominance of other cleaners, information on the best cleaning conditions remains limited: cleaners are currently being pre-checked for performance and to test the local corrosion behavior.

A short note is required at this point concerning the use of chemical cleaning with simultaneous mechanical support. In some cases, where heavy removable contamination is present on the magnesium surface, chemical cleaning assisted by additional mechanical treatment is recommended. The most effective and widely-used methods are combinations of barrel abrading or ultrasonic cleaning with either the alkaline cleaners described in the previous subsections or light acidic cleaners/activators: these provide reasonably high levels of performance and efficiency.

4.3.3 Rinsing

The rinsing step is often overlooked but has important effects in the case of magnesium surfaces. With standard water or distilled DI water, dilution effects occur. The electrical conductivity should be below 1500 μS/cm to avoid ion mobility and to meet the following requirements:

- removal of adhesive bath solutions;
- minimizing drag-in to subsequent zones;
- DI water to optimize the removal of salts;
- tap water rinse < 1000 μS/cm;
- DI water rinse < 50 μS/cm.

After rinsing, the level of contamination should be minimized and a clean metal surface should be available. Long rinse durations must be avoided because prolonged exposure to DI water weakens paint adhesion and promotes the formation of magnesium oxide (MgO). Pre-treated magnesium substrates should reach a peak metal temperature (PMT) of 100°C during a drying process to remove all humidity. Under these conditions, the corrosion protection and paint adhesion of organic coated material will show significant improvement.

4.3.4 Activation

We should first examine what is meant by 'activation'. In principle for magnesium alloys in the context of corrosion prevention, it means the preparation of a homogeneous and oxide (MgO) free metal surface which is highly suitable for post-processing and finishing. Acid pickling often equates to activation, but this is not always the case. There are two methods of activation with useful potential for magnesium. The first and most important is the use of fluxing compounds that wet and dissolve oxides, which in most cases are the standard chemicals. A perfect activator such as this has low viscosity at working temperature and a high wetting ability, disperses easily and is homogenous. Chemical activators can generally be classified into three groups according to the active substance:

- mineral acids: pickling,
- organic acids: pickling and complexing and
- fluorides: complexing and pickling.

The first of these has already been discussed in the section on inorganic acid pickle processes. A further example is the use of hydrochloric acid (HCl) prior to cerium-based conversion coating treatments (Brunelli *et al.*, 2005). The efficiency of organic acetic pickling on sol-gel coatings has been assessed by Supplit *et al.* (2007). The most popular and effective treatment agents are fluorides acting as separate activators. An example of standard activation by HF before plating is described by Liu *et al.* (2011), while the very detailed study by Zhao *et al.* (2007) demonstrates the influence of immersion into fluorides such as NH_4HF_2, $NaHF_2$ or NaF as a safe and very effective alternative for Cr-free conversion coatings. The use of these fluorides depends on the etching performance requirements and on their ability to remove the oxide and alloy components and contaminations. In the classical Dow process, for example, a mixture of phosphoric acid and NH_4HF_2 is used for activation (Hu *et al.*, 2005). Typically, etching rates of between 5 and 30 g/m^2 are preferred depending on the alloy, the impurity content and the oxidation state.

The second method of activation is mechanical, specifically shear or pressure (e.g., ultrasonic pressure). This method is clearly not very popular, nor particularly suitable for corrosion protection from a cleaning perspective, but does have interesting potential for other magnesium applications, for example, its use as a powdered hydride former in fuel cells (Lukashev *et al.*, 2008), and for soldering. However, since this chapter focuses on corrosion issues, we leave the question of mechanical activation aside. Since activation is usually the last step before finishing by coating or chemical conversion processes, it must be carried out with great care and attention, taking into account the subsequent processing steps and their requirements, as well as the quality of the material. Figure 4.6 provides a summary of the surface states of the various treatment steps carried out on technical magnesium, including a preview of conversion coated (chrome free and chromated) magnesium.

4.3.5 Corrosion performance without finishing

In most cases cleaning is simply one stage in a process chain. However, it is worthwhile examining what occurs when further finishing is not carried out after cleaning. Purely mechanical treatments do not usually have a positive effect on the corrosion performance. The work of Song and Xu (2010) clearly demonstrates this for sand blasting on AZ31 components. Mechanical cleaning, then, is in fact a conditioning rather than true cleaning process. The significant improvement in corrosion resistance due to impurity reduction brought about by acid pickling has been discussed in the previous sections. Nevertheless, the subsequent surface quality is subject to significant variation (Supplit *et al.*, 2007). Color variations, which are compound dependent, and roughness problems (i.e. lack of homogeneity) occur, making post-processing indispensable.

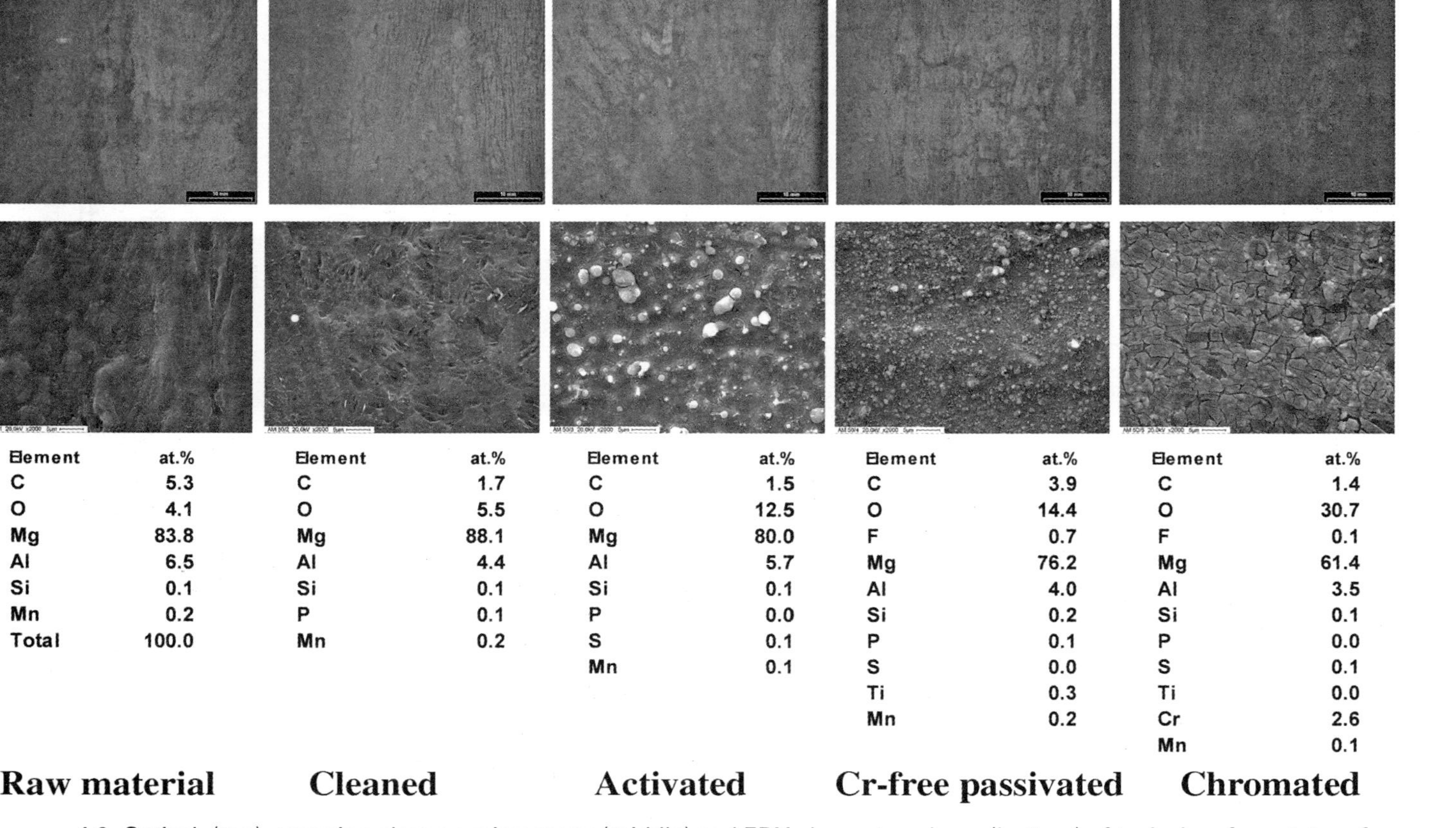

Raw material	
Element	at.%
C	5.3
O	4.1
Mg	83.8
Al	6.5
Si	0.1
Mn	0.2
Total	100.0

Cleaned	
Element	at.%
C	1.7
O	5.5
Mg	88.1
Al	4.4
Si	0.1
P	0.1
Mn	0.2

Activated	
Element	at.%
C	1.5
O	12.5
Mg	80.0
Al	5.7
Si	0.1
P	0.0
S	0.1
Mn	0.1

Cr-free passivated	
Element	at.%
C	3.9
O	14.4
F	0.7
Mg	76.2
Al	4.0
Si	0.2
P	0.1
S	0.0
Ti	0.3
Mn	0.2

Chromated	
Element	at.%
C	1.4
O	30.7
F	0.1
Mg	61.4
Al	3.5
Si	0.1
P	0.0
S	0.1
Ti	0.0
Cr	2.6
Mn	0.1

4.6 Optical- (top), scanning electron microscopy (middle) and EDX element analyses (bottom) of typical surface states of technical magnesium for different treatment processes (with kind permission of Henkel).

Table 4.3 Expected corrosion performance after single cleaning for typical treatments

Method	Acting process	Performance compared to untreated surface
Sand blast	Micro stress, cracks and contamination	↘
'Slurry' process	Impurity removal and conversion	↗
Wire brushing	Contamination	↘
Flat brushing	Secondary material (soil, etc.) removal	→
Acid pickle	Impurity removal and compound formation	↗
Alkaline cleaning	Material (soil, oxides, etc.) removal	→
Neutral cleaning	Removal of soil, debris, grease, etc.	→
Solvent cleaning	Water-repellent film with temporary protection	↗

Alkaline and neutral cleaners, for example, do not have a significant direct effect on the corrosion properties of magnesium, since they do not affect the metallic character of the magnesium surface. The main benefit that they offer is pre-conditioning thanks to their ability to provide wettable, uniform and homogenous surfaces. Solvent (emulsion) cleaners based on organic compounds can affect the surface by forming a thin protective film, which is then removed by the subsequent finishing treatments. This film provides temporary protection, particularly against exposure to water, which is useful for short term storage. Table 4.3 provides an overview of changes in corrosion performance due to cleaning observed through experience.

4.4 Application of cleaning solutions to achieve the desired finish

There is a large number and variety of commercial and non-commercial cleaning solutions available; the surface treatments do of course differ in many aspects, but there are some general rules that should always be taken into account for magnesium pre-treatments. Prior to any conversion, anodizing and other finishing treatments, it should be ensured that the surfaces are homogenous (with homogeneity provided by mechanical machining or grinding, for example), free of oils, other contaminating compounds and residues of lubricants, and of course dry. Rinsing should be carried out very carefully and very quickly between every processing step, as mentioned above. Contact with more noble metallic elements must be prevented and the duration of inventory times minimized.

Many other aspects should certainly be taken into account, but in most cases these are very specific to particular applications and will not be discussed in detail individually here. Because of the reliability of many of the traditional and well-known chemical treatments, a short list is provided below, showing suggested cleaning steps (discounting rinsing, which is always required) for the most common technical processes and providing some general guidelines:

Anodizing, as in HAE or Dow 17 Process
Solvent degrease /machine → vapor degrease (TCE) → caustic soda → chromic etch → HAE / DOW
Ni-P -Electroless plating
Machine → alkaline cleaning (NaOH + ····.) → nitric acid etches → HF activation → plating
Cu – plating like the Norsk-Hydro or WCM process
Machine → solvent degrease (e.g. acetone) and/or alkaline cleaning (NaOH + ····.) → oxalic acid pickle → fluoride activation (for WCM) → Zinc immersion → Cu plating
Zn -Phosphate conversion
Machine → alkaline cleaning (e.g. KOH) → Ti-phosphates (like activation) → phosphating bath
Phosphate- permanganate conversion
Machine → solvent degreases (e.g. acetone) → alkaline cleaning (NaOH + Na_3PO_4 + ····.) → conversion
Microarc oxidation – MAO
Machine → ultrasonic solvent degrease (e.g. acetone) → electrolyte bath

It is clear that this list provides only a small sample of the possible variations in processing steps. The requirements differ according to the process requirements; nevertheless, the major aspects of conditioning as described in the previous sections, remain the same. If the suggestions above are followed, the systems are effective for post-processing purposes even under moderate exposure.

4.4.1 Surface cleaning, pre-conditioning and other corrosion prevention techniques

Some major trends have emerged as a result of previous studies and during our earlier discussions on the various possibilities for handling magnesium surfaces, its requirements and the choice of treatment methods for the huge number of processing and finishing techniques (a nice summary is

Table 4.4 Alternative methods for the preparation of magnesium surfaces

Method	Acting process	Benefits/ disadvantage	Reference
Laser cleaning	Laser induced evaporation of dirt	Excellent adhesion/ time consuming	Li *et al.*, 2008
Dry ice blasting	Thermal and kinetic energy lead to damage/remove of surface near region	Excellent adhesion/ oxide formation	Spur *et al.*, 1999
Atmospehric plasma induced cleaning	Ion sputtering and surface material removal of some nm	Dry process and fast	Horn *et al.*, 2009
Sputter etching	Argon ion sputtering	Very clean surfaces/ vacuum process	Hoche *et al.*, 2003

also available in Gray and Luan (2002)). The first trend, which we consider indispensable, is avoiding the use of chromium compounds and the identification and adoption of adequate alternatives. Due to legislation on safety and environmental grounds, the use of substances such as hexavalent chromium in Europe has been limited to special aerospace applications: see EU directive RoHS (Restriction of Harmful Substances 2002/95/EC). Working alternatives that are commercially available have already been discussed; however their use has yet to become widely accepted.

A wholly different approach comes from the bare materials side. The development of new alloys with high tolerance to impurities tolerances could help to make 'excessive cleaning' less necessary. The first investigations in this regard were carried out by Blawert *et al.* (2010), who successfully increased the copper tolerance in a secondary alloy. A third trend is particularly important from an economic perspective. A variety of 'all-in-one' solutions (though of course degrease and pickle have to be separated) have been developed to minimize the number of processing steps by mixing tensides, builders, inhibitors and so on. There are a huge number of patents surrounding this particular issue; we will therefore not discuss it in detail here.

In addition to the traditionally used treatments a large number of new and in particular 'dry' methods have been developed or are under development for the preparation and activation of surfaces for finishing without the use of treatments such as acid baths. Some possibilities are given in Table 4.4. However, the use and functionality of these new methods is not guaranteed, and due to geometric or procedural limitations their application is still restricted to specific processes. In the future, it can be assumed that these methods will become increasingly viable and their cost will reduce. If they can achieve the same level of efficiency as traditional treatments they will constitute an excellent alternative, particularly as they pose a much lower risk to health and the environment.

Since functional materials are a central issue in magnesium research, some new aspects and challenges regarding alloys and coating systems should also be considered. New concepts of surface modification and conditioning are becoming increasingly important; these involve tailored functionalized surfaces for top coats or degradation, or even more specifically for impurity control. These new concepts, in turn, require new surface preparation methods that are specifically tailored to requirements without striving for perfect corrosion-protected flat surfaces. One very wide field in this regard is biotechnology (Peng *et al.*, 2010), where magnesium is used as an implant material, and *in vivo* corrosion is a key issue. In the biotechnological context, the surface pre-treatment takes into account the requirements of a controlled degradation process in a biological environment.

4.5 Conclusion and sources of further information and advice

The chapter first described the processing steps for the typical surface treatments of mechanical conditioning, chemical cleaning and activation of magnesium alloy-based components. An overview and analysis of the cleaning ability of different classes of cleaner was then provided, followed by some general remarks on handling the pre-treatments in the context of both standard industrial processes and scientific investigations. New aspects of corrosion protection and corrosion science demonstrate the need for innovative conditioning processes, leading to ongoing developments in the field.

The ASTM D 2651 standard (ASTM, 2008) and the recommendations of the International Magnesium Association are still the main reference point and state of the art for the preparation of magnesium alloy surfaces. Further information can be found on the websites of the various companies working with magnesium, such as Magnesium Elektron, or on coating technology websites which will not be cited here. Commercial suppliers including Henkel, Aeromagnesium/Chemetall, MacDermid and AHC, among others, offer a wide range of products to address the issue of pre-conditioning magnesium alloy surfaces, the compositions of which are not publicly available. Finally, it should be stressed that although this chapter provides the fundamentals of cleaning magnesium alloy surfaces and gives guidelines and advice, end users should ultimately carry out their own tests to identify the best course of action for their product.

4.6 References

ASTM (2008). D2651 - 01 Standard Guide for Preparation of Metal Surfaces for Adhesive Bonding. *Magnesium Alloys*. West Conshohocken, PA: ASM International.

Avedesian, M. and Baker, H. (1999). *Magnesium and magnesium alloys*. Materials Park, OH: ASM International.

Bharadwaj, M. D., Sundarraj, S. and Tiwari, S. M. (2007). Micro-Galvanic Corrosion Issues in AM Series Magnesium Alloys. *Materials Technology*. TMS.

Blawert, C., Fechner, D., Höche, D., Heitmann, V., Dietzel, W., Kainer, K. U., Zivanovic, P., Scharf, C., Ditze, A., Grobner, J. and Schmid-Fetzer, R. (2010). 'Magnesium secondary alloys: Alloy design for magnesium alloys with improved tolerance limits against impurities'. *Corrosion Science*, **52**, 2452–2468.

Brunelli, K., Dabalà, M., Calliari, I. and Magrini, M. (2005). 'Effect of HCl pre-treatment on corrosion resistance of cerium-based conversion coatings on magnesium and magnesium alloys'. *Corrosion Science*, **47**, 989–1000.

Eliezer, D. and Alves, H. (2002). *Corrosion and oxidation of magnesium alloys*. Wiley Online Library.

Friedrich, H. E. and Mordike, B. L. (2006). *Magnesium technology: metallurgy, design data, applications*. Berlin: Springer Verlag.

Gray, J. E. and Luan, B. (2002). 'Protective coatings on magnesium and its alloys – a critical review'. *Journal of Alloys and Compounds*, **336**, 88–113.

Hoche, H., Scheerer, H., Probst, D., Broszeit, E. and Berger, C. (2003). 'Development of a plasma surface treatment for magnesium alloys to ensure sufficient wear and corrosion resistance'. *Surface and Coatings Technology*, **174–175**, 1018–1023.

Horn, K., Pfuch, A. and Schmidt, J. (2009). *Neue Methode für einen effektiven Korrosionsschutz auf Magnesium*, Galvanotechnik (9/2009), Leuze Verlag, Bad Saulgau, S. 1976–1979.

Hu, B., Yu, G., Chen, J., Li, Y. and Ye, L. (2005). 'Investigation on a Non-cyanide Plating Process of Ni-P Coating on Magnesium Alloy AZ91D'. *Journal of Materials Science & Technology*, **21**(03), 301–306.

Jia, S. Q., Jia, S. S. and Yao, J. (2007). 'Electroless nickel-plating on die cast magnesium alloy AZ91D'. *Transactions of Nonferrous Metals Society of China*, **17**, s866–s870.

Kurze, P. (2006). Corrosion and Surface Protections. *In:* Friedrich, H. E. and Mordike, B. L. (eds.) *Magnesium technology: Metallurgy, design data, applications*. Berlin: Springer Verlag.

Li, H., Costil, S., Liao, H., Coddet, C., Barnier, V. and Oltra, R. (2008). 'Surface preparation by using laser cleaning in thermal spray', *Journal of Laser Application*, **20**(1), 12–22.

Liang, C., Zheng, R., Huang, N. and Xu, L. (2009). 'Conversion coating treatment for AZ31 magnesium alloys by a phytic acid bath'. *Journal of Applied Electrochemistry*, **39**, 1857–1862.

Liu, M., Uggowitzer, P., Schmutz, P. and Atrens, A. (2008). 'Calculated phase diagrams, iron tolerance limits, and corrosion of Mg-Al alloys'. *Journal of the Minerals, Metals and Materials Society*, **60**, 39–44.

Liu, Q., Liu, K., Han, Q. and Tu, G. (2011). 'Surface pretreatment of Mg alloys prior to Al electroplating in TMPAC-AlCl3 ionic liquids'. *Transactions of Nonferrous Metals Society of China*, **21**(9), 2111–2116.

Lukashev, R., Yakovleva, N., Klyamkin, S. and Tarasov, B. (2008). 'Effect of mechanical activation on the reaction of magnesium hydride with water'. *Russian Journal of Inorganic Chemistry*, **53**, 343–349.

Makar, G. and Kruger, J. (1993). *Corrosion of magnesium*. London: Maney Publishing.

Nwaogu, U. C., Blawert, C., Scharnagl, N., Dietzel, W. and Kainer, K. U. (2009). 'Influence of inorganic acid pickling on the corrosion resistance of magnesium alloy AZ31 sheet'. *Corrosion Science*, **51**, 2544–2556.

Nwaogu, U. C., Blawert, C., Scharnagl, N., Dietzel, W. and Kainer, K. U. (2010). 'Effects of organic acid pickling on the corrosion resistance of magnesium alloy AZ31 sheet'. *Corrosion Science*, **52**, 2143–2154.

Peng, Q., Huang, Y., Zhou, L., Hort, N. and Kainer, K. U. (2010). 'Preparation and properties of high purity Mg–Y biomaterials'. *Biomaterials*, **31**(3), 398–403.

Sharma, A. K., Uma Rani, R. and Giri, K. (1997). 'Studies on anodization of magnesium alloy for thermal control applications'. *Metal Finishing*, **95**, 43–51.

Sharma, A. K., Uma Rani, R. and Mayanna, S. M. (2001). 'Thermal studies on electrodeposited black oxide coating on magnesium alloys'. *Thermochimica Acta*, **376**, 67–75.

Singh, S. S., Roy, A., Lee, B. and Kumta, P. N. (2011). 'Aqueous deposition of calcium phosphates and silicate substituted calcium phosphates on magnesium alloys'. *Materials Science and Engineering: B*, **176**, 1695–1702.

Skar, J. I., Sivertsen, L. K. and Öster, J. M. (2004). 'Chrome-free conversion coatings for magnesium die castings – a review', Proceedings from ICEPAM 2004, Holmenkollen Park Hotel, June 16–18, 2004.

Song, D., Jing, X., Wang, J., Lu, S., Yang, P., Wang, Y. and Zhang, M. (2011). 'Microwave-assisted synthesis of lanthanum conversion coating on Mg–Li alloy and its corrosion resistance'. *Corrosion Science*, **53**, 3651–3656.

Song, G.-L. and Xu, Z. (2010). 'The surface, microstructure and corrosion of magnesium alloy AZ31 sheet'. *Electrochimica Acta*, **55**, 4148–4161.

Song, R., Blawert, C., Dietzel, W. and Atrens, A. (2005). 'A study on stress corrosion cracking and hydrogen embrittlement of AZ31 magnesium alloy', *Materials Science and Engineering: A*, **399**(1–2), 308–317.

Spur, G., Uhlmann, E. and Elbing, F. (1999). 'Dry-ice blasting for cleaning: process, optimization and application'. *Wear*, **233–235**, 402–411.

Supplit, R., Koch, T. and Schubert, U. (2007). 'Evaluation of the anti-corrosive effect of acid pickling and sol–gel coating on magnesium AZ31 alloy'. *Corrosion Science*, **49**, 3015–3023.

Zhang, W. X., Jiang, Z. H., Li, G. Y., Jiang, Q. and Lian, J. S. (2008). 'Electroless Ni-P/Ni-B duplex coatings for improving the hardness and the corrosion resistance of AZ91D magnesium alloy'. *Applied Surface Science*, **254**, 4949–4955.

Zhao, M., Wu, S., An, P. and Luo, J. (2007). 'Influence of surface pretreatment on the chromium-free conversion coating of magnesium alloy'. *Materials Chemistry and Physics*, **103**, 475–483.

5

Surface processing and alloying to improve the corrosion resistance of magnesium (Mg) alloys

G.-L. SONG, General Motors Corporation, USA and
Z. XU, Meda Engineering, USA

DOI: 10.1533/9780857098962.2.110

Abstract: Corrosion is a surface degradation process determined by an alloy's surface state, composition and microstructure. This chapter presents an overview of surface processing and alloying techniques that can improve the corrosion resistance of Mg alloys. It is found that some surface engineering processes can significantly change the surface state, composition and microstructure of Mg alloys. For example, surface mechanical and chemical cleaning can effectively remove surface contaminations; burnishing can modify the surface layer's microstructure including grain size and grain orientation; surface alloying can produce a metallic layer on top of the Mg alloy base material. All these surface engineering processes lead to enhanced corrosion resistance of Mg alloys without affecting their bulk properties.

Key words: magnesium, corrosion, alloying, surface treatment.

5.1 Introduction

Magnesium (Mg) is one of the promising lightweight engineering metallic materials that can find many applications particularly in the auto industry.[1,2] Owing to increasing energy costs, using high strength and low density Mg alloys to build light vehicles has become an important approach for improving fuel efficiency. However, the use of Mg alloys is limited by their high surface electrochemical activity or high susceptibility to corrosion attack.[3–13] Mg and its alloys are among the most active metals in aqueous solution. Their galvanic corrosion susceptibility is significant if they are in contact with other engineering metals.

Mg alloys can also suffer severe corrosion damage due to the micro-galvanic effect caused by heavier metal impurities, such as iron- and copper-containing compounds,[11,14] in the alloys. It is well-known that the presence of a very small amount of Fe, Ni, Co or Cu can significantly increase the corrosion rate of a

Mg alloy. Studies have shown that the corrosion rate can be accelerated by a factor of 10 to 100 when the concentrations of the impurities are above their tolerance limits.[15,16] Besides impurities, a Mg alloy is not uniform in terms of the composition,[17–21] microstructure[19,22–30] and crystal orientation.[27,30–32] All of these will make a contribution to the micro-galvanic effect, causing corrosion damage of a Mg alloy.

In principle, corrosion is a surface chemical process. All the surface related factors may influence the corrosion behavior of a metal to a great degree. The surface activity of a metal is closely associated with the composition and microstructure of the metal as well as their distributions in its surface layer. A suitable modification of these surface factors may reduce the surface activity, resulting in enhanced corrosion resistance of a metal.

5.1.1 Surface and applications

The surface of a metal is different from its bulk in lattice structure, microstructure or even composition, which means that the surface has physical and chemical properties different from the bulk material. Surface properties determine many applications of a metal. The functional performance and service life of an engineering component not only rely on the bulk material properties, but are also determined by its surface characteristics. For example, the behavior of a wear resistant component is greatly dependent on its surface hardness. The fatigue failure of a material may start from its surface after repeated rolling or sliding. A rough or heterogeneous surface can result in an uneven distribution of stress or locally concentrated stress to trigger initiation of a crack from the surface, which may propagate into the bulk, resulting in an ultimate fracture failure.

Apart from these mechanical functions and performance, the corrosion behavior of a metal is very sensitively influenced by surface conditions, since corrosion is a result of the reaction between the metal surface and its environment. Techniques that can passivate metal surfaces, such as passivation,[33–36] surface conversion and coating,[37–39] have been used in many practical applications to prevent corrosion. Furthermore, surface quality also concerns decoration and finishing of an article. Components manufactured in automobile industries are required to go through various painting and coating processes. Most coating failures can be directly attributed to poor surface preparation of the substrate. Surface conditions or quality therefore have an important effect on the practical applications of a metallic component in industry. For an active metal with a highly reactive surface, like Mg or its alloys, control and modification of its surface state has become one of the most important research and engineering topics.

5.1.2 Surface modification

Basically, the surface and bulk chemical composition, phase constituents and crystal parameters of an alloy depend on how the material is formed, processed and heat-treated. It is unlikely that surface composition and microstructure can be altered through modification of the normal alloy production and heat-treatment processes without changing the properties of the bulk material. Unfortunately, a composition or microstructure that can yield great mechanical properties does not usually exhibit excellent corrosion resistance. Conversely, undesired corrosion performance often results when the composition and microstructure of a material are modified for better mechanical properties. For example, a Mg alloy usually has a high strength/weight ratio, which is determined by its bulk properties. However, its surface is too active for many practical applications and engineers have to look for Mg alloys that have passivated surfaces. Therefore, a surface technique that can modify the surface to enhance the corrosion resistance without changing the bulk from its original excellent bulk properties would be highly desirable in practical applications of Mg alloys.

Several techniques have been developed and used for this purpose. For example, coating is one of the simplest approaches that can effectively passivate the surface without reducing the bulk performance of a material. Similar techniques include anodizing, conversion, plating, etc. All these surface processes result in a distinct surface layer on the top of the bulk and thus adhesion is often a big concern in their practical applications. A surface layer that naturally evolves from the bulk without a clear interface would be a solution to the adhesion problem. Such an interface-free surface layer can only be obtained by modifying the composition or microstructure in the surface layer of a metal. If such a surface layer can be obtained without altering the bulk material, and if it can still exhibit the desired mechanical properties, then the material will have significantly expanded applications in practice. It should be noted that there are at least two different surface treatments that can produce such an interface-free surface layer: (1) surface processing and (2) surface alloying. They will be reviewed and discussed in the following sections.

5.2 Surface processing techniques to improve corrosion resistance

Surface processing in this chapter refers to a chemical or physical engineering process on a metal surface that can result in a surface layer different in composition or microstructure but not separated from the bulk by a distinct interface. It is one of the easiest and most inexpensive engineering approaches that can improve the corrosion resistance of a Mg alloy.

Surface processing techniques are usually classified into mechanical and chemical groups. Mechanical methods include abrasive and non-abrasive processes. Sample abrading can be done through polishing, grinding, blasting, abrasive-finishing, burnishing, etc. Non-abrasive surface treatments include stirring, and spraying. Mechanical surface processing is usually a preliminary procedure for further surface modification. It should be noted that, in addition to the surface material removal, a mechanical process may also deform the surface layer, modifying its microstructure. Different surface processes can lead to different degrees of surface cleanliness, surface deformation, micro-stress storage, phase transformation and/or microstructure change, and thus various surface corrosion behaviors.

The chemical surface process is normally known as surface cleaning/etching. It is usually implemented in acidic solutions. Organic solvents can also be used to remove oil, grease and wax. However, cleaning by alkaline bath is preferred for Mg, as Mg is more stable at high pH values. For example, aqueous solutions containing 50 g/L NaOH, 10 g/L Na_3PO_4 and 1 g/L surface-active agent with a pH value of 11 are recommended[40] to clean Mg surfaces. Immersion cleaning can be aided by polarization at a temperature around 80–95°C. Acid pickling can reduce surface impurity levels and enhance the corrosion resistance of a Mg alloy.[41] Supplit *et al.*[42] used four different acid pickling processes on Mg AZ31 alloy, and found that nitric acid only led to a slight improvement in corrosion resistance owing to the removal of contaminations on the surface; hydrofluoric and phosphoric acid further improved the corrosion resistance; acetic acid showed the best results although it had higher weight loss during pickling. Therefore, acid pickling should be used with caution in cleaning Mg based alloys.

Mechanical and chemical methods can be combined in practice. Which process is more suitable or effective depends on the shape and specific characteristics of the individual component. For example, Hadzima *et al.*[43] investigated the corrosion rate of Mg AZ21 whose surfaces were ground, ground + phosphated and ground + shot-peened + phosphated. They found that shot-peening pre-treatment resulted in increased corrosion resistance of a phosphated Mg surface. Generally speaking, surface processing can more or less change the surface composition and/or microstructure, and thus it can either improve the corrosion resistance or deteriorate the corrosion behavior of a material. In this chapter, only those processes that can result in enhanced corrosion performance are reviewed and discussed.

5.3 Effect of surface processing and corrosion resistance

Surface processing can lead to various changes in surface composition and microstructure, thus influencing the corrosion resistance of a Mg alloy.

5.3.1 Surface cleaning

Acid etching, alkaline cleaning, grinding and blasting can remove surface impurities, although their cleaning effects are different in practice.[14]. For example, after the surface of an as-received AZ31 sheet sample was ground and acid pickled, its corrosion resistance significantly differentiates in 5 wt.% NaCl solution. According to their surface morphologies, weight loss results and hydrogen evolution measurements, the corrosion resistance of these differently surface-treated samples can be roughly ranged in the following order: acid cleaned > ground >> as-received > sandblasted (see Plate I in the colour section between pages 124 and 125). The as-received AZ31 is badly corroded; the sandblasted sample has the most severe corrosion damage; the ground sample shows much less corrosion; while the corrosion damage on the acid cleaned specimen is hardly seen from the picture. In other words, surface grinding and acid cleaning significantly improve the corrosion resistance of AZ31 while sandblasting deteriorates its corrosion performance.

The main purpose of grinding, cleaning and sandblasting is surface purification. The surface of as-received AZ31 is usually contaminated by Fe impurities during the sheet rolling process. Both grinding and acid cleaning can successfully remove the contaminated surface layer including Fe-containing particles there (Fig. 5.1a–5.1c). Sandblasting is also common in manufacturing to remove dirt and stain on metal surfaces. However, if the sand used to blast steel is repeatedly recycled and reused on a Mg alloy, it will contaminate the alloy surface and thus significantly deteriorate its corrosion performance, because Fe impurity from steel mixed with the sand will be deposited on the Mg alloy surface. Surface contamination by sandblasting has been confirmed by elemental mapping. The Fe impurity content on the surface of AZ31 was not reduced by sandblasting. Instead, more Fe contamination was detected after sand blasting (see Fig. 5.1d).[14]

It should be noted that polishing or grinding is a costly process. It is also difficult to polish or grind an uneven surface. Acid cleaning, on the

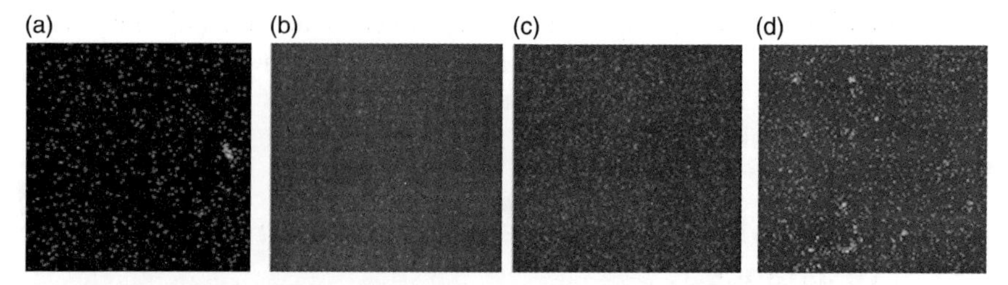

5.1 Fe mapping on AZ31 sample surfaces: (a) as-received, (b) abraded, (c) acid cleaned, (d) sandblasted. The white particles shown in the mapping are Fe-containing impurities.

other hand, is more popular in industries as an alternative surface cleaning technique. It even leads to a surface with slightly higher corrosion resistance than grinding (Plate I). Moreover, mechanical processing does not normally lead to enrichment of alloying elements in the surface layer. Chemical dipping or cleaning, if properly controlled, may result in selective dissolution of a certain element or phase. A multi-phase Mg alloy, such as AZ91 or AM60, has a less corrosion resistant matrix phase and a more resistant secondary phase. It is proposed that a controlled chemical or electrochemical etching process can produce a surface layer rich in resistant secondary phase and thus evidently improve the corrosion performance of these Mg alloys.[44]

5.3.2 Surface roughness and microstructure

Surface processing can change surface roughness. Mechanical processing may even introduce micro-stress in the surface layer. Both roughness and residual stress can influence the corrosion behavior of a metal. Generally speaking, a rougher surface of a material has worse corrosion performance;[45–47] corrosion initiation and propagation is less likely to happen on a smoother surface. Hilbert *et al.*[47] found that surface roughness is an important parameter affecting the corrosion resistance of stainless steel. Koutsomichalis *et al.*[48] demonstrated that the surface of irradiated magnesium AZ31B exhibited a wavy topography and its corrosion behavior was better than an untreated sample. The hydrogen evolution measurements of coarsely, regularly and finely polished AZ31 (in Fig. 5.2) also show that corrosion activity increases with surface roughness.[14]

Surface roughness is related to real surface area. However, the latter's influence on corrosion rate is relatively less significant than residual micro-stress which may also be closely associated with surface roughness. High surface roughness resulting from mechanical processing always accompanies some micro-stress stored in the surface layer. Under certain conditions, the surface roughness to some extent reflects the level of residual micro-stress. For example, a roughly polished or rolled surface usually means a higher level of surface micro-stress, and the latter normally has a contribution to an increase in corrosion rate. Figure 5.2 shows that the corrosion damage on the AZ31 sample surface abraded with coarse SiC paper follows the grinding tracks while the surface ground with fine SiC paper does not have a clearly orientated corrosion pattern. This is because the grinding tracks store residual micro-stress, which accelerates the corrosion. The detrimental effect of surface roughness and residual stress on corrosion implies that mechanical surface processing techniques should be used carefully for the purpose of improving corrosion resistance of a Mg alloy. Ideally, a surface

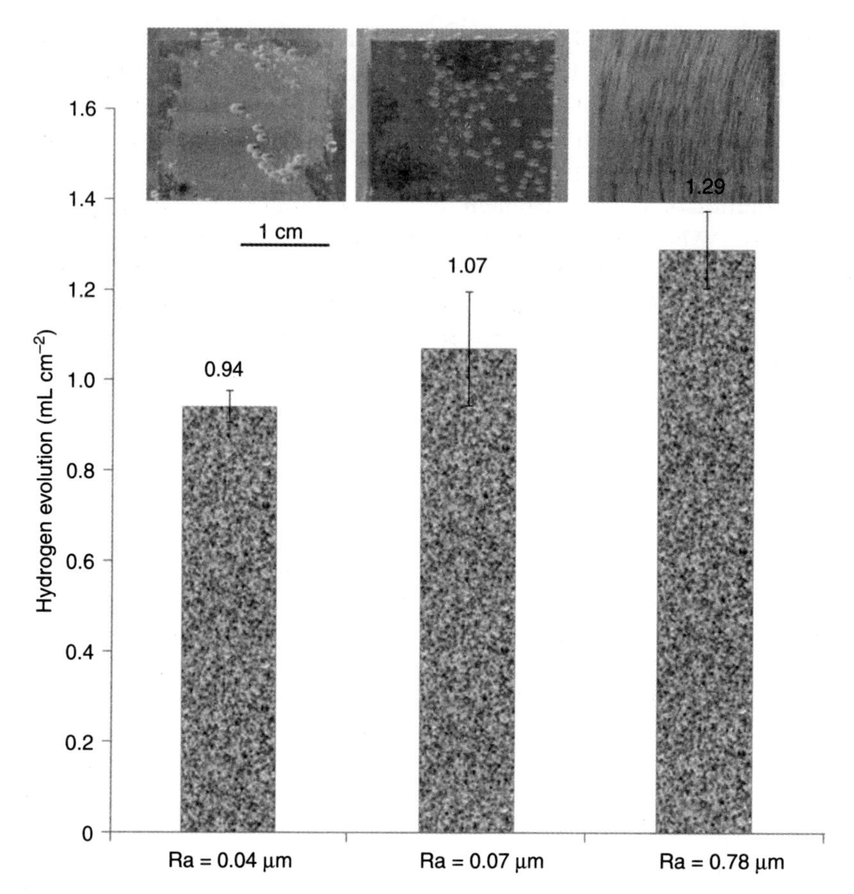

5.2 Hydrogen evolution rates of Mg AZ31 alloy with different surface roughness. The inserted pictures show corrosion behavior of finely, regularly and coarsely polished AZ31 alloy immersed in 5 wt% NaCl.

mechanical treatment can clean and smooth the surface without introducing micro-stress in the surface layer.

5.3.3 Grain orientation

It is interesting that mechanical processing can actually change the grain orientations in the surface layer. Figure 5.3 shows the texture evolution of AZ31 caused by burnishing.[49] The relative heights of the peaks corresponding to the basal plane (0002) increase significantly after both dry and cryogenic burnishing, suggesting that basal textures have been created on the burnished surface, which should have an effect on the corrosion behavior of the AZ31.

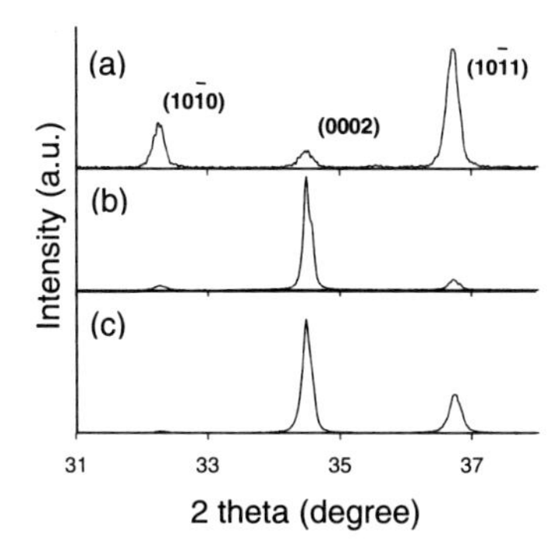

5.3 X-ray diffraction spectra of surfaces: (a) as-ground before burnishing; (b) after dry and (c) cryogenic burnishing.[49]

In fact, the preferential grain orientation in the surface layer by a compression process can be predicted based on the texture of hot-rolled AZ31 sheet. The basal slip system is the easiest one dominating the deformation of Mg and Mg alloy at room temperature.[50,51] This deformation mechanism determines that grains will have a preferential orientation after deformation (Plate II); the basal planes in the alloy turn to the direction parallel to the rolling surface. The burnishing process has a similar grain orientation behavior in the surface layer as the hot-rolling.

Different crystal planes have different chemical activities and thus different corrosion resistance. The atomic density in a crystallographic structure can vary significantly from one plane to another. The hexagonal close packed (hcp) basal plane (0001) has the highest atomic density (1.13×10^{19} atom/m^2), followed by the prismatic planes ($11\bar{2}0$) and ($01\bar{1}0$) that have atomic densities 6.94×10^{18} and 5.99×10^{18} atom/m^2, respectively.[52] A close packed plane has a relatively high binding energy due to its high atomic coordination. Therefore, the atomic packing density of the exposed crystallographic plane can affect the corrosion process of Mg, in which Mg atoms break the bonds, leave their crystallographic lattice and get into the solution.

For example, a polished pure Mg ingot sample was immersed in 0.1 M HCl solution. After 15 h, the close packed crystallographic plane had dissolved more slowly than the prismatic planes.[32] The relationship between grain orientation and corrosion resistance of Mg also holds in an alkaline environment. It has been found that in a $Mg(OH)_2$ saturated aqueous solution, the corrosion potential of a grain with basal orientation is about 70 mV more positive than that of another grain close to the prismatic orientation, and the corrosion current densities of the basal and prismatic planes are about 0.1 and 0.5 mA/cm^2, respectively.[53] Similar results have also been reported for AZ31 Mg alloy.[14,27,30] After 3 h immersion in an

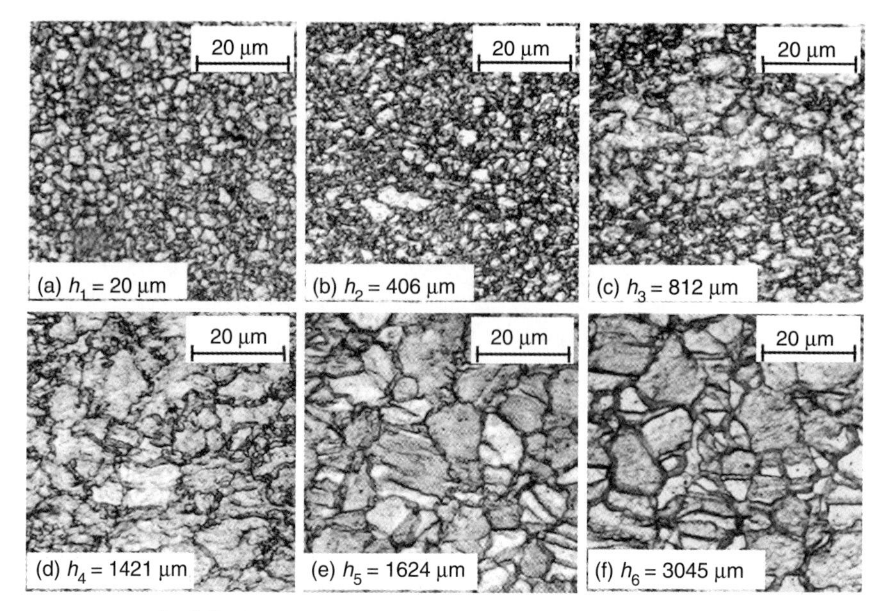

5.4 (a–f) Typical microstructures after dry burnishing at different depths $h_1 - h_6$ from the surface.[49]

5 wt.% NaCl acidic solution, the cross-section surface which is dominated by the prismatic planes was found to have dissolved much faster than the rolling surface that mainly consists of the basal planes. In neutral 5 wt.% NaCl solution, the corrosion rates of the rolling surface and cross-section are around 0.01 and 0.02 mg/cm^2/h, respectively, according to hydrogen evolution measurements.[54,55]

5.3.4 Grain size

A compression machining process can even change the grain size of the surface layer. For example, surface burnishing has been found to dramatically reduce the grain size of AZ31 in the surface deformation zone.[56] The cross-sectional microstructure images at different depths after dry burnishing are shown in Fig. 5.4. The corrosion rate of the AZ31 after burnishing becomes significantly lower.

Grain size is an important parameter which determines mechanical properties and corrosion behavior of a material. Song *et al.*[57] reported that AZ91D alloy with ultra-fine grained α-phase matrix grains and refined β-phase particles has a significantly lower corrosion rate. The secondary phase which varies with grain size is a critical factor determining the corrosion behavior of this multi-phase Mg alloy. The corrosion rate of Mg AZ31 also varies with grain size after heat-treatment.[14,30,58]

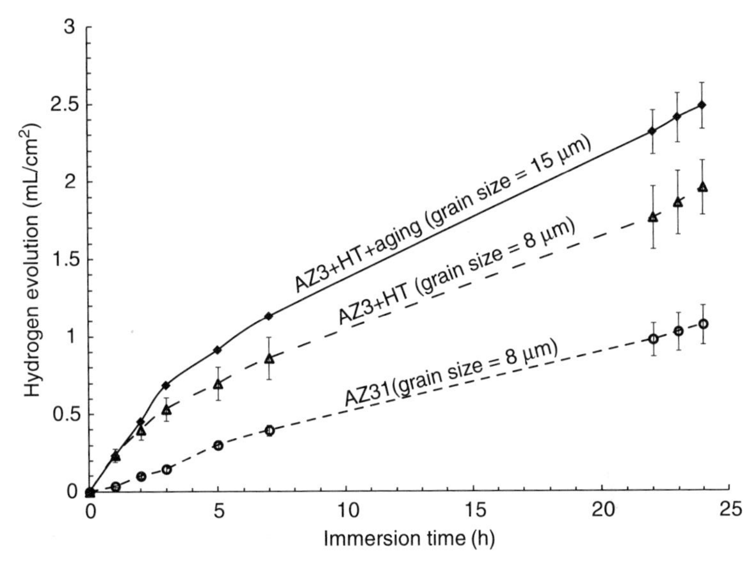

5.5 Hydrogen evolution rates of Mg AZ31 alloy samples with different grain sizes after heat-treatment. Heat-treatment conditions: HT-450°C in air for 10 min; HT + aging – 450°C in air for 5 h.

However, grain size is not a critical factor affecting the corrosivity (see Fig. 5.5). Samples that have similar grain size do show very different corrosion resistance as indicated by hydrogen evolution rates. This suggests that other factors may be involved in the variation of grain size during heat treatments and these factors can actually influence the corrosion behavior. In fact, in addition to the grain size effect, the influence of the secondary phase and the alloying element in the solid solution on corrosion cannot be neglected.

5.3.5 Intermetallic particles and crystal defects

The intermetallic phase in a Mg alloy can play a 'dual-role' in corrosion, depending on its continuity.[11] If the phase is discontinuous, it will deteriorate the corrosion performance. Otherwise, it is a corrosion barrier. Intermetallic particles normally refer to a small amount of isolated phases, which cannot form a continuous network to block the development of corrosion from grain to grain; instead they accelerate the corrosion of the matrix phase through galvanic effect. These particles can be impurity-containing compounds, such as AlMn(Fe), a typical intermetallic in commercial AZ31.[21,59] The tiny intermetallic precipitates randomly distributed in the alloy are cathodic to the matrix phase of AZ31 because of their relatively positive corrosion potentials.[60] The presence of these precipitates can accelerate the dissolution of the

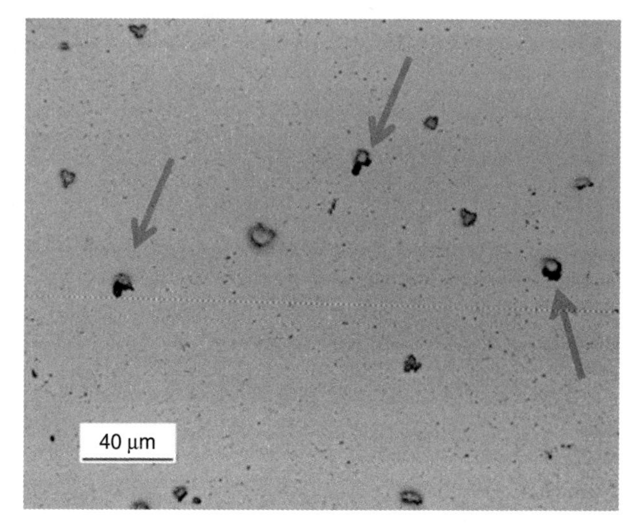

5.6 Corrosion morphology of the rolling surface of AZ31 after 10 min of heat-treatment at 450°C and 5 min of immersion in 5 wt% NaCl solution.

surrounding matrix phase due to galvanic effect. Figure 5.6 is a photo showing the corrosion damage (indicated by arrows in the photo) of the matrix phase next to the particles in the rolling surface of a heat-treated AZ31 sample after 5 min immersion in 5 wt.% NaCl solution. It confirms that the presence of cathodic AlMn(Fe) particles is detrimental to the corrosion of AZ31.

Crystallographic defects mainly refer to grain boundaries, dislocations and twins in the matrix phase. In theory, they should be preferentially corroded due to their higher chemical activity than the bulk. For example, after 6 days of immersion in 5 wt.% NaCl, the exposed grain boundaries and twins appear to be dissolved slightly deeper (as marked by arrows in Fig. 5.7).

The presence of the intermetallic particles and crystal defects can accelerate the corrosion of Mg alloys. Since surface mechanical processing can undoubtedly modify these parameters, the variation of these factors should not be overlooked in considering the effect of surface mechanical processing on corrosion behavior of a Mg alloy. It should be noted that crystallographic defects (grain boundaries, twins and dislocations) and intermetallic particles can evolve with grain size during a surface deformation or heat-treatment process and finally influence the corrosion performance of a Mg alloy. Therefore, grain size, crystal defect and intermetallic particle are not independent factors.

Figure 5.8 shows that when the grain size of AZ31 is large (e.g. >10 μm), the corrosion rate is relatively high; but if the grain size is small (e.g. <10 μm),

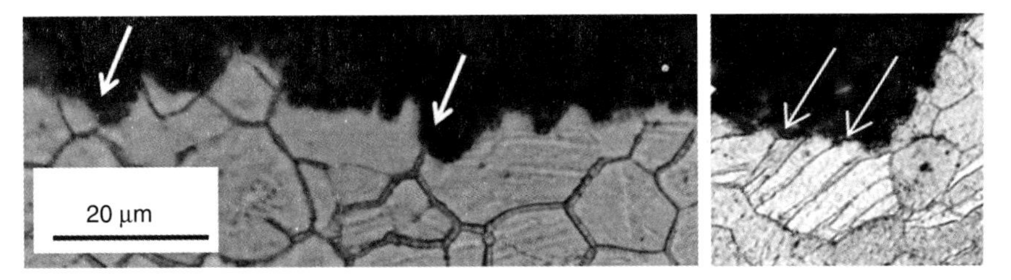

5.7 Cross-section images of corroded surfaces of heat-treated AZ31 sheet samples after 6 days of immersion in 5 wt% NaCl solution.

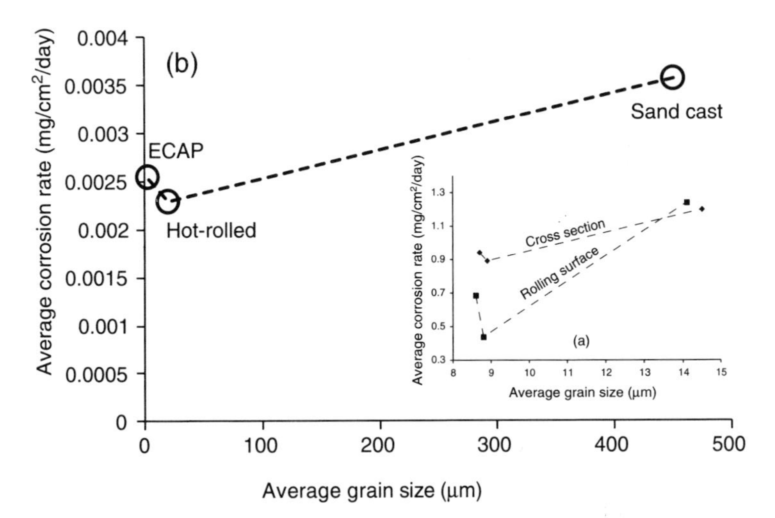

5.8 Dependence of corrosion rate on grain size: (a) (the insert) hot-rolled AZ31 sheet samples in 5 wt% NaCl solution,[30] (b) squeeze cast, hot-rolled and ECAP AZ31 in a SBF solution ($CaCl_2$ 0.14 g/L, KCl 0.4 g/L, NaCl 8.0 g/L, $MgSO_4$ 0.1 g/L, KH_2PO_4 0.06 g/L, $NaHPO_4$ 0.05 g/L and D-Glucose 1.0 g/L).[61]

the corrosion rate increases with decreasing size.[53] The non-monotonous relationship between the grain size and the corrosion rate of AZ31 suggests that not only the density of crystal defects but also the amount of precipitated particles are responsible for the corrosion behavior. The detrimental effect of crystallographic defects (dislocations, grain boundaries and twins) on corrosion performance interprets the slightly worsened corrosion resistance when grain is small (grain size <10 μm). After heat-treatment, or in case of squeeze casting, while grains are growing significantly, the number of particles also increases. The latter variation deteriorates the corrosion performance of AZ31.

In summary, the dependence of corrosion rate of AZ31 on grain size is a result of two effects: (1) the preferential dissolution at crystallographic defects and (2) the accelerated corrosion next to tiny intermetallic precipitates. When a Mg alloy has relatively large grains or fewer crystallographic defects, deformation processing will introduce crystallographic defects, such as grain boundaries, twins and dislocations, and reduce the size and amount of the intermetallic particles by making them dissolve in the defects. (The solid solubility of alloying elements is higher in the distorted defect lattice, and thus more particles will be dissolved.) In this case, corrosion rate decreases with grain refinement. After grains become small enough and most particles have been dissolved in the defects, further grain refinement can only increase the densities of the crystallographic defects without further reducing the amount of intermetallic particles. Consequently, the corrosion rate of AZ31 increases with further grain refinement. Heat-treatment has a contrary effect on the crystallographic defects and intermetallic particles, and hence the corrosion behavior.

In practice, surface over-processing is unnecessary for corrosion resistance improvement. For example, a normal dry surface burnishing treatment can significantly improve the corrosion resistance of AZ31 in 5 wt.% NaCl solution. Although cryogenic burnishing may further refine the grain size in the surface layer, it is actually not helpful in corrosion performance improvement.[49] Figure 5.8b also indicates that equal-channel angular processing (ECAP) does not further improve the corrosion resistance of AZ31 in comparison with hot-rolling, although ECAP can more effectively refine grains.

5.3.6 Burnishing to improve corrosion resistance

Since corrosion resistance is determined by the surface layer, some surface engineering processes, such as machining, may improve the corrosion performance through modifying the surface microstructure. Burnishing is one of the practical surface plastic deformation techniques. Pu *et al.*[49,56,62] have demonstrated that severe plasticity burnishing can significantly improve corrosion resistance of AZ31 by refining grains and forming strong basal texture in the surface layer. The roller burnishing can also reduce surface roughness, increase hardness and introduce beneficial compressive residual stresses. As shown in Fig. 5.9, after 7 h immersion, the burnished samples are about 40% less active than the normally ground one (measured by hydrogen evolution rate). Figure 5.10 schematically illustrates a few possible surface compression friction treatments. It is believed that these compressive deformation processes can refine the grains and change the crystal orientation in the surface layer for improved corrosion performance.

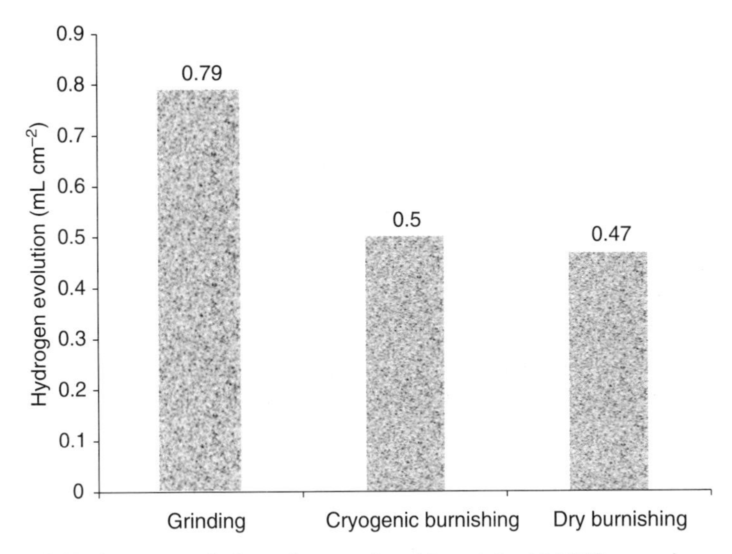

5.9 Hydrogen evolution of ground and burnished AZ31B samples processed in 5 wt% NaCl solution for 7 h.[49]

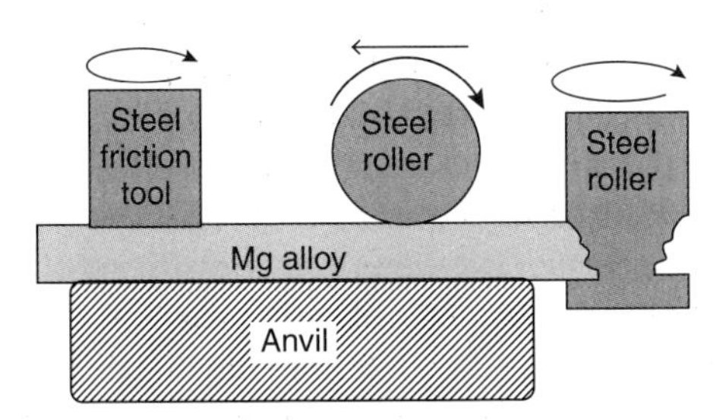

5.10 Schematic illustration of surface compression friction processes to modify the grain size and orientation for corrosion performance improvement.

5.4 Alloying and corrosion resistance

Mechanical processing mainly changes the microstructure of a surface layer. Chemical cleaning may result in some alloying elements or phases enriched on the surface. However, neither one can introduce a new alloying element into the surface layer of the substrate. Surface alloying is a process that can form a new surface alloy layer. As the layer is formed in the substrate, it normally does not have the adhesion problem that a plated, sprayed or deposited coating on the substrate usually has.

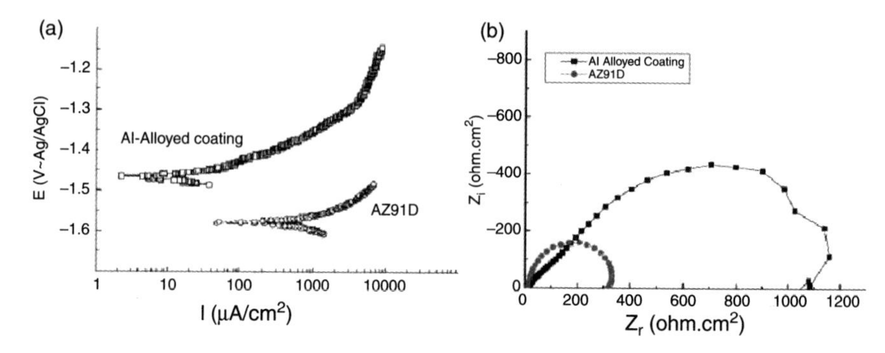

5.11 (a) Polarization and (b) electrochemical AC impedance spectra of coated and uncoated AZ91D in 5 wt% NaCl solution.[63]

5.4.1 Hot-diffusion

Hot-diffusion alloying is a direct approach to surface alloying.[63] Alloying elements diffuse from the environment into the surface layer at a high temperature to form a new alloy layer there.

Since the β-phase of a Mg-Al alloy is much more corrosion resistant than the matrix, a surface layer consisting of a large amount of β-phase would have much improved corrosion resistance. It has been reported[63] that an AZ91 Mg alloy coupon covered with a paste-like mixture of Al powder and ethylene glycol, heated to let the Al diffuse from the paste into AZ91D has significantly improved corrosion resistance in 5 wt.% NaCl solution (see Fig. 5.11). X-ray diffraction results suggest the major constituent of the surface layer is β-phase. The Al alloyed surface not only provides a corrosion barrier of β-phase, but also increases the Al content in the matrix phase. The continuous β-phase and Al-rich zone along underneath the surface layer can effectively stop corrosion penetration.

A similar study has also been carried out by Zhu *et al.*[64] who alloyed AZ91D surface with Zn. The Zn alloyed surface is about 55 microns in thickness, composed of Mg_7Zn_3, Mg and Zn. The Zn alloyed sample corrodes much slower than its counterpart without surface alloying. In fact, surface alloying of a Mg alloy can also occur if the Mg alloy is dipped in a melted alloying metal. This has been proposed and filed by Song *et al.*[65] Alloyed surfaces should have much higher surface micro-hardness, which is beneficial for wearing resistance.

5.4.2 Ion implantation

Alloying elements can also be introduced into the surface layer of a Mg alloy through implantation. Ion implantation is a well-known process to improve

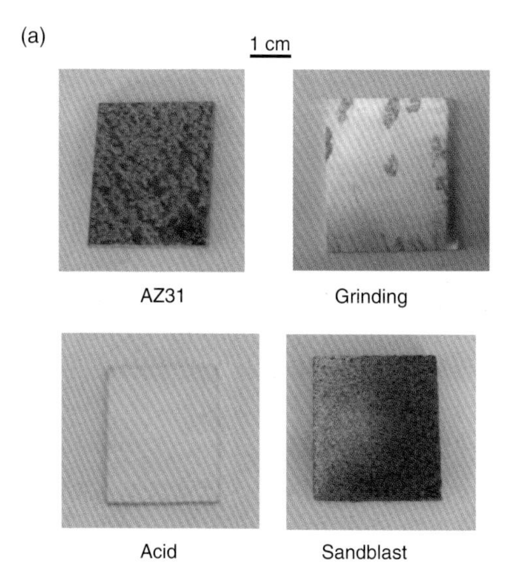

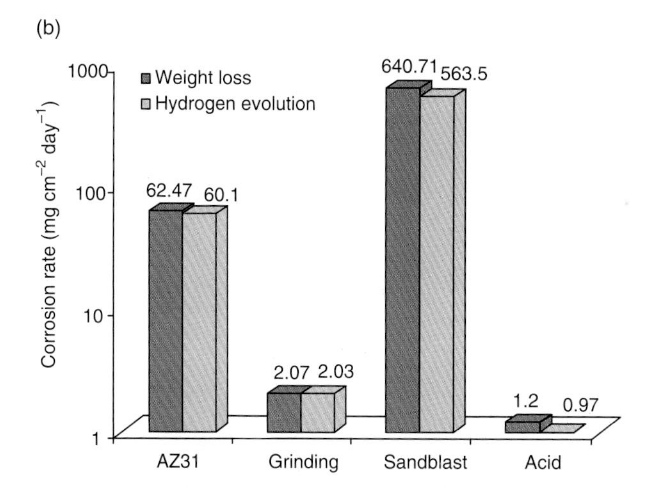

Plate I (a) Surface appearance of corroded AZ31 samples with various surface treatments after 1 h immersion test in 5 wt% NaCl; (b) corresponding corrosion rates of these samples calculated from weight loss and hydrogen evolution measurements.

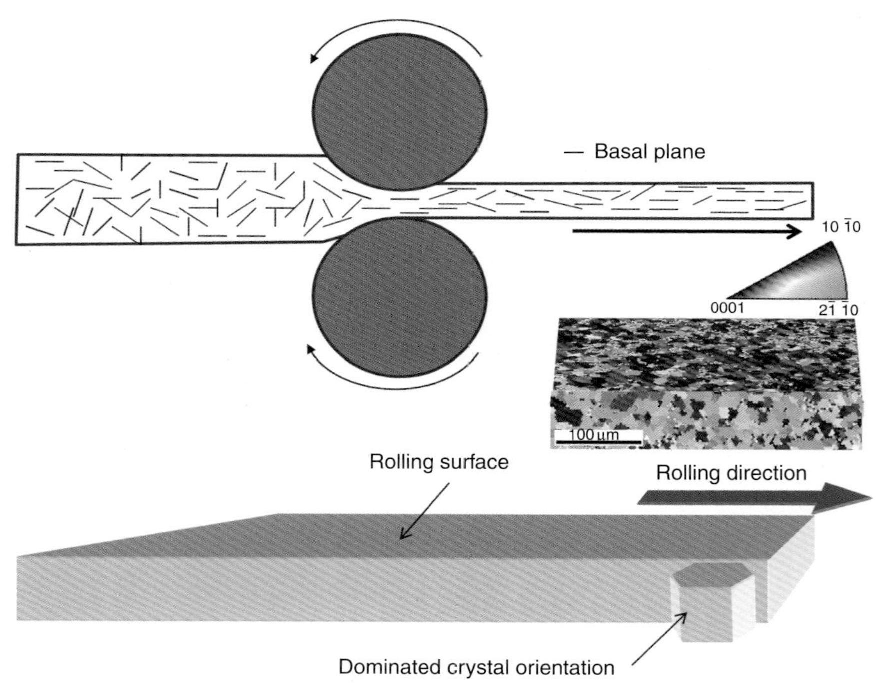

Plate II Schematic illustration of the change in grain orientation (basal planes) of AZ31 during a hot-rolling process and EBSD detected crystal orientations of hot-rolled AZ31 sheet.

Mg surface properties.[66,67] Owing to the formation of Magnesium nitrides, nitrogen implantation can enhance the corrosion resistance of Mg based alloys. Corrosion resistance of AZ91D Mg alloy by nitrogen ion implantation has been observed in 5 wt.% NaCl solution by Nakatsugawa *et al.*[68] Ion implantation-improved corrosion resistance has been found when doses were larger than 1×10^{16} ions/cm^2; at doses of 5×10^{16} ions/cm^2, the corrosion rate and pitting depth were are their minimum. Hoche *et al.*[67] modified the Mg alloys by means of nitrogen ion implantation to form a 600 nm thick nanocrystalline Mg_3N_2 film, which led to some improved corrosion resistance.

Aluminum and oxygen ion implantation was also employed to modify the surface of an Mg-Y-RE alloy by Zhao *et al.*[69] Al and O ion implantation produced an Al_2O_3-containing protection layer which improved the corrosion resistance of the Mg-Y-RE alloy. Titanium ion implantation was recently conducted on Mg AZ91D alloy.[70] An intermixed layer was produced and the surface oxidized films were mainly composed of titanium oxide. It was found that the titanium ion implantation significantly shifted the open circuit potential of Mg AZ91D to a more positive region and improved the corrosion resistance of AZ91D. This phenomenon was attributed to the compact surface oxide film structures and enhanced reoxidation process on the implanted surface. Overall, all the ion implantation techniques appear to be promising in generating a corrosion resistant surface layer for Mg based alloys.

5.4.3 Possible surface alloying systems

Currently, surface alloying to improve the corrosion resistance of a Mg alloy is limited to a great extent by (1) the solid solubility of the alloying elements in Mg and (2) the difficulty in introducing a new element into the substrate through normal surface engineering techniques. With the rapid development of surface processing technology, formation of any alloyed surface layers will soon become possible. It is important to know what, in an alloying system in the surface layer, is beneficial to the corrosion resistance enhancement for a Mg alloy. Such knowledge is a basis for future development of surface alloying technology.

Mg and Mg alloys present very negative corrosion potentials and can be actively dissolved in aqueous solutions.[21,25,71,72] Adding strong passivating elements like Cr[73,74] and Ti[24,75,76] to pure Mg could possibly improve its passivity. Titanium is much lighter and thus considered first. However, according to the thermodynamic phase diagrams, the solubility of Mg in Ti, and vice versa, is very small. Moreover, the much higher melting point of Ti than Mg is also a big problem in alloying. Conventional equilibrium metallurgical approach cannot successfully lead to a Mg-Ti solid solution phase due to their low solid solubility. For example, Candan *et al.* could only add 0.2–0.5 wt.% of Ti in Mg AZ91 alloy.[70]

A non-equilibrium physical vapor deposition (PVD) deposition technique can substantially extend the solubility of Ti in Mg to form Mg-Ti alloys. In PVD, the big melting point difference between Ti and Mg is no longer an issue.[77,78] In addition, a homogenous element distribution and low impurity levels, which are difficult to achieve by conventional casting, can easily be obtained, since PVD is a non-equilibrium process. Recently, Bohne *et al.*[79] deposited Mg-Ti alloys (maximum 60 at % Ti content) on Si and Mg AM50 substrates by Ion Beam Sputtering (IBS) and investigated their corrosion properties. Blawert *et al.*[80] synthesized supersaturated Mg alloys Mg-Al, Mg-Ti and Mg-Sn on Si and Mg substrates and suggested that a uniform passive film $Mg(OH)_2$ is responsible for the improved corrosion properties. These studies suggest that the non-equilibrium techniques can successfully make Mg alloy with a desired passivating element to achieve higher corrosion resistance.

For example, Mg-Ti thin-film alloys with different Ti concentrations can be deposited on a quartz substrate.[78] Using quartz instead of Si or Mg eliminates the substrate effect during electrochemical and corrosion tests. To investigate the effect of Ti on the electrochemical corrosion behavior of a Mg-Ti alloy, polarization curve measurements were employed (see Fig. 5.12). It is found that with increasing Ti content, the polarization curve moves towards a more positive potential and low current density regime. The higher corrosion potential and lower current density can be attributed to the substitution of Mg by Ti. Ti has excellent resistance to many highly corrosive environments. In theory, greater Ti content in Mg should lead to a smaller corrosion current density and high corrosion resistance. This example suggests that once the introduction of alloying elements into the substrate to form an alloyed surface layer no longer becomes a hurdle, a strong passive element, like Ti,

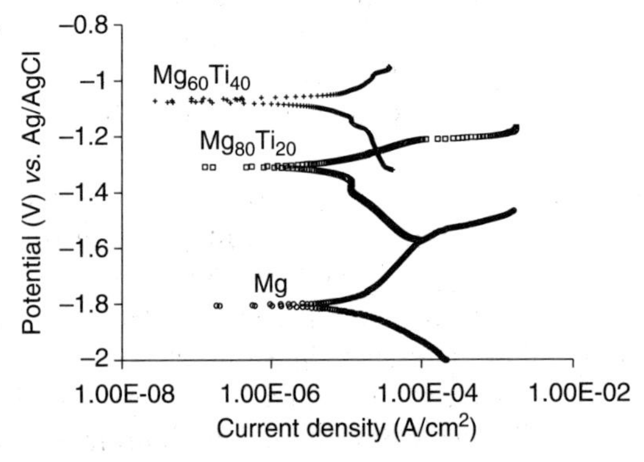

5.12 Polarization curves of Mg_xTi_{100-x} alloys in 0.1 M NaCl solution.

should be one of the alloying element candidates and has the potential to significantly improve the corrosion resistance of a Mg alloy.

5.5 Conclusions and future trends

Both surface processing and alloying can improve the corrosion resistance of a Mg alloy. These corrosion prevention approaches have the following characteristics:

- Both surface processing and alloying do not significantly change the alloy bulk properties.
- Surface processing is one of the easy and inexpensive surface treatments that can enhance the corrosion performance of a Mg alloy. Surface alloying is relatively difficult and costly, but more effective in corrosion resistance improvement.
- Both surface processing and alloying can be easily combined in production or manufacturing procedures. For example, surface cleaning is an essential step before coating. Hot-diffusion can be conducted during heat-treatment of a Mg alloy part. Therefore, the additional cost of these surface techniques in industry cannot be too high.
- Different from coating or plating, surface processing and alloying is a modification of the surface layer in the substrate. The composition and microstructure gradually change from the bulk to the surface and there is no distinct interface between them. Therefore, adhesion is not an issue and localized corrosion damage risk is very low even though the surface is locally damaged.
- Some additional surface properties, such as high hardness, attractive appearance and other functional features may also be obtained with a processed or alloyed surface.

However, surface processing and alloying have some limitations. For example, compared with other typical corrosion prevention techniques, such as coating, the improvement in corrosion resistance by surface processing and alloying is not so evident. Desired surface processing and alloying can only occur under carefully controlled conditions, which may not be easily realized on some Mg alloy parts in practice.

Further developments of surface processing and alloying techniques may be expected in the following areas:

- Introduction of innovative surface engineering techniques to generate more corrosion resistant surface layers more easily.
- Combinations of chemical, mechanical and alloying processes to obtain better surface layers.

- Application of some post-treatments (e.g. conversion and coating) on processed or alloyed surfaces to further improve the corrosion resistance and also to obtain some additional surface properties.

5.6 Acknowledgements

The authors would like to thank Dr Zhengwen Pu, Dr Daad Haddad and Professor Liqun Zhu for their support in relevant studies and experiments.

5.7 References

1. Ghali E. Magnesium and magnesium alloys. Uhlig's Corrosion Handbook (2000):793.
2. Avedesian MM and Baker H. ASM Specialty Handbook: Magnesium and Magnesium Alloys (1999).
3. Song G, Atrens A, Wu X and Zhang B. Corrosion behaviour of AZ21, AZ501 and AZ91 in Sodium Chloride. *Corrosion Science* (1998);**40**:1769.
4. Song G, Atrens A and Dargusch M. Influence of microstructure on the corrosion of diecast AZ91D. *Corrosion Science* 1998;**41**:249.
5. Li Y, Song G, Lin H and Cao C. Study on the relationship between the corrosion interface structure and negative difference effect for pure magnisuim. *Corrosion Science and Protection Technology* (1999);**11**:206.
6. Song G and St John D. Corrosion performance of magnesium alloys MEZ and AZ91. *International Journal of Cast Metals Research* (2000);**12**:327.
7. Song G. Investigation on corrosion of magnesium and its alloys. *Journal of Corrosion Science and Engineering* (2003);**6**.
8. Song G and Atrens A. Understanding magnesium corrosion. A framework for improved alloy performance. *Advanced Engineering Materials* (2003);**5**:837.
9. Song G, Johannesson B, Hapugoda S and St John D. Galvanic corrosion of magnesium alloy AZ91D in contact with an aluminium alloy, steel and zinc. *Corrosion Science* (2004);**46**:955.
10. Shi Z, Song G and Atrens A. Influence of the β phase on the corrosion performance of anodised coatings on magnesium-aluminium alloys. *Corrosion Science* (2005);**47**:2760.
11. Song G. Recent progress in corrosion and protection of magnesium alloys. *Advanced Engineering Materials* (2005);**7**:563.
12. Jia JX, Song G and Atrens A. Influence of geometry on galvanic corrosion of AZ91D coupled to steel. *Corrosion Science* (2006);**48**:2133.
13. Song G and Atrens A. Recent insights into the mechanism of magnesium corrosion and research suggestions. *Advanced Engineering Materials* (2007);**9**:177.
14. Song GL and Xu Z. The surface, microstructure and corrosion of magnesium alloy AZ31 sheet. *Electrochimica Acta* (2010);**55**:4148.
15. Hillis JE and Murray RW. The effects of heavy metal contamination on Magnesium corrosion performance. SAE Technical paper # 830523. Detroit, 1983.
16. Hillis JE and Murray RW. Finishing alternatives for high purity Magnesium alloys. SDCE 14th International Die Casting Congress and Exposition, vol. # G-T87-003. Toronto, 1987.

17. Lunder O, Lein JE, Aune TK and Nisancioglu K. Role of Mg17Al12 phase in the corrosion of Mg alloy AZ91. *Corrosion* (1989);**45**:741.
18. Nakatsugawa I, Kamado S, Kojima Y, Ninomiya R and Kubota K. Corrosion of magnesium alloys containing rare earth elements. *Corrosion Reviews* (1998);**16**:139.
19. Spassov T and Koster U. Microstructure, microhardness and corrosion behavior of rapidly solidified magnesium based Mg-Ni-(Y, MM) alloys. *Zeitschrift Fur Metallkunde* (2000);**91**:675.
20. Ye HZ and Liu XY. In situ formation behaviors of Al8Mn5 particles in Mg-Al alloys. *Journal of Alloys and Compounds* (2006);**419**:54.
21. Pardo A, Merino MC, Coy AE, Arrabal R, Viejo F and Matykina E. Corrosion behaviour of magnesium/aluminium alloys in 3.5 wt.% NaCl. *Corrosion Science* (2008);**50**:823.
22. Shi Z, Song G and Atrens A. Corrosion resistance of anodised single-phase Mg alloys. *Surface and Coatings Technology* (2006);**201**:492.
23. Ben-Haroush M, Ben-Hamu G, Eliezer D and Wagner L. The relation between microstructure and corrosion behavior of AZ80 Mg alloy following different extrusion temperatures. *Corrosion Science* (2008);**50**:1766.
24. Rousselot S, Bichat MP, Guay D and Roue L. Structure and electrochemical behaviour of metastable Mg50Ti50 alloy prepared by ball milling. *Journal of Power Sources* (2008);**175**:621.
25. Zhao MC, Liu M, Song G and Atrens A. Influence of the β-phase morphology on the corrosion of the Mg alloy AZ91. *Corrosion Science* (2008);**50**:1939.
26. Izumi S, Yamasaki M and Kawamura Y. Relation between corrosion behavior and microstructure of Mg-Zn-Y alloys prepared by rapid solidification at various cooling rates. *Corrosion Science* (2009);**51**:395.
27. Song GL, Mishra R and Xu Z. Crystallographic orientation and electrochemical activity of AZ31 Mg alloy. *Electrochemistry Communications* (2010);**12**:1009.
28. Song GL and Xu Z. Improving corrosion performance of AZ31B Mg alloy sheet by surface polishing. *Magnesium Technology* (2010):**181**.
29. Marlaud T, Malki B, Henon C, Deschamps A and Baroux B. Relationship between alloy composition, microstructure and exfoliation corrosion in Al-Zn-Mg-Cu alloys. *Corrosion Science* (2011);**53**:3139.
30. Song GL and Xu Z. Effect of microstructure evolution on corrosion of different crystal surfaces of AZ31 Mg alloy in a chloride containing solution. *Corrosion Science* (2012);**54**:97.
31. Abayarathna D, Hale EB, O'Keefe TJ, Wang YM and Radovic D. Effects of sample orientation on the corrosion of zinc in ammonium sulfate and sodium hydroxide solutions. *Corrosion Science* (1991);**32**:755.
32. Liu M, Qiu D, Zhao MC, Song G and Atrens A. The effect of crystallographic orientation on the active corrosion of pure magnesium. *Scripta Materialia* (2008);**58**:421.
33. Ashton RF and Hepworth MT. Effect of crystal orientation on the anodic polarization and passivity of zinc. *Corrosion* (1968);**24**:50.
34. Song G and Cao Cn. On the linear response of a passivated metallic electrode to potential step perturbation. *Corrosion Science* (1992);**33**:413.
35. Song GL, Cao CN and Lin HC. Effects of AC-modulated passivation and post-treatment on composition and stability of passive films. *Corrosion* (1993);**49**:271.

36. Song G, Cao CN and Chen SH. A study on transition of iron from active into passive state. *Corrosion Science* (2005);**47**:323.
37. Hinton BRW, Arnott DR and Ryan NE. Cerium conversion coatings for the corrosion protection of aluminium. *Materials Forum* (1986);**9**:162.
38. Gonzalez-Nunez MA, Nunez-Lopez CA, Skeldon P, Thompson GE, Karimzadeh H, Lyon P and Wilks TE. A non-chromate conversion coating for magnesium alloys and magnesium-based metal matrix composites. *Corrosion Science* (1995);**37**:1763.
39. Chong KZ and Shih TS. Conversion-coating treatment for magnesium alloys by a permanganate-phosphate solution. *Materials Chemistry and Physics* (2003);**80**:191.
40. Kurze P. Magnesium technology: metallurgy, design data, applications, (2006).
41. Nwaogu UC, Blawert C, Scharnagl N, Dietzel W and Kainer KU. Influence of inorganic acid pickling on the corrosion resistance of magnesium alloy AZ31 sheet. *Corrosion Science* (2009);**51**:2544.
42. Supplit R, Koch T and Schubert U. Evaluation of the anti-corrosive effect of acid pickling and sol-gel coating on magnesium AZ31 alloy. *Corrosion Science* (2007);**49**:3015.
43. Hadzima B and Bukovinová L. Electrochemical characteristics of shot-peened and phosphatized AE21 magnesium alloy. *Acta Metallurgica Slovaca* (2011);**17**:228.
44. Sachdeva D, Kumar AM and Song G. Surface Treatment of Mg Alloys for Corrosion Protection. vol. 12/969710. US 12/969710, Dec. 16 2010.
45. Rammelt U and Reinhard G. The influence of surface roughness on the impedance data for iron electrodes in acid solutions. *Corrosion Science* (1987);**27**:373.
46. Burstein GT and Pistorius PC. Surface roughness and the metastable pitting of stainless steel in chloride solutions. *Corrosion* (1995);**51**:380.
47. Hilbert LR, Bagge-Ravn D, Kold J and Gram L. Influence of surface roughness of stainless steel on microbial adhesion and corrosion resistance. *International Biodeterioration and Biodegradation* (2003);**52**:175.
48. Koutsomichalis A, Saettas L and Badekas H. Laser treatment of magnesium. *Journal of Materials Science* (1994);**29**:6543.
49. Pu Z, Song GL, Yang S, Outeiro JC, Dillon Jr OW, Puleo DA and Jawahir IS. Grain refined and basal textured surface produced by burnishing for improved corrosion performance of AZ31B Mg alloy. *Corrosion Science* (2012);**57**:192.
50. Obara T, Yoshinga H and Morozumi S. Slip system in magnesium. *Acta Metallurgica* (1973);**21**:845.
51. Lou XY, Li M, Boger RK, Agnew SR and Wagoner RH. Hardening evolution of AZ31B Mg sheet. *International Journal of Plasticity* (2007);**23**:44.
52. Konig U and Davepon B. Microstructure of polycrystalline Ti and its microelectrochemical properties by means of electron-backscattering diffraction (EBSD). *Electrochimica Acta* (2001);**47**:149.
53. Song G-L. Effect of the texture on the corrosion behavior of Mg AZ31 alloy. *Journal of the Minerals, Metals and Materials Society* submitted.
54. Song G-L. Effect of tin modification on corrosion of AM70 magnesium alloy. *Corrosion Science* (2009);**51**:2063.
55. Song G and Song S. A possible biodegradable magnesium implant material. *Advanced Engineering Materials* (2007);**9**:298.
56. Pu Z, Outeiro JC, Batista AC, Dillon OW, Puleo DA and Jawahir IS. Enhanced surface integrity of AZ31B Mg alloy by cryogenic machining towards improved

functional performance of machined components. *International Journal of Machine Tools and Manufacture* **56**:17.

57. Song D, Ma AB, Jiang JH, Lin PH, Yang DH and Fan JF. Corrosion behaviour of bulk ultra-fine grained AZ91D magnesium alloy fabricated by equal-channel angular pressing. *Corrosion Science* (2011);**53**:362.
58. Krajewski PE, Kim S, Carter JT and Verma R. Magnesium Sheet: Automotive Applications and Future Opportunities. *Korean Institute of Metals and Materials: Trends in Metals & Materials Engineering* (2007);**20**:60.
59. Aung NN and Zhou W. Effect of grain size and twins on corrosion behaviour of AZ31B magnesium alloy. *Corrosion Science* (2010);**52**:589.
60. Jönsson M, Thierry D and LeBozec N. The influence of microstructure on the corrosion behaviour of AZ91D studied by scanning Kelvin probe force microscopy and scanning Kelvin probe. *Corrosion Science* (2006);**48**:1193.
61. Wang H, Estrin Y, Fu H, Song G and Zúberová Z. The effect of pre-processing and grain structure on the bio-corrosion and fatigue resistance of magnesium alloy AZ31. *Advanced Engineering Materials* (2007);**9**:967.
62. Pu Z, Yang S, Song GL, Dillon Jr OW and Puleo DA, Jawahir IS. Ultrafine-grained surface layer on Mg-Al-Zn alloy produced by cryogenic burnishing for enhanced corrosion resistance. *Scripta Materialia* (2011);**65**:520.
63. Zhu L and Song G. Improved corrosion resistance of AZ91D magnesium alloy by an aluminium-alloyed coating. *Surface and Coatings Technology* (2006);**200**:2834.
64. Zhu L, Liu H, Li W and Song G. Zinc alloyed coating on AZ91D magnesium alloys. *Beijing Hangkong Hangtian Daxue Xuebao/Journal of Beijing University of Aeronautics and Astronautics* (2005);**31**:8.
65. Song G-L and Powell B. Method of forming a coated article including a magnesium alloy. vol. 12/431726. US 12/969710, Apr. 28 2009.
66. Peng XD, Edwards DS and Barteau MA. Reactions of O2 and H2O with magnesium nitride films. *Surface Science* (1988);**195**:103.
67. Höche D, Blawert C, Cavellier M, Busardo D and Gloriant T. Magnesium nitride phase formation by means of ion beam implantation technique. *Applied Surface Science* (2011);**257**:5626.
68. Nakatsugawa I, Martin R and Knystautas EJ. Improving corrosion resistance of AZ91D magnesium alloy by nitrogen ion implantation. *Corrosion* (1996);**52**:921.
69. Zhao Y, Wu G, Pan H, Yeung KWK and Chu PK. Formation and electrochemical behavior of Al and O plasma-implanted biodegradable Mg-Y-RE alloy. *Materials Chemistry and Physics* (2012);**132**:187.
70. Candan S, Unal M, Koc E, Turen Y and Candan E. Effects of titanium addition on mechanical and corrosion behaviours of AZ91 magnesium alloy. *Journal of Alloys and Compounds* (2011);**509**:1958.
71. Makar GL and Kruger J. Corrosion of magnesium. *International Materials Reviews* (1993);**38**:138.
72. Song G, Atrens A, Stjohn D, Nairn J and Li Y. The electrochemical corrosion of pure magnesium in 1 N NaCl. *Corrosion Science* (1997);**39**:855.
73. Vilarigues M, Alves LC, Nogueira ID, Franco N, Sequeira AD and da Silva RC. Characterisation of corrosion products in Cr implanted Mg surfaces. *Surface & Coatings Technology* (2002);**158**:328.
74. Samulevičien M, Miečinskas P, Leinartas K, Grigucevičien A, Kalinauskas P, Jasulaitien V, Jušknas R and Juzeliūnas E. Corrosion resistance of nanocrystalline

Mg-Cr alloys deposited by magnetron sputtering. *Materials Chemistry and Physics* (2011);**126**:898.

75. Baldwin KR, Bray DJ, Howard GD and Gardiner RW. Corrosion behaviour of some vapour deposited magnesium alloys. *Materials Science and Technology* (1996);**12**:937.
76. Mitchell T, Diplas S and Tsakiropoulos P. Characterisation of corrosion products formed on PVD in situ mechanically worked Mg-Ti alloys. *Journal of Alloys and Compounds* (2005);**392**:127.
77. Olk CH and Haddad DB. Growth and structure of a combinatorial array of mixed-phase magnesium-aluminum thin-film alloys. *Applied Physics A: Materials Science and Processing* (2007);**88**:249.
78. Song G-L and Haddad D. The topography of magnetron sputter-deposited Mg-Ti alloy thin films. *Materials Chemistry and Physics* (2011);**125**:548.
79. Bohne Y, Blawert C, Dietzel W and Mändl S. Formation and corrosion properties of Mg-Ti alloys prepared by physical vapour deposition. *Plasma Processes and Polymers* (2007);**4**:S584.
80. Blawert C, Heitmann V, Dietzel W, Störmer M, Bohne Y, Mändl S and Rauschenbach B. Corrosion properties of supersaturated magnesium alloy systems. *Materials Science Forum* (2007);**539–543**:1679.

6

Laser treatment to improve the corrosion resistance of magnesium (Mg) alloys

J. DUTTA MAJUMDAR and I. MANNA,
Indian Institute of Technology, Kharagpur, India

DOI: 10.1533/9780857098962.2.133

Abstract: Magnesium and its alloys are promising materials for automotive, aerospace and bio-implant applications due to low density (1.81 g/cm^3), very low elastic modulus (45 GPa) and high tensile yield strength (200 MPa). However, they corrode too quickly in aggressive environments and possess poor wear resistance. Wear and corrosion are essentially surface dependent degradation which may be improved by suitable modification of surface microstructure and composition. Laser as a source of coherent and monochromatic radiation may be applied for tailoring the surface microstructure and composition. In the present contribution, a detailed overview of laser surface engineering of magnesium-based alloys is presented. Finally, the future scope of application of laser surface engineering of magnesium-based alloys is presented with examples.

Key words: magnesium, laser, surface engineering, wear, corrosion.

6.1 Introduction

Magnesium is an alkaline metal with the atomic number 12 and the 8th most abundant element in the earth's crust.[1] Magnesium alloys are lightweight (two thirds that of aluminum), high strength and possess exceptional vibration damping characteristics, as well as being easy to machine and fabricate.[1] The metal is produced either by electrolytic reduction of $MgCl_2$ (obtained from seawater, brine deposits or salt lakes) or by chemical reduction of magnesium oxides (MgO) (obtained principally from seawater or dolomite) by Si in the form of ferrosilicon.[2] Magnesium has a hexagonal crystal structure and the ability to form a solid solution with a variety of elements, particularly those which are of commercial importance including Al, Zn, Li, Ce, Ag, Zr and Th because of a favorable size factor (atomic diameter 0.320 nm).

Commercially used magnesium alloys are fabricated by (a) liquid and semi-solid melting processes and (b) wrought forming processes.[3] Many magnesium alloys have been developed to provide a range of properties and characteristics to meet the needs of a wide variety of applications. Among them, there are two major classes—one containing aluminum as the principal alloying ingredient, the other containing zirconium. The alloys containing aluminum are strong and ductile, and have excellent resistance to atmospheric corrosion.[3] Zirconium is a potential grain refiner, however, is usually not added for the magnesium alloys containing aluminum. Hence, the alloys containing rare-earth or yttrium are suitable for applications at temperatures up to a maximum of 300°C within this class. The alloys not containing rare-earth or yttrium contains zinc as the principal alloying element and are strong, ductile, and tough. Recently developed magnesium alloys include: AZ91E (8.3–9.7% Al, 0.13% Mn, 0.35–1.0% Zn, 0.50% Sim, 0.10% Cu, 0.03% Ni, 0.30% max. other, balance Mg) developed by sand and permanent mold castings method and AZ91D (8.3–9.7% Al, 0.15% Mn, 0.35–1.0% Zn, 0.10% Si, 0.005% Fe, 0.1030% Cu, 0.002% Ni, 0.02% max. other balance Mg), AM60B (5.5–6.5% Al, 0.25% Mn, 0.22% Zn, 0.10% Si, 0.005% Fe, 0.10% Cu, 0.002% Ni, 0.003% max. other, balance Mg), AM50A 4.5–5.3% Al, 0.28–0.50% Mn, 0.0005–0.0015%Be, 0.20% Zn, 0. 0.05% Si, 0.008% Cu, 0.004% Fe, 0.001% Ni, 0.01% max. other, balance Mg, and AS41B (4.17%Al–0.87%Si–0.41%Mn–0.01%Zn–0.002%Fe–0.001%Cu–0.001%Ca, balance Mg) developed by die casting technique.

In spite of a large expansion of magnesium alloys, the corrosion resistance of these alloys, especially pitting and galvanic corrosion, is of major concern, particularly when the environment contains acid or salt.[4–12] Figure 6.1 shows the Pourbaix diagram schematically representing the stability of MgO and hydroxides (OH) as a function of the pH of the environment.[13] It should be noted that magnesium hydroxide formed through the interaction of magnesium with its environment is stable in the basic range of pH values, but not in the neutral or acid ranges. The immunity region of the diagram is well below the region of water stability; as a result, in neutral and low pH environments magnesium dissolution is accompanied by hydrogen evolution. In basic environments, passivation is possible as a result of the formation of a $Mg(OH)_2$ layer on the metal surface. Since the films that form on unalloyed magnesium are slightly soluble in water, they do not provide long-term protection.[14]

The influence of alloying additions on the corrosion of commercially pure magnesium is presented in Fig. 6.2.[15] Six of the elements included in Fig. 6.2 (aluminum, manganese, sodium, silicon, tin and lead), as well as thorium, zirconium, beryllium, cerium, praseodymium and yttrium, have little if any deleterious effect on the basic saltwater corrosion performance of pure magnesium when present at levels exceeding their solid solubility or up to a maximum of 5%.[4] Cadmium, zinc, calcium and silver have mild-to-moderate accelerating effects on corrosion rates, whereas iron, nickel, copper and cobalt have extremely deleterious effects because of their low solid solubility limits and their ability to serve as active cathodic

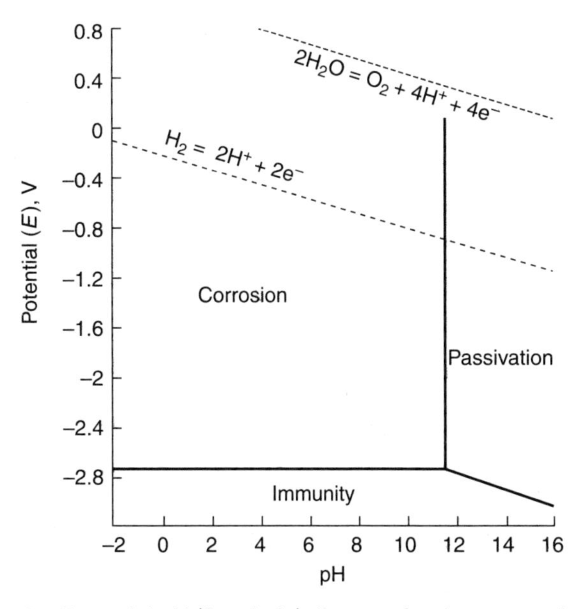

6.1 Potential-pH (Pourbaix) diagram for the magnesium and water system at 25°C, showing the theoretical domains of corrosion, immunity and passivation.[13]

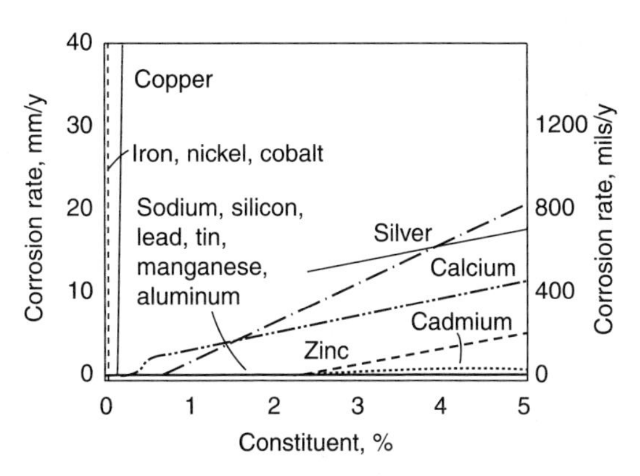

6.2 Effect of alloying and contaminant metals on the corrosion rate of magnesium as determined by alternate immersion in 3% NaCl solution.[15]

sites for the reduction of water at the sacrifice of elemental magnesium. Furthermore, in magnesium alloys, grain boundaries are usually cathodic compared to the grains. Corrosion attack is concentrated on the anodic area adjoining the boundary until, eventually, metallic particles surrounded by the attack fall out of the matrix.[16]

Magnesium and its alloys are promising materials for components in the aerospace, automobile and biomedical sectors.[17–19] Magnesium is employed extensively in aircraft engines, airframes and landing wheels due to its good machinability, high strength to weight ratio in casting and stiffness/density ratio in the wrought forms, combined with factors such as good elevated temperature, fatigue and impact properties, always with good machinability.[20,21] Due to their net shape cast ability and ease of component integration, magnesium and its alloys are extensively used in the automotive sector as steering wheels, steering columns and seat risers which take advantage of magnesium's high strength-to-density ratio and excellent ductility, combined with attractive energy absorbing characteristics.[22]

Corrosion is essentially a surface dependent degradation which may be minimized by suitable modification of surface microstructure and composition.[5] Gray and Laun[6] discussed the possible coating techniques used for improving the corrosion resistance of magnesium and its alloys. Laser as a source of coherent and monochromatic radiation may be applied for tailoring the surface microstructure and composition for improving corrosion resistance of magnesium and its alloys.[7] In this chapter, a detailed overview of the application of laser surface engineering in improving the corrosion resistance of magnesium and its alloys are presented.

6.2 Surface engineering for combating corrosion

As corrosion is a surface dependent degradation, it may be mitigated by suitable modification of the surface microstructure and/or composition. Corrosion of magnesium and its alloys may be minimized by the application of a conventional barrier or noble materials coating using a numbers of techniques. The easiest ways of enhancing the corrosion resistance of magnesium and its alloys include different types of conversion films, such as chromate, phosphate and anodizing.[23–28] Among them, chromate conversion technique is the most effective process and has been widely exploited in industry due to its excellent adhesion and corrosion resistance. However, the Cr^{6+} in chromate bath is a highly toxic carcinogen and is gradually facing prohibition. The development of an environmentally friendly process[29] is necessary due to increasingly stringent environmental protection laws currently in effect or being proposed.

Plating on magnesium and alloy surfaces has been shown to be useful in applications. However, due to the high chemical activity of magnesium, a strong replacement reaction occurs in the plating reprocess accompanied by a large amount of hydrogen evolution. The plating films, therefore, have weak adhesion to magnesium alloys. Generally, pretreatment is complicated for plating on magnesium alloys. Thus, standard ASTM B 480–1988 was established for preparation of magnesium alloys in electroplating.[30]

Many other protective technologies such as surface heat treatment,[31] vapor deposition[32–34] and ion implantation[35,36] have been invented. They are excellent alternatives with respect to environmental impact. However, these technologies usually involve more capital investment and power consumption. Further, the corrosion and adhesion properties of these coatings on magnesium are not yet satisfactory.

6.2.1 Laser as a source of heat

Light amplification by stimulated emission of radiation or laser is a coherent, convergent and monochromatic beam of electromagnetic radiation with wavelength ranging from ultraviolet to infrared.[37] Laser can deliver very low (~mW) to extremely high (1–100 kW) focused power with a precise spot size/dimension and interaction/pulse time (10^{-3}–10^{-15} s) on to any kind of substrate through any medium.[37–40] Laser has a wide-ranging application from printer to pointer, surgery to spectroscopy, isotope separation to invisible surveillance and medical to material treatment.[41,42] The important properties that enable and justify the use of laser in such a wide spectrum of applications are (a) spatial and temporal coherence (i.e. phase and amplitude are unique), (b) low divergence (parallel to the optical axis), (c) high continuous or pulsed power density and (d) monochromaticity. Table 6.1 summarizes the commercially available lasers used for materials processing applications and their main areas of application. Depending on the type of laser and wavelength desired, the laser medium could be solid, liquid or gaseous. Different laser types are commonly named according to the state or the physical properties of the active medium. Consequently, there are glass or semiconductor, solid state, liquid and gas lasers. Gas can be further subdivided into neutral atom lasers, ion lasers, molecular lasers and excimer lasers. The typical commercially available lasers are (a) solid state or glass laser – Nd:YAG, (b) semiconductor laser – AlGaAs, GaAsSb and GaAlSb, (c) dye or liquid lasers – solutions of dyes in water/alcohol and other solvents, (d) neutral or atomic gas lasers – He-Ne, Cu or Au vapor, (e) ion lasers – argon (Ar^+) and krypton (Kr^+) ion, (f) molecular gas lasers – CO_2 or CO and (g) excimer laser –XeCl, KrF, etc. The wavelengths of the currently available lasers cover a wide spectral range from the far infrared to the soft X-ray.

6.2.2 Laser surface engineering

Laser, as a source of intense heat, is capable of heating, melting and evaporating materials from the surface.[42] Laser may be applied as a source of heat to modify the microstructure and/or composition of the near surface region

Table 6.1 Commercially available lasers and their industrial applications

Laser	Discovery	Commercialization	Application
Ruby	1960	1963	Metrology, medical applications, inorganic material processing
Nd-Glass	1961	1968	Length and velocity measurement
Diode	1962	1965	Semiconductor processing, biomedical applications, welding
He-Ne	1962		Light-pointers, length/velocity measurement, alignment devices
Carbon dioxide	1964	1966	Material processing – cutting/joining, atomic fusion
Nd-YAG	1964	1966	Material processing, joining, analytical technique
Argon ion	1964	1966	Powerful light, medical applications
Dye	1966	1969	Pollution detection, isotope separation
Copper	1966	1989	Isotope separation
Excimer	1975	1976	Medical application, material processing, coloring
Free electron laser	1971	1997	Medical surgery, surface modification of polymer

and offers advantages in terms of economy, precision, flexibility and novelty (thermodynamic and kinetic) of processing and improvement in the surface dependent properties of interest.[37–40]

Laser processing is characterized by an extremely fast heating/cooling rate (10^4–10^{11} K/s), very high thermal gradient (10^6–10^8 K/m) and ultra-rapid resolidification velocity (1–30 m/s).[43–45] These extreme processing conditions often develop an exotic microstructure and composition in the near surface region with large extension of solid solubility and formation of metastable, including nanocrystalline and amorphous, phases. Laser surface engineering involves tailoring the performance of a component by microstructural modification (laser transformation hardening, laser surface melting), microstructures and composition (laser surface alloying) and compositional change (laser assisted chemical vapor deposition, pulsed laser deposition, etc.).

6.2.3 Laser surface melting for combating corrosion

In laser surface melting (LSM), the surface of the substrate is heated to its melting point with a high power laser beam and rapidly solidified and aimed at (a) refinement of surface microstructure, (b) homogenization of composition and (c) dissolution of precipitates.[46] In the past, laser surface melting was successfully applied to improve corrosion resistance of iron and steel.[47–50] Lo *et al.*[47,48] showed that laser surface melting of 440°C martensitic stainless steel was more effective in improving the corrosion resistance (in NaCl solution) through carbide refinement.

The possibility of inter-granular corrosion was minimized or suppressed by laser surface melting of sensitized AISI 304 stainless steel by eliminating the carbides formed during sensitizing treatment and subsequent homogenization.[49,50] Kwok *et al.*[51] observed that laser surface melting improved the corrosion resistance of high speed steel (HSS) in 0.6 M NaCl and 0.5 M $NaHCO_3$ solution. The enhancement in the corrosion resistance was attributed to the combined effects of dissociation and refinement of large carbides and the increase in Cr, Mo and W in solid solution. Conde *et al.*[52] studied the effect of laser surface melting on the corrosion behavior of two austenitic steels (AISI 310 and AISI 304), one ferritic (AISI 430) and one martensitic (AISI 420) steel processed under different conditions. It was observed that corrosion resistance was improved by laser surface melting and depended on the laser processing parameters. Laser surface melting of SAE 52100 steel improved its general and pitting corrosion resistance in a 3.56 wt.% NaCl solution; this was attributed to microstructural refinement and homogenization.[53]

Several attempts have been made to understand the effect of laser surface melting on corrosion resistance of magnesium and its alloys. However, precautions need to be undertaken so that a suitable shroud is provided to avoid oxidation during melting.[46] It was observed that laser surface melting was effective in refining the microstructure and improving corrosion resistance of magnesium and its alloys.[46] Kattamis[54] reported that laser surface melting improved the corrosion resistance of JK60 wrought Mg alloy which was attributed to the refinement of grains and grain boundary precipitates. A similar effect of rapid solidification on enhancement of corrosion resistance was reported in laser surface melted Mg-Li alloy (MA21), Mg-Al and Mg-Zn-Al alloys containing Si, Y and lanthanides.[55–57] The corrosion resistance of laser surface melted as-cast ACM720 (7.0 wt.% Al, 2.0 wt.% Ca, 0.5 wt.% Sn, 0.3 wt.% Mn, 0.1 wt.% Sr, balance Mg) alloy was significantly improved due to the absence of the second phase Al_2Ca at the grain boundary; microstructural refinement; and extended solid solubility, particularly of Al, owing to rapid solidification.[58] However, the corrosion behavior of the melted surface was found to depend on the microstructure and composition of the substrate, melt zone and the defect content on the surface. Abbas *et al.*[59] compared the corrosion behavior of laser surface melted AZ31, AZ61 and WE43 alloys. Figure 6.3 shows the effect of laser surface melting on the average corrosion rates in 5 wt.% sodium chloride solution of pH 10.5 at 20°C. The corrosion rate is reduced by about 30%, 66% and 87% for the AZ31, AZ61 and WE43 alloys, respectively. The corrosion rate of the WE43 alloy was significantly reduced such that the melted layer was not consumed in the test period.

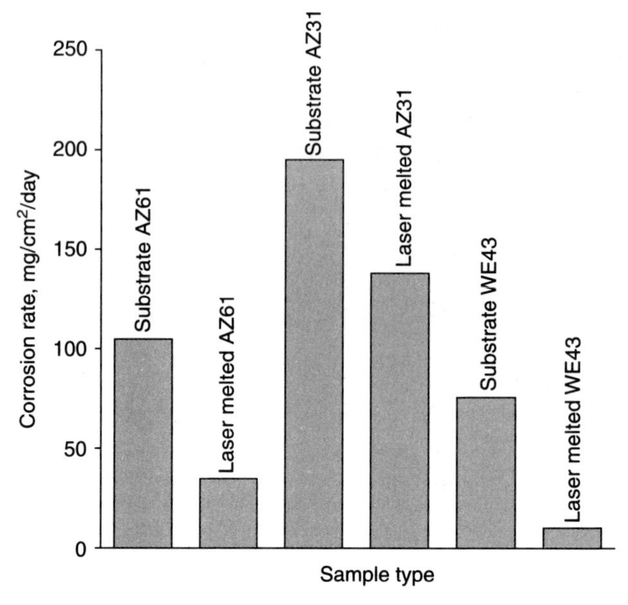

6.3 Average corrosion rates of AZ31, AZ61 and WE43 alloys, before and after LSM, during immersion for 10 days in 5 wt.% sodium chloride solution of pH 10.5 at 20°C.[59]

However, with increased corrosion rates, the melted layers of the AZ31 and AZ61 alloys were consumed during the test period, with subsequent attack proceeding in the underlying alloy (Fig. 6.3). The improved behavior is associated with the refinement of microstructure and microstructural uniformity. This observation was further supported by Singh Raman *et al.*[60] on laser surface melted AZ31 and AZ61 in 0.001 N NaCl solution. In the same group, however, it was observed that laser surface melting did not improve the corrosion resistance of ZE41 (4%-Zn, 1%-RE).[61]

Detailed studies on kinetics and mechanisms of corrosion resistance enhancement were undertaken in laser surface melted MEZ (a Mg-Zn alloy).[62] Laser surface melting was carried out using a 10 kW continuous wave (CW) CO_2 laser (Model: Rofin Sinar, RS 10 000) having a beam diameter of 4 mm with Ar as shrouding environment (at a flow rate of 6 L/min) to avoid oxidation during lasing. Figure 6.4 shows the scanning electron micrograph of the cross section of the laser surface melted MEZ alloy, lased with a power of 2 kW and scan speed of 200 mm/min. A significant refinement in microstructure is achieved in the melt zone as compared to the as-received substrate. In addition, the precipitates of Mg/Zr/Ce-rich compound observed in as-received MEZ substrate were partly remelted causing supersaturation of the matrix with alloying element. This was due to rapid cooling during solidification, and partly intermixed with molten metal and

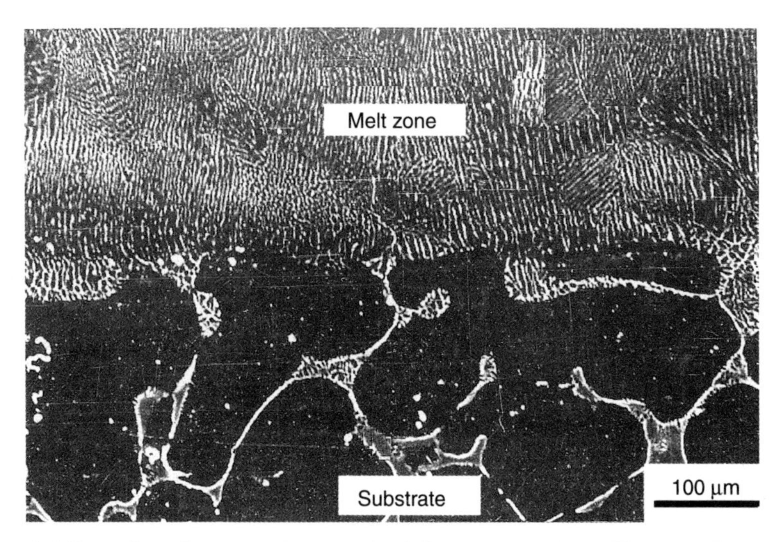

6.4 Scanning electron micrograph of the cross section of laser surface melted MEZ sample (lased with a power of 2 kW and scan speed of 200 mm/min). The defect free interface with significant grain refinement in the laser-melted zone may be noted.[62]

uniformly dispersed as fine precipitates in the melt zone. The melted zone-substrate interface is crack/defect-free and well compatible with practically with no noticeable amount of heat-affected zone.

Figure 6.5 compares the kinetics of pit formation (cumulative area fraction of pits as a function of time) of as-received (plot a) *vis-a-vis* laser surface melted (with a power of 2 kW and scan speed of 200 mm/min) (plot b) MEZ samples subjected to standard immersion tests in a 3.56 wt.% NaCl solution.[62] From Fig. 6.5 it is evident that both the extent and rate of pitting are significantly reduced following laser surface melting. In as-received MEZ, pit formation starts as soon as 30 min after immersion in contrast to laser surface melted MEZ where visible pits are observed after 12 h of exposure. Furthermore, the rate of pitting is initially faster and decreases after 30 h of immersion time in as-received MEZ. The kinetics of pitting is much slower in laser surface melted MEZ. The corrosion behavior of the MEZ samples was evaluated by potentiostatic polarization. The calculated values of corrosion voltage, corrosion current and corrosion rate are presented in Table 6.2.[62] A detailed analysis of the corrosion data in Table 6.2 reveals that laser surface melting has significantly reduced the corrosion rate to 0.133 mpy as compared to 6.12 mpy of the substrate in a 3.56 wt.% NaCl of MEZ solution. Furthermore, the polarization resistance, which is a measure of corrosion resistance (dE/di), was also found to increase significantly as a result of laser surface melting. A detailed study

Table 6.2 Summary of corrosion rate and polarization resistance of MEZ (as-received *vis-à-vis* laser remelted) in a 3.56 wt.% NaCl solution[62]

Sample	*OCP* mV(SCE)	E_{corr} mV(SCE)	i_{corr} (mA/cm²)	Corrosion rate (mpy)	Polarization resistance (mΩcm²)
MEZ	−1525	−1530	69	6.12	15.07
Remelted	−1185	−1190	1.5	0.133	899.9

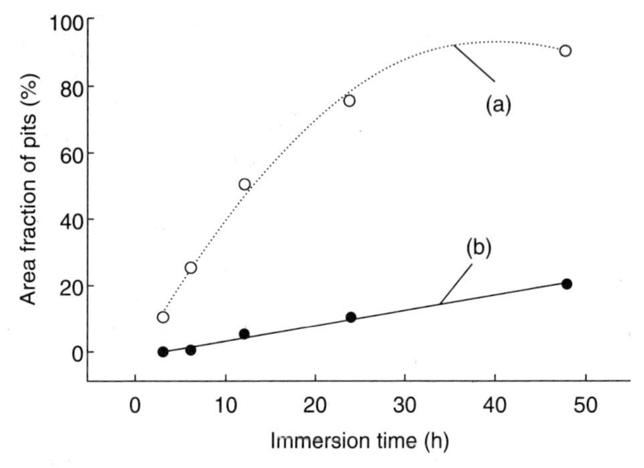

6.5 Kinetics of pit formation (in terms of area fraction of pits as a function of immersion time) in (a) as-received and (b) laser surface melted (with a power of 2 kW, scan speed of 200 mm/min) MEZ samples followed by immersion test in a 3.56 wt.% NaCl solution.[62]

of the mechanism of corrosion shows that the enhanced corrosion resistance of laser remelted specimens is attributed to the combined influence of grain refinement, dissolution of intermetallic phases and retention of alloying elements (RE elements) in extended solid solution. The fine grain boundary precipitates seem more effective in anchoring and retaining the $Mg(OH)_2$ film in laser surface melted samples.

Hino *et al.*[63] investigated the removal of anodic oxidation coatings of magnesium alloy by laser processing and its effect on its corrosion resistance. The area on the removed anodized coating under appropriate laser processing condition showed the excellent corrosion resistance as well as good conductivity. Excimer laser treatment of peak-aged WE43 alloy (4.1%Y-2.3%Nd-1.0%HRE-0.5Zr) led to dissolution of cathodic $Mg_{12}Nd$ particles and microstructural homogenization causing a significant improvement in corrosion resistance.[64] Yue *et al.*[65] studied the effect of laser surface melting on corrosion behavior in the same 3.56% NaCl solution for a Mg-SiC (17 vol.%) composite using KrF excimer laser with Ar and N_2 as shrouding environment. A considerable improvement in corrosion resistance was achieved primarily through microstructural refinement when melted using Ar shroud and it was also reported that surface melting in N_2 atmosphere improved the corrosion resistance further because of the formation of magnesium nitrides.

6.2.4 Laser shock processing for combating corrosion

Laser shock processing (LSP) involves rapid irradiation of the component surface with a laser (at a power density of ~10^{12} W/m^2) resulting in the generation of a shock wave (due to a volume expansion of the plasma plume formed on the surface) and subsequent alteration of microstructure/state of stress. The effect of the shock wave may be enhanced when it is propagated through water. This novel laser processing has the capability of improving material hardness and fatigue strength in a number of alloys. LSP has also been proposed as a competitively alternative technology to classical treatments for improving fatigue, corrosion cracking and wear resistance of metallic materials. It generates a high strength impact wave and then induces compressive residual stress of several hundred MPa by exposing metallic samples to high power density (GW/cm^2), short pulse (ns level) laser beam.[66–68] The previous studies have indicated that LSP virtually eliminates the occurrence of stress corrosion cracking (SCC) and crack-growth in austenitic stainless steels.[69] Zhang *et al.*[70] investigated the effects of LSP (using a Q-switched Nd:glass laser with a wavelength of 1064 nm) on the SCC susceptibility of AZ31B Mg alloy. Residual stress distribution as a function of depth was assessed using X-ray diffraction technology. Figure 6.6 shows the residual stress variations with different laser impacts as functions of the distance from the surface on the AZ31B Mg alloy.[70] (1: as-received, 2: one impact, 3: two impacts and 4: 4 impacts) It can be seen that the residual stress remarkably changes from tensile to compressive after LSP, and the minimum values are located at the surface. It is well known that the compressive residual stresses near the specimen surface are generated due to the local plastic deformation after LSP.[71] From Fig. 6.6 it is evident that there is introduction of compressive residual stress both in x (s_y) and y direction (s_y) on the surface, the magnitude of which is maximum at the surface (varying from –120 to –220 MPa) and increases with increasing no. of impacts. The residual stress does not vary much with direction. Furthermore, from Fig. 6.6 it may also be concluded that the depth of the compressive residual stress from the surface remains unchanged with different laser impacts. This was explained that the behavior was mainly because of the fact that the depth of the plastic deformation does not change with change in number of impact.[70]

6.3 Laser surface alloying for combating corrosion

Laser surface alloying causes melting of alloy ingredients (either pre-deposited or simultaneously added in the form of powder or wire) along with a part

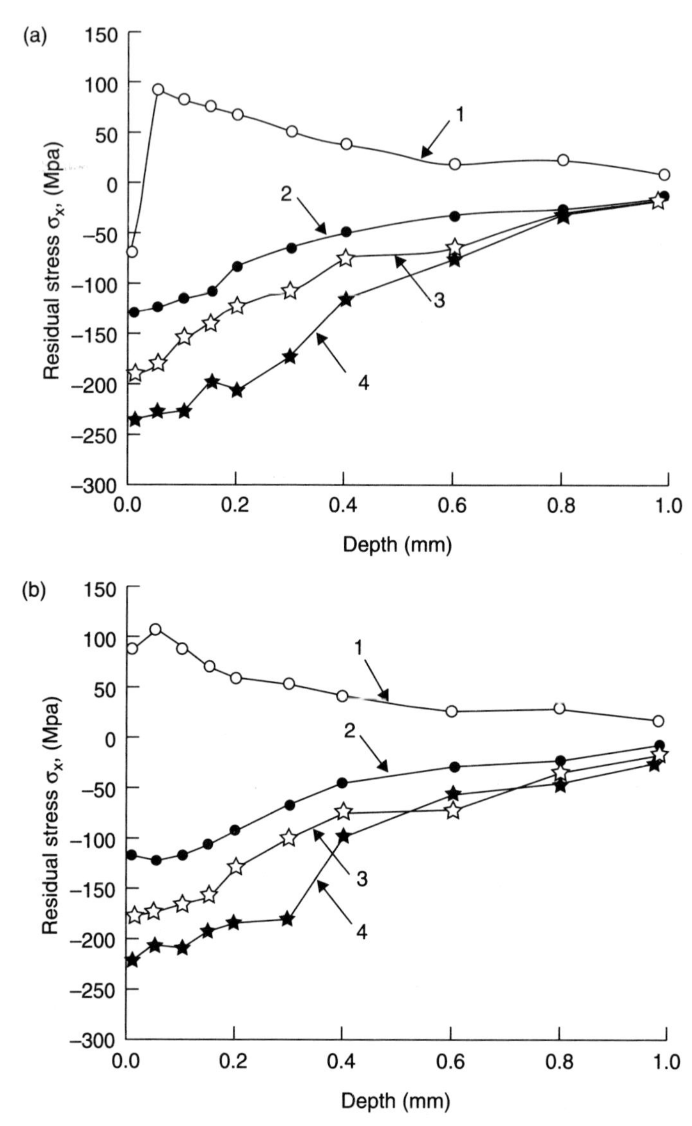

6.6 Variation of the residual stress along (a) x (σ_x) and (b) y (σ_y) direction with different laser impacts (1: as-received, and with 2: one impact, 3: 2 impacts, 4: 4 impacts) as a functions of the distance from the surface of AZ31B Mg alloy.[70]

of the underlying substrate to form an alloyed zone for improving the surface dependent engineering properties of the substrate.[46] Figure 6.7 illustrates the scheme of laser surface alloying with a CW laser. It includes three major parts: a laser source with a beam focusing and delivery system, a microprocessor controlled sweeping stage where the specimen is mounted for lasing and

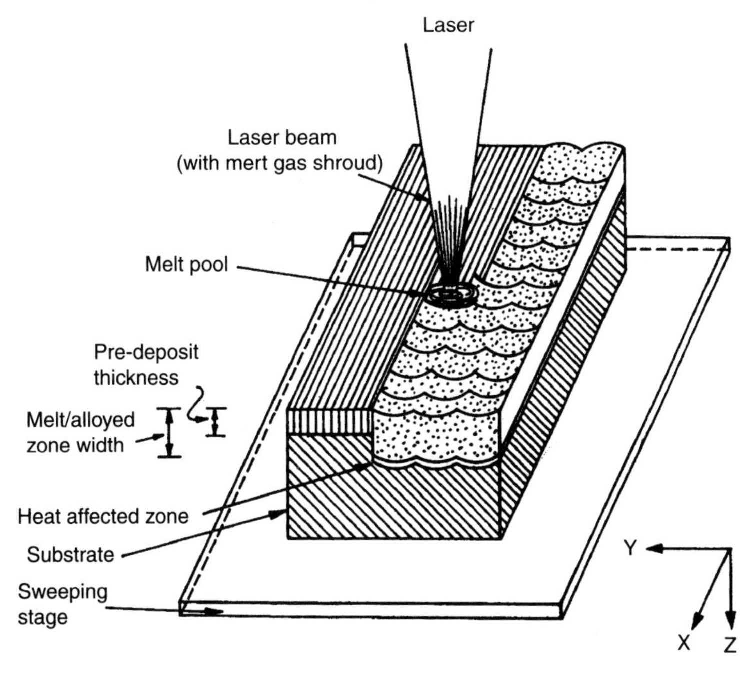

6.7 Schematic of laser surface alloying process using a pre-placed powder.[46]

the arrangement for delivering the alloying ingredients in the form of powder. The process includes melting of the alloy ingredient along with part of substrate, subsequent intermixing and rapid solidification to form the alloyed zone confined to a very shallow depth from the surface. A 20–30% overlap of the successive molten/alloyed track is intended to ensure microstructural/compositional homogeneity of the laser-treated surface in the laser-alloyed zone that has a composition distinctly different from that of the underlying base substrate. The sweeping stage (x–y or x–y–z–θ) allows laser irradiation of the intended area of the sample surface at an appropriate rate and interaction time/frequency. The depth, chemistry, microstructure and associated properties of the alloyed zone depend on the suitable choice of laser/process parameters *which are:* incident power/energy, beam diameter/profile, interaction time/pulse width, pre- or co-deposition thickness/composition and relevant physical properties like reflectivity, absorption coefficient, thermal conductivity, melting point and density. Though laser surface alloying is effective in enhancing corrosion resistance of magnesium and its alloys, a proper choice of alloying elements and laser processing conditions is essential to ensure a defect-free alloyed zone with improved corrosion resistance. Laser surface alloying has been successfully applied to improve corrosion resistance of ferrous and non-ferrous materials.[72–81] Ni-P coating provides corrosion protection in steel.

Laser surface melting of the same was found to reduce the corrosion resistance in comparison to as-deposited Ni-P due to the alloying of Fe with Ni-P.[72] Khanna *et al.*[73] coated stainless steel powder on mild steel substrates using a plasma coating technique and melted it with a CW CO_2 laser. The porous coating thus obtained was modified using laser surface alloying. The oxidation behavior of the plasma-coated as well as the laser-treated materials was studied in air at 1073 K. As expected, the oxidation rate decreased considerably after plasma coating compared to that of the stainless steel, but there was a significant improvement in the oxidation resistance after laser treatment of the coated surface. The adherence of the scale, tested using an acoustic emission technique and thermal cycling, showed significant improvement.

Dutta Majumdar and Manna[74] investigated the effect of lasersurface melting and alloying of plasma spray-deposited Mo on AISI 304 stainless steel on its corrosion resistance. The optimum processing zone corresponding to the formation of a defect-free and homogeneous microstructure was derived. Potentiodynamic anodic polarization tests of the substrate and laser surface alloyed samples in 3.56 wt.% NaCl solution (both in forward and reverse potential) showed that the critical potential for pit formation (E_{PP1}) was significantly (2–3 times) improved from 75 mV(SCE) in the substrate to 550 mV(SCE) after laser surface alloying.[74] E_{PP2} has also been found to be nobler in as-lased specimens than in the stainless steel substrate. Pillai *et al.*[75] have obtained enhanced oxidation resistance (at 873 K for up to 200 h) of plain carbon steel following laser surface alloying with Al due to presence of several intermetallic phases like $Al_{13}Fe_4$ and Al_2Fe_2.

Tang *et al.*[76] observed that laser surface alloying of manganese-nickel-aluminum bronze with Al was more effective in enhancing corrosion and cavitation erosion resistance in a 3.56 wt.% NaCl solution than laser surface melting. By employing polarization resistance measurements under quiescent and cavitating conditions, the contributions of erosion-induced corrosion and corrosion-induced erosion were determined.

Almeida *et al.*[77] carried out laser surface alloying of Al with Cr using the powder injection method with a laser beam power of 2000 W and 5 mm/s scan speed. Cr content in the alloyed zone was 4.3 wt.%. Similarly, Ferreira *et al.*[78] also reported that crevice corrosion in Al 7175-T7351 alloy could be significantly reduced by laser surface alloying with Cr.

Thermal barrier coatings could also be effective for improving oxidation resistance of stainless steel.[79] However, such coatings are brittle and hence must be applied with a favorable composition gradient like applying a bond coat (Ni-22Cr-14Al-1Y) between the topcoat (ZrO_2 + 7.5 wt.% Y_2O_3) and stainless steel substrate before single or multiple pass LSM treatments. Oxidation tests at 1200°C indicated that double pass LSM was more effective in imparting a better oxidation resistance.[79] Similar attempts to improve oxidation resistance have been made with Incoloy 800H by laser surface alloying with Al[80] and by laser surface cladding of 316 L stainless steel with

Fe-Cr-Al-Y alloy coatings.[81] In all these cases, it appears that the degree of enhancement of oxidation resistance by LSE primarily depends on stability, adherence, imperviousness and strength of the complex oxide scale comprising multiple oxides on top of the base metal.

Laser surface alloying was successfully used to improve aqueous corrosion resistance and wear resistance of magnesium and its alloys. Galun *et al.*[82] studied laser surface alloying of commercial pure Mg with Al, Ni, Si, Cu, Al + Si, Al + Cu and Al + Ni and found a maximum improvement in wear resistance in laser-alloyed Mg with Al + Ni. On the other hand, Al is the only alloying element which offered maximum resistance to pitting corrosion, in particular, at Al content >20 at%, there was no visible pit formation.[83]

Addition of Mn with Al was found to be beneficial in enhancing the corrosion resistance further. A detailed study of the effect of laser surface alloying of a magnesium alloy, MEZ (RE 2%, Zn 0.5%, Mn 0.1%, Zr 0.1%) with Al + Mn (at different ratios) on microstructure, mechanical properties and electrochemical properties was made by Dutta Majumdar *et al.*[84] Laser surface alloying of MEZ with Al + Mn (in weight ratios of 3:1 and 1:3) was carried out with a 10 kW CW CO_2 laser (Model: Rofin Sinar, RS 10 000) having a beam diameter of 4 mm (focal point 30 mm above the surface) with Ar as shrouding gas to avoid oxidation during lasing.

Figure 6.8 shows the scanning electron micrograph of the cross section of laser-alloyed MEZ with (a) 76Al + 24Mn and (b) 45Al + 55Mn lased with a power of 2.5 kW, scan speed of 200 mm/min and powder feed rate of 20 mg/s.[84] The microstructures primarily consist of two distinct regions containing mainly the intermetallics of Al + Mn and Al + Mg, respectively. The upper region is enriched in alloying contents having a very large volume fraction of intermetallics and the lower region is substrate enriched with dispersion of intermetallic precipitates. It is relevant to mention that the morphology and the area fraction of the intermetallic phases were different near to the surface as compared to the near-interface region. Moreover, it is evident from Fig. 6.8 that the interface is adherent and defect-free with a negligible heat-affected zone. However, it is relevant to note that this kind of homogeneous and defect-free microstructure with graded composition is formed only for a very narrow range of laser/process parameters. As evidence, see Fig. 6.9 (a, b) which shows the variation of Al and Mn content in the alloyed zone with depth for laser surface alloyed (lased with a power of 2.5 kW, scan speed of 200 mm/min and powder feed rate of 20 mg/s) MEZ with (a) 76Al + 24Mn and (b) 45Al + 55Mn, respectively.[84] From Fig. 6.9 it is relevant that Al and Mn content in the alloyed zone is maximum near to the surface and gradually decreases with depth for both the cases of laser surface alloyed MEZ with (a) 76Al + 24 Mn and (b) 45Al + 55Mn. This kind of microstructure is beneficial in reducing residual stress distribution with maximum enhancement in hardness and corrosion resistance property on the top surface due to maximum enrichment of alloying content near to the surface

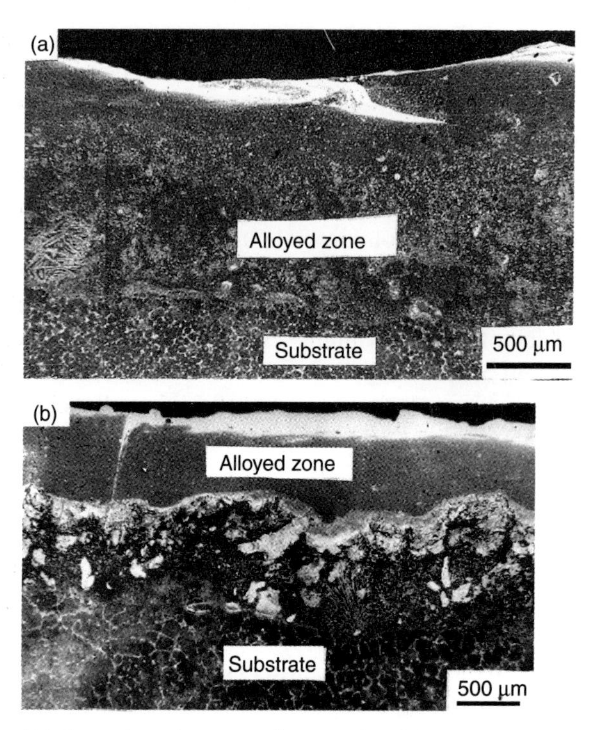

6.8 Scanning electron micrograph of the cross section of laser alloyed MEZ with (a) 76Al + 24Mn and (b) 45Al+55Mn lased with a power of lased with a power of 2.5 kW, scan speed of 200 mm/min and powder feed rate of 20 mg/s.[84]

region. However, the average alloying content on the surface and its distribution with depth were found to vary with laser parameters. It has been found that increasing laser power decreases the average alloying element content in the alloyed zone mainly because of increased dilution. On the other hand, increasing scan speed also decreases the average composition distribution mainly because of lower powder input at a low interaction time.

A detailed corrosion behavior of as-received and laser surface alloyed MEZ is summarized in Table 6.3.[84] From Table 6.3 it is evident that laser surface alloying of MEZ with Al and Mn has remarkably improved the corrosion resistance in a 3.56 wt.% NaCl solution. However, alloying with 45Al + 55Mn is more effective in enhancing corrosion resistance of MEZ substrate. As a comparison, a corrosion study of the laser remelted sample is also shown in Table 6.3. It is interesting to note that surface grain refinement and solutionizing of RE element in the matrix phase cause improvement in corrosion resistance of the remelted region.[84] A detailed analysis of the microstructure and phase analysis of the corroded film shows that the surface scale of as-received MEZ followed by immersion for 72 h in 3.56 wt.% NaCl solution mainly consists of $Mg(OH)_2$

Table 6.3 Summary of the parameters defining the corrosion behavior of as-received and laser surface alloyed MEZ under different conditions in a 3.56% NaCl solution[84]

Sample	OCP (mV)	E_{corr} (mV)	I_{corr} (mA)	Corrosion rate (mpy)
MEZ (as-received)	−1528	−1531	1.714	1521.32
MEZ (laser remelted)	−1185	−1181	0.875	777.34
Laser surface alloyed with 76Al + 24Mn	−1445	−1448	0.286	254.5
Laser surface alloyed with 45Al + 55Mn	−1441	−1437	0.230	204.32

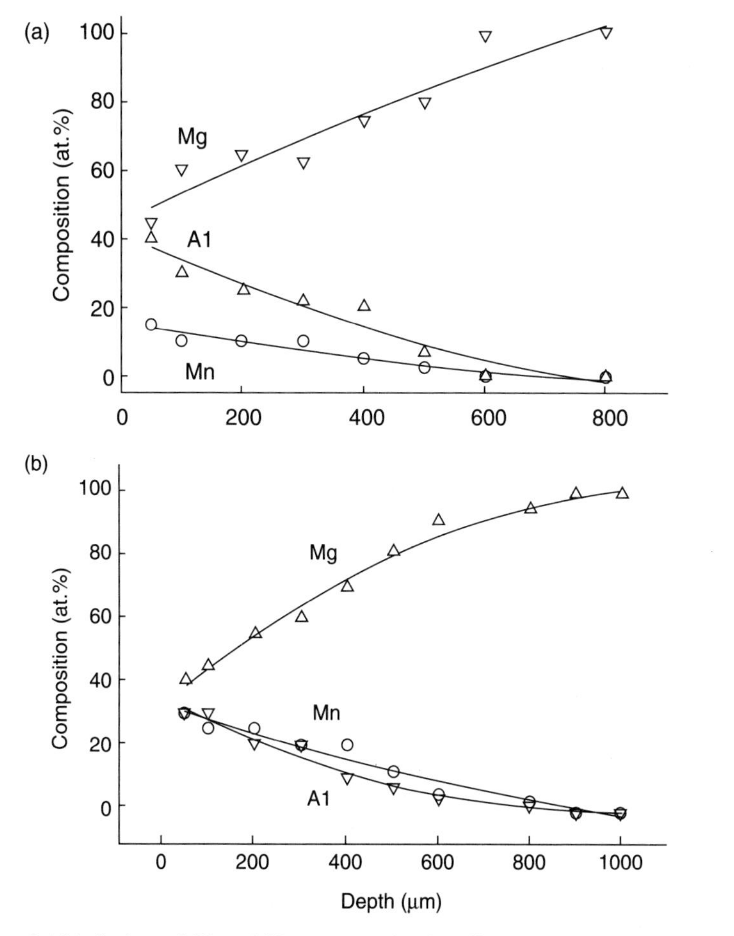

6.9 Variation of Al and Mn content in the alloyed zone with depth for laser surface alloyed (lased with a power of 2.5 kW, scan speed of 200 mm/min and powder feed rate of 20 mg/s) MEZ with (a) 76Al + 24Mn and (b) 45Al + 55Mn, respectively.[84]

with the presence of some $MgCl_2$. On the other hand, in laser surface alloyed MEZ with 76Al + 24Mn and 45Al + 55Mn, the main corrosion products are Mn_3O_4 and Al_2O_3. Hence, it may be concluded that $Mg(OH)_2$ was the main corrosion product in MEZ, which was responsible for a higher degree of corrosion. Better corrosion resistance of the laser surface alloyed specimens is attributed to the presence of adherent and stable films of Mn_3O_4 and Al_2O_3 in the corrosion film of the laser surface alloyed specimens, making it more stable in an aqueous environment.

In another attempt, laser surface alloying of a magnesium alloy, Mg-9Al-0.9Zn (AZ91) with nickel has been carried out with an objective of improving corrosion resistance.[85] Laser surface alloying was carried out on a sand-blasted magnesium alloy (AZ91) substrate by melting with a 10 kW CW CO_2 laser (Model: Rofin Sinar, RS 10 000) having a top-hat beam profile (of 4 mm beam diameter and focal point 30 mm above the surface) and simultaneous addition of nickel with Ar as shrouding gas to avoid oxidation during lasing. Laser surface alloying was carried out under the following optimum process parameters: applied power of 1.5–4 kW, scan speed of 100–400 mm/min and powder feed rate of 16 mg/s.

Figure 6.10a–c shows the scanning electron micrographs of the (a) top surface, (b) at a depth of 500 μm from the surface and (c) near-interface region of the alloyed zone formed in laser surface alloyed AZ91 with Ni lased with a power of 3 kW, scan speed of 300 mm/min and powder feed rate of 20 mg/s. Application of a high scan speed refines the microstructure considerably. Furthermore, the primary phase was found to solidify along a particular direction and the shape changes as compared to the same when processed at a low scan speed. As observed in Fig. 6.10, the volume fraction of primary precipitates decrease with increase in depth. The area fraction of precipitates was also found to decrease (along depth) with increase in scan speed. From the microstructural observation it is clear that under those laser conditions a graded microstructure formed with a maximum volume fraction of intermetallic precipitates at the near surface region, decreasing with depth from the surface. A detailed X-ray diffraction profile shows that the microstructure of as-received AZ91 mainly consists of Mg and intermetallics of Mg and Al ($Mg_{17}Al_{12}$). Laser surface alloying of AZ91 with Ni, on the other hand, leads to the formation of intermetallics of Mg and Ni (predominantly Mg_2Ni) with some undissolved nickel and Mg. The corrosion resistance of as-received AZ91 and laser surfaced AZ91 with Ni was evaluated to understand the Ni alloying in a 3.56 wt.% NaCl solution. Average corrosion potential (E_{corr}) observed for laser surface alloyed AZ91 was considerably nobler, –0.946 V (SCE) as compared to as-received substrate, –1.523 V(SCE). Similarly, the corrosion rate of laser surface alloyed AZ91 with Ni was significantly reduced (0.000252 mm/year) when compared to as-received substrate (17.95 mm/year). Hence, surface alloying with Ni is beneficial in improving the corrosion resistance of AZ91.

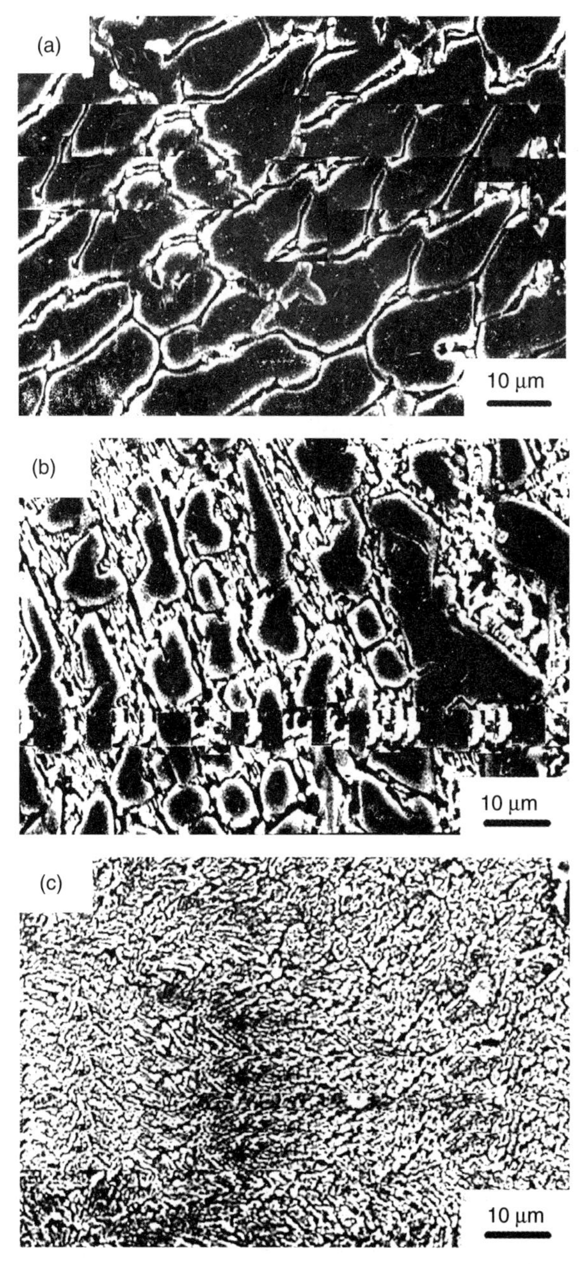

6.10 Scanning electron micrographs (SEI) of the cross section of laser surface alloyed AZ91 with Ni lased with power 3 kW, scan speed 300 mm/min and powder rate of 20 mg/s (a) at the surface; (b) at a distance of 500 μm from surface; (c) near to solid-liquid interface.

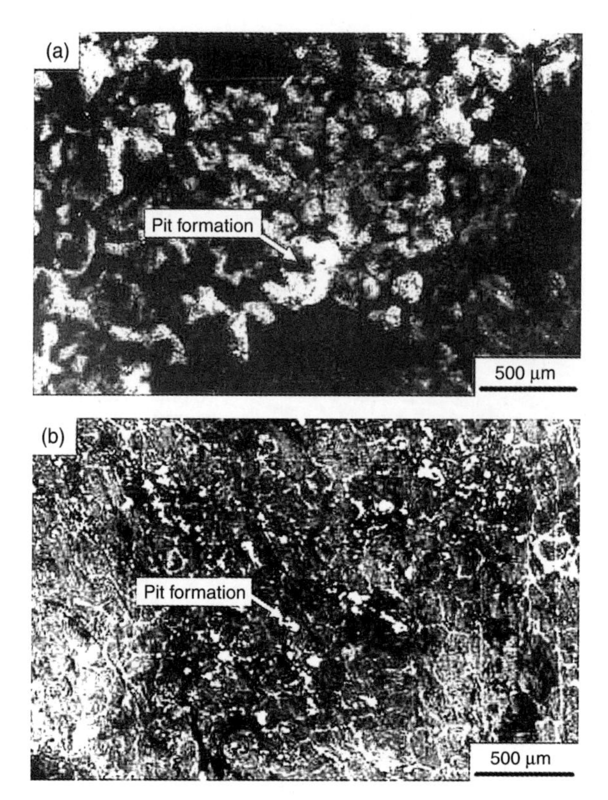

6.11 Scanning electron micrographs of the corroded surface of (a) as-received and (b) laser surface alloyed AZ91 with Ni followed by immersion in 3.56% NaCl solution for 24 h.

A detailed analysis of the corroded surface of as-received *vis-à-vis* laser surface alloyed AZ91 with Ni was undertaken to investigate the extent of corrosion and its mechanism. Figure 6.11a and 6.11b shows the scanning electron micrographs of the corroded surface of (a) as-received and (b) laser surface alloyed AZ91 with Ni followed by immersion in 3.56% NaCl solution for 24 h. A close comparison between Fig. 6.11a and 6.11b shows that in as-received AZ91 alloy almost 75% of the surface area is pitted, and there is interconnection of individual pits to form a large pitted area, with the dispersion of corroded products in a discontinuous fashion. On the other hand, in laser surface alloyed AZ91 with Ni, small pits are present. The area fraction of pits was 25–30%. The volume fraction of the corroded product was much less compared to as-received AZ91. Figure 6.12 a and 6.12 b shows the X-ray diffraction profiles of the corroded surface of (a) as-received and (b) laser surface alloyed AZ91 with Ni (lased with a power of 2.5 kW, scan speed of 400 mm/min and powder rate of 20 mg/s) followed by immersion in 3.56% NaCl solution for 24 h. It is observed that

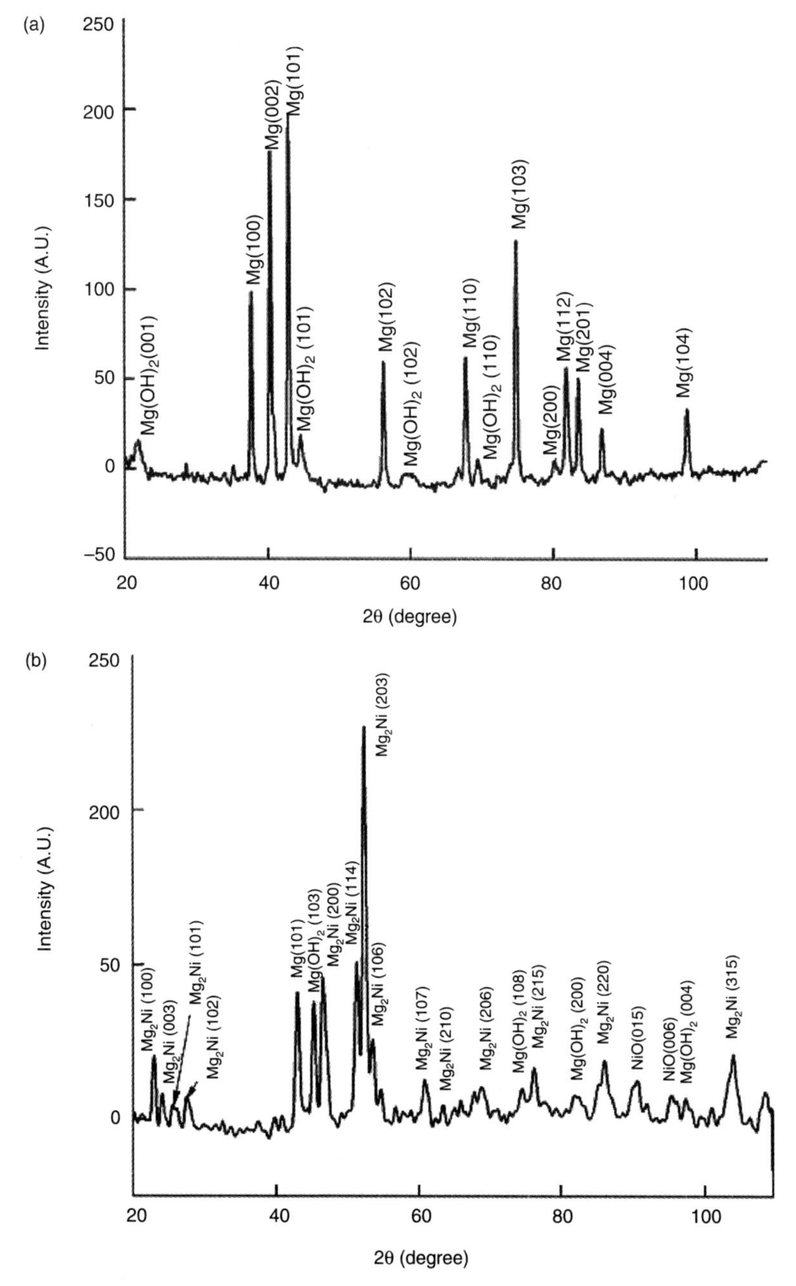

6.12 X-ray diffraction profiles of the corroded surface of (a) as-received and (b) laser surface alloyed AZ91 with Ni (lased with a power of 2.5 kW, scan speed of 400 mm/min and powder rate of 20 mg/s) followed by immersion in 3.56% NaCl solution for 24 h.

in as-received AZ91, there is the presence of a mostly $Mg(OH)_2$ phase, along with the presence of a few Mg peaks. On the other hand, on the corroded surface of laser surface alloyed AZ91 with Ni, there is the presence of strong Mg_2Ni peaks with a few $Mg(OH)_2$ peaks revealing a very low extent of corrosion in laser surface alloyed AZ91 with Ni. The improved corrosion resistance of laser surface alloyed AZ91 with Ni is attributed to surface coverage by Mg_2Ni which is electrochemically more noble than Mg, as a result of which the corrosion potential of the surface shifts towards the nobler direction. Furthermore, laser surface alloyed AZ91 has an almost negligible corrosion product compared to the as-received surface. It is relevant to mention that, even though there is minor galvanic attack at the interface between the Mg_2Ni and Mg, it cannot propagate in a lateral direction due to the presence of Mg_2Ni, which is electrochemically noble.

6.4 Laser surface cladding for combating corrosion

Laser surface cladding involves melting of the clad material with a high power laser and applying the molten layer onto the substrate surface to cause solidification of the clad layer with a minimum dilution at the interface.[46] A large number of attempts have been made to study the effect of laser surface cladding on corrosion behavior of magnesium and its alloys. The clad compositions were based on alloys which have a positive influence on corrosion resistance of magnesium and its alloys and whose melting point does not differ largely from the substrate. Surface cladding of Mg-Zr alloy onto magnesium and its alloys was found to improve corrosion resistance.[86,87] Cladding of Mg + Al (in the weight ratio of 1:4) was developed on ZM5 substrate by pre-placement of precursor powder and subsequent laser melting.[88] The corrosion resistance of laser clad was superior to that of the untreated ZM5 substrate due to the refined microstructure, the decrease in the anode to cathode area and the extended solid solubility of Al in the Mg matrix. The primary corrosion mechanism was reported to be galvanic corrosion due to the presence of $Mg_{17}Al_{12}$ phase which acts as cathodic phase to the α phase.

Al + Si cladding was also found to improve corrosion resistance of magnesium-based substrate. The microstructure and corrosion behavior of laser surface clad Al-Si on ZE41 was studied by Volovitch *et al.*[89] The clad layer was found to be composed of an Al-Mg matrix and dendrite precipitates of Mg_2Si. In the matrix there was also formation of $Mg_{17}Al_{12}$. intermetallics in a Mg matrix. The presence of different phases was responsible for galvanic corrosion and decrease of corrosion resistance in the interfacial area between coats. However, the coated surface showed similar corrosion potentials in spite of its different matrix composition. Figure 6.13 shows the mechanism of corrosion in the presence of Mg_2Si phase.[89] From Fig. 6.13 it may be concluded that the mechanism of corrosion involves two stages:

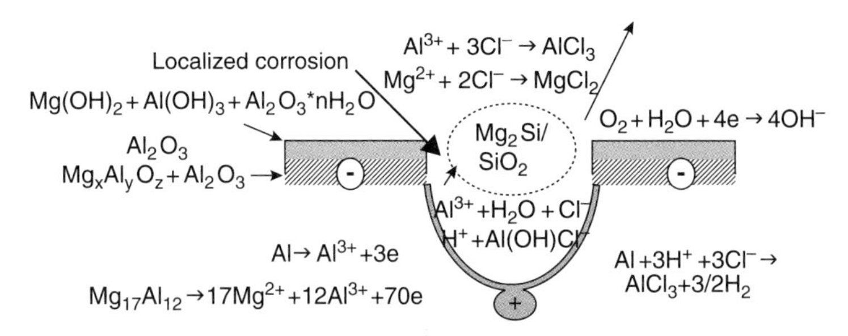

6.13 Schematic showing the localized corrosion mechanism of intermetallic matrix in the presence of Mg_2Si particles.[89]

(1) an initial dissolution of anodic Mg_2Si particles followed by (2) pitting in the formed crevices. The initial dissolution of anodic Mg_2Si particles pitting decreases the local pH in the pits and increases the local pH at the surface. The acid character of the solution in the pits and crevices corresponds with formation of SiO_2 and the basic character of the surface causes the corrosion potentials of Al and $Mg_{17}Al_{12}$ phases in NaCl to be very close and hence, the corrosion potentials of the two matrixes become nearly the same. The proposed mechanism corresponds well with the experimental observations and the mechanisms of localized corrosion observed for Al alloys in the chloride media described in the literature.

A graded multi-layer coating of Ni/Cu/Al was successfully deposited on magnesium substrates using a high power CW CO_2 laser.[90] A metallurgical bond was achieved at the interfaces between Mg-Al, Al-Cu and Cu-Ni layers. Significant mixing had occurred in the Al layer: a considerable amount of Mg and Cu had entered into the Al layer, and as a result, some complex Mg-Al-Cu ternary phases were formed. Nevertheless, Mg was not found in the upper portion of the Cu layer. With regard to the Ni layer, the objective of obtaining a single phase structure of (Ni) solid solution was achieved. Figure 6.14 shows the corrosion behavior of as-clad and as-received substrate tested using potentiodynamic anodic polarization.[90] The corrosion potential (E_{corr}) of the laser-clad graded coating was about −580 mV, which was 1020 mV more noble than the uncoated material. In addition, the laser-clad specimen exhibited a large span of passivity of over 500 mV, in contrast to the magnesium substrate which showed no sign of passivation.

Magnesium-based metal matrix composites (MMCs) are of great interest for structural applications where there is an obvious need for weight saving.[91,92] However, one of the major obstacles for wider usage in the automotive and aerospace industries is the inherently poor corrosion resistance of the material. Laser surface cladding of Al-Cu alloy on SiC dispersed magnesium-based composite improved its corrosion potential as well as the breakdown potential.[93] Moreover, the corrosion current was reduced by at

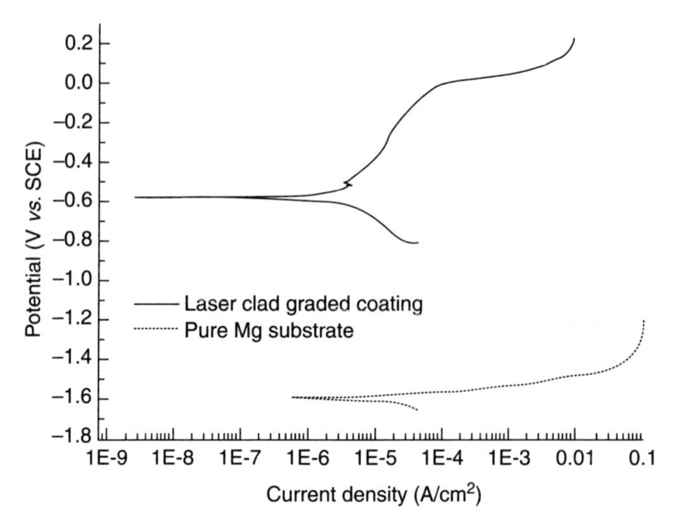

6.14 Potentiodynamic polarization behavior of as-clad and as-received substrate using potentiodynamic anodic polarization test.[90]

least one order of magnitude. Laser surface cladding of an Al-Zn mixture was also produced on ZK60/SiC composite by thermal spraying and subsequent laser cladding. The corrosion potential of the laser-clad sample was about 300 mV higher than that of the as-received sample and the corrosion rate was three orders of magnitude lower than as-received substrate.[94] In an another attempt, laser surface cladding of an Al-Si eutectic alloy on ZK60/SiC composite was successfully produced by using a two-step method: thermal spraying then laser cladding.[95]

Amorphous metal is usually an alloy rather than a pure metal. The alloys contain atoms of significantly different sizes, leading to low free volume (and therefore up to orders of magnitude higher viscosity than other metals and alloys) in their molten state. The viscosity prevents the atoms moving enough to form an ordered lattice. The material structure also results in low shrinkage during cooling and resistance to plastic deformation. The absence of grain boundaries, the weak spots of crystalline materials, leads to better resistance to wear and corrosion. Amorphous metals, while technically glasses, are also much tougher and less brittle than oxide glasses and ceramics. However, it is brittle and surface cladding of amorphous materials is capable of significantly improving corrosion resistance. Laser surface cladding has been successfully attempted in order to develop amorphous coating on magnesium and its alloys.[96–98] Huang *et al.*[96–98] developed magnesium-based ($Mg_{65}Cu_{25}Y_{10}$/SiC), copper based ($Cu_{47}Ti_{34}Zr_{11}Ni_8$) and zirconia based ($Zr_{55}Al_{10}Ni_5Cu_{30}$ + 20 wt.% SiC) amorphous composite coatings on AZ91D magnesium alloy by laser cladding using pre-placement of mixed

Table 6.4 Summary of the corrosion behavior of as-received and laser-clad AZ91D samples in a 3.5% NaCl solution[96–98]

Sl. No.	Coating composition	E_{corr} (V/SCE)	Corrosion rate (mm/year)	References
1	AZ91D	−1.46	8.59	Huang *et al.*[96]
2	$Mg_{65}Cu_{25}Y_{10}$ + SiC	−1.34	1.63	Huang *et al.*[97]
3	$Cu_{47}Ti_{34}Zr_{11}Ni_{8}$ + SiC	−1.29	2.56	Huang *et al.*[98]
4	$Zr_{55}Al_{10}Ni_{5}Cu_{30}$ + 20wt.% SiC	−1.33	2.13	Huang *et al.*[98]
5	$Zr_{65}Al_{7.5}Ni_{10}Cu_{17.5}$	−0.480		Huang *et al.*[98]

powders and laser melting. In all the coatings, there was formation of mainly amorphous phase with a few crystalline peaks. Table 6.4 compares the corrosion behavior of the coatings in a 3.56 wt.% NaCl solution and its comparison with the base material.[96–98] From Table 6.4 it may be noted that in all the coatings, there is significant improvement in corrosion resistance both in terms of shifting corrosion potential to the noble direction and corrosion rate. A close comparison of corrosion behavior of different coatings shows that there is a marginal difference in corrosion potential with changes in coating composition and a maximum shift in corrosion potential is achieved in copper based amorphous coating ($Cu_{47}Ti_{34}Zr_{11}Ni_{8}$ + SiC). Comparison of corrosion rate, on the other hand, shows that a maximum decrease in corrosion rate is achieved when a magnesium-based ($Mg_{65}Cu_{25}Y_{10}$/SiC) amorphous coating is applied. It was concluded that the improvement in corrosion resistance is attributed to the presence of the amorphous phase in the coatings.

6.5 Conclusions and future trends

Laser surface engineering is an emerging technique for the modification of microstructures and composition of the near surface region of magnesium and its alloys for tailoring electrochemical properties. Laser surface melting is the simplest treatment for compositional homogenization, grain refinement and solubility extension of the alloying ingredients. Hence, laser surface melting improves corrosion resistance magnesium-based alloys where, alloying constituents contribute positively to the enhancement of corrosion resistance. However, process parameters need to be optimized for the minimization of residual stress level on the surface. Reported research on laser surface alloying of magnesium and its alloys for combating corrosion is limited. Fewer attempts have been made on laser surface alloying of magnesium and its alloys; this is mainly due to their low melting temperature, there being less control over average composition/its distribution and the high reactivity of magnesium. However, laser surface alloying develops a defect-free microstructure and improves the corrosion resistance if it is designed properly and performed under optimum parameters. A large number of reports is available in the literature on the application of laser surface cladding in improving wear resistance. Clad compositions are selected based on their electrochemical properties. As the composition of the clad layer does not change after laser processing, similar or improved properties are usually achieved in the clad surface as compared to the bulk composition. When there is a large difference in melting temperature between the clad layer and base metal a graded layer is developed by multi-step processing.

In the present chapter an attempt has been made to review the recent developments and highlight the important issues concerning them. However, the individual influence of residual stress, grain size and metastable microstructures following laser surface engineering on corrosion resistance needs to be quantified and understood for the optimization of process parameters intended for achievement of maximum corrosion resistance in the treated surface. Furthermore, the role of process parameters on the grain orientation, its distribution and its effect on corrosion resistance is another area where basic understanding needs to be developed. The role of the excimer and diode lasers with shorter wavelengths in surface engineering might introduce superior properties on the surface and hence, sustained efforts are needed to establish the microstructural evolution and scope of improvement of corrosion resistance following laser surface engineering with those new lasers.

6.6 Acknowledgement

Partial financial supports from Department of Sci. & Technol., N. Delhi; German Academic Exchange Service, N. Delhi; Naval Research Board, N. Delhi; and Council of Scientific and Industrial Research, N. Delhi, for this work are gratefully acknowledged. The authors are grateful to Professor B. L. Mordike, Dr R. Galun, Dr Andreas Weisheit and Professor K. U. Kainer, their collaborators, for their cooperation and help during the research work related to laser surface engineering of magnesium and its alloys.

6.7 References

1. R. S. Busk, *Handbook of Materials Selection* (Ed.: Myer Kutz), ISBN 0-471-35924-6, John Wiley & Sons, Inc., New York: 2002, pp. 259–265.
2. A. M. Cameron, A. V. Hatten and V. G. Aurich, Proceeding of Magnesium Technology Conference (London, 1987), vol. **119**, pp. 7–17.
3. C. S. Roberts, *Magnesium and Its Alloys*, John Wiley & Sons, Inc., New York: 1960, pp. 3–5.
4. M. R. Bothwell, *The Corrosion of Light Metals*, John Wiley & Sons Inc., New York: 1967.

5. M. G. Fontana (ed.), *Corrosion Engineering*, McGraw-Hill, New York: 1987, p. 71.
6. J. E. Gray and B. Luan, *Journal of Alloys and Compounds*, **336** (2002) 88–113.
7. W. M. Steen (ed.), *Laser Materials Processing*, Springer-Verlag, London: 2003, p. 1.
8. G. L. Makar and J. Kruger, *Journal of the Electrochemical Society*, **137** (1990) 414–421.
9. M. E. Straumanis and B. K. Bathia, *Journal of the Electrochemical Society*, **110** (1963) 357–360.
10. W. J. James, M. E. Straumanis, B. K. Bathia and J. W. Jhonson, *Journal of the Electrochemical Society*, **110** (1963) 1117–1120.
11. G. R. Hoey and M. Cohen, *Journal of the Electrochemical Society*, **105** (1958) 245–250.
12. G. L. Song and A. Atrens, *Advanced Engineering Materials*, **1** (1999) 11–33.
13. M. Pourbaix, *Atlas of Electrochemical Equilibria in Aqueous Solutions*, National Association of Corrosion Engineers, 1974, p. 139.
14. H. P. Godard, W. P. Jepson, M. R. Bothwell and R. L. Lane (eds), *The Corrosion of Light Metals*, John Wiley & Sons, New York: 1967, p. 283.
15. J. D. Hanawalt, C. E. Nelson and J. A. Peloubet, *Trans. AIME*, **147** (1942) 273–299.
16. L. L. Shreir, *Corrosion*, George Newness, London: 1965.
17. E. F. Emley, *Principles of Magnesium Technology*, Pergamon Press, Oxford: 1966.
18. *Magnesium Materials, A Publication of Magnesium Elecktron* (1993), p. 1.
19. J. E. Harris, *Light Metals: Science and Technology, Proceeding of International Symposium* (Switzerland, 1985), p. 225.
20. D. Elizer, E. Aghion, F. H. (SAM) Froes, *Advanced Performance Materials*, **5** (1998) 201–212.
21. Magnesium Electron Limited, Magnesium Alloy Database, MATUS Databases, Engineering Information Co. Ltd., 1992.
22. S. Sugimoto, 53rd Annual Magnesium Congress (Ube City, Japan, 1996), p. 38.
23. T. Yoshinori, Surface treatment of magnesium or magnesium alloy. JP patent 60132460, 1986-12-22.
24. K. Hezbert and L. De, Surface treatment of magnesium alloys. US patent 2428749, 1947.
25. M. Osamu, Y. Ataru, S. Saburo and Y. Gohara, Surface treatment of magnesium or magnesium alloy. JP patent 54057110, 1980.
26. L. M. Leuzinger, Conversion of coating of magnesium alloys surfaces. CA patent 726661, 1966.
27. P. L. Hagans, Method for providing a corrosion resistant coating for magnesium containing materials. US patent 4569699, 1986.
28. F. P. Heller, Method of forming a chromate conversion coating on magnesium. CA patent 603580, 1960.
29. I. Sadao, N. Masahiko and S. Yokichi, Composition and process for treating magnesium-containing metals and product therefrom. US patent 63763596A, 1997.
30. ASTM B 480. In: *Standard Guide for Preparation of Magnesium and Magnesium Alloys for Electroplating*. 1988.
31. H. Suzuki, K. Sato, K. Okazaki, T. Ito and H. Kawaguchi, Surface treatment method for magnesium base material, and method for manufacturing magnesium shaped article. JP patent 2004204957, 2006.

32. T. Yamanishi, H. Tsubakino, A. Yamamoto and Y. Takatani, Method for surface treating Mg alloy. JP patent 11367394, 2001.
33. K. Inagawa, N. Tani and S. Fujimoto, Method and device for surface treatment of Mg alloy member. JP patent 2004048142, 2005.
34. S. Fujimoto, N. Tani and K. Inagawa, Surface treatment method and surface treatment apparatus for Mg alloy member. JP patent 2004067175, 2005.
35. X. M. Wang, X. Q. Zeng, G. S. Wu and S. S. Yao, Modification method of ion implantation on magnesium alloy surface. CN patent 200610027164.9, 2006.
36. X. M. Wang, X. Q. Zeng, G. S. Wu and S. S. Yao, Ion implantation method to improve oxidation resistance on magnesium alloy surface. CN patent 200610027991.8, 2006.
37. J. F. Ready, D. F. Farson and T. Feeley (ed.), *LIA Handbook of Laser Materials Processing*, Springer-Verlag, Berlin: 2001.
38. W. M. Steen and K. Watkins (ed.), *Laser Material Processing*, Springer-Verlag, New York: 2003, p. 1.
39. B. L. Mordike, *Materials Science and Technology*, (eds. R. W. Cahn, P. Haasen, E. J. Kramer), vol.15, VCH, Weinheim: 1993, p. 111.
40. J. Mazumdar, *Lasers for Materials Processing* (ed. M. Bass), North Holland Pub. Co., New York: 1983, p. 113.
41. N. N. Rykalin, A. Uglov and A. Kokora, *Laser Machining and Welding*, MIR, Moscow: 1978.
42. S. T. Picraux and D. M. Follstaedt, *Laser-Solid Interactions and Transient Thermal Processing of Materials* (eds. J. Narayan, W. L. Brown and R. A. Lemons), North-Holland, New York: 1983, p. 751.
43. P. A. Molian, *Surface Modification Technologies – An Engineers Guide*, (ed. T. S. Sudarshan), Marcel Dekker Inc., New York: 1989, p. 421.
44. C. W. Draper and J. M. Poate, *International Metals Reviews*, **30** (1985) 85–108.
45. C. W. Draper and C. A. Ewing, *Journal of Materials Science*, **19** (1984) 3815.
46 J. Dutta Majumdar and I. Manna, *International Materials Reviews*, **56** (2011) 341–388.
47. K. H. Lo, F. T. Cheng and H. C. Man, *Surface and Coatings Technology*, **173** (2003) 96–104.
48. K. H. Lo, F. T. Cheng, C. T. Kwok and H. C. Man, *Materials Letters*, **58** (2003) 88–93.
49. Q. Y. Pan, W. D. Huang, R. G. Song, Y. H. Zhou and G. H. Zhang, *Surface & Coating Technology*, **102** (1998) 245–255.
50. N. Parvathavarthini, R. V. Subbarao, S. Kumar, R. K. Dayal and H. S. Khatak, *Journal of Materials Engineering and Performance*, **10** (2001) 5–13.
51. C. T. Kwok, F. T. Cheng and H. C. Man, *Surface & Coating Technology*, **202** (2007) 336–348.
52. A. Conde, R. Colac, R. Vilar and J. de Damborene, *Materials & Design*, **21** (2000) 441.
53. J. Dutta Majumdar, A. K. Nath and I. Manna, *Surface & Coatings Technology*, **204** (2010) 1321–1325.
54. T. Z. Kattamis, in: K. Mukherjee and J. Mazumder (Eds.), *Lasers in Metallurgy*, The Metals Society of AIME, Warrendale, PA: 1981, p. 1.
55. R. K. Kalimullin, V. V. Valuev and A. T. Berdnikov, *Metal Science and Heat Treatment*, **28** (1986) 668.

56. C. F. Chang, S. K. Das and D. Raybould, in: P. W. Lee, R. S. Carbonara (Eds.), *Rapidly Solidified Materials*, ASM, Metals Park, Ohio: 1986, p. 129.
57. C. F. Chang, S. K. Das, D. Raybould and A. Brown, *Metal Powder Report*, **41** (1986) 302.
58. A. K. Mondal, S. Kumar, C. Blawert and N. B. Dahotre, *Surface & Coatings Technology*, **202** (2008) 3187–3198.
59. G. Abbas, Z. Liu and P. Skeldon, *Applied Surface Science*, **247** (2005) 347–353.
60. R. K. Singh Raman, S. Murray and M. Brandt, *Surface Engineering*, **23** (2007) 107–111
61. P. Chakraborty Banerjee, R.K. Singh Raman, Y. Durandet and G. McAdam, *Materials Science Forum*, **618–619** (2009) 263–268.
62. J. Dutta Majumdar, R. Galun, B. L. Mordike and I. Manna, *Materials Science and Engineering: A*, **A361** (2003) 119–129.
63. M. Hino, Y. Mitooka, K. Murakami, K. Nishimoto and T. Kanadani, *Materials Science Forum*, **654–656** (2010) 1960–1963.
64. L. F. Guo, T. M. Yue and H. C. Man, *Journal of Materials Science*, **40** (2005) 3531–3533.
65. T. M. Yue, A. H. Wang and H. C. Man, *Scripta Materialia*, **38** (1998) 191–198.
66. H. Zhang, C. Y. Yu, *Materials Science and Engineering: A*, **257** (1998) 322–327.
67. C. S. Montross, T. Wei, L. Ye, G. Clark and Y.-W. Mai, *International Journal of Fatigue*, **24** (2002) 1021–1036.
68. Y. X. Hu and Z. Q. Yao, *Surface & Coating Technology*, **202** (2008) 1517–1525.
69. Y. Sano, M. Obata, T. Kubo, N. Mukai, M. Yoda, K. Masaki and Y. Ochiet, *Materials Science and Engineering: A*, **417** (2006) 334–340.
70. Y. Zhang, J. You, J. Lu, C. Cui, Y. Jiang and X. Ren, *Surface & Coatings Technology*, **204** (2010) 3947–3953.
71. P. Peyre and R. Fabbro, *Optical and Quantum Electronics*, **27** (1995) 1213–1229.
72. T. R. Jervis, M. Nastasi, A. J. Griffin, Jr, T. G. Zocco, T. N. Taylor and S. R. Folton, *Surface & Coating Technology*, **89** (1997) 158–164.
73. A. S. Khanna, R. K. Singh Raman, E. W. Kreutz and A. L. E. Terrance, *Corrosion Science*, **33** (1992) 949–958.
74. J. Dutta Majumdar and I. Manna, *Materials Science and Engineering: A*, **A267** (1999) 50–59.
75. S. R. Pillai, P. Shankar, R. V. Subba-Rao, N. B. Sivai and S. Kumaravel, *Materials Science and Technology*, **17** (2001) 1249–1252.
76. C. H. Tang, F. T. Cheng and H. C. Man, *Surface & Coating Technology*, **200** (2006) 2606–2609.
77. A. Almeida, P. Petrov, I. Nogueira and R. Vilar, *Materials Science and Engineering: A*, **A303** (2001) 273–280.
78. M. G. Ferriera, R. Li and R. Vilar, *Corrosion Science*, **38** (1996) 2091–2094.
79. K. C. Chang, W. J. Wei and C. Chen, *Surface & Coatings Technology*, **102** (1998) 197–204.
80. A. Gutierrez and J. Damborenea, *Oxidation of Metals*, **47** (1997) 259–275.
81. K. Nagarathnam and K. Komvopoulos, *Metallurgical and Materials Transactions A-Physical Metallurgy and Materials Science*, **27A** (1996) 381–390.
82. R. Galun, A. Weisheit and B. L. Mordike, *Corrosion Reviews*, **16** (1998) 53–73.
83. R. Galun, A. Weisheit and B. L. Mordike in: *Magnesium Alloys and Their Applications*, B. L. Mordike and K. U. Kainer (Eds.), Werkstoff-Informationsge sellschaft mbh, Germany: 1998, p. 439.

84. J. Dutta Majumdar, B. L. Mordike, R. Galun, T. Maiwald and I. Manna, *Lasers in Engineering*, **12** (2002) 147–169.
85. J. Dutta Majumdar and I. Manna, *Scripta Materiallia*, **62** (2010) 579–581.
86. R. Subramanian, S. Sircar and J. Mazumda, *Journal of Materials Science*, **26** (1991) 951–956.
87. A. A. Wang, S. Sircar and J. Mazumda, *Journal of Materials Science*, **28** (1993) 5113–5122.
88. C. J. Chen, M. C. Wang, D. S. Wang, R. Jin and Y. M. Liu, *Corrosion Engineering, Science and Technology*, **42** (2007) 130–136.
89. P. Volovitch, J. E. Masse, A. Fabre, L. Barrallier and W. Saikaly, *Surface & Coatings Technology*, **202** (2008) 4901–4914.
90. T. M. Yue, T. Li, *Surface & Coatings Technology*, **202** (2008) 3043–3049.
91. T. E. Wilks, in *Proceedings of Metal Matrix Composites*, Detroit, 28 Feb.–3 Mar. 1994, p. 49.
92. J. F. King, T. E. Wilks and G. D. Wardlow, in *Proceedings of Magnesium – A Material Advancing to the 21st Century*, Japan, June 1996, p. 77.
93. T. M. Yue, Z. Mei and H. C. Man, *Journal of Materials Science*, **20** (2001) 1479–1482.
94. Z. Mei, L. F. Guo and T. M. Yue, *Journal of Materials Processing Technology*, **161** (2005) 462–466.
95. T. M. Yue, A. H. Wang and H. C. Man, *Scripta Materialia*, **40** (1999) 303–311.
96. K.-J. Huang, H.-G. Liu and C.-R. Zhou, *Advanced Materials Research*, **143–144** (2011) 758–762.
97. K.-J. Huang, L. Yan, C. Wang, C. S. Xie and C. R. Zhou, *Transactions of Nonferrous Metals Society of China*, **20** (2010) 1352–1355.
98. K.-J. Huang, H.-G. Liu and C.-R. Zhou, *Advanced Materials Research*, **179–189** (2011) 757–761.

7

Micro-arc oxidation (MAO) to improve the corrosion resistance of magnesium (Mg) alloys

B. L. JIANG and Y. F. GE,
Xi'an University of Technology, China

DOI: 10.1533/9780857098962.2.163

Abstract: Micro-arc oxidation (MAO) is a novel and attractive surface engineering process for magnesium (Mg) alloys. In an electrolytic bath with high electric energy, the surface of a magnesium alloy can be converted into a dense and hard ceramic oxide coating. The MAO process can be used in diverse applications as a corrosion control technique. Due to its simplified pre-treatment, superior corrosion resistance performance and environmentally safe coating process, MAO technology has emerged as an important alternative to anodizing techniques in certain areas. This chapter presents a review of MAO techniques for Mg alloys from the scientific, technological and application points of view.

Key words: micro-arc oxidation (MAO), magnesium alloys, coating, microstructure, corrosion resistance.

7.1 Introduction

The use of magnesium alloys in aircraft, aerospace and automobile applications has increased steadily in recent years because of the low density and high strength-to-weight ratio of magnesium. However, magnesium has a very negative standard electrode potential resulting in poor corrosion resistance, thereby limiting its application. This issue can be addressed, to a limited extent, by alloying, but one of the most effective ways to prevent corrosion is to apply chemical conversion coatings to the Mg substrate. Various chemical conversion surface treatments such as chrome pickling or anodizing have been used on Mg alloys to give limited corrosion protection in mild corrosive environments. Also the traditional surface processes for Mg alloys, which contain chrome, are being replaced with more environmentally safe surface treatments because of environmental concerns (Gray and Luan, 2002). A new anodizing technology, micro-arc oxidation (MAO), has emerged in the field of anodizing for light alloys in recent years. The

MAO process is an anodic plasma chemical surface treatment that forms a ceramic oxide coating on Al, Mg or Ti alloys. For Mg alloys, the MAO treatment can enhance corrosion and wear resistance properties, or confer various other functional properties including anti-friction, thermal protection, optical and dielectric, as well as a pre-treatment to provide load support for top coating (Curran and Clyne, 2005, 2007).

The micro-arc sparking phenomenon in anodizing was initially observed by scientists in the former Soviet Union, particularly in the current Ukraine and Russia, about 40 years ago. Further investigation led to MAO technology being developed for military purposes. The United States started to explore this technology in the 1970s. Scientific information on this technology has spread around the world since the late 1980s (Yerokhin *et al.*, 1999). Based on the fact that is difficult to catch and analyze the instant discharge event, understanding of the discharge nature is still limited due to the scarcity of experimental proof, which also leads to the use of various terms in modern scientific literature such as 'plasma electrolytic oxidation' (PEO) (Yerokhin *et al.*, 1998a, 1998b; Arrabal *et al.*, 2009), 'microplasma oxidation' (MPO) (Rudnev *et al.*, 1998; Timoshenko and Magurova, 2000; Xin *et al.*, 2001), 'anodic spark deposition' (ASD) (Brown *et al.*, 1971; Wirtz *et al.*, 1991) and 'anodic oxidation by spark deposition' (ANOF) in Germany (Krysmann *et al.*, 1971, 1984; Districh *et al.*, 1984; Kurze *et al.*, 1986). In the meantime, scientific research in many countries including Russia, China, Japan, UK, Germany and Australia has greatly contributed to the understanding of MAO mechanisms and developing novel functionally structured MAO coatings. Companies such as Keronite (UK), Magoxid-coat (Germany) and Micro-plasmic (USA) have also developed commercial applications of MAO coatings for the improvement of corrosion resistance properties of Mg alloys. As a fascinating technology, MAO is attracting increasing attention from both academic institutions and many industries.

This chapter aims to give an overview of the MAO technique from the scientific, technological and application points of view. Discussion on the latest developments will enable engineering designers to realize the potential of the MAO technique on surface treatment for Mg alloy components and stimulate research into developing new coating materials for advanced applications of Mg alloys.

7.2 Micro-arc oxidation fundamentals

Micro-arc oxidation (MAO) is a novel surface engineering process for Mg alloys. In an electrolytic bath with high electric energy, the surface of a Mg alloy can be converted into a dense and hard ceramic oxide coating. A typical schematic of the MAO coating system is shown in Fig. 7.1. The stainless

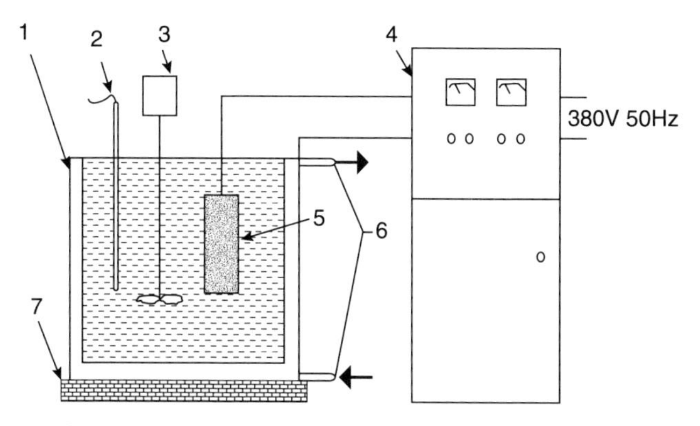

7.1 Schematic of the MAO coating system: 1 stainless steel tank (cathode); 2 thermocouple; 3 mixer; 4 AC power supply unit; 5 specimen (anode); 6 cooling system; 7 insulating plate.

steel tank acts as the cathode, the specimen as the anode, with the thermocouple, mixer and cooling system also contained in the tank. The plasma is discharged by an external power supply in a slightly alkaline electrolyte near the surface of the specimen. The oxygen plasma generated causes partial surface melting and ultimately the formation of a ceramic oxide coating. The phenomenology, principles, process and advantages of the MAO technique and microstructure of the MAO coating on Mg alloys will be introduced in more detail in this section.

7.2.1 Key phenomena in the MAO process

It is well known that the electrolysis of aqueous solutions is accompanied by a number of electrode processes (Fig. 7.2). In particular, the liberation of gaseous oxygen and/or metal oxidation occurs on the anodic surface. Depending on the electrolyte chemical activity in relation to the metal, the oxidation process can lead either to surface dissolution or to oxide film formation. Liberation of gaseous hydrogen and/or cation reduction can also occur on the cathodic surface. When a 'conventional' electrolytic process is studied (e.g. electroplating, electrochemical machining, anodizing, etc.), the electrode processes are usually considered in the framework of a simplified model, where the electrode–electrolyte interface can be represented by a two phase system (i.e. metal–electrolyte or oxide–electrolyte couple) with a single phase boundary consisting of a double-charged layer. Concurrent by-product processes (such as gas liberation) are either neglected or taken into account using special correction factors, for example 'current yield' or 'electrode shielding' coefficients (Yerokhin *et al.*, 1999). However, as will be shown below, such a simplification is not always justifiable since, under

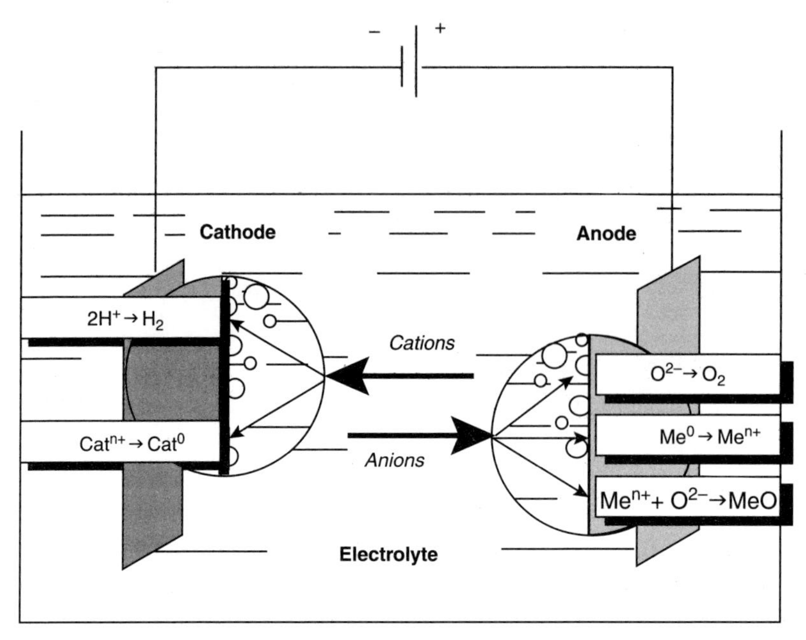

7.2 Electrode processes in electrolysis of aqueous solutions (Yerokhin *et al.*, 1999).

certain conditions, the results obtained from the treatment are influenced considerably by the processes that occur in the gaseous environment surrounding the electrode and/or in its surface layers.

In a MAO treatment, the micro discharges occur on Mg alloy when the applied voltage exceeds the dielectric breakdown potential, then the oxide ceramic coating will be formed on substrate. The evolution of the micro-discharge aspect as a function of the treatment time is depicted in Fig. 7.3 corresponding to surface view pictures of the Mg alloy sample during MAO process. For these views, the integration time of the camera was set between 8 and 10 ms. The MAO process involves the creation of a plasma discharge around a Mg alloy specimen submersed in an electrolyte under high voltage (Gerasimov *et al.*, 1994; Guo *et al.*, 2005). From the micro-arcing phenomenon during the MAO process (Fig. 7.3), it can be seen that the sparks become bright and concentrated on the surface of the Mg alloy specimen with increased processing time.

These processes affect the characteristic current–voltage profile of the electrochemical system shown in Fig. 7.4 (Yerokhin *et al.*, 1999). A 'type-a' current–voltage plot represents a metal–electrolyte system with underlying gas liberation on either the anode or cathode surface; 'type-b' represents a system where oxide film formation occurs (Tchernenko *et al.*, 1991; Duradzhy and Parsadanyan, 1998). At relatively low voltages the kinetics of

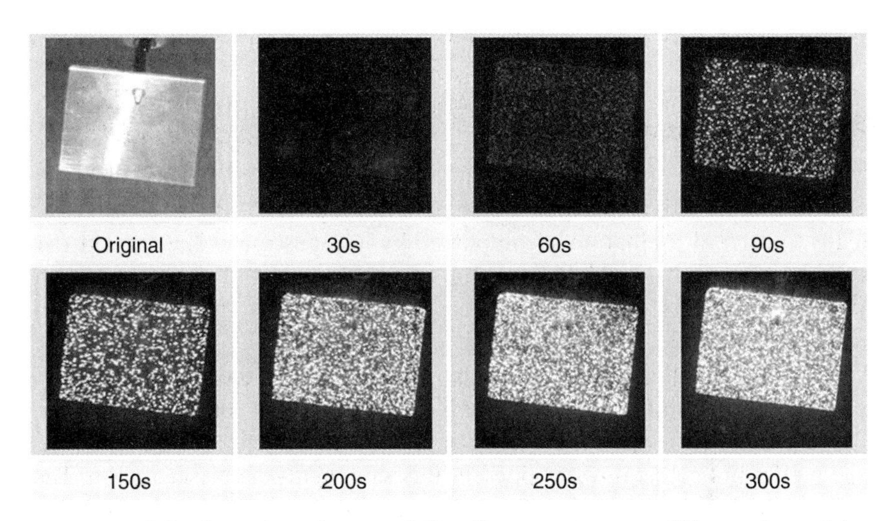

7.3 Surface view pictures of Mg alloys specimen at different time of the MAO process.

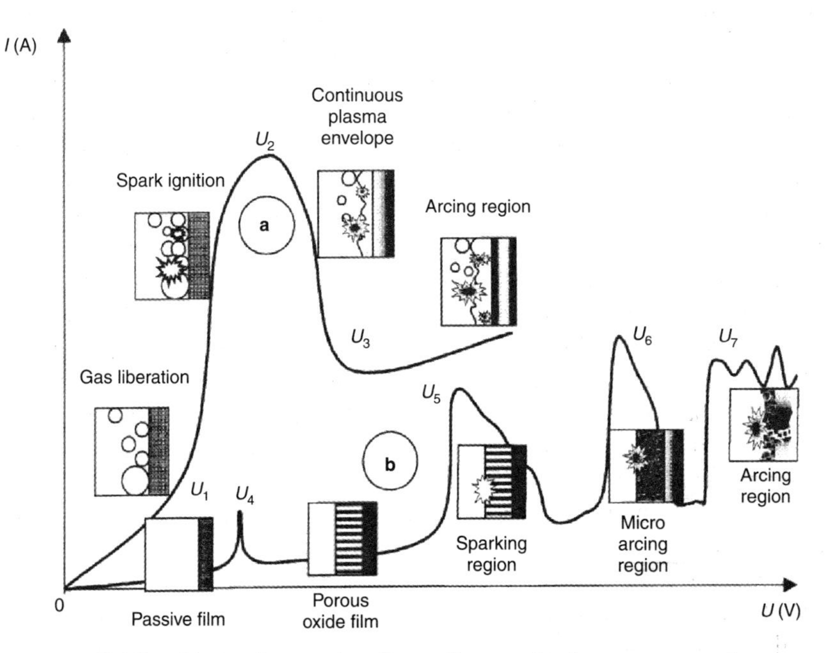

7.4 Two kinds of current–voltage diagram for the processes of plasma electrolysis: discharge phenomena are developed (a) in the near-electrode area and (b) in the dielectric film on the electrode surface (Yerokhin *et al.*, 1999).

the electrode processes for both systems conform to Faraday's law and the current–voltage characteristics of the cell vary according to Ohm's law. Thus, an increase in voltage leads to a proportional rise in the current (region '0–U_1' in the type-a system and '0–U_4' in the type-b system). However, beyond a certain critical voltage, the behavior of a particular system may change significantly. For a type-a system in the region U_1–U_2, a potential rise leads to current oscillation accompanied by luminescence. The current rise is limited by a partial shielding action of gaseous reaction products (O_2 or H_2) over the electrode surface. In areas where the electrode remains in contact with the liquid, however, the current density continues to rise, causing local boiling (ebullition) of the electrolyte adjacent to the electrode. Upon progression to point U_2 the electrode is enshrouded by a continuous gaseous vapor plasma envelope of low electrical conductivity. Almost all of the voltage across the cell is now dropped in this thin, near- electrode region. The electric field strength within this region therefore reaches a value between 10^6 and 10^8 V/m, which is sufficient for initiation of ionization processes in the vapor envelope. The ionization phenomena appear initially as a rapid sparking in scattered gaseous bubbles and then transform into a uniform glow distributed throughout the vapor plasma envelope. Due to the hydrodynamic stabilization of the vapor envelope in the region U_2–U_3, the current drops and, beyond point U_3, the glow discharge transforms into intensive arcing accompanied by a characteristic low frequency acoustic emission.

The behavior of type-b systems is more complicated. Firstly, the passive film previously formed begins to dissolve at point U_4, which, in practice, corresponds to the corrosion potential of the material. Then, a porous oxide film grows in the region of re-passivation (U_4–U_5), across which most of the voltage drop now occurs. At point U_5, the electric field strength in the oxide film reaches a critical value beyond which the film is broken through due to impact or tunneling ionization (Ikonopisov, 1977; Monterro *et al.*, 1987). In this case, small luminescent sparks are observed to move rapidly across the surface of the oxide film, facilitating its continued growth. At point U_6, the mechanism of impact ionization is supported by the onset of thermal ionization processes and slower, larger arc-discharges arise. In the region U_6–U_7, thermal ionization is partially blocked by negative charge build-up in the bulk of the thickening oxide film, resulting in discharge decay shorting of the substrate. This effect determines the relatively low power and duration of the resultant arc-discharges, that is micro discharges, which are (somewhat misleadingly) termed 'micro-arcs'. Owing to this 'micro-arcing', the film is gradually fused and alloyed with elements contained in the electrolyte. Above the point U_7, the micro-arc discharges occurring throughout the film penetrate through to the substrate and (since negative charge blocking effects can no longer occur) transform into powerful arcs which may cause destructive effects such as thermal cracking of the film (Tchernenko *et al.*, 1991).

In practice, a number of the electrode processes described above may occur concurrently over adjacent areas of the electrode surface. The simple two phase electrode–electrolyte model normally encountered in conventional electrolysis must therefore be replaced by a more complex four phase system (metal–dielectric–gas–electrolyte) with a number of possible phase boundaries, particularly when considering electrochemical systems running above the critical voltages of U_1 and U_5. Two phases of low conductivity are formed (i.e. dielectric and gas), where the main voltage drop is concentrated. Since the resistance of these phases varies continuously, it is very difficult to discern in what phase the ionization phenomena are initiated (Krysmann *et al.*, 1984). Thus, the division of electrochemical systems into the two types is not distinct.

7.2.2 Principles and mechanisms of the MAO process

Regardless of what kind of electrolyte systems, electrical system or electrical parameter control modes (constant current or constant voltage) are applied, the basic formation mechanism of an MAO coating on a Mg alloy is similar. Taking the constant current mode as an example, Fig. 7.5 gives a schematic of a typical voltage vs. time plot obtained from MAO processing

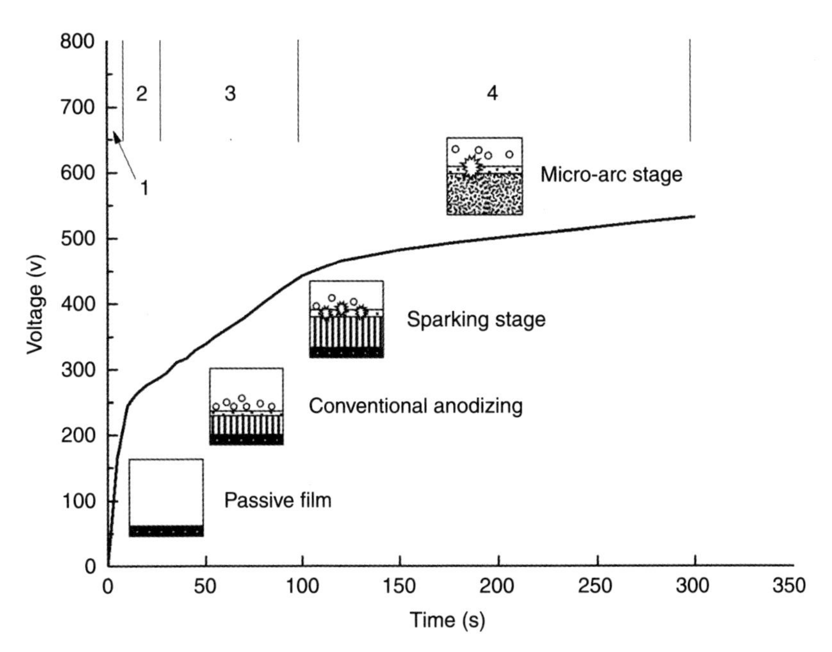

7.5 Schematic voltage *vs.* time plot representation of different stages and discharge phenomena of MAO process on AZ31B Mg alloy.

of an AZ31B Mg alloy in a silicate based electrolyte indicating the different stages of the MAO process and the discharge phenomenon. It is well known that a very thin, natural, passive film exists on substrate metal surfaces which could provide a very limited protective effect. As the applied voltage increases, a large number of gas bubbles are produced; this is the traditional anodizing stage resulting in the formation of a porous insulation film with a columnar structure perpendicular to the substrate (Mizutania *et al.*, 2003). When the voltage exceeds a certain threshold (i.e. breakdown voltage), dielectric breakdown occurs in some scattered weak regions across the insulating film accompanied by the phenomenon of spark discharge. In this case, a large number of fine, uniform, white sparks are generated on the sample surfaces which result in the formation of a large number of small uniform microspores (Chigrinova *et al.*, 2001).

The formation and evolution of the coating on a Mg alloy substrate during the MAO process in alkaline electrolyte can be described in four stages corresponding to the voltage vs. time plot (Fig. 7.5). The first stage is the dissolution of the Mg alloy substrate, resulting in the formation of a passive film on the surface comprising $Mg(OH)_2$ and MgO. Depending on the electrolyte used, other compounds can be incorporated into the passive film as well.

$$Mg \rightarrow Mg^{2+} + 2e^- \quad [7.1]$$

$$Mg^{2+} + 2OH^- \rightarrow Mg(OH)_2 \quad [7.2]$$

$$Mg(OH)_2 \rightarrow MgO + H_2O \quad [7.3]$$

Mizutani *et al.* (2003) have observed $Mg(OH)_2$ and MgO on anodized coatings of Mg alloy formed at 60 V under non-sparking conditions.

The second stage is marked by the beginning of sparks on the surface of the Mg alloy due to the breakdown of the passive film, thus termed the breakdown voltage. The breakdown voltage is characteristic for a given electrolyte system and substrate, depending on their composition and conductivity of electrolyte. Upon breakdown of the passive film on the surface of the Mg alloy, a vigorous gas evolution can be observed. The following reactions occur at the anode surface:

$$2H_2O \rightarrow 2H_2 + O_2 \quad [7.4]$$

$$2Mg + O_2 \rightarrow 2MgO \quad [7.5]$$

The sparking characteristics and intensity vary from electrolyte to electrolyte and they are also dependent on the processing parameters such as applied current density, frequency, etc. At the beginning of the second stage, the sparks are very fine and normally possess a very short lifetime.

With increase in voltage and time, and thus with the growth of the oxide ceramic coating on the surface of the Mg alloy, the sparks grow in size and number, which is described as the third stage. The increase in voltage during the first three stages in the MAO process is generally quite rapid. During the final stage, where the increase in voltage with time is very marginal, the sparks grow much bigger and have a much longer lifetime compared to the earlier stages.

Figure 7.6 shows a schematic of the key mechanisms of the MAO process (Gupta *et al.*, 2007). As discussed in the previous section, during the stable plasma regime the work piece is surrounded by a continuous gas envelope. High voltage between the electrodes leads to a concentration of positive ions in the electrolyte, in the close proximity of the anode, mostly on the surface of the gas bubbles. Thus, a very high positive charge is present in close proximity to the work piece. This results in high localized electric field strength between the cathode and the positive charges. It was reported earlier that in electrolytic plasma processing, plasma layer electric field strength can reach 10^5 V/m or higher (Yerokhin *et al.*, 1999; Tynrin and Pogrehnjak, 2001). When such a high electric field is reached, gas space inside the bubbles is ionized and a plasma discharge is initiated. Figure 7.6a shows the magnified view of the surface of the work piece. Single bubble that is surrounded by plasma shown to illustrate the concept. In reality, the bubble is surrounded by numerous plasma bubbles. The temperature of plasma can reach as high as 2000°C locally. Based on anodic plasma in the MAO process, Klapkiv (1998) suggested that the discharge plasma temperature corresponds to $6–7 \times 10^3$ K. This high temperature plasma bubble is surrounded by relatively cool electrolyte (at the boiling point of water), thereby resulting in cooling of the plasma. Finally, the bubble implodes on the metal surface (Fig. 7.6b–d). Belevantsev *et al.* (1998) have described the presence of bubbles containing negative oxygen ions in anodic electrolytic plasma processing. Furthermore, they have described the extinction of discharge as a consequence of expansion and cooling of the bubbles. The plasma discharge duration is expected to be 10^{-6} s for each individual event. The entire surface of the anode may not be covered by a continuous plasma layer but by a limited quantity of discrete plasma discharges, which take place at any one instant in time. Yerokhin *et al.* (2003) studied the characteristics of discharge in anodic electrolytic plasma processing.

Two phenomena are expected on the work piece surface due to the implosion of a plasma bubble. Firstly, the positive ions, which are concentrated around the bubble, are accelerated directly to the anode surface, the characteristics of which can be described as similar to an avalanche. Secondly, as the bubble implodes, the stored energy is released into the gas

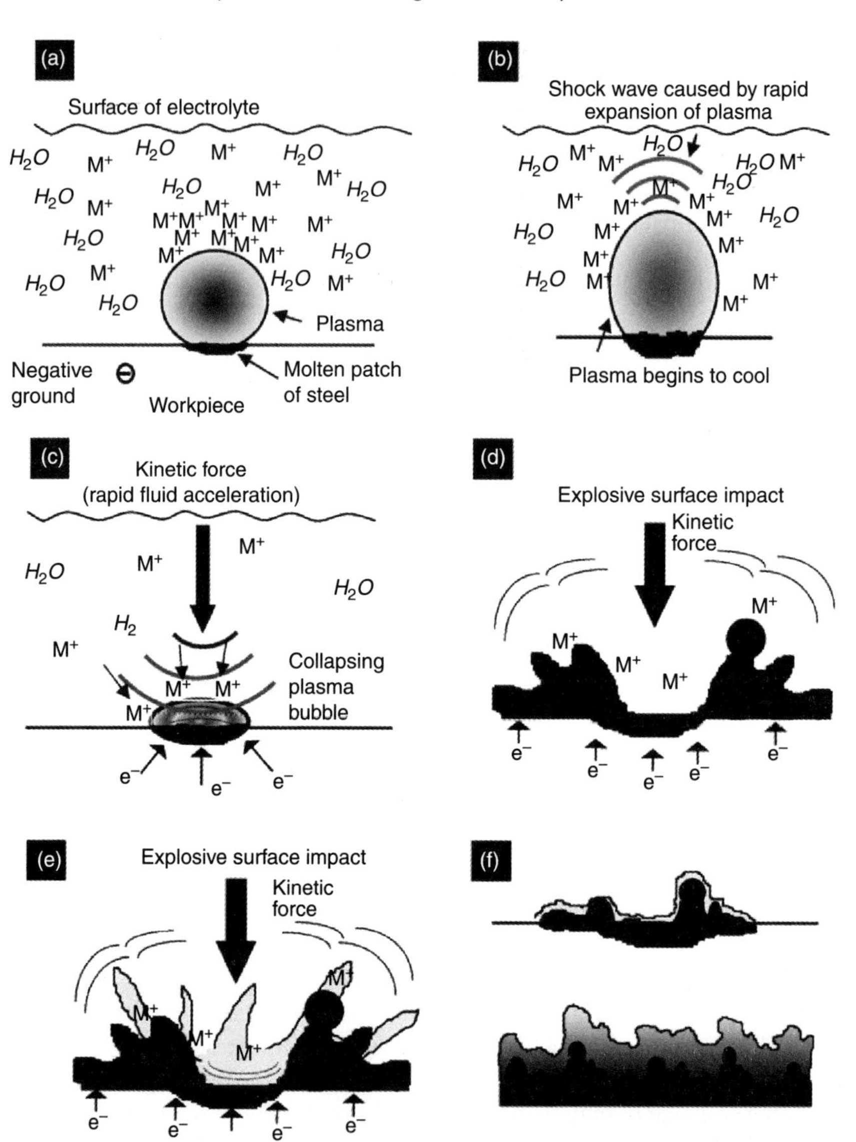

7.6 Schematic of MAO process mechanism: (a) plasma bubble on the surface of the work piece, (b) shockwave production by the cooling plasma bubble, (c) collapsing plasma bubble and cleaning, (d) collapsing plasma bubble and creation of micro-crater, (e) collapsing bubble deposits ions in case of coating, and (f) increasing coating thickness with processing time (Gupta *et al.*, 2007).

layer and kinetic energy is transferred to and from the liquid layer to the surface of the work piece. This energy can be very high, similar to cavitations (having pressure of the order of several hundred MPa or greater), and the ions to be deposited, which are initially accelerated by the void created by the imploding bubble, are further assisted and moved to the anode surface by this kinetic energy. This leads to deposition of metal ions present in the electrolyte onto the surface of the work piece (Fig. 7.6e and 7.6f). The movement of ions, in the MAO process, takes place mainly by ion acceleration through the plasma and ion bubble absorption transport as the bubble collapses. Furthermore, both of these modes of transport eliminate the phase boundary diffusion layer that is generally present in the conventional electroplating processes. MAO technology is a dynamic system, where the electrolyte moving through the reactor at a high rate leads to rapid transport of ions to the plasma layer. Tynrin *et al.* (2001) have also reported hydrodynamic transport of ions from the bulk to the work piece in similar electrolytic plasma systems. A combination of hydrodynamic flow and efficient ion transport mechanisms during the MAO coating process leads to high deposition rates.

7.2.3 Advantages of MAO technology

The MAO technique for Mg alloys has many advantages (Patel and Saka, 2001): (i) a wide range of coating properties, including good wear resistance, corrosion resistance and other functional properties are conferred (such as thermo optical, dielectric, thermal barrier); (ii) no deterioration of the mechanical properties of the substrate material is caused because of negligible heat input; (iii) high metallurgical bonding strength is measured between the coating and substrate; (iv) there is the possibility of processing parts with complex geometric shape or large size; (v) equipment is simple and easy to operate; (vi) cost is low, as it has no need of vacuum or gas shielding conditions; (vii) the technique is ecologically friendly, as alkaline electrolytes are employed and no noxious exhaust emission is involved in the process, meeting the requirement of environmentally friendly surface modification technology.

Compared with conventional anodizing and hard anodizing for Mg alloys, MAO coating exhibits many excellent characteristics (Wu *et al.*, 2007). Although conventional anodizing in strongly alkaline solutions can produce coatings up to several microns thick, these are still too thin to provide effective protection against wear and corrosion, and are therefore used mainly for decoration. The MAO process, derived from conventional anodizing but enhanced by spark discharge events when the applied voltage exceeds the critical value of the insulator film, can generate a thicker oxide ceramic coating with excellent properties, for example high hardness, good wear and corrosion properties, as well as excellent bonding strength with the substrate, compared with conventional anodizing methods. The MAO process has demonstrated great success in offering improved surface oxidation treatment of Mg, Al and Ti alloys, replacing the conventional acid based anodizing processes and/or conversion treatments, which contain hexavalent chrome and other environmentally hazardous substances. These positive features of MAO technology should encourage a widespread exploration of its potential applications.

7.2.4 Microstructure and composition of MAO coating

The microstructural features of MAO coatings depend on the processing conditions, and processing time is one of the critical parameters for growth of MAO coatings, which to a large extent decides the microstructural features of these layers. Figure 7.7 shows an overview of the surface appearance of a typical MAO treated AZ31B Mg alloy specimen under different processing times using scanning electron microscopy (SEM). The SEM images display typical morphology of the MAO coating surface with agglomeration of uniformly distributed oxide particles, inter-oxide particle gaps, micro-cracks due to shrinkage effect and micro-pores formed through discharge channels. The SEM images of a MAO coating with 5–8 μm thickness obtained from silicate electrolyte at short processing time (5 min) in Fig. 7.7a show fine and uniformly distributed pores on the surface of the AZ31B Mg alloy. On the other hand, the surface images of a 12–15 μm thick MAO coating in Fig. 7.7b show large, coarse oxide deposits with large pores on the surface. Both images in Fig. 7.7 display very similar surface morphology, and the size of micro-pores increased with processing time.

Characteristically, all cross-sectional micrographs of MAO coating on Mg alloys show that there is no uniform coating thickness, and thickness can range from 5 to 100 μm. Figure 7.8 shows a schematic drawing of a typical cross-section of MAO coating produced using silicate based electrolyte on Mg alloy (Blawert *et al.*, 2005). Generally, there is porosity throughout the whole MAO coating, but from the fractured cross sections four different regions can be identified according to the amount of porosity. Region one is the outermost surface with the crater structure on top. The ceramic coating is enforced by a large number of discharge channels reaching from the crater surface towards region two. This second region is visible in all specimens and can be identified by a remarkable band of cavities in a depth of 20–40 μm from the surface. Most of the visible surface pores (craters with discharge channels) seem to end in this band of cavities. Underneath this band, the third region has an extension from 20 up to 80 μm. Here, the ceramic oxide coating is denser, and only a small number of larger pores or cavities are visible. The fourth region is the interface which is very rough. It appears as if there is a thin layer of less than 1 μm thickness with a finer and denser structure in the interface region.

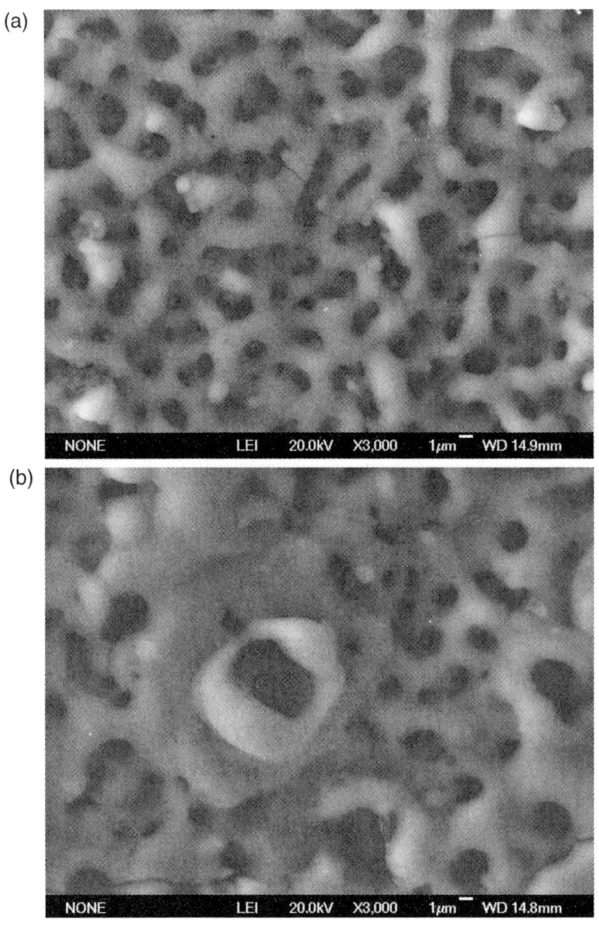

7.7 SEM shows the surface micrographs of MAO coating on AZ31B Mg alloy under different processing time: (a) 5 min; (b) 10 min.

The phase composition of the MAO coating on an Mg alloy is mainly influenced by the electrolyte composition; the energy intensity during the discharge is also considered to play a role. A large number of research publications on MAO processing for Mg alloys using silicate based electrolyte have shown the formation of Mg_2SiO_3 as the major phase (Liang *et al.*, 2009; Srinivasan *et al.*, 2009). In addition, the MAO coatings from silicate based electrolyte were reported to also contain MgO, $MgAl_2O_3$ and MgF_2, depending on the alloy, processing conditions and additives employed (Wang *et al.*, 2005, 2006). Liang *et al.* (2009) reported the presence of only MgO in the MAO coating obtained from a sodium silicate based electrolyte. Figure 7.9 displays the X-ray diffraction (XRD) pattern acquired from 10 μm thickness

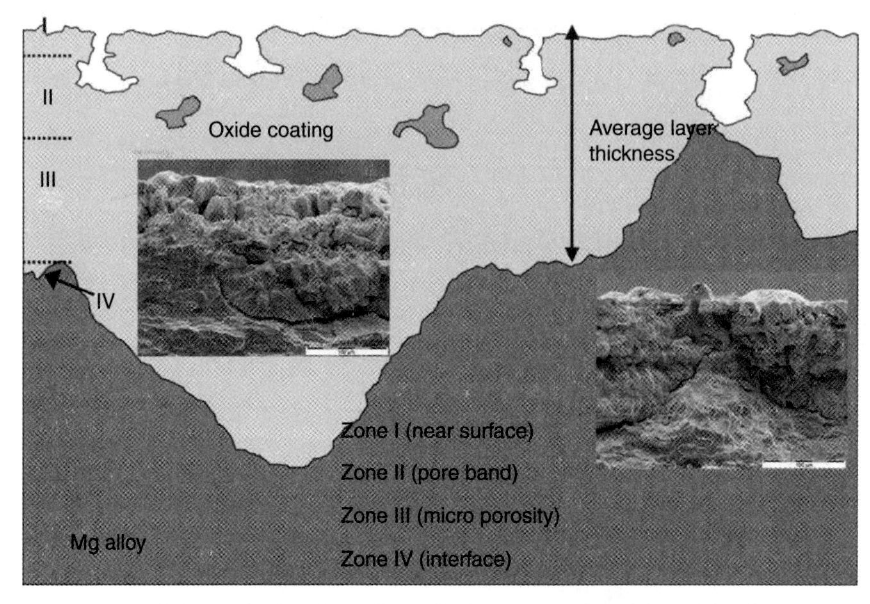

7.8 Schematic drawing of a typical cross-sectional found for the ceramic oxide coating on Mg alloy; inserted are SEM micrographs showing typical features of the layer (Blawert *et al.*, 2005).

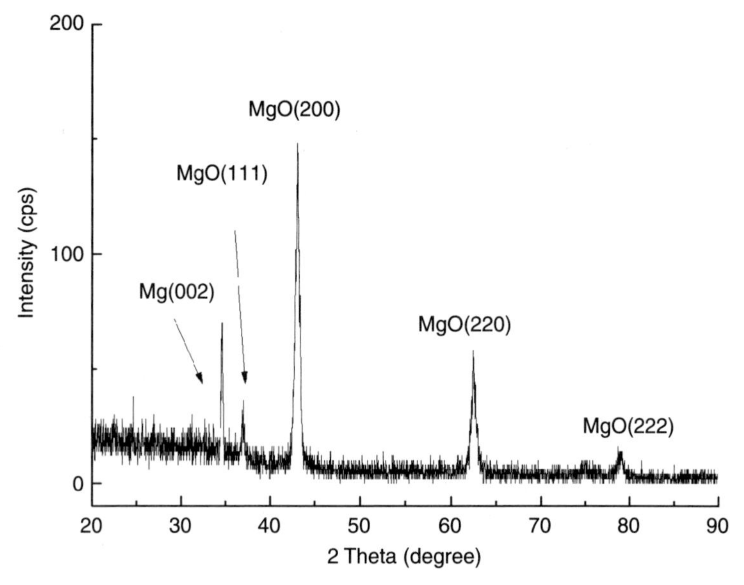

7.9 XRD patterns of MAO coating on AZ31B Mg alloy in a sodium silicate based electrolyte.

MAO coating on AZ31B Mg alloy in a sodium silicate based electrolyte. The results indicate that the MAO coatings are mainly composed of MgO, and some of the substrate peaks are Mg. The intensity of the Mg peaks depends on the layer thickness: with increasing coating thickness less information is obtained from the substrate.

To further study the electronic states, the chemical binding state of the elements silicon (Si) and oxygen (O) on the surface of MAO coating for AZ31B Mg alloy were determined by X-ray photoelectron spectroscopic (XPS) analysis (Fig. 7.10). It can be seen from Fig. 7.10a that the O_{1s} spectra of MAO coating contain two distinct peaks, one is in agreement with XPS data for MgO (O_{1s} = 530.9 eV), and the other may correspond to $MgSiO_3$ or H_2O (O_{1s} = 532.8 eV), suggesting that O in ceramic coating exists in the form of MgO, $MgSiO_3$ and H_2O. As shown in Fig. 7.10b, the peak at 102.5 eV of the Si_{2p} spectrum is in excellent agreement with XPS data for $MgSiO_3$, indicating that part of the silicon exists in the form of silicate (Ge *et al.*, 2010). The peak at 101.4 eV accords with SiO_x, resulting from the reaction between $MgSiO_3$ and carbon dioxide in atmospheric. So the results of XRD and XPS both show that there is mainly MgO in the MAO coating on AZ31B Mg alloy.

7.3 Techniques for the MAO process

Both intrinsic factors (electrolyte composition and pH value) and extrinsic factors (power source types, electrical parameters and electrolyte temperature) affect the formation and microstructure of an MAO coating on Mg alloys. The composition and concentration of electrolyte and electrical parameters used during the MAO process play a crucial role in obtaining the desired ceramic coating with special phase components and microstructure.

7.3.1 Power sources

A typical treatment unit for the MAO process consists of an electrolyser and a high power electrical source (Fig. 7.1). The electrolyser is usually a water cooled bath made of stainless steel, which also serves as the cathode. It is placed on a dielectric base and confined in a grounded steel frame. The electrolyser incorporates electrolyte mixing and recycling, as well as electrical interlocks. Various types of power source can be used to bring about plasma electrolysis. According to the applied electrical regime they can be classified into various types of power sources, such as DC sources (Kuhn, 2003; Verdier *et al.*, 2005; Shi *et al.*, 2006), DC-pulsed sources (Nie *et al.*, 1996; Zhao *et al.*, 2005; Han *et al.*, 2007), AC sources (Magurova and Timoshenko, 1995; Yerokhin *et al.*, 1999; Timoshenko and Magurova, 2000; Arrabal *et al.*, 2008;

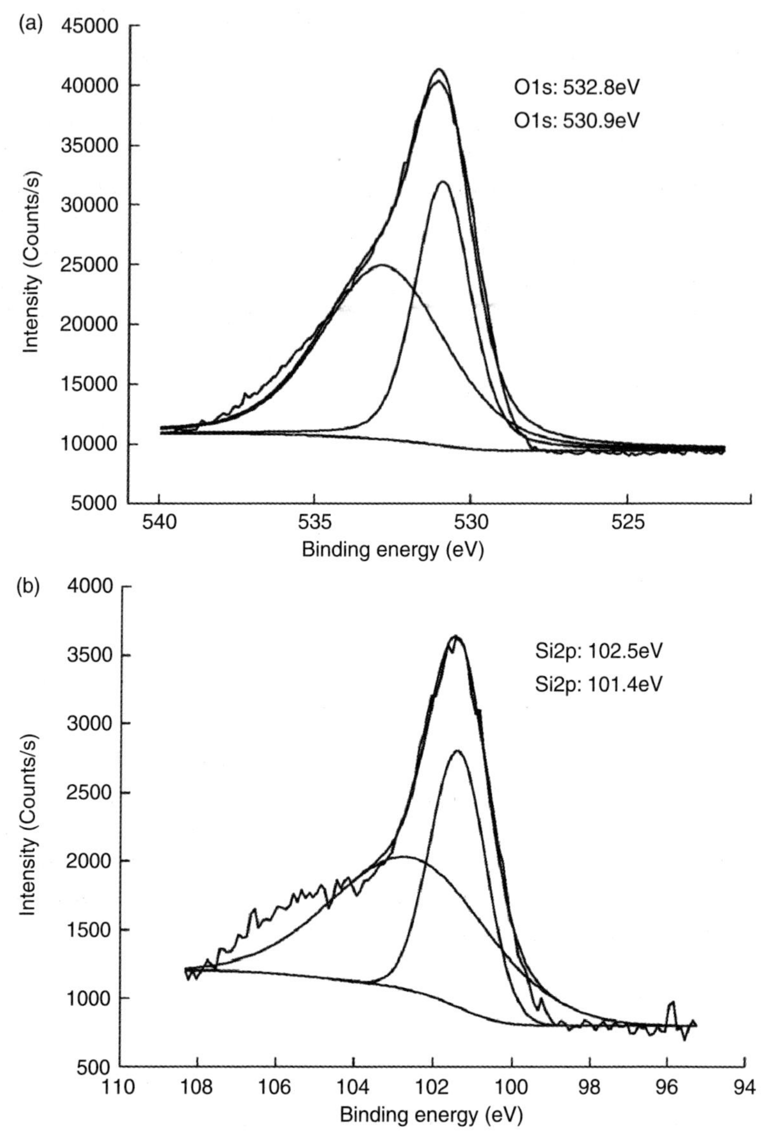

7.10 High-resolution spectra at the (a) O_{1s} and (b) Si_{2p} region of the surface on MAO coating (Ge *et al.*, 2010).

Wang *et al.*, 2009) and bipolar pulsed sources (Belov *et al.*, 2002; Timoshenko and Magurova, 2005; Yerokhin *et al.*, 2005; Jin *et al.*, 2006; Jaspard-Mécuson *et al.*, 2007) distinguished by different electrical regimes that have been applied to produce oxide ceramic coating on Mg alloys. The adjustable flexibility of electrical regimes enables one to regulate the surface discharge characteristics over a wide range, which is closely related to the coating growth, microstructure and associated properties.

The DC power sources are normally based on a bridge circuit and allow the application of galvanostatic or potentiostatic regimes of direct current. However, the ability to control the MAO process is limited, because of difficulties in regulating the surface discharge characteristics. For such reasons DC power sources are used only for simply shaped components and to prepare a thin MAO coating on Mg alloys. More recently, the invention of DC-pulsed and AC sources has greatly supported the rapid development and practical applications of the MAO technique. DC-pulsed sources allow control of the discharge duration by adjustment of the pulse duty cycle. In this way, the power energy can be used more efficiently with reduced energy consumption induced by interval discharge, and the corresponding microstructure and composition of the MAO coating can be tailored. A larger current density output is needed for the higher applied voltage (generally 800 V) due to the additional polarization caused by the creation of a charged double layer. Consequently, extensive industrial applications are severely limited. In the case of an AC source, additional polarization of the electrode can be avoided, whilst maintaining the ability to control the discharge duration by setting the half-wave and full-wave using a diode rectifier circuit. The full-wave power source is inclined to accelerate the growth of the anodic films, and the half-wave power source mainly contributes to the uniformity and fineness of the MAO coating (Wang *et al.*, 2009). However, the limitations in the power (typically 10 kW) and current frequency (mains frequency only) are the principal disadvantages of these sources that restrict commercial application.

Recently, a novel pulse source, which can supply higher power and a wide range of frequencies, has been developed. It is attracting great interest for industrial production due to the strong appeal of obtaining a much more compact coating with a relatively low energy consumption. Yerokhin *et al.* (2005) made a comparison between aluminum oxide layer properties obtained by 50 Hz AC and bipolar pulsed modes of the MAO power source, confirming that the bipolar pulsed process is able to improve the coating morphology, particularly by reducing the thickness of the porous outer layer. Investigations from the literature (Belov *et al.*, 2002; Timoshenko and Magurova, 2005; Jin *et al.*, 2006; Jaspard-Mécuson *et al.*, 2007) have all shown the importance of the pulse initiation delay and duration using bipolar pulsed sources at high frequency (up to kHz range). The significantly better properties of the bipolar pulsed samples can be attributed to the higher frequency current pulses, which enable the creation of shorter and more energetic micro-discharge events.

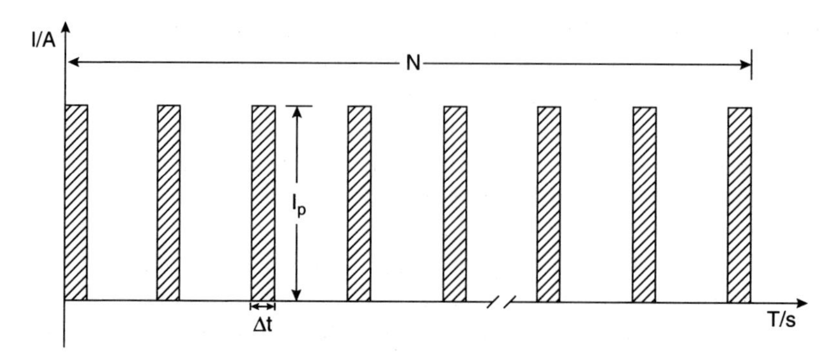

7.11 Schematic of pulse output of MAO power supply unit (I_p: current intensity of plus; Δt: pulse-on time; *N*: pulse numbers).

7.3.2 Effects of electrical parameters

The larger amount of the porous outer layer of the MAO coating leads to lower properties, and thus worse performance. Much effort has been devoted to limiting or suppressing the growth of the porous layer during MAO treatment. One way to achieve this requirement is to optimize the electrolyte composition. Another approach consists of using special current regimes such as pulses of current combined with optimal electrical parameters (Liang *et al.*, 2007; Wei *et al.*, 2007). A typical pulsed source power supply unit, as an example, and the effects of electrical parameters (voltage, current, duty cycle and frequency) on the microstructure of the MAO coating formed on Mg alloys are discussed. The corresponding parameters and waveforms of the pulsed power source are given in Fig. 7.11. The pulse parameters (voltage or current, pulse-on time and number of pulses) can be adjusted independently of each other using respective electronic amplifiers.

The sparking discharge duration and intensity depend on the pulse-on time and energy, respectively. The single pulse energy E_p is defined as:

$$E_p = \int_0^{\Delta t} U_p I_p \mathrm{d}t \qquad [7.6]$$

where U_p is the pulse voltage, I_p is the pulse current and Δt is the pulse-on time. Therefore, a change in the pulse cycle can regulate the surface discharge characteristics, which are responsible for the growth, microstructure and phase composition of MAO coating on Mg alloys.

Taking the MAO process for AZ31B Mg alloy in a sodium silicate based electrolyte as an example, only one parameter at a time was changed in order to detect the evolution of coating characteristics. Table 7.1 summarizes the effect of pulse parameters on the growth and microstructure of MAO

Table 7.1 Effect of pulse parameters on growth and microstructure of MAO coatings formed on AZ31B Mg alloy

Electrical parameters	Coating thickness	Surface morphology	Phase composition
Current density (A/dm^2): 0.5, 1, 2, 4, 8	Growth with increasing current density: 5–18 μm	Roughness growth	Slight phase transformation
Frequencies (Hz): 500, 1000, 1500, 2000	Growth with increasing frequencies: 12–19 μm	Roughness decrease	
Duty cycle: 3%, 6%, 9%, 12%, 15%	Growth with increasing duty cycle: 8–21 μm	Roughness decrease	
Voltage (V): 250, 300, 350, 400, 450	Growth with increasing voltage: 7–18 μm	Roughness growth	

coatings. It has been established from Table 7.1 that coating growth rate and surface morphologies depend mainly on the single pulse energy of the discharge. When increasing pulse voltage (reducing frequency, increasing duty cycle or increasing current density), the single pulse discharge energy rises, thus the coating product mass increases by a single pulse, finally resulting in increasing growth rate of the coating. In addition, increase of the single pulse discharge energy leads to rapidly rising temperature in the local discharge zones by heat accumulation, which induces a much stronger hot plasma effect. As a result, larger size pores are formed on the coating surface leading to high surface roughness.

7.3.3 Effects of electrolyte

The intrinsic effects of the electrolyte during the MAO process can be summarized as follows: (i) first and most important, promoting metal passivation to form a thin insulating film, which is a necessary prerequisite for dielectric breakdown to induce spark discharge; (ii) as the medium for conducting current, transmitting the essential energy needed for anode oxidizing to occur at the metal–electrolyte interface; (iii) providing the oxygen source in the form of oxysalt needed for oxidation; (iv) finally and also interestingly, allowing components present in the electrolyte to be incorporated into the coatings, further modifying or improving the properties of the MAO coating.

Table 7.2 MAO coating phase constituents formed on Mg alloys in complex electrolytes

Reference	Electrolyte composition	Substrate	Phase composition of coating
Liang *et al.*, 2009	Na_2SiO_3 (10.0 g/L) KOH (1.0 g/L)	AM50	MgO, $MgSiO_4$
Liang *et al.*, 2009	Na_3PO_4 (10.0 g/L) KOH (1.0 g/L)	AM50	MgO, $Mg_3(PO_4)_2$
Ding *et al.*, 2007	Na_2SiO_3 (10.0 g/L) KOH (1.0 g/L) Na_2WO_4 (6.0 g/L)	AM60B	MgO, $MgSiO_4$, $MgAl_2SiO_4$
Liang *et al.*, 2005	KOH (1.0 g/L) Na_3PO_4 (10.0 g/L) $NaAlO_2$ (8.0 g/L)	AM60B	MgO, $MgAl_2O_4$

To meet the prerequisite for dielectric breakdown, additives to promote strong metal passivation (such as silicates, aluminates and phosphates) are widely used as basic constituents of the electrolytes. The three groups have the following advantages: (i) they allow the sparking voltage to be reached easily, thus saving time; (ii) components present in the electrolyte (such as SiO_3^{2-}, AlO_2^- and PO_4^{3-}) are easily incorporated into the coatings by poly-reactions and deposition, thus increasing the coating growth rate; (iii) use of environmentally friendly and inexpensive electrolytes produces good wear and corrosion resistant coatings which are beneficial for the commercial producer.

It has been proved by experiments that simple alkaline electrolytes are unfeasible for commercialization of the process, because of lower coating growth rates and very high energy consumption. Thus, a complex electrolyte composition is more desirable for investigation into commercial applications (Ding *et al.*, 2007; Luo *et al.*, 2009). Examples from the published information on the coating phases formed on Mg alloys using complex electrolytes (mainly containing strong passive silicates) are summarized in Table 7.2.

As well as the limit of greater power consumption, the short service lifetime of the electrolyte is another obstacle and challenge for industrial applications of the MAO process on Mg alloys, which also affects the reproducibility of coating and increases costs by having to replenish the electrolyte. In fact, almost all the published literature involves the development and optimization of electrolytes (including composition and concentration) for desirable coating properties, while issues around stability of the electrolyte restrict extensive applications. Therefore, selection of reasonable stabilizers and further optimization of the electrolyte composition to improve long term stability is still an important research direction.

7.4 Corrosion resistance properties of MAO coating

Among light alloys, Mg is the most active and thus sensitive to corrosion, but MAO processing can significantly improve the corrosion resistance properties of Mg alloys in mild environments and/or for short durations. The corrosion resistance of the MAO coating in aggressive environments and for long term exposures is dictated by many factors, including the coating composition, thickness and defect levels. This section discusses how the MAO process can be applied in such areas as improving traditional corrosion resistance and preventing galvanic corrosion and stress corrosion cracking (SCC) of Mg alloys.

7.4.1 Traditional corrosion resistance

The corrosion performance of the MAO coating on Mg alloys has been studied extensively by many researchers. The corrosion resistance of a silicate electrolyte and a phosphate electrolyte MAO coating on AZ91D Mg alloy to 5 wt.% NaCl aqueous solution was reported by Cai *et al.* (2006). Optical macrograph (Fig. 7.12) shows the macroscopic appearance of the samples after 120 h total immersion test. It can be seen that MAO coating provided effective protection to the substrates of AZ91D Mg alloy and the anti corrosion performance is enhanced significantly. Furthermore, it is clear that the coating produced by silicate electrolyte is more corrosion resistant than those produced by phosphate electrolyte.

The electrochemical test and hydrogen evolution method were used to evaluate the corrosion resistance of AZ31B Mg alloy substrate and different thickness of MAO coating in 5 wt.% NaCl aqueous solution. Figure 7.13 shows the potentiodynamic polarization plots and corrosion rate that were calculated based on the ideal gas law through the hydrogen evolution method for 120 h. It is well known that the high corrosion potential (E_{corr}) and low corrosion current density (I_{corr}) of the sample suggests that it exhibits a low corrosion rate and a good corrosion resistance. The results shown in Fig. 7.13 demonstrate that the sample with MAO coating has a higher E_{corr} (−1.362 V) than the substrate (−1.511V). The I_{corr} of the substrate and different thickness MAO coatings calculated from Fig. 7.13 are 3.83×10^{-4} A/cm^2, 3.21×10^{-9} A/cm^2 and 5.44×10^{-9} A/cm^2, respectively. The results showed that the I_{corr} of MAO coating was about five orders of magnitude lower than that of the AZ31B Mg substrate. And the hydrogen evolution method results also show that the MAO coating could decrease the corrosion rate so improving the corrosion resistance of AZ31B Mg alloy.

7.12 The appearance of the samples after total immersion test in 5 wt.% NaCl aqueous solution for 120 h: (a) AZ91D; (b) silicate MAO coating (c) phosphate MAO coating (Cai *et al.*, 2006).

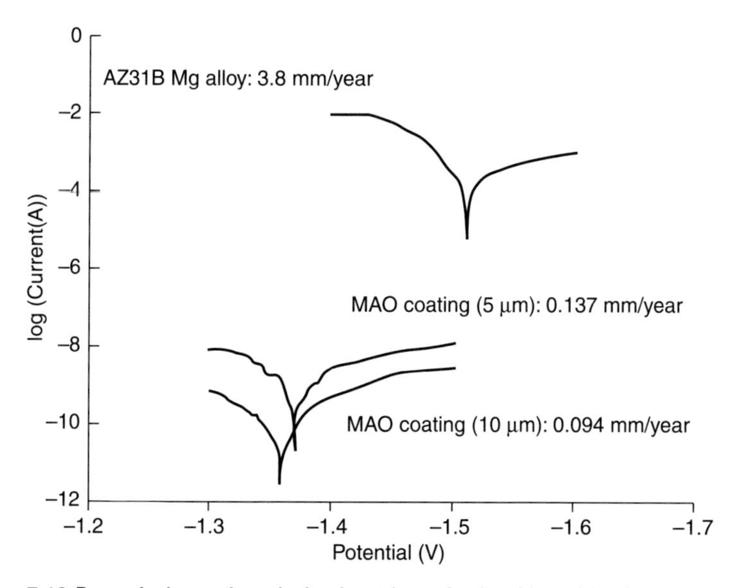

7.13 Potentiodynamic polarization plots obtained in 5% NaCl aqueous solution (scanning rate 0.1 mV/s, potentials measured against NHE) for the untreated substrate and MAO coating. The corrosion rates shown in this graph are calculated from the results of hydrogen evolution method.

7.4.2 Galvanic corrosion resistance

Mg alloys are susceptible to galvanic corrosion due to excessive levels of heavy metal or flux contamination, and to poor design and assembly practices. Galvanic corrosion is a major obstacle to the coupled use of Mg alloys with other metals. If these metals, such as iron, nickel and copper, have a low

7.14 The appearance of different treatment AZ31B Mg alloy connect with LY12 aluminum alloy after 120 h slat spay test using 5 wt.% NaCl aqueous solution: (a) substrate; (b) MAO coating.

hydrogen over-voltage, they can serve as efficient cathodes, consequently causing severe galvanic corrosion. MAO coating has proved effective in insulating or blocking direct contact between Mg alloys and other metals, thus eliminating galvanic corrosion (Gordienko *et al.*, 1993; Xue *et al.*, 2007; Barchiche *et al.*, 2008; Ghasemi *et al.*, 2008). The salt spray test was used to evaluate the galvanic corrosion resistance of AZ31B Mg alloy (substrate and MAO coated) connected with LY12 aluminum alloy. Figure 7.14a shows the surface morphology of AZ31B substrate connected with aluminum alloys after 120 h salt spray test. It finds that the corrosion on the Mg alloy is very severe, and there are many deep corrosion pits on the sample surface. It can be seen from Fig. 7.14b, when connected with the aluminum alloy for 120 h salt spray test, the ceramic coating on MAO treated AZ31B Mg alloy is without obvious corrosion pits. The results show that MAO treatment could prevent the galvanic corrosion of Mg alloy when connected with LY12 aluminum alloy.

7.4.3 Stress corrosion cracking

Stress corrosion cracking (SCC) is an extremely dangerous type of corrosion damage in engineering service of equipment, vessels, etc. SCC in Mg alloys has been generally attributed to one of two groups of mechanisms: continuous crack propagation by anodic dissolution at the crack tip or discontinuous crack propagation by a series of mechanical fractures at the crack tip (Winzer *et al.*, 2005). Thus there are two types of model for SCC: the dissolution models and the brittle fracture models. The former includes a preferential attack model, film rupture model, tunneling theory, etc; the latter involves cleavage processes and hydrogen embrittlement (HE) theory.

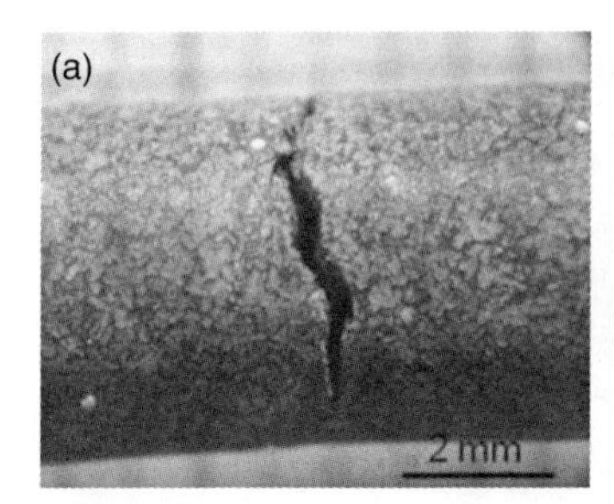

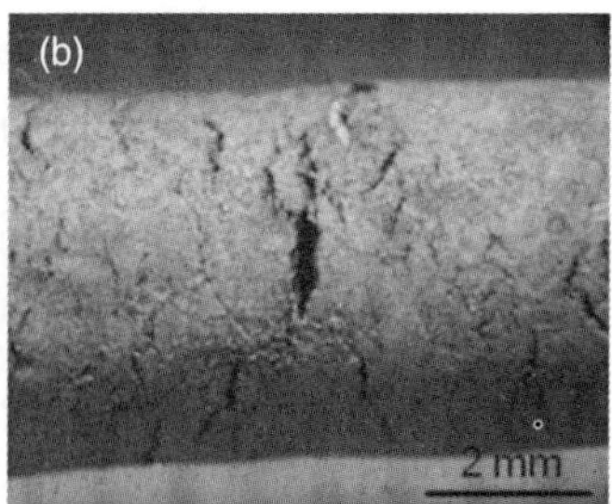

7.15 Optical macrographs showing the surface appearance of the SSRT tested (ASTM D1384 solution) on AM50 Mg alloy: (a) substrate; (b) MAO coated (Srinivasan *et al.*, 2008b).

Information on the environmentally assisted cracking behavior of MAO coated Mg alloys is limited. Srinivasan *et al.* (2008a, 2008b) have assessed the tensile strength and SCC of a cast AM50 Mg alloy with and without a silicate based MAO coating, by performing slow strain-rate tensile (SSRT) tests in air and in ASTM D1384 solution. The mechanical properties (SSRT tests in air) of this MAO coated Mg alloy were found to be slight lower than those of untreated alloy. Even though the general corrosion resistance of the MAO coated alloy was superior, the MAO coating could not eliminate the SCC susceptibility of the alloy in the tests performed at strain rates of 10^{-6} and 10^{-7} s^{-1}. However, the MAO coating could delay the failure under both conditions (Fig. 7.15). The optical macrographs of the specimens presented in Fig. 7.15a and 7.15b give an account of the type and extent of damage on these two specimens. The untreated specimen apparently has a large crack, which would have caused the damage, while the MAO coated specimen is observed to have developed multiple cracks. The size of the cracks observed in this specimen were quite large compared to those observed in the MAO coated specimen tested in air. It is quite evident that the MAO coating, despite being thin, was cracking under conditions of constant/continuous straining, thus developing numerous micro-cracks. Thus, on account of the synergistic dissolution and straining at these defect sites, the cracks opened up wider and became longer, leading to SCC. The formation of numerous cracks and their widening is a plausible reason for the higher elongation values registered for the coated sample. The MAO treatment performed well for a relatively longer test duration (in SSRT tests) on Mg alloys, but could not obviate the SCC process.

7.4.4 Corrosion mechanism of MAO coating

It is well known that MAO coatings on Mg alloys are porous in nature, and the extent of porosity and other defects such as cracks can influence the

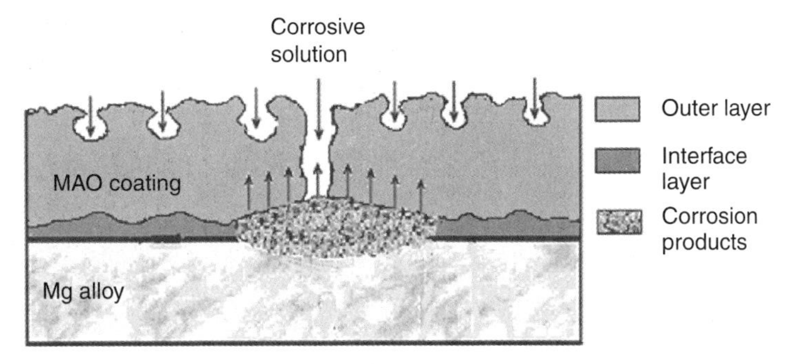

7.16 Schematic representation of the corrosion process of MAO coating on Mg alloys.

corrosion behavior. Figure 7.16 schematically represents the corrosion process of the MAO coating. The structural imperfection of the MAO coating enlarges the effective surface area immersed in the corrosive medium, while the large micro-pores act as passages for the corrosive ions entering into the MAO coating. Since the electrolyte can reach the interface and spread on the surface, the corrosion product (magnesium hydroxide) formed at the interface grows to such a level that it exerts a stress which lifts/damages the MAO coating. The lift-off of the coating in these areas results in exposure of the substrate underneath to the electrolyte, when the pitting corrosion forms on the surface of the MAO coated Mg alloy.

7.5 Applications, industrial MAO processing and exploration of new processes

Besides the extensive scientific research around the world, the MAO process for Mg alloys may find special applications in a variety of industries such as automotive, aerospace, sports, military, etc., since this technology has a number of unique advantages. Various research has been undertaken to promote the industrial application of MAO techniques during the last decade.

7.5.1 Applications

Because the MAO process produces a thick, well bonded, ceramic coating on a variety of reactive Mg alloys, it can be used for a broad range of applications such as improving wear and corrosion resistance. The primary application could be the replacement of heavier metallic alloys or the more expensive light metal composite materials required by the aerospace and automotive

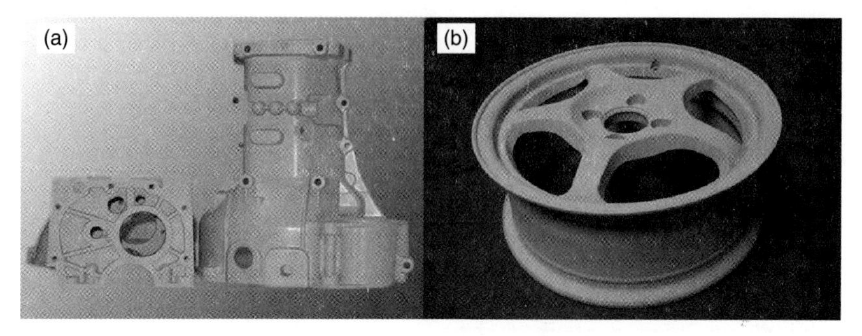

7.17 MAO treated Mg alloys components: (a) gearbox casing; (b) wheel boss.

industries. Figure 7.17 shows MAO treated Mg alloy components (gearbox casing and wheel boss) used in the automotive industry.

The porous ceramic coatings resulting from MAO can act as an excellent paint base because of their strong affinity to paints. They can be used alone or as a base for further build-up of a coating system, such as promoting the adhesion of paint and powder coating to the Mg alloy substrate. Furthermore, the MAO process can be utilized as pre-treatment in duplex processes involving electroplating, physical vapor deposition (PVD) or other coatings (Mu and Han, 2003; Liu and Gao, 2006). For example, thin, hard PVD layers can be deposited onto a thick diffusion coating fabricated by MAO to provide good adhesion and load support on low alloy substrates. The high heat resistance of the MAO coating may be valuable in manufacturing. Also, the technique can also be used as an alternative to conventional induction hardening, owing to its characteristic of rapid heating, which can be used to create a high temperature difference between the substrate surface and the interior.

7.5.2 Industrial MAO process

MAO coating technology on Mg alloys is readily available on a commercial basis. Commercial processes such as Keronite (http://www.keronite.com), Magoxid (http://www.aimt-group.com), Tagnite (http://www.tagnite.com) and others have been developed. As any other technology for commercial application, MAO technology has its limitations. It requires high voltage (up to 1000 V) and power source capacity (even larger than 1 MW) to run a MAO process for large scale production. This means high energy consumption and a dangerous production environment as well as restrictions in production scale and part size. As a consequence, the capital investments and production costs of the MAO process are considerably higher than conventional anodizing technologies. So there is lots of work needed to improve and popularize the MAO process for industrial application.

7.5.3 New process exploration

With the development of extensive scientific and industrial research, there are some new areas of exploration for processing and applying MAO technology, such as combined processes (electrophoresis or powder) with MAO for enhanced corrosion resistance properties of Mg alloys, and low energy consumption of MAO power source.

Combined processes with MAO for enhanced properties

The pores which exist on the MAO coating surface pose a negative effect on corrosion properties. However, the coatings can be used as pretreated subsurface layers to mechanically support surface layers produced by other processes (Curran and Clyne, 2006). This will result in duplex surface systems with combined property improvements and can extend application areas. Taking advantage of the beneficial surface connected pores, a wide variety of compounds, including paints, lubricants, sol-gels and polymers (such as electrophoresis or powder) can be impregnated into the pores on an MAO coating to achieve duplex coatings to enhance corrosion properties of Mg alloys. Porosity in the porous MAO coating can effectively improve adhesion of the top layer to the subsurface layer.

Using the MAO composited electrophoresis coat (E-coat) on Mg alloys as an example, Fig. 7.18 shows the cross-sectional micrograph of MAO composited E-coat on AZ31B Mg alloy. It can be seen that the E-coat adheres tightly with the MAO coating, and the open pores on the surface of the MAO coating are filled in nicely with the E-coat material. The salt spray test was used to compare the corrosion resistance of MAO coating and MAO composited E-coat on Mg alloy. The corrosion-lost weight of MAO treated Mg alloy after 300 h corrosion is as high as ~ 70 g/m^2, and the ceramic coating is totally damaged. But the corrosion-lost weight of the sample with MAO composited E-coat after 800 h salt spray test remains almost negligible, and the surface morphology change is indistinct. Also the adhesion test results show that E-coat adheres well with MAO coating without peeling off. And this excellent adhesion between the E-coat and MAO coating comes from the infiltration of the E-coat into the surface pores of the MAO layer. After curing, E-coat forms a strong bond with the MAO coating. The new process exploration that the MAO composited E-coat could greatly improve the corrosion resistance of Mg alloys.

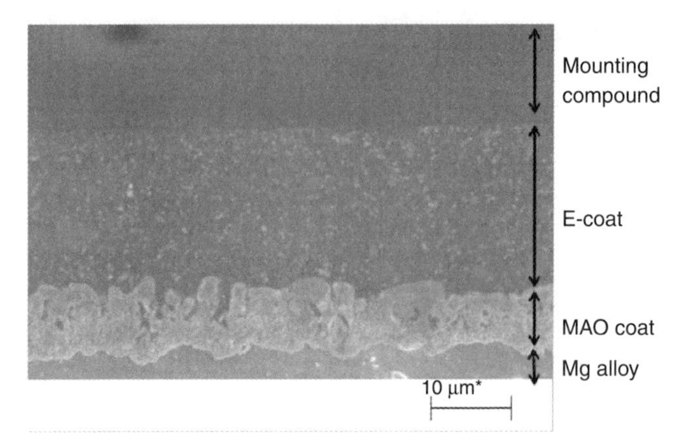

7.18 Cross-sectional micrograph of MAO composited E-coat on AZ31B Mg alloy.

Low energy consumption of MAO power source

As discussed in Section 7.5.2, one of limitations for commercial MAO technology is that energy consumption for MAO process in large scale production is too high. Owing to the generation of a large quantity of heat in a MAO process, a high chilling capacity is required to cool the electrolyte, so lots of electricity is needed to warm and cool the electrolyte recurrently. As a consequence, the production costs of the MAO process are considerably higher than conventional anodizing technologies. In order to improve and popularize the MAO process for industrial application, the exploration of MAO power sources with low energy consumption is very important.

The Xi'an University of Technology (XAUT) in P.R.China has made significant efforts in its MAO coating facility. The researchers in XAUT have developed the MAO power source using pulse current anodizing, and the pulse current, pulse numbers and pulse-on time can be adjusted independently. From the schematic representation of pulse current changed with the constant electron flux shown in Fig. 7.19, it can be seen that under constant average current, the intensity of pulse current could increase with decreased pulse numbers and pulse-on time. The results of computation show that energy consumption of the MAO process on Mg alloys using this new power source is decreased by 40% compared with the normal AC power source.

7.6 Future trends

MAO is a very attractive, environmentally friendly surface engineering technique for Mg alloys. MAO treatment utilizes micro discharges to

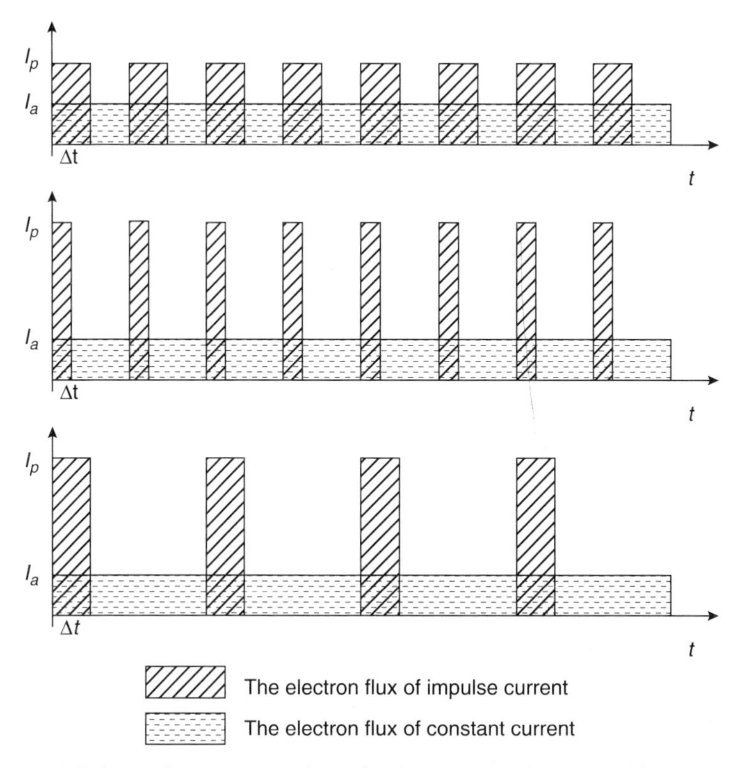

7.19 Schematic representation of pulse current changed with the constant electron flux (I_a: average current; I_p: pulse current; Δt: pulse-on time).

convert the surface of Mg alloys into a variety of oxide ceramic coatings with different properties using mainly alkaline based, chromate free solutions, which can effectively improve not only the corrosion resistance but also wear resistance. Furthermore, the MAO coatings have good adhesion to the substrate, thermal stability, thermal shock resistance and good optical properties, which offer a wide range of property improvements for Mg alloys in their applications. In particular, MAO is a unique technique which can be used to fabricate functional coatings for specific applications.

In spite of those advantages, the porous structure of MAO coating on Mg alloys could influence the corrosion resistance under harsh conditions and allow galvanic corrosion when in contact with other metals. Thus, combined processes with MAO are necessary for Mg alloys to retain long term stability under such conditions. However, this can be used to advantage as well, because the porosity structure offers good adhesion if used as pretreatment for paint or powder coating. Currently, the MAO process is in a transition phase from research to commercial application, mainly focused on the corrosion and wear protection of Mg alloys. It is necessary to further study the fundamentals of the MAO technique to advance scientific understanding and to develop new functional MAO coatings. Exploration of MAO process power source units with low energy consumption is also significant for widening possible commercial application.

7.7 Acknowledgements

Support from National Natural Science Foundation of China (Grant NO. 51071121), as well as from the National Key Technologies Research and Development Program of China (Grant No. 2011BAE22B05) is gratefully acknowledged.

7.8 References

Arrabal R, Matykina E, Skeldon P and Thompson G E (2009), 'Coating formation by plasma electrolytic oxidation on ZC71/SiC/12p–T6 magnesium metal matrix composite', *Appl Surf Sci*, **255**, 5071–5078.

Arrabal R, Matykina E, Viejo F, Skeldon P, Thompson G E and Merino M C (2008), 'AC plasma electrolytic oxidation of magnesium with zirconia nanoparticles', *Appl Surf Sci*, **254**, 6937–6942.

Barchiche C E, Rocca E and Hazan J (2008), 'Corrosion behaviour of Sn–containing oxide layer on AZ91D alloy formed by plasma electrolytic oxidation', *Surf Coat Tech*, **202**, 4145–4152.

Belevantsev V I, Terleeva O P, Markov G A, Shulepko E K, Slonova A I and Utkin V V (1998), 'Micro–plasma electrochemical processes', *Zashch Met*, **34**, 468–484.

Belov N A, Zolotarevsky V S and Shatrov A S (2002), 'Construction material based on aluminum and method for producing parts from said material', *WO 02/22902*, publ. 21.03.02.

Blawert C, Heitmann V, Dietzel W, Nykyforchyn H M and Klapkiv M D (2005), 'Influence of process parameters on the corrosion properties of electrolytic conversion plasma coated magnesium alloys', *Surf Coat Tech*, **200**, 68–72.

Brown S D, Kuna K J and Van T B (1971), 'Anodic spark deposition from aqueous solution of $NaAlO_2$ and Na_2SiO_3', *J Am Ceram Soc*, **54**(8), 384–390.

Cai Q Z, Wang L S and Wei B K (2006), 'Electrochemical performance of microarc oxidation films formed on AZ91D magnesium alloy in silicate and phosphate electrolytes', *Surf Coat Tech*, **200**, 3727–3733.

Chigrinova N M, Chigrinova V E and Kukharev A A (2001), 'Formation of coatings by anodic microarc oxidation and their operation in thermally–stressed assemblies', *Powder Metall Met C+*, **40**, 213–220.

Curran J A and Clyne T W (2005), 'The thermal conductivity of plasma electrolytic oxide coatings on aluminium and magnesium', *Surf Coat Tech*, **199**, 177–183.

Curran J A and Clyne T W (2006), 'Porosity in plasma electrolytic oxide coatings', *Acta Mater*, **54**, 1985–1993.

Curran J A, Kalkanc H, Magurova Y and Clyne T W (2007), 'Mullite–rich plasma electrolytic oxide coatings for thermal barrier applications', *Surf Coat Tech*, **201**, 8683–8687.

Ding J, Liang J, Hu L H, Hao J and Xue Q (2007), 'Effects of sodium tungstate on characteristics of microarc oxidation coatings formed on magnesium alloy in silicate–KOH electrolyte', *T Nonferr Metal Soc*, **17**, 244–249.

Districh K H, Krysmann W and Kurze P (1984), 'Structure and properties of ANOF layers', *Cryst Res Technol*, **19**(1), 93–99.

Duradzhy V N and Parsadanyan A S (1998), 'Metal heating in electrolytic plasma', *Shtiintsa*, **125**, 1235–1239 (in Russian).

Ge Y F, Jiang B L, Yang Z Y and Li Y L (2010), 'Microstructure and corrosion resistance behavior of composite micro–arc oxidation and SiO_2 coatings on magnesium alloys', *Adv Mater Res*, **160–162**, 1834–1838.

Gerasimov M V, Nikolaev V A and Shcherbakov A N (1994), 'Microplasma oxidation of metals and alloys', *Metallurgist*, **38**(7–8), 179.

Ghasemi A, Raja V S, Blawert C, Dietzel W and Kainer K U (2008), 'Study of the structure and corrosion behavior of PEO coatings on AM50 magnesium alloy by electrochemical impedance spectroscopy', *Surf Coat Tech*, **202**, 3513–3518.

Gordienko P S, Rudnev V S, Orlova I I, Kurnosova A G, Zavidnaya A G, Rudnev A S and Tyrin V I (1993), 'Vanadium–containing anode–oxide films on aluminium alloys', *Zashchita Metallov*, **5**, 739–742.

Gray J E and Luan B (2002), 'Protective coatings on magnesium and its alloys – A critical review', *J Alloy Compd*, **336**, 88–113.

Guo H F, An M Z, Xu S and Huo H B (2005), 'Formation of oxygen bubbles and its influence on current efficiency in micro–arc oxidation process of AZ91D magnesium alloy', *Thin Solid Films*, **485**, 53–58.

Gupta P, Tenhundfeld G and Daigle E O (2007), 'Electrolytic plasma technology: Science and engineering – An overview', *Surf Coat Tech*, **201**(21), 8746–8760.

Han I, Choi J H, Zhao B H, Baik H K and Lee I S (2007), 'Changes in anodized titanium surface morphology by virtue of different unipolar DC pulse waveform', *Surf Coat Tech*, **201**, 5533–5536.

Homepage: AIMT AHC Group, Inc. at http://www.aimt-group.com

Homepage: Keronite International Ltd at http://www.keronite.com

Homepage: Technology Applications Group, Inc. at http://www.keronite.com

Ikonopisov S (1977), 'Theory of electrical breakdown during formation of barrier anodic films', *Electrochim Acta*, **22**(10), 1077–1082.

Jaspard-Mécuson, Czerwiec T, Henrion G, Belmonte T, Dujardin L, Viola A and Beauvir A (2007), 'Tailored aluminium oxide layers by bipolar current adjustment in the plasma electrolytic oxidation (PEO) process', *Surf Coat Tech*, **201**, 8677–8682.

Jin F, Paul K C, Xu G D, Zhao J, Tang D L and Tong H H (2006), 'Structure and mechanical properties of magnesium alloy treated by micro–arc discharge oxidation using direct current and high–frequency bipolar pulsing modes', *Mat Sci Eng A*, **435–436**, 123–126.

Klapkiv M D (1998), 'State of an electrolytic plasma in the process of synthesis of oxides based on aluminum', *Mater Sci*, **31**(4), 494–499.

Krysmann W, Kurze P and Districh K H (1984), 'Process characteristics and parameters of anodic oxidation by spark deposition (ANOF)', *Cryst Res Tech*, **19**(7), 973–979.

Krysmann W, Kurze P, Districh K H and Schneider H G (1971), 'Process characteristics and parameters of anode oxidation by spark discharge (ANOF)', *Corros Sci*, **11**, 411–417.

Kuhn A (2003), 'Plasma anodizing of magnesium alloys', *Met Finish*, **101**, 44–50.
Kurze P, Krysmann W and Schneider H G (1986), 'Application field of ANOF layers and composites', *Cryst Res Technol*, **21**(12), 1603–1609.
Liang J, Guo B, Tian J, Liu H, Zhou J, Liu W and Xu T (2005), 'Effects of $NaAlO_2$ on structure and corrosion resistance of microarc oxidation coatings formed on AM60B magnesium alloy in phosphate–KOH electrolyte', *Surf Coat Tech*, **199**, 121–126.
Liang J, Hu L T and Hao J C (2007), 'Improvement of corrosion properties of micro-arc oxidation coating on magnesium alloy by optimizing current density parameters', *Appl Surf Sci*, **253**, 6939–6945.
Liang J, Srinivasan P B, Blawert C, Störmer M and Dietzel W (2009), 'Electrochemical corrosion behaviour of plasma electrolytic oxidation coatings on AM50 magnesium alloy formed in silicate and phosphate based electrolytes', *Electrochim Acta*, **54**, 3842–3850.
Liu Z M and Gao W (2006), 'A novel process of electroless Ni–P plating with plasma electrolytic oxidation pretreatment', *Appl Surf Sci*, **253**, 2988–2991.
Luo H H, Cai Q Z, Wei B K, Yu B, He J and Li D J (2009), 'Study on the microstructure and corrosion resistance of ZrO_2–containing ceramic coatings formed on magnesium alloy by plasma electrolytic oxidation', *J Alloys Compd*, **474**, 551–556.
Magurova Y V and Timoshenko A V (1995), 'The effect of a cathodic component on AC microplasma oxidation of aluminum alloys', *Zashch Met*, **31**, 414–418.
Mizutania Y, Kim S J, Ichinob B R and Okidob M (2003), 'Anodizing of Mg alloys in alkaline solutions', *Surf Coat Tech*, **169–170**, 143–146.
Monterro J, Fernandez M and Albella J M (1987), 'Pore formation during the breakdown process in anodic Ta_2O_5 films', *Electrochim Acta*, **32**(1), 171–174.
Mu W Y and Han Y (2008), 'Characterization and properties of the MgF_2/ZrO_2 composite coatings on magnesium prepared by micro–arc oxidation', *Surf Coat Tech*, **202**, 4278–4284.
Nie X, Hao Q K and Wei J M (1996), 'A novel modification technique for metal surface', *J Wuhan Univ Technol*, **11**(1), 28–35.
Patel J L and Saka N (2001), 'Microplasmic coatings', *Am Ceram Soc Bull*, **80**(4), 27–29.
Rudnev V S, Yarovaya T P and Konshin V V (1998), 'Microplasma oxidation of an aluminum alloy in aqueous solutions containing sodium cyclohexaphosphate and nitrates of lanthanum and europium', *Russ JE lectrochem*, **34**(6), 510–516.
Shi Z, Song G and Atrens A (2006), 'Influence of anodizing current on the corrosion resistance of anodized AZ91D magnesium alloy', *Corros Sci*, **48**, 1939–1959.
Srinivasan P B, Blawert C and Dietzel W (2008a), 'Effect of plasma electrolytic oxidation coating on the stress corrosion cracking behaviour of wrought AZ61 magnesium alloy', *Corros Sci*, **50**, 2415–2418.
Srinivasan P B, Blawert C, Dietzel W and Kainer K U (2008b), 'Stress corrosion cracking behaviour of a surface–modified magnesium alloy', *Scripta Mater*, **59**, 43–46.
Srinivasan P B, Liang J, Blawert C, Störmer M and Dietzel W (2009), 'Effect of current density on the microstructure and corrosion behavior of plasma electrolytic oxidation treated AM50 magnesium alloy', *Appl Surf Sci*, **255**, 4212–4218.
Tchernenko V I, Snezhko L A and Papanova I I (1991), 'Coatings by anodic spark electrolysis', *Khimiya*, **68**, 568–573 (in Russian).

Timoshenko A V and Magurova Y V (2000), 'Application of oxide coatings to metals in electrolyte solutions by microplasma methods', *Rev Metal Madrid*, **36**(5), 323–330.

Timoshenko A V and Magurova Y V (2005), 'Investigation of plasma electrolytic oxidation processes of magnesium alloy MA2–1 under pulse polarisation modes', *Surf Coat Tech*, **199**, 135–140.

Tynrin Y N and Pogrehnjak A D (2001), 'Electric heating using a liquid electrode', *Surf Coat Tech*, **142–144**, 293–299.

Verdier S, Boinet M, Maximovitch S and Dalard F (2005), 'Formation, structure and composition of anodic films on AM60 magnesium alloy obtained by DC plasma anodizing', *Corros Sci*, **47**, 1429–1444.

Vijh A K (1971), 'Sparking voltage and side reactions during anodization of valve metals in terms of electron tunneling', *Corros Sci*, **11**, 411–417.

Wang Y Q, Wu K and Zheng M Y (2006), 'Effects of reinforcement phases in magnesium matrix composites on microarc discharge behavior and characteristics of microarc oxidation coatings', *Surf Coat Tech*, **201**, 353–360.

Wang Y Q, Zheng M Y and Wu K (2005), 'Microarc oxidation coating formed on SiCW/AZ91 magnesium matrix composite and its corrosion resistance', *Mater Lett*, **59**, 1727–1731.

Wang X M, Zhu L Q, Li W P, Liu H C and Li Y H (2009), 'Effects of half–wave and full–wave power source on the anodic oxidation process on AZ91D magnesium alloy', *Appl Surf Sci*, **255**, 5721–5728.

Wei C B, Tian X B, Yang S Q, Wang X B, Fu K Y and Chu P K (2007), 'Anode current effects in plasma electrolytic oxidation', *Surf Coat Tech*, **201**, 5021–5024.

Winzer N, Atrens A, Song C, Ghali E. Dietzel W, Kainer K U, Hort N and Blawert C (2005), 'A critical review of the stress corrosion cracking (SCC) of magnesium alloys', *Adv Eng Mater*, **7**(8), 659–693.

Wirtz G P, Brown S D and Kriven W M (1991), 'Ceramic coatings by anodic spark deposition', *Mater Manuf Process*, **6**(1), 87–115.

Wu K, Wang Y Q and Zheng M Y (2007), 'Effects of microarc oxidation surface treatment on the mechanical properties of Mg alloy and Mg matrix composites', *Mat Sci Eng A*, **447**, 227–232.

Xin S G, Jiang Z H, Wang F P, Wu X H, Zhao L C and S himizu T (2001), 'Effect of current density on Al alloy microplasma oxidation', *J Mater Sci Technol*, **17**(6), 657–660.

Xue W B, Wang C, Tian H and Lai Y C (2007), 'Corrosion behaviors and galvanic studies of microarc oxidation films on Al–Zn–Mg–Cu alloy', *Surf Coat Tech*, **201**, 8695–8701.

Yerokhin A L, Lyubimov V V and Ashitkov R V (1998b), 'Phase formation in ceramic coatings during plasma electrolytic oxidation of aluminium alloys', *Ceram Int*, **24**(1), 1–6.

Yerokhin A L, Nie X, Leyland A, Matthews A and Dowey S G (1999), 'Plasma electrolysis for surface engineering', *Surf Coat Tech*, **122**, 73–93.

Yerokhin A L, Shatrov T A, Samsonov V, Shashkov P, Leyland A and Matthews A (2004), 'Fatigue properties of Keronite coatings on a magnesium alloy', *Surf Coat Tech*, **182**, 78–84.

Yerokhin A L, Shatrov T A, Samsonov V, Shashkov P, Pilkington A, Leyland A and Matthews A (2005), 'Oxide ceramic coatings on aluminium alloys produced by a pulsed bipolar plasma electrolytic oxidation process', *Surf Coat Tech*, **199**, 150–157.

Yerokhin A L, Snizhko L O, Gurevina L, Leyland A, Pilkington A and Matthews A (2003), 'Discharge characterization in plasma electrolytic oxidation of aluminium', *J Phys D: Appl Phys*, **36**, 2110–2120.

Yerokhin A L, Voevodin A A, Lyubimov V V, Zabinski J and Donley M (1998a), 'Plasma electrolytic fabrication of oxide ceramic surface layers for tribotechnical purposes on aluminium alloys', *Surf Coat Tech*, **110**(3), 140–146.

Zhang R F, Wang F Y, Hu C Y, Li W K, Xiang J H, Duo S W, Li M S and He X M (2007), 'Research progress in sealing treatment of anodic coatings formed on magnesium alloys', *J Mat Eng*, **11**, 82–86.

Zhao H, Chen J and Liu Z (2005), 'Microstructure and mechanical properties of ceramic coating on AM50 magnesium alloy by microarc oxidation', *Mater Sci Forum*, **88**, 693–696.

8
Anodization of magnesium (Mg) alloys to improve corrosion resistance

S. A. SALMAN, Nagoya University, Japan and Al-Azhar University, Egypt and M. OKIDO, Nagoya University, Japan

DOI: 10.1533/9780857098962.2.197

Abstract: Anodizing is one of the most widely used surface treatments for magnesium and its alloys. In this electrolytic oxidation process, the surface of a metal is converted to an anodic oxide/hydroxide film with desirable protective, decorative or functional properties. This chapter describes the fundamentals of anodizing treatment, with an emphasis on the difference between magnesium anodizing and aluminum anodizing. The chapter then discusses the most widely used commercial anodic treatments and describes how the treatment conditions affect the corrosion resistance and the physical and mechanical properties of the anodic coating that forms on magnesium alloys. A critical review of recent patents and advances in the anodizing of magnesium is given. After reviewing the literature, the chapter explains how surface treatment can lead to more widespread use of magnesium alloys in a range of industrial applications.

Key words: anodizing, anodic film, Mg alloys, corrosion resistance.

8.1 Introduction

Magnesium is the lightest of all structural metal materials in practical use, with a density equivalent to two-thirds that of Al, one-third that of Zn and one quarter that of steel. Magnesium alloys are therefore recognized as alternatives to Al alloys and steel in reducing the weight of structural materials. Furthermore, magnesium alloys have high strength-to-weight ratios, excellent castability, machinability, weldability, thermal stability and good damping capacity. These properties make magnesium alloys an attractive target in a number of applications, especially in fields where weight reduction is critical or there are particular technical requirements. Magnesium has several applications, including in automobiles, computer parts, mobile phones, sporting goods, hand-held tools, household equipment and aerospace components. Magnesium alloys could also contribute towards meeting future requirements of the automotive industry to make cars lighter and thus improve fuel efficiency and reduce emissions, as the use of magnesium alloys would significantly decrease the weight of automobiles without sacrificing structural strength.

Pure magnesium is rarely used in industrial engineering applications. However, to improve its mechanical and other properties, alloying elements are added, most commonly aluminum, manganese, zinc, zirconium, silicon, calcium and rare earth elements. In appropriate amounts, these additives enhance the anticorrosion and mechanical properties of magnesium alloys. Aluminum has the most favorable effect on magnesium: it improves strength and hardness, increases the freezing range and makes the alloy easier to cast. However, a major obstacle to the widespread use of magnesium alloys is poor corrosion resistance: magnesium alloys are highly susceptible to corrosion attack, particularly in wet environments. Therefore, selecting appropriate alloying elements and finding the best alloy design constitute the first step to improve the anticorrosion property of magnesium alloys. Further surface treatment of magnesium and its alloys is important in meeting several industrial specifications. Electrochemical plating, conversion coating, anodizing, ceramic coating, organic coating, gas-phase deposition and laser surface alloying are among the most common surface treatments applied to magnesium alloys. Of these, anodizing produces a non-conductive, dense, durable, abrasion-resistant and corrosion-resistant film. Furthermore, adjusting anodizing parameters such as applied current and temperature and applying a dye treatment after anodizing makes it possible to control the film thickness and decorative appearance. Previously described specifications of anodic films suggest that anodizing treatment can be used to meet most industrial requirements.

8.1.1 Corrosion behavior of magnesium

Poor corrosion resistance is a major obstacle to the widespread use of magnesium alloys in many applications because it has higher chemical and electrochemical activity than other structural metals such as steel and Al alloys. Magnesium alloys readily corrode in the presence of only a small amount of aggressive ions in the environment. Kurze (2006) reported the corrosion susceptibility of magnesium and its alloys in both organic and inorganic substances.

With regard to inorganic substances, magnesium is not suitable for use in acid media other than hydrofluoric acid medium, water other than distilled water, salts of halogen acids, aqueous solutions of sulfates, ammonium hydroxide or mercury salts. As for organic substances, magnesium is not suitable for use with methanol, glycerol, glycol, formaldehyde or any organic acids. Magnesium alloys often serve as sacrificial anodes on structures such as buried pipelines and steel piles. There are two primary reasons for the poor corrosion resistance of magnesium alloys. The first is internal galvanic corrosion by second phases or impurities. Second, the hydroxide film on magnesium is much less stable than the passive films that form on metals such as Al alloys and stainless steels (Emley, 1966; Makar and Kruger, 1989). Although the standard reduction potential of magnesium has been given as −2.37 V *vs* NHE (Bard *et al.*, 1985), its actual corrosion potential is usually −1.7 V *vs* NHE in dilute chloride solutions. The difference between the theoretical standard potential and the actual corrosion potential is attributed to the formation of a surface film of $Mg(OH)_2$ or perhaps MgO. In addition, the measured potential corresponds to the mixed potential for Mg dissolution and hydrogen gas evolution in aqueous solutions (Makar and Kruger, 1993; Song and Atrens, 1999).

Magnesium dissolution in aqueous environments generally proceeds by electrochemical reaction with water to produce magnesium hydroxide and hydrogen gas according to reaction [8.1].

$$Mg + 2H_2O \rightarrow Mg(OH)_2 + H_2 \qquad [8.1]$$

Most studies on the kinetics of reaction [8.1] have concluded that the rate of attack at a pH of less than about 11 is controlled by the diffusion of reactants or products through the surface film. As corrosion proceeds, the pH of the metal surface increases because of the formation of $Mg(OH)_2$, which has an equilibrium pH of about 11. This film provides some corrosion protection over a wide pH range. However, the presence of damaging electrolyte species and impurities in the metal hinders the formation of the film (Pourbaix, 1974). The thermodynamics that govern the formation of the film are described by the Pourbaix (potential–pH) diagram given in Fig. 8.1. Although we consider here the formation of MgO, the diagram given by Pourbaix indicates that the lines correspond to $Mg(OH)_2$ Pourbaix explained that this is because $Mg(OH)_2$ is thermodynamically more stable than MgO in the presence of water. In the figure, the ringed number lines divide the diagram into three regions: a region of corrosion (dissolved Mg^{2+}), a region of immunity (unreacted metal e.g. Mg), and a region of passivation (formation of passive film e.g. $Mg(OH)_2$). The immunity region in the diagram is well below the region of water stability. In neutral and low-pH environments, magnesium dissolution is accompanied by hydrogen evolution. In basic environments, the surface of the magnesium alloy is passivated by the formation of an $Mg(OH)_2$ film. Because the magnesium oxide and hydroxide films that form on unalloyed magnesium are slightly soluble in water, they do not provide long-term protection. When chloride, bromide and/or sulfate are present in the environments, the surface films break down. Likewise, as carbon dioxide (CO_2) in air acidifies water, the films are not stable.

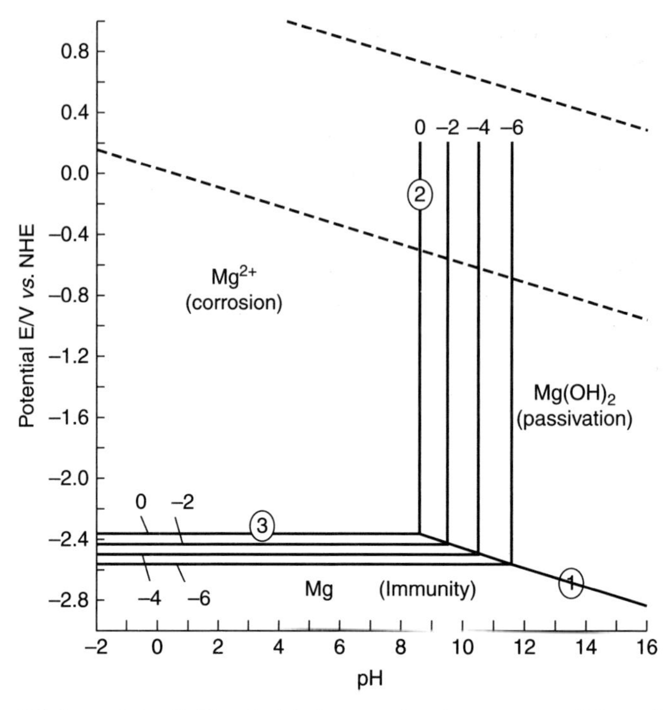

8.1 Potential–pH (Pourbaix) diagram for the system of Mg and water at 25°C. (Source: G. L. Makar, J. Kruger, 'Corrosion of magnesium', International Materials Review, 1993. Reproduced with permission from Maney Publishing.)

8.1.2 Surface treatment of magnesium alloys

In recent years, many surface treatments have been developed for the protection of magnesium alloys. A schematic diagram of the mixed potential is presented in Fig. 8.2. The anodic polarization represents the dissolution of magnesium, and the cathodic polarization curves are assumed to represent cathodic hydrogen evolution through water reduction. The anticorrosion property can be improved by suppressing the anodic or cathodic reactions. Available methods for suppressing the anodic reaction from point A to point B are film passivation, anodizing, conversion coating and painting. The cathodic reaction can be suppressed from point A to point C through surface treatment that improves the hydrogen overpotential and the removal of any undesirable impurities (on the cathode side) that readily generate hydrogen. Various surface treatment techniques can be employed to improve the surface properties of magnesium such as wear resistance, corrosion resistance and hardness, as well as decorative appearance. Table 8.1 lists the main surface treatments used for magnesium alloys categorized according to wet and

Table 8.1 Commonly applied wet and dry methods for magnesium surface treatment

Dry process	Wet process
Thermal spray and cold spray coating (Al/SiC, Al-Si)	Electroplating (Zn, Cu)
Laser surface cladding (WC,TiC)	Electroless plating (Nickel)
Physical vapour deposition (PVD) (TiN, Al_2O_3)	Anodization (HAE, Dow 17)
Chemical vapour deposition (CVD) (Al coating)	Plasma electrolytic oxidation (Tagnite, Keronite)
Solid diffusion coating (Al powder coating)	Sol-gel coating (silica, ceria)
	Organic /polymer coating (painting, powder coating)

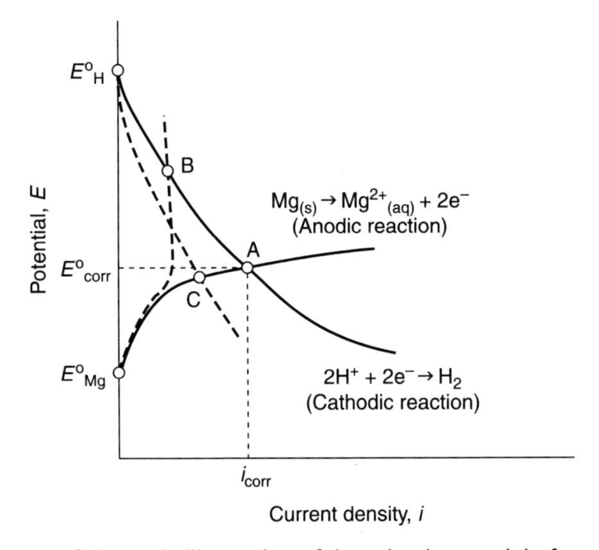

8.2 Schematic illustration of the mixed potential of magnesium.

dry methods. Dry methods are usually environmentally friendly, and can be used to achieve a decorative appearance and accurate coating. However, use of such methods is limited because of their high cost and inability to treat parts with complex shape and large size.

By contrast, wet methods are economical, easy to use and offer various valuable surface properties. Figure 8.3 presents the wet processes most widely used commercially to protect magnesium alloys against corrosion in terms of film thickness and the film-formation rate. Chromate conversion coating is characterized by its ability to produce a very thin, uniform and reliable coating with high productivity. This coating is widely used in

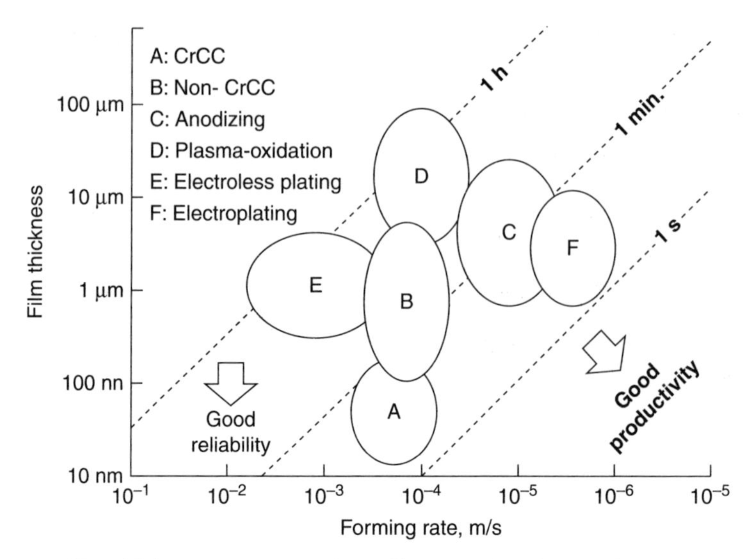

8.3 Film thickness on magnesium alloys produced by various wet processes as a function of the film-formation rate.

industry because of its desirable properties: it provides excellent corrosion protection through the corrosion-inhibiting properties of the hexavalent chromium present in the film and through the physical barrier of the film itself. Other attractive properties of this type of coating include its adhesiveness, uniformity, hardness and wear resistance.

In this type of treatment, magnesium is oxidized to magnesium ions, and chromium is partially reduced to its trivalent state [8.2]. The magnesium ions produced react further with the chromate and form a protective film on the magnesium surface [8.3] and [8.4] (Skar and Albright, 2002).

$$3Mg + Cr_2O_7^{2-} + 14H^+ = 3Mg^{2+} + 2Cr^{3+} + 7H_2O \quad [8.2]$$

$$2Mg^{2+} + Cr_2O_7{}^{2-} + 2H_2O = 2MgCrO_4 + 2H^+ \quad [8.3]$$

$$2Cr^{3+} + Cr_2O_7^{2-} + 3H_2O = 2Cr(OH)\left(CrO^4\right) + 4H^+ \quad [8.4]$$

However, changing environmental regulations and pollution prevention requirements have led to a significant push to find a new alternative to the use of poisonous Cr^{6+}, with increasing attention directed towards developing an environmentally friendly surface treatment with desirable properties. Ideally, the surface treatments must have compatibility and adhesiveness with substrate metals, as well as a ‘function’ and ‘response’ to the external environment. As non-chromate conversion coatings produce a thicker film with a lower formation rate compared with chromate conversion coatings, a great deal of research interest has been directed towards developing non-chromate conversion coatings such as stannate, phosphate, zinc/phosphate, phosphate/permanganate, manganese/phosphate, cerium and rare earth conversion coatings. In particular, Shashikala *et al.* (2008) compared the anticorrosion properties of various chemical conversion coatings, including cerium oxide, stannate, chromate and galvanic black anodizing coatings on AZ31B magnesium alloy. In polarization and electrochemical impedance studies, the corrosion resistance of the coatings was found to have the order galvanic black anodizing > chromating > cerium oxide coating > stannate coating. Electroplating and electroless plating have been applied to magnesium alloys because of the possibility of producing a multifunctional, variously physiognomic and economical coating (Li-ping *et al.*, 2010). They reported that most electroless baths contain toxic chemicals such as chromium, cyanide and fluoride with short service life. So, more research is needed to develop environmentally friendly solutions. Furthermore, it is difficult to get a uniform coating for the complex shapes and small diameter holes in electroplating.

Anodizing is recognized as one of the most promising surface treatments for magnesium alloys. Anodizing can produces a relatively thick, dense, hard, adherent, abrasion-resistant and durable film to improve one or more surface properties, including chemical, mechanical, electrical or optical properties. Anodizing treatment can also be used to achieve a number of cosmetic effects, either with thick porous coatings that can absorb dyes or with thin transparent coatings that add interference effects to reflected light. Anodizing is also used to prevent galling of threaded components and to make dielectric films for electrolytic capacitors. It is generally accepted that the best anticorrosion properties for magnesium and its alloys is achieved by anodizing (Ostrovsky, 2005; Zhang *et al.*, 2008).

8.2 Anodization treatment

Anodizing treatment is an electrolytic oxidation process in which the surface of a metal is converted to a film with desirable protective, decorative or functional properties. The process is called 'anodization' because the metal to be treated serves as the anode of an electrical circuit. The process has been applied to various metals and alloys, among them steel alloys, Al alloys, Ti alloys and Zn alloys. According to Hugh (1974), the anodizing treatment was first disclosed in a patent applied for in 1923 by G. D. Bengough using chromic acid solution. Shortly afterwards, in 1924, a Tokyo firm applied for intellectual property protection of a process of aluminum anodizing in solutions of oxalic acid. Then, in the mid-1940s, the Dow Chemical Company developed Dow 17, the first anodized magnesium coating. Since World War II, anodizing treatment has developed considerably because of the greater availability and use of the metal and the expiry of earlier patents. The HAE anodizing treatment, developed by Harry A. Evangelides in the mid-1950s, is also widely used (Manoj and Nai, 2010). The second half of the twentieth century saw great interest in magnesium anodizing, with much research carried out and many patent applications lodged. In recent years, anodizing in an environmentally friendly electrolyte has received considerable attention. In the latter part of this chapter, we present the most important researches and patents in the field of magnesium anodizing.

8.2.1 Main differences between aluminum and magnesium anodizing

Aluminum is among the metals which exhibit a sharp increase in corrosion rate with increasing pH. The corrosion rate of aluminum alloys increases very dramatically, changing by almost two orders of magnitude between pH 8 and 10. This increase is virtually independent of counter-ion and can be attributed to the formation of soluble aluminum hydroxide products (Tomashov, 1966; Pourbaix, 1974; Vujicic and Lovrecek, 1985). Consequently, the Al surface requires additional surface treatments such as conversion coating, painting, chemical vapor deposition or physical vapor deposition to improve its corrosion resistance and other surface properties. However, Al metal usually does not have good adherence to the coatings obtained with these kinds of treatment (Yakovleva *et al.*, 2002). Therefore, anodic oxidation in certain electrolytes is required to form porous anodic films that should have good adherence for further coating layers. Aluminum anodizing is an electrochemical method of converting aluminum into aluminum oxide (Al_2O_3) at the surface of the item being coated. Although several metals can be anodized, including aluminum, titanium and magnesium, only aluminum anodizing has found widespread use in industry; it is used in almost every industry that employs aluminum because of the wide variety of coating properties that can be produced through variations in the process (Pernick, 1987). Aluminum can be anodized in a wide variety of electrolytes, employing varied operating conditions including the concentration and composition of the electrolyte, presence of additives, temperature, voltage and current. The metal and many of its alloys are anodized in such acids as boric (H_3BO_3), oxalic ($H_2C_2O_4$), phosphoric (H_3PO_4) and sulfuric (H_2SO_4) under conditions whereby an oxide forms on the surface. The oxides that form in H_3BO_3 are relatively thin and non-porous in nature, but they are not suitable for protecting aluminum against corrosion in aggressive environments. The oxides formed in H_3PO_4 or H_2SO_4 are thicker and very porous. The porosity is reduced in a second step in which the oxide

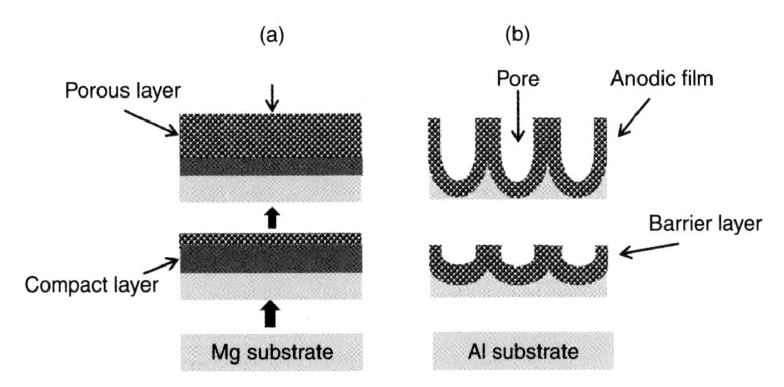

8.4 Schematic illustration of the anodic film structure produced on magnesium (a) and aluminum (b) alloys.

is sealed in steam, boiling water or an aqueous solution containing nickel acetate ($Ni(C_2H_3O_2)_2$) or other salts (Leidheiser, 1987).

Magnesium alloys may be subjected to anodizing treatment in the same way as aluminum alloys; however, it is not possible to obtain the same level of protection as for aluminum-based alloys, because the degree of coverage for MgO is 0.81 (i.e., the Pilling–Bedworth ratio is less than 1). For this reason, the anodic oxide coating that forms on magnesium alloys is porous and rough, with a coarse surface structure. This coarse structure permits corrosive substances to infiltrate down to the base metal and cause its deterioration. The anodic finish, once applied, must therefore be sealed to make it corrosion-proof. A further property of magnesium oxide finishes is their relative lack of hardness in comparison with aluminum oxide, which negatively affects the wear resistance (Kurze, 2006). The schematic diagram in Fig. 8.4 shows the growth of the anodic films on magnesium and aluminum alloys. Figure 8.4a shows that the anodic film on magnesium consists of a compact layer and a porous layer. The growth of the anodic film on the aluminum alloy is shown in Fig. 8.4b. The thickness of the anodic film and the diameter of the pores are dependent on the process conditions. Table 8.2 summarizes some differences between aluminum and magnesium anodizing.

8.2.2 Procedures of magnesium anodizing

Pre-treatment processes

The procedures for magnesium anodizing are presented in Fig. 8.5. The pre-treatment processes, including both mechanical and chemical methods, are important in creating a work piece with the desired surface characteristics. Mechanical methods employed in pre-treatment include grinding, polishing,

Table 8.2 Differences between aluminium and magnesium anodizing

	Magnesium anodizing	Aluminium anodizing
Naturally formed oxide layer	The naturally formed oxide layer on Mg surfaces is very loose and cannot offer an effective resistance to corrosion.	The naturally formed oxide layer on Al surfaces is stable and provides little corrosion resistance under normal atmospheric conditions.
Electrolyte pH	High pH. Alkaline solutions preferred.	Low pH. Acidic solutions preferred.
Electrolyte Temperature	Thickness decreases at high temperature and a compact film is obtained.	Thickness decreases at high temperature and the film becomes porous with bad mechanical properties.

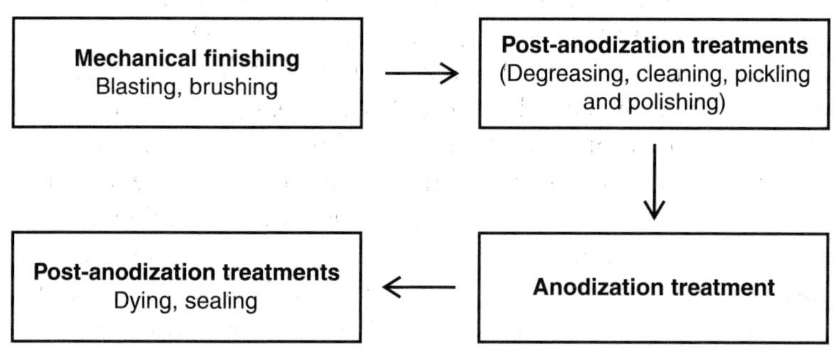

8.5 Steps in magnesium anodizing.

buffing, blasting and brushing; however, dry blasting is usually avoided because of the cathodic-particle contamination that arises when employing typical blasting media (Habashi, 1998). Chemical pre-treatments such as degreasing and pickling are employed to remove oxides, oil, impurities and any undesired materials from the substrate surface before anodizing treatment. Alkaline degreasing, organic degreasing, alkali pickling and acid pickling are typical anodizing pre-treatment methods for magnesium alloys.

Anodic oxide film formation

The anodizing step is the main part of the process. Whether other steps are carried out or not depends on the required specifications of the anodic films. In contrast to the case for chemical conversion, the properties of anodic coatings depend on several factors, such as the composition of the substrate, applied voltage, electrolyte composition and electrolyte temperature. Anodizing treatment can be accomplished by controlling either the voltage

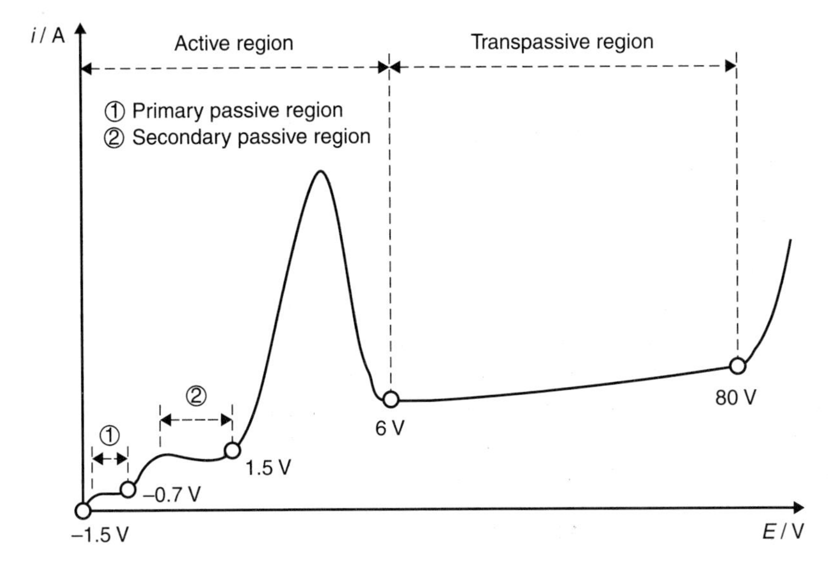

8.6 Anodic polarization curve of AZ31 magnesium alloy in 1 M NaOH alkaline solution.

or the current. It has been found that the use of pulsed current is critical and, as the rate of anodic film formation is low when alternating current (AC) is applied, direct current (DC) is preferred (Dolan, 2003). Under voltage control, the current drops with treatment as the insulating oxide film grows. Under current control, the voltage increases with treatment time to maintain the applied current density while the isolating oxide film grows. The anodic behavior of Mg in NaOH solutions in a wide range of potentials was first described in the studies of Emley, Huber and Evangelides, cited in Kim and Okido (2003). Huber showed the relationship between the applied voltage and the characteristics of the anodic film that formed on Mg in 1 M NaOH. At voltages of up to 3 V, the current density remained low, and a light-grey protective film of $Mg(OH)_2$ formed. At intermediate voltages (i.e. 3–20 V), oxygen evolved and a thick, dark film of $Mg(OH)_2$ was found. Above 20 V, a thin protective coating was again produced. The formation of a compact anodic film has been shown to be limited by the breakdown phenomenon accompanied by intensive sparking (above 50 V). Similar behavior was described by Yaniv and Shick and later by Zengnan *et al.* when anodizing Mg in fluoride solutions and by Takaya when anodizing Mg-Mn alloy in potassium hydroxide (KOH) solution (Khaselev and Yahalom, 1998).

Figure 8.6 shows the anodic polarization curve of AZ31 magnesium alloy in NaOH alkaline solution. The curve can be divided into two parts: an active region and transpassive region. The active region can be further divided into a primary passive region and secondary passive region. The anodic dissolution of Mg alloys begins at –1.55 V *vs.* Ag/AgCl reference electrode and the current density increases with the anodic overpotential. This region mainly corresponds to the anodic process, that is, the formation of the anodic oxidation product (Mg^{2+}). The magnesium dissolution and the formation of magnesium ion occur according to reaction [8.5].

$$Mg \rightarrow Mg^{2+} + 2e \qquad [8.5]$$

The rate of increase in current density of this active–passive metal is significantly limited by the shift of the anodic overpotential in the more positive direction because of the initial formation of a passive film on the electrode surface. Film formation progressively reduces the oxidation current and finally leads to passivation at −1.2 to −0.7 V (primary passivity). The Mg^{2+} produced from the dissolution reaction reacts with the hydroxyl ions (OH^-) to form $Mg(OH)_2$, as described by reaction [8.6].

$$Mg^{2+} + 2OH \rightarrow Mg(OH)_2 \qquad [8.6]$$

However, the current density subsequently increases with anodic overpotential at an applied potential at −0.7 V because of the breakdown of the primary passivation film. Subsequently, the alloy becomes passive once again (secondary passivity) at 0.2–1.5 V. This secondary passive state occurs through the direct oxidation of MgO on the electrode surface (Cai *et al.*, 2009).

$$Mg + H_2O \rightarrow MgO + 2H^+ + 2e \qquad [8.7]$$

With the potential increasing above 1.5 V, the current density increases again because of oxygen evolution on the electrode surface, which can be clearly observed with the naked eye. Thick and rough anodic film forms at 3 V, and the surface becomes porous at an applied potential of 6.2 V. With a further increase in the potential, the anodic current remains almost constant because of the equilibrium of dissolution and film formation. At a high applied potential of 80 V, the temperature at the specimen surface increases and the following reaction occurs at the film/electrolyte interface.

$$Mg(OH)_2 \rightarrow MgO + H_2O \qquad [8.8]$$

Feitknecht and Braun (1967) demonstrated that these structures can be converted into each other by either hydration or dehydration. Anodizing is accompanied by intensive sparking and oxygen evolution. Therefore, the following reaction occurs at the film/electrolyte interface.

$$Mg(OH)_2 \rightarrow Mg^{2+} + \frac{1}{2}O_2^- + H_2O + 2e^- \qquad [8.9]$$

Post-treatments

Anodic film properties such as porosity, corrosion resistance, wear resistance and color can be achieved by applying an appropriate seal or dye. Coloring of anodized films can be achieved by employing one of the following methods (Mittal, 1995; Brace, 1997; Gray and Luan, 2002).

- Absorption of organic or inorganic dye into the film after anodizing.
- Electrolytic deposition of inorganic metal oxides and hydroxides into the anodic pores. This process can be carried out with a combination of organic dyes, whereby metal is deposited deep into the pores electrolytically and dye is then distributed throughout the coating nearer the surface.
- Integral color anodizing by adding organic constituents to anodizing electrolyte that decomposes during the process and produces particles that become trapped in the film as it grows.
- Interference coloring, which involves control of the pore structure to produce color through the interference of light reflected from the top and bottom of the pore.

Sealing treatment of the porous surface layer of the anodic film is necessary to achieve abrasion- and corrosion-resistant anodic films. Researchers have developed numerous sealing techniques for anodic films on magnesium alloys, the most commonly used of which are hot-water sealing, steam sealing and sodium silicate sealing. Kim *et al.* (2004) investigated the effects of the temperature of sealing solution and sealing time on the morphology and corrosion properties of AZ91 magnesium alloys and found that the number of micropores and microcracks decreased with increasing water temperature. The suggested reason for this finding is that the bonding strength of OH^- ions increases with the increasing sealing temperature of distilled water. The surface micropores and microcracks disappear with volume expansion when sealing in distilled water at 100°C, which improves the corrosion resistance. The development of a new sealing treatment for the production of high-performance anodic films is currently attracting much attention. Hsiao and Tsaia (2005) reported the effect of heat treatment on the electrochemical properties of anodic films formed on AZ91D magnesium alloy. The results show that the cooling rate affects the microstructural evolution of AZ91D magnesium alloy after solution heat treatment at 440°C for 20 h in N_2 atmosphere. The highest polarization resistance in 3.5 wt.% NaCl solution is obtained for the solution-annealed and air-cooled AZ91D Mg alloy.

8.3 Effects of anodizing parameters

The anodizing process parameters have a significant influence on the properties of the anodic films formed on magnesium alloys. The effect of the most important anodizing parameters on the anodic films properties were shown throughout this section.

8.3.1 Applied potential and current density

Salman *et al.* (2010) investigated the effect of the anodizing potential on the surface morphology and corrosion property of the anodic film that forms on AZ31 magnesium alloy. Figure 8.7 shows the surface morphologies of the anodic films that form at various applied potentials.

Anodizing at 3 V produces a very rough, grey surface, as shown in Fig. 8.7a. Anodizing at 10 V produces a porous structure, with a pore diameter of 0.84 μm, as shown in Fig. 8.7b. Figure 8.7c and 8.7d show that the surface becomes smoother as the anodizing potential increases from 20 to 70 V. Intense sparking is observed at anodizing potentials of 80 and 100 V. Initially, the film formed by sparking is a porous microstructure with circular or elliptical pores, as shown in Fig. 8.7e and 8.7f. The anodic film that forms at 100 V has a thickness of 1.2 μm and is white in color; it has the lowest hydrogen overpotential and the best corrosion resistance compared with films anodized at lower potential. Chaia *et al.* (2008) investigated the effect of applied current densitics on the microstructure and

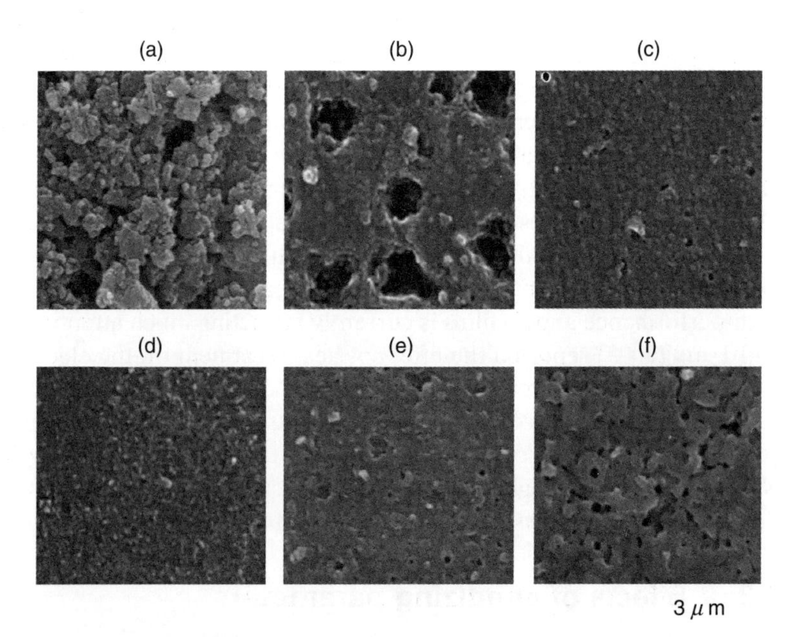

8.7 Surface morphologies of AZ31 magnesium alloy after anodizing at (a) 3 V, (b) 10 V, (c) 20 V, (d) 70 V, (e) 80 V and (f) 100 V.

corrosion properties of the anodic film that forms in alkaline solution. They found that with current density less than 20 mA/cm^2, a rough surface is obtained on the anodizing film, and when the current density reaches 20 mA/cm^2, a smooth surface is observed. With further increases of the current density, more rough areas with more pores are observed on the film surface. A possible reason is the rapid movement of sparking, leading to a greater number of pores and smaller particles. The thickness of the anodic film, especially the thickness of the compact layer, is increased by increasing the applied current, which improves the anticorrosion property. Among the selected oxysalts, sodium silicate contributes to the formation of an anodic film with the best anticorrosion property. Different concentrations of sodium silicate lead to the formation of various anodic film structures.

Shi *et al.*, (2006) concluded that the thickness, composition, and microstructure are dependent on the anodizing current density and that they affect the corrosion performance of the anodized coatings. Furthermore, the current density is ideally high at the start of anodizing so that an initial anodized coating is obtained quickly. In the later stages, when sparking becomes too intensive or too localized, a low current density is appropriate, leading to a less porous coating.

8.3.2 Effect of anodizing time

Many authors have reported that the anodizing time greatly affects the anodic film formation. However, the effect of anodizing time depends on anodizing conditions, such as the applied potential, electrolyte type and additives. Wu *et al.*, (2007) investigated the anodizing of AZ31 in NaOH with boric acid, sodium tetraborate and various additives. They found that with the prolongation of the anodizing time, the primary oxide film formed evolves into a porous outer layer and thickens with time, while a more compact inner layer simultaneously forms on the Mg alloy matrix. The current transients for AZ31 magnesium alloy in 1 M NaOH solution are shown in Fig. 8.8. The current density differs with different anodizing potentials in the range 3–100 V. At a potential of 3 V, the current density does not flow immediately when the potential is applied. With time, the current density increases because of the dissolution of magnesium and consequently the formation of the anodic film. The current density reaches its maximum value (0.2 A/cm^2) at about 90 s and then remains constant until the end of the anodizing process. By contrast, with a higher anodizing potential in the range 10–100 V, the current flows immediately and decreases with time. Anodizing at 10 or 20 V produces the lowest current density. An anodizing potential of 70–100 V produces nearly the same structure of current transients. At 100 V, the anodizing process is accompanied by intense sparking.

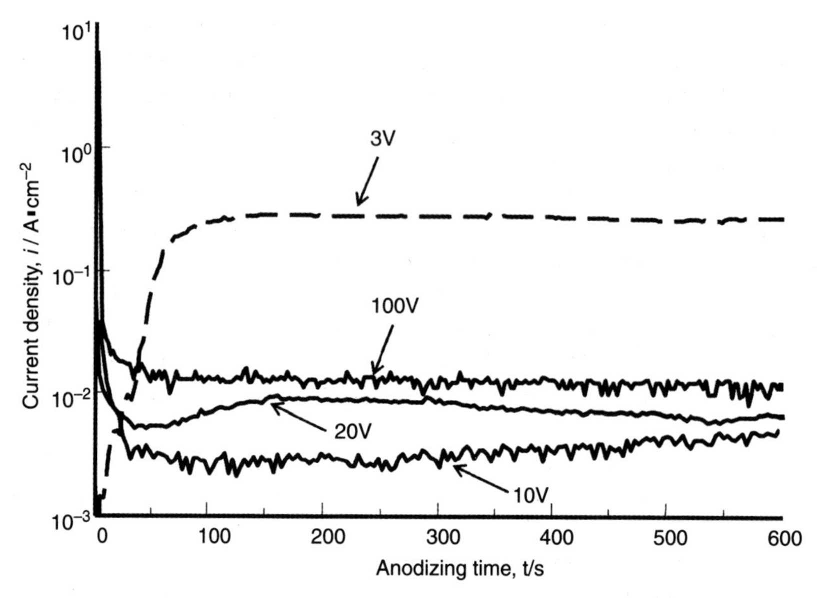

8.8 Current changes during constant-potential anodizing of AZ31 magnesium alloy in 1 M NaOH solution at 298 K.

Consequently, the current density sharply decreases and becomes approximately constant at about 30 s. The decrease in the current density value at an anodizing potential of 10–100 V is related to the formation of a more protective anodic film in comparison with the case for low-potential anodizing. Figure 8.9 shows the surface morphology and cross-section of the anodized films. The anodic film has a porous structure with pore size of about 0.2 μm. The surface morphology does not change with anodizing time, although the film thickness does. A thin anodic film partially forms at a very early point in the anodizing process, and there is then a notable increase in the film thickness. The thicknesses of the anodic films are 0.42, 0.9 and 1.9 μm at anodizing times of 0.5, 1.0 and 2 s, respectively. The thickness of the anodic films continues to increase with time, reaching a maximum value (2.3 μm) at an anodizing time of 30 s. As the anodizing time increases further from 30 to 600 s, the thickness of the film decreases to 1.22 μm. The anticorrosion property was examined employing the anodic polarization technique. Table 8.3 summarizes the E_{pit} values of treated and untreated specimens. The pitting potential of all anodized specimens is less negative than that of the untreated specimens. This indicates that the anticorrosion property of anodized specimens at different times is better than that of untreated specimens. E_{pit} values of specimens after treatment for 2, 30, 200 and 600 s are very close, which indicates that there is no notable change in the anticorrosion property after anodizing for 2 s.

Table 8.3 Pitting potential determined from the anodic polarization curve at various times

Anodizing time (s)	E_{pit} *vs.* Ag/AgCl reference electrode
0.0	−1.56 V
0.5	−1.52 V
1.0	−1.52 V
2.0	−1.43 V
30	−1.47 V
200	−1.4 V
600	−1.4 V

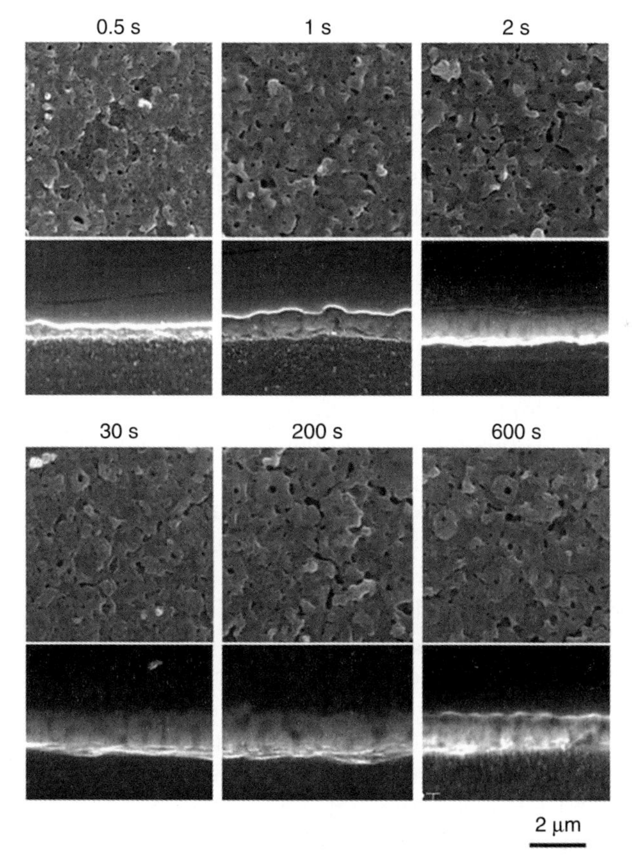

8.9 Surface morphologies and cross-sections of AZ31 magnesium alloy after anodizing for various times at 100 V.

8.3.3 Solution temperature

Increasing the electrolyte temperature decreases the microhardness and wear resistance of the anodic oxide film on an aluminum alloy (Aerts *et al.*, 2007). This is mainly attributed to the increase in porosity in the outer region of the oxides. Tomita (1980) also noted that when the temperature rises above 30°C, the film formation rate decreases and a powder layer forms on the surface because of the redissolution of the formed film or substrate metal. This occurs because, while the film formation rate increases with the rise in temperature, the rate of dissolution also increases. At low temperatures (below 5°C) a dense and poorly stainable film forms, the limited applications of which mean it is not desirable.

In the case of magnesium anodizing, Sharma *et al.* (1993) reported that both the current density and the voltage tend to be higher at higher operating temperatures because of higher conductivity and an increase in the rate of dissolution of the alloy in the electrolyte at higher temperatures. The film-formation rate was found to increase with increase in the electrolyte temperature and applied current density for electrolyte composed of ammonium bifluoride, sodium dichromate and orthophosphoric acid (Sharma *et al.*, 1997). The relative increase in film growth, however, is rapid in the initial state and slowdown gradually, approaching an almost constant value at electrolyte temperatures above 82°C and applied current densities above 50 A/ft. Zhang *et al.* (2002) examined the effect of the electrolyte temperature for KOH with carbonate, silicate and borate additives. The results can be explained by the conclusion that anodic film formation and dissolution both increase with an increase in the solution temperature. However, the increase in the dissolution rate is dominant. Therefore, the destruction of the old film occurs more rapidly than the formation of the new film, leading to a reduction in the thickness of the anodic film.

Chaia *et al.* (2008) studied the effect of the solution temperature on the anodic film that forms on AZ31 magnesium alloy in alkaline solution with borate and silicate additives. The increase in solution temperature has a negative effect on the anticorrosion property. The poor corrosion resistance is attributed to a decrease in the film formation rate and the formation of large holes during the sparking process. On the other hand, the effect of the solution temperature on the anodic film that forms in alkaline-based solution with calcium hydroxide additive at low potential was reported by Salman *et al.* (2008). Figure 8.10 shows that by increasing the temperature, a smooth and compact anodic film can be obtained. The increase in temperature affects the formation and crystallization of magnesium hydroxide, thus improving the anticorrosion properties of the anodic film. In conclusion, the solution temperature greatly affects the formation rate and porosity of the film. Increasing the solution temperature during spark-free anodizing could

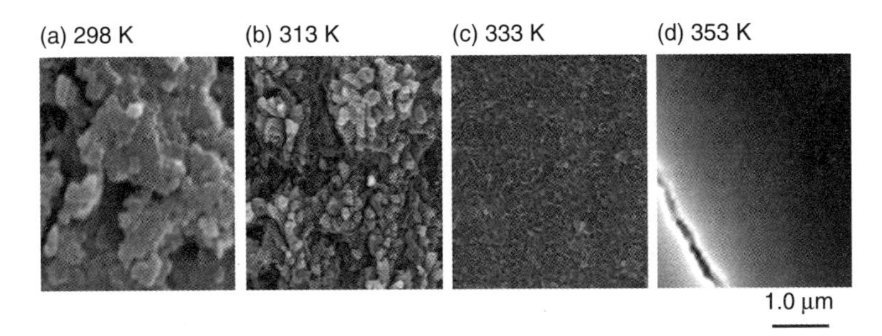

8.10 (a–d) Surface morphologies after anodizing for 30min at 3V in 1M NaOH with 35mM $Ca(OH)_2$ solution at different temperatures. S. Salman, R. Ichino and M. Okido, 'Influence of calcium hydroxide and anodic solution temperature on corrosion property of anodizing coatings formed on AZ31 Mg alloys', Surface Engineering, 2008. Reproduced with permission from Maney Publishing.

improve the surface morphology and anticorrosion property. On the other hand, sparking anodizing may have a negative effect on the anticorrosion property of the anodic film.

8.3.4 Electrolyte type and additives

Throughout the anodizing process, the content of electrolyte plays an important role. The effects of oxysalts such as Na_3PO_4, $NaAlO_2$, Na_2MoO_4, Na_2SiO_3 and $Na_2B_4O_7$ on the surface morphology and structure of the anodic films that form on AZ31 magnesium alloy in alkaline solution are shown in Fig. 8.11. During anodizing in 1 M NaOH solution, the anodic films grow preferentially in a certain direction. Adding oxysalts affects this phenomenon. Sodium phosphate, sodium aluminate and sodium borate have similar effects on the surfaces of films. The surface structure of the anodic film formed in the solutions containing these three oxysalts are smooth with deep cracks. The addition of molybdate produces a rough anodic film with many little holes. With anodizing in a bath solution containing sodium silicate, irregular movements of sparks were observed. These spark lead to formation of a white anodic film with rough surface.

Lin and Fu (2006) investigated the effects of F^- and PO_4^{3-} ions on the anodizing behavior and microstructure of AZ31 magnesium alloy. The addition of F^- induces intense sparking resulting from the dielectric breakdown of a compact layer composed of fluorine, oxygen and magnesium. The fluorine species are detected mainly in the compact layer adhered to magnesium substrate. PO_4^{3-} ions have a minor effect on the potential response

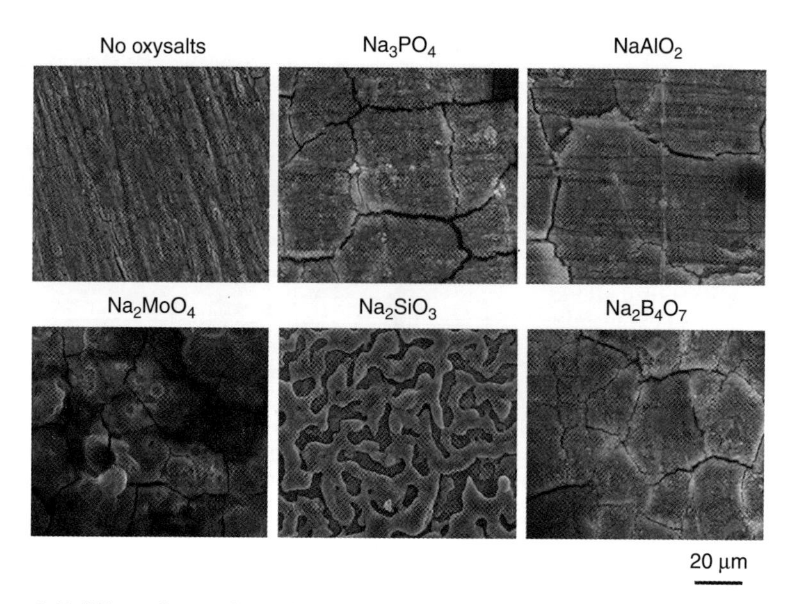

8.11 Effect of oxysalt additives on the surface morphology of the anodic film that forms on AZ31 magnesium alloy.

of the cell. However, they form a thicker magnesium hydroxide film. In spark-free anodizing, phosphorus species are distributed uniformly in the anodic film, presumably in the form of $Mg_3(PO_4)_2$. However, in a solution in which sparking occurs, PO_4^{3-} ions in the solution promote the formation of numerous caves. The phosphorus species are detected mainly in the walls and bases of such caves.

The effects on the anodic film properties of $Al(NO_3)_3$ and/or Na_2SiO_3 added to 3 M KOH + 0.21 M Na_3PO_4 + 0.6 M KF solution were investigated by Hsiao, Tsung, and Tsai (2005). Their results show that the anodic film is mainly composed of MgO. Adding Na_2SiO_3 and/or 0.15 M $Al(NO_3)_3$ enhances the polarization resistance of the anodic film because of the formation of Al_2O_3 and $MgAl_2O_4$ in the silicate-containing anodized film, accompanied by silicate formation.

Organic electrolyte, which differs from usual aqueous solutions, was used to improve anodic film properties. Asoh and Ono (2003) investigated the anodizing of magnesium in 1 M/dm^3 triethylamine/ethylene glycol solution. The water content significantly affects the growth behavior, appearance and incorporated organic species of anodic films. In the case of electrolyte having water content below 10% or above 50%, the appearances are yellowish white and whitish grey, respectively. In the water-content range between 10% and 40%, compact barrier-type films are obtained regardless of substrate purity. The surface appearance of the obtained films is transparent and enamel-like, in contrast to the films that usually form in aqueous solutions. Incorporating organic species improves the anticorrosion property of the anodic film.

8.4 Commercial magnesium anodizing processes

Various anodizing technologies have been developed to protect Mg alloys from corrosion. The electrolytes employed are mainly aqueous solutions of inorganic substances, used individually or in mixture. Anions such as aluminate, silicate, vanadate.etc. can be used as electrolyte additives in order to enhance the properties of the anodic films. Other methods use a strong base substance such as sodium hydroxide or potassium hydroxide in order to produce a large amount of the alkaline hydroxyl (OH^-) ions required to perform the anodic reaction. In this section, some of the most commonly applied anodizing methods were described.

8.4.1 Galvanic anodizing

Galvanic anodizing (Dow 9 process) produces a protective layer on all types of magnesium alloys. The color of the anodic films varies from dark brown to black and has good characteristics for a paint base with no appreciable dimensional change. Because the coating makes an excellent paint base, itis often used for cameras, optical devices and electrical components, as well as other components that require a non-reflective surface. Sharma *et al.* (1993) reported that a galvanic anodic coating of Mg-Li alloy produced a highly corrosion-resistant film with excellent adhesion in a Scotch tape peel test.

Caustic anodizing treatment has long been used for all magnesium alloys that do not require close dimensional tolerance. The method is as follows. The electrolyte is composed of 240 g/l NaOH, 83 ml/l $(HOCH_2CH_2)_2O$, and 2.5 g/l $Na_2C_2O_4$. The temperature of the solution should be between 73.9°C and 79.9°C. Anodizing is carried out for 15–25 min using 6–24 V AC or 6 V DC. The film produced is immersed in solution containing sodium acid fluoride and sodium dichromate.

8.4.2 Dow 17 process

Dow Chemical has established many surface treatments for protecting magnesium alloys against corrosion including Dow 1 (chrome pickling), Dow 7 (dichromate application) and Dow 9 (galvanic anodizing) (Licari, 2003). However, Dow 17 is the most versatile of the surface treatments established by the company. It was developed in the mid-1940s and can be applied with either alternating or direct current to all forms and alloys of magnesium. The color of the Dow 17 coating varies between light and dark green depending on the coating thickness. This process produces a two-phase, two-layer coating. The first layer is deposited at a lower voltage and results in a thin (approximately 5 μm), light-green coating. The overlayer forms at a higher voltage, is thick (approximately 30.4 μm) and dark green with good abrasion resistance, good paint-base properties and corrosion resistance. The corrosion resistance of AZ91D alloy treated with this technique was subjected to a three-year atmospheric exposure test, and superior corrosion resistance compared with tested conversion coatings was reported (Umehara *et al.*, 2000).

The composition, structure and growth mechanism of anodic films on pure magnesium and magnesium alloy AZ91D produced employing the Dow 17 process have been studied by Ono (1998). The study confirmed that the film has a thin barrier layer at the interface with an overlying cylindrical-pore structure. The authors assume that film growth on pure magnesium proceeds by the formation of magnesium fluoride and magnesium oxyhydroxide at the metal–film interface and dissolution of the film at the base of the pores. The crystallization of magnesium fluoride and $NaMgF_3$ proceeds in the overlying porous layer. The films that form on AZ91D alloy are uneven, most likely because of the presence of intermetallic Mg-Al particles located at grain boundaries and the surface porosity. However, the growth mechanism of the coating is similar and a porous cell structure with crystalline particles of MgF_2 and $NaMgF_3$ was also found in these films. Nevertheless, the chromate additive in the electrolyte will restrict the future use of this treatment. Figure 8.12 shows scanning electron microscope photographs of the surface of the coating layer formed by Dow 17, HAE and Tagnite coatings. These photographs provide an excellent comparison of pore sizes on each of the coatings.

8.4.3 HAE process

The HAE process, which was named after its inventor Harry A. Evangelides, was first published by Pitman–Dunn Laboratories, Frankfort Arsenal, Philadelphia, USA, in 1950. The HAE treatment is effective for all forms and alloys of magnesium alloy. Anodizing treatment is carried out in aqueous alkaline solution with aluminum oxide and phosphate, fluoride and manganite ions. The treatment produces a two-phase coating as in the Dow 17 process. A light subcoating film with thickness of 5–10 μm is produced at low voltage. At higher voltage, a dark brown, thicker film of 25–80 μm is produced (Kurze, 2006). Both thick and thin HAE anodic layers have a porous structure. The thick HAE coating is very rough and relatively hard, but it is also very brittle. Both thick and thin layers have resistance to corrosion and abrasion. Upon sealing, the HAE treatment provides an excellent anticorrosion property. The dark brown coating is hard with good

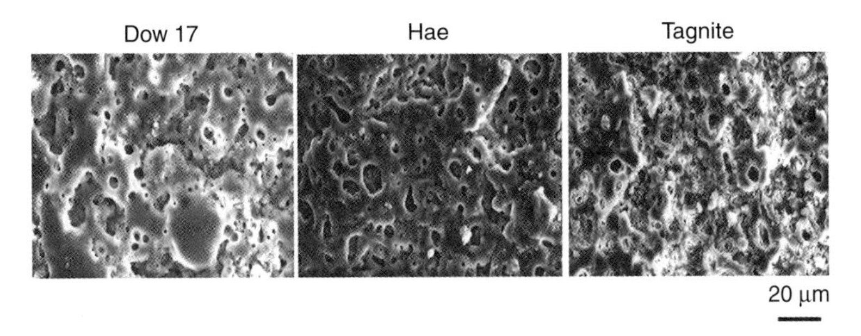

8.12 Surface morphologies of the coating layer on ZE41A magnesium alloy formed by the Tagnite, HAE and Dow 17 coatings. Reproduced with permission from Tagnite®.

abrasion resistance. However, it can adversely affect the fatigue strength of the underlying magnesium, particularly if the underlying layer is thin. The anticorrosion property of AZ91D treated with this technique is superior to those of conversion coatings.

8.4.4 Tagnite treatment

Technology Applications Group invented the Tagnite Coating System, a chromate-free anodic treatment that can be used for all types of magnesium alloys (Kurze, 2006). The coating of the magnesium alloy is carried out in a two-step process. In the first step, magnesium alloy is immersed in an aqueous solution of magnesium fluoride and/or magnesium ammonium fluoride. The second step is anodic plasma treatment in an aqueous electrolytic solution of hydroxides, fluorides, fluorosilicates and silicates having a pH of at least about 12.5. This process provides a superior coating with higher abrasion and corrosion resistance compared with the HAE and Dow 17 processes (Bartak *et al.*, 1995). The process uses high voltages above 300 V, which lead to the formation of thicker porous coatings (Eltron Research & Development, 2009). The coating has a white to grayish-white appearance, and consists of crystalline MgO, $MgSiO_3$, and MgF_2. Two classes of layers (5–10 μm and 20–25 μm) can be produced through the Tagnite treatment (Kurze, 2006). Although this anodizing process is much less hazardous than the HAE and Dow 17 processes, concerns persist about the use of fluorides, which are a potential environmental and health hazard (Eltron Research & Development, 2009).

8.4.5 Anomag process

The Anomag process was developed and licensed by Magnesium Technologies Licensing, Ltd (Auckland, New Zealand) (Beals *et al.*, 2003).

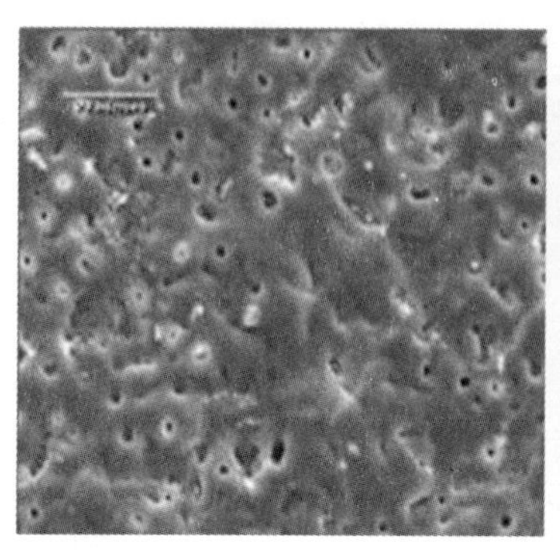
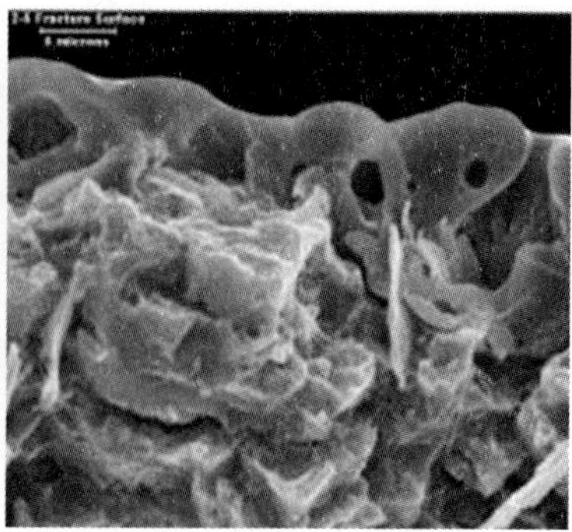

8.13 Surface morphology and cross-section of the coating layer on magnesium alloy formed by Anomag coating. Reproduced with permission from Anomag®.

The electrolyte used in this treatment is composed of ammonia and sodium ammonium hydrogen phosphate ($NaNH_4$ $HPO_4.4H_2O$) (Barton, 1998). Three classes of layers (3–8 μm, 10–15 μm and 20–25 μm) can be produced with a gentle spark-free plasma discharge on the substrate. The coatings produced are characterized by good corrosion resistance, wear resistance and fatiguc behaviors, and low cost (Kurze, 2003). Saijo *et al.* (2005) investigated the the corrosion resistance and structure of the Anomag coating and Dow 17 coating. The Anomag coating consists of two layers. The lower layer is compact and is formed at the boundary between the base metal and the coating. The upper layer has an irregular porous microstructure, shown in Fig. 8.13. The report suggested that the excellent corrosion resistance of the Anomag coating is attributed to this thin compact layer. This anodizing process is relatively environmentally friendly, ammonia is compound although there are concerns about the naturally produced ammonia gas, which must be removed from the exhaust gas and wastewater produced (Kurze, 2006).

8.4.6 Magoxide-coat treatment

The Magoxide process was invented in Russia and developed by AHC (AHC Oberflächentechnik GmbH) in Germany (Abyazani, 2011). The Magoxide process produces a hard anodic coating on magnesium base substrates using an external power supply to produce discharges in alkaline solutions mainly consisting of phosphate, borate, silicate, aluminate or fluoride anions. The oxygen plasma generated causes partial short-term surface melting and ultimately the formation of an oxide-ceramic layer (Kruze *et al.*, 1995) Various film thicknesses can be produced using this method; for example, thin films (5 μm) provide an excellent primer base for paint, and thicker films (25–30 μm) have excellent corrosion resistance and very good wear resistance

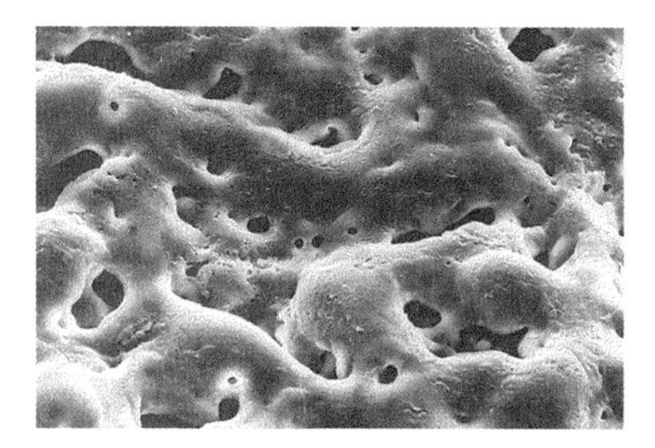

8.14 Surface morphology of the coating layer on magnesium alloy formed by Magoxide coating. Reproduced with permission from Springer Science & Business Media, © 2006, Kurze, P., Corrosion and surface protection. In H. E. Friedrich and B. L. Mordike, Magnesium Technology.

(Kurze, 2006). The coating consists of three layers: a thin layer lying directly on the metal surface; a middle layer with compact structure acting as a barrier against corrosion; and an outer porous layer that is a good base for post-treatments such as painting and polymer coating (Gray and Luan, 2002). Figure 8.14 shows the SEM image of this outer porous layer.

8.4.7 Keronite surface treatment

The Keronite process was invented in Russia and released commercially by Keronite Ltd (Cambridge, UK). This process applies an alternating pulsed voltage high enough to produce a plasma discharge in an aqueous solution on the surface of light metals such as aluminum, magnesium and titanium (Shelley, 2009). Complex compounds can be synthesized inside the high-voltage breakthrough channels that form across the growing oxide layer. These compounds are composed of oxides of both the substrate material and electrolyte-containing elements (Yerokhin *et al.*, 1999). Figure 8.15 shows that the Keronite surface seems smooth with less microporosity and the cross-section image shows that the anodized film adheres to the substrate with a close interfacial connection.

These complex mixtures of ceramic phases are highly resistant to a wide range of chemicals, and thus provide excellent corrosion protection for magnesium alloys. The coatings show excellent corrosion protection in the salt spray test according to ASTM B117 for 2000 h. Keronite treatment can provide magnesium with a surface that is harder and more wear-resistant than the surface of magnesium produced by hard anodizing and even than the surfaces of common hard materials such as hard steel, sand and glass

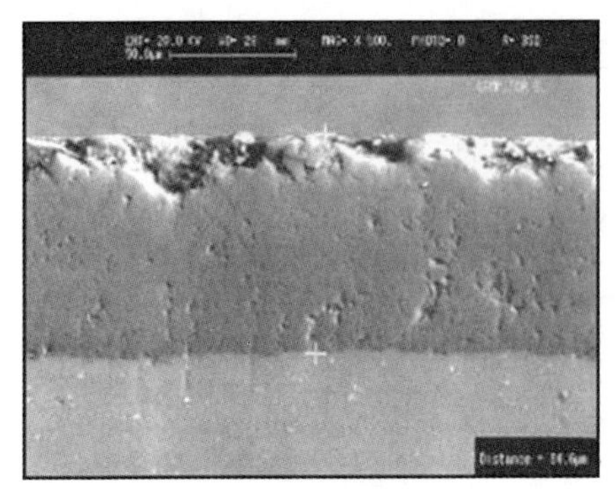

Keronite cross section

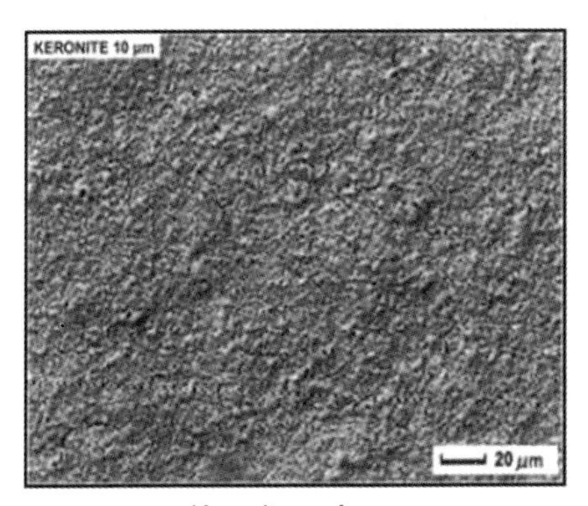

Keronite surface

8.15 Surface morphology and cross-section of the coating layer on magnesium alloy formed by Keronite. Reproduced with permission from Keronite®, Keronite copyright © 2011.

(Keronite, 2011). The improved Keronite process provides dense and uniform ceramic oxide layers with a fine-grain microstructure, which has high fatigue strength. However, the fatigue strength of the base metal can be reduced by hard anodizing, especially in the case of thicker films (Ghali and Revie, 2010).

8.5 Other methods and recent developments in anodizing treatment of magnesium

Many patents related to the anodizing of magnesium alloys have been approved in recent years. (Kozak, 1980) investigated the anodizing treatment for alkali metal silicate alone or for an admixture with an aqueous alkaline solution (e.g., sodium potassium or lithium hydroxide). After pickling in hydrofluoric acid for 3–10 min, anodizing is carried out at 150–350 V for 1–5 min. The anodic coating has greater resistance to corrosion compared with coatings produced with previously developed anodizing processes. An anodizing process for the formation of a chemically stable and hard spinel compound of MgO-Al_2O_3 on magnesium surfaces was reported in a patent by Kobayashi and Takahata (1985). The anodizing bath consists of alkali hydroxide, aluminate and at least one boron, phenol, sulfate or iodine compound. The coating is white in color; however, dyes can be strongly adsorbed by the surface and the resulting color is highly resistant to decolorization.

The anodic oxidation coating that forms on the surface of magnesium or its alloys when using the anodizing solution contains silicate, carboxylate and alkali hydroxide has a superior combination of properties, especially with respect to corrosion resistance, abrasion resistance and ornamental properties, as compared with previous anodic oxidation coatings. Other additives to the previously mentioned components further improved the properties of the coating. Table 8.4 lists these additives along with the effects of insufficient or excess amounts of each component on the anodizing treatment (Kobayashi *et al.*, 1988).

The use of an electrolytic solution containing ammonia was reported in a patent by Barton (1998). The use of such an electrolytic solution alters the manner in which the anodizing occurs to provide a coating on the magnesium material without spark formation. The additives aluminate, phosphate and fluoride produce magnesium aluminate, magnesium phosphate and magnesium fluoride, respectively. Anodizing using ammonium phosphate compounds gives significant corrosion resistance to the coating. However, because of the volatile and corrosive nature of ammonia, special care and appropriately designed equipment are required to prevent the ammonia escaping from an electrolytic bath into the workplace (Dolan, 2003).

Anodizing using pulsed current and relatively low average voltages has been carried out in various solutions. The anodizing solutions employed may contain phosphate, permanganate, silicate, zirconate, vanadate, titanate, hydroxide, alkali metal fluoride and/or complex fluoride, optionally with other components present. Anodizing has been carried out in an aqueous solution of alkali hydroxide, sodium silicate, and ethylene glycol at constant current density. The resulting anodic film consists of a porous layer and a barrier layer. Insertion of carbon, including magnesium and silicate, into the anodic film leads to excellent anticorrosion properties, which are 10 times better than those of the anodic film produced employing the HAE method (Akihiro and Sakai, 2009; Yabuki and Sakai, 2008). Anodizing has also been carried out using 2–12 A/dm^2 DC in an aqueous solution comprising water-soluble inorganic hydroxide, phosphorus, oxygen-containing anions, and surfactant with other additives. The anodic coating was found to have a corrosion resistance of less than 1% area of corrosion on a flat surface after 560 h exposure in a 5% NaCl salt spray test according to ASTM B117 (Ostrovsky, 2010). Kim *et al.* (2011) investigated

Table 8.4 Effects of the amounts of additives on the formation of anodic film

Additives	Insufficient amount	Excess amount
Silicate (e.g. sodium silicate) 50–100 g/L	It is difficult to form an oxide coating of high quality on the surface of magnesium or its alloys	Allows the precipitation of other components dissolved in the anodizing solution and, thus, is undesirable
Carboxylate (e.g. oxalic acid) 40–80 g/L	Invisible effect on the anodic film	No further advantage
The alkali hydroxides (e.g. sodium hydroxide) 60–120 g/L	High decomposition voltage and formation of rough anodic coating	Flowing of an excessive current in the anodizing bath and the dissociation voltage of magnesium or its alloy will not reach the desired level
Fluoride (e.g. sodium fluoride) 2–20 g/L	The quality of the anodic oxidation coating will be deteriorated	Concentrated sparks on the surface, causing difficulties in the operation and dissolving fluoride in the solution
Phosphate (e.g. trisodium phosphate) 10–30 g/L	Affects the formation rate of the coating films	Porous anodic films, which will cause the lowering of abrasion resistance
Borate (e.g. sodium metaborate) 10–40 g/L	Undesired level of anodic film thickness	An excess amount will not dissolve in the solution

Source: Kobayashi *et al.*, 1988.

the anodizing treatment in an aqueous alkaline solution with sodium carbonate, potassium fluoride and other additives using 3 A/dm^2 DC. Degreasing in a strong alkaline solution with other additives was performed prior to the anodizing treatment, and no corrosion was observed after 240 h in a salt spray test.

8.6 Applications of magnesium alloys

Magnesium alloys, which are the most light-weight metal, have always been attractive to designers as promising structural materials. Furthermore, magnesium alloys have many other desired properties such as high strength-to-weight ratios, excellent castability, machinability, weldability, and thermal stability, and good damping capacity. These has been a major factor in the widespread use of magnesium alloy castings and wrought products. Improvements in mechanical properties and corrosion resistance have led to greater interest in magnesium alloys for aerospace, automobile and many other applications, especially in fields where weight reduction is required.

8.6.1 Automobile industry

The use of magnesium alloys in the automobile industry has increased recently, with the aim of reducing car weights and consequently decreasing fuel consumption and CO_2 emissions. According to the European Automobile Manufacturers' Association, automakers will need to reduce CO_2 emissions in new cars to 120 g/km by 2012–2015. Furthermore, a new objective of 95 g/km has been set for 2020 (Errico *et al.*, 2011). Achieving optimum corrosion protection using an anodizing treatment and solving other problems related to magnesium alloys will increase the amount of magnesium used in automotive applications. According to Czerwinski (2008), magnesium was first used in the automotive industry in Germany in 1919, and Volkswagen made considerable developments after World War II. The American automobile industry uses 70 000 tonnes/year of magnesium. Examples of the amounts used per car are based on 2001 data: up to 26 kg in a full-sized GM van, up to 17 kg in a GM minivan, 14 kg in a Ford F-150 truck, 9.5 kg in a Buick Park Avenue and 6 kg in a Chrysler minivan. The Volkswagen Passat and Audi A4 and A6 each use 14 kg, and the Mercedes SLK Roadster uses 8 kg. Japanese automobile manufacturers have already used magnesium alloys in various car parts such as steering wheels, seat frames and transmission cases. Figure 8.16 illustrates car parts for which magnesium is used and their positions in the car (Watarai, 2006).

8.6.2 Aerospace industry

Magnesium fuel tanks, propellers and engine crank-cases were used in German aircraft in the early 1930s. The trend continued through the 1940s, when the B-47 jet bomber had a total of 5.5 tonnes of magnesium sheets, extrusions and casting. Magnesium was intensively used as a skin material on the wings and tails, jet engine pods, cowlings, gun enclosures and doors. At present, commercial aircraft have very few magnesium parts. Helicopters contain some cast components, such as gearboxes, seats and pedals. There are cast and wrought magnesium parts in satellites. The military industry uses much more magnesium, including in wheels, radar antennae, radar bases, covers and missile elements (Czerwinski, 2008). The DOW 17 and HAE anodizing coating do not confer the required protection properties (Manoj and Nai, 2010). By contrast, the Tagnite coating provides superior corrosion resistance for magnesium sand castings, die-castings, extrusions and forgings, and has therefore been used on a production basis in the aerospace industry

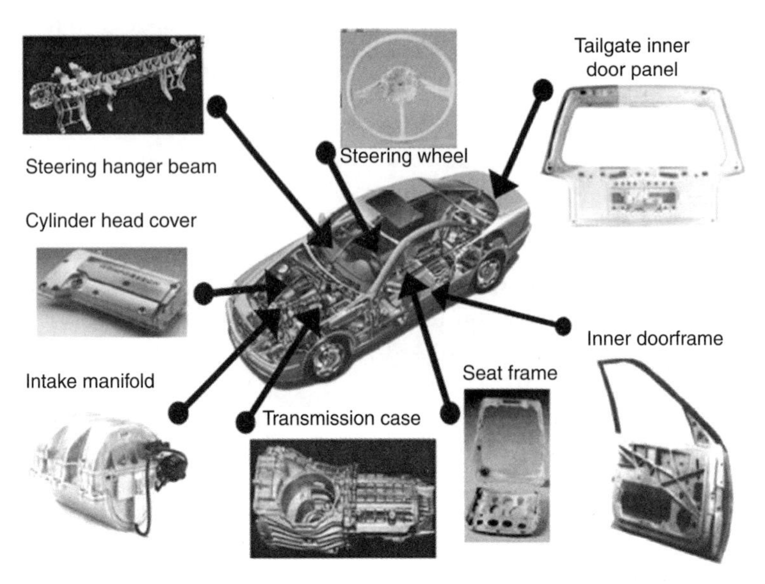

8.16 List of some magnesium car parts and their positions in the car. Reproduced with permission of Professor Kamado, Nagaoka University of Technology, Japan.

for more than a decade. Many applications require paint and Tagnite is an excellent base for all kinds of paint. In the aerospace industry, the coating is used for gearboxes, transmission cases and covers, and sump and oil pump housings (Tagnite, 2011). Keronite treatment is now approved by FAA, US Army and US NavAir as a surface pre-treatment for housings. A durable coating with long service intervals could be obtained, after combining with a class-leading topcoat system (Keronite, 2011).

8.6.3 Electronics industry

Because of its low weight and good mechanical and electrical properties, magnesium can contribute towards overcoming design challenges for the production of small, light, portable electronic equipment. Therefore, there is great interest and rapid expansion in the employment of magnesium alloys in the electronics industry. The first digital video camera to use magnesium was the Sony DCR-VX1000, whose design objectives included reduction of noise and vibration, prevention of heat build-up inside the machine, electromagnetic shielding, high strength, low weight and small size (Ohshima, 1996). Today, there is growing interest in using magnesium in manufacturing electronic devices such as cameras, projectors, notebook computers, portable digital audio/video players, television cabinets and mobile phones. Many leading electronics manufacturers, including Sony, Toshiba, Panasonic, Sharp,

Canon, JVC, Hitachi, Minolta, Nikon, NEC, Ericsson, IBM and Compaq, have used magnesium in their products (Magnesium Elektron, 2011).

8.6.4 Other applications

To use magnesium alloy as a magnetic shielding material, Akimoto and Akimoto (2006) introduced a magnetoresistance effect element. Fine powder magnetic material and/or a magnetic material precursor were poured into or adhered inside pores and/or a layer of porous anodic oxide film. The surface photocatalytic activity was improved employing the same method by filling the minute pores of the film with the photocatalyst (Akimoto, 2005).

Compared with commonly used metallic biomaterials, magnesium alloys have many outstanding advantages because of their attractive biological properties, including that they are essential for human metabolism, biocompatible and biodegradable. Through anodizing treatment, a hydroxyapatite coating was deposited on the surface of metallic implants to improve biocompatibility and to decrease the degradation rate in the physiological environment (Zhao *et al.*, 2010).

8.7 Future trends

The light-weight, high strength, good castability and machinability make magnesium alloys a promising engineering material for electronics, automotive and aerospace industries. However, many technical challenges hamper the widespread use of magnesium alloys in industry. In particular, the high cost of magnesium production, insufficient creep resistance, poor flammability resistance and poor cold workability are among the major problems to be overcome when using magnesium alloys. Nevertheless, susceptibility to corrosion is the main characteristic limiting its application. Overcoming such limitations therefore requires the development of low-cost, hard, wear-resistant, corrosion-resistant and aesthetic anodic coatings. Including high-redox materials such as Ce^{4+}/Ce^{2+}, Sn^{4+}/Sn^{2+}, MnO^{4-}/Mn^{2+} and MoO^{4-}/Mo^{2+} in the anodizing solution will produce an anodic coating with self-healing properties. Much more research is needed to understand the surface chemistry and reaction kinetics of the anodizing treatments.

8.8 Sources of further information and advice

Most of the references cited in this chapter are specialist handbooks and review articles that cover the subject of magnesium anodizing. The following books provide a comprehensive collection of engineering information on magnesium alloy published in recent years:

M. M. Avedesian and H. Baker, ASM Specialty Handbook: Magnesium and Magnesium Alloys, ASM International, Novelty, Ohio, USA, 1999.

E. F. Horst, B.L. Mordike, Magnesium Technology. Metallurgy, Design Data, Application, Springer-Verlag, Berlin-Heidelberg, 2006.

K. U. Kainer (Ed.), *Magnesium Alloys and Technologies* (F. Kaiser, Trans.). Betz-Druck GmbH Darmstadt: Wiley-VCH, 2003.

Guang-Ling Song (Ed.), Corrosion of magnesium alloys, Woodhead Publishing Limited.

Magnesium Technology Proceedings (Minerals, Metals & Materials Society)

8.9 References

Abyazani, A. G. (2011). Contribution to understanding the formation process and corrosion protection of the PEO coating on AM50 magnesium alloy. Doctoral dissertation, Faculty of Natural and Materials Sciences, Clausthal University, Germany

Aerts, T., Dimogerontakis, T., Graeve, I. D., Fransaer, J. and Terryn, H. (2007). Influence of the anodizing temperature on the porosity and the mechanical properties of the porous anodic oxide film. *Surface &Coatings Technology*, **201**, 7310–7317.

Akihiro, Y. and Sakai, M. (2009). Anodic films formed on magnesium in organic, silicate-containing electrolytes. *Corrosion Science*, **51**, 793–798.

Akimoto, M. (2005). *Patent No. 103505*. Japan.

Akimoto, K. and Akimoto, M. (2006). *Patent No. 152360*. Japan.

Asoh, H. and Ono, S. (2003). Anodizing of magnesium in amine-ethylene glycol electrolyte. *Materials Science Forum*, **419–422**, 957–962.

Bard, A., Parsons, R. and Jordan, J. (1985). *Standard Potentials in Aqueous Solution*. New York: Marcel Dekker.

Bartak, D. E., Lemieux, B. E. and Woolsey, E. R. (1995). *Patent No. US 5470664*

Barton, T. F. (1998). *Patent No. 5792335*. United States.

Beals, R., LeBeau, S., Roberto, O. and Shashkov, P. (2003). Advances in thixomolding magnesium alloys part II. *TMS (The Minerals, Metals and Materials Society)*, 283–288.

Brace, A. (1997). Seventy years of sulphuric acid anodizing. *Transactions of the Institute of Metal Finishing*, **75**(5), B101–B106.

Cai, Z., Lu, D., Li, W., Liang, Y., and Zhou, H. (2009). Study on anodic oxidation of magnesium in 6 M KOH solution by alternative current impedance. *International Journal of Hydrogen Energy*, **34**(1), 467–472.

Chaia, L., Yua, X., Yanga, Z., Wang, Y. and Okido, M. (2008). Anodizing of magnesium alloy AZ31 in alkaline solutions with silicate under continuous sparking. *Corrosion Science*, **50**(12), 3274–3279.

Czerwinski, F. (2008). *Magnesium Injection Molding*. New York: Springer Science+Business Media.

Dolan S. E. (2003). *Patent No. 0075453*. United States.

Eltron Research & Development. (2009). *Low-cost, Environmentally Friendly Magnesium Anodization*. Nautilus Court South, Boulder, USA. www.eltronresearch.com/

Emley, E. (1966). *Principles of Magnesium Technology*. London: Pergamon Press.
Errico, F. D., Fare, S. and Garces, G. (2011). The next generation of magnesium-based material to sustain the intergovernmental panel on climate change policy. *Magnesium Technology 2011*. TMS annual meeting 2011, 26 April 2011, Wiley Online Library, pp. 19–24.
Feitknecht, W. and Braun, H. (1967), The mechanism of the hydration of magnesium oxide with water vapour. *Helvetica Chimica Acta*, **50**, 2040.
Ghali, E. and Revie, W. (2010). *Corrosion Resistance of Aluminum and Magnesium Alloys: Understanding, Performance, and Testing*. Hoboken, New Jersey: John Wiley & Sons, Inc.
Gray, J. E. and Luan, B. (2002). Protective coatings on magnesium and its alloys—a critical review. *Journal of Alloys and Compounds*, **336**(1–2), 88–113.
Habashi, F. (1998). *Alloys: Preparation, Properties, Applications*. Weinheim, Germany: Wiley-VCH.
Hsiao, H.-Y. and Tsaia, W.-T. (2005). Effect of heat treatment on anodization and electrochemical of AZ91D magnesium alloy. *Journal of Materials Research*, **20**(10), 2763–2771.
Hsiao, H.-Y., Tsung, H.-C. and Tsai, W.-T. (2005). Anodization of AZ91D magnesium alloy in silicate containing electrolytes. *Surface and Coatings Technology*, **199**(2–3), 127–134.
Hugh, G. C. (1974). The coloration of aluminium. *Coloration Technology*, **5**(1), 49–64.
Keronite (2011). Retrieved 11.21.2011, from Keronite: www.keronite.com.
Khaselev, O. and Yahalom, J. (1998). The anodic behavior of binary Mg-Al alloys in KOH-aluminate solutions. *Corrosion Science*, **40**(7), 1149–1160.
Kim, H. G., Seo, K. H., Lee, J. E. and Tae, S. (2011). *Patent No. 0114497*. United States.
Kim, S.-J. and Okido, M. (2003). The electrochemical properties and mechanism of formation of anodic oxide films on Mg-Al alloys. *Bulletin of the Korean Chemical Society*, **24**(8), 975–980.
Kim S. J., Kim J. I., Okido M. (2004). Sealing Effects of Anodic Oxide Films Formed on Mg-Al Alloys. *Korean Journal of Chemical Engineering*, **21**(4), 915–920.
Kobayashi, W. and Takahata, S. (1985). *Patent No. US4551211*.
Kobayashi, W., Uehori, K. and Furuta, M. (1988). *Patent No. US4744872*.
Kozak, O. (1980). *Patent No. 4184926*. United States.
Kurze, P. (2003). Corrosion and corrosion protection of magnesium. In K. U. Kainer, *Magnesium Alloys and Technologies* (F. Kaiser, Trans.). Betz-Druck GmbH Darmstadt: Wiley-VCH.
Kurze, P. (2006). Corrosion and surface protection. In H. E. Friedrich and B. L. Mordike, *Magnesium Technology*. Berlin Heidelberg: Springer.
Leidheiser, H. (1987). Fundamentals of corrosion protection in aqueous solutions. *In Metals Handbook 9th Edition, Vol. 13, Corrosion*, Metals Park, OH: ASM International.
Licari, J. J. (2003). *Coating Materials for Electronic Applications*. Norwich, NY: William Andrew.
Lin, C. S. and Fu, Y. C. (2006). Characterization of anodic films on AZ31 magnesium alloys in alkaline solutions containing fluoride and phosphate anions. *Journal of the Electrochemical Society*, **153**(10), B417–B424.
Li-ping, W., Zhao, J.-J., Xie, Y.-P. and Yang, Z.-D. (2010). Progress of electroplating and electroless plating on magnesium alloy. *Transactions of Nonferrous Metals Society of China*, **20**, s630–s637.

Magnesium Elektron (n.d.). *Magnesium in ICT*. Retrieved 11.29.2011 from http://www.magnesium-elektron.com/markets-applications.asp?ID=12.

Makar, G. and Kruger, J. (1989). Corrosion studies of rapidly solidified magnesium alloys. *Journal of the Electrochemical Society*, **137**(2), 414–421.

Makar, G. L. and Kruger, J. (1993). Corrosion of magnesium. *International Materials Review*, **38**(3), 138–153.

Manoj, G. and Nai, S. M. (2010). *Magnesium, Magnesium Alloys, and Magnesium Composites*. New Jersey: John Wiley & Sons, Inc.

Mittal C K. (1995). Chemical Conversion and Anodized Coatings. *Transactions of the Metal Finishers Association of India*, **4**, 227–231.

Ohshima, E. (1996). Application of die-cast magnesium to AVCC. In Proc. Conference International Magnesium Association, Ube city, Japan. IMA, 1303 Vincent Place, Mclean, Virginia 221001 USA, 1–16.

Ono, S. (1998). Surface phenomena and protective film growth on magnesium and magnesium alloys *Metallurgical Science and Technology*, **16**(1–2), 91–104.

Ostrovsky, I. (2005). *Patent No. US6875334.*

Ostrovsky, I. (2010). *Patent No. US7780838.*

Pernick, J. (1987). Aluminum anodizing. *In Metals Handbook 9th Edition, Vol. 13, Corrosion*, Metals Park, OH: ASM International.

Pourbaix, M. (1974). *Atlas of Electrochemical Equilibria in Aqueous Solutions*. Houston, TX: National Association of Engineering Corrosion.

Saijo, A., Hino, M., Hiramatsu, M. and Kanadani, T. (2005). Environmental friendly anodizing on AZ91D magnesium alloys and coating characteristics. *Acta Metallurgica Sinica (English Letters)*, **18**(3), 411–415.

Salman, S., Ichino, R. and Okido, M. (2008). Influence of calcium hydroxide and anodic solution temperature on corrosion property of anodising coatings formed on AZ31 Mg alloys. *Surface Engineering*, **24**(3), 242–245.

Salman, S. A., Mori, R., Ichino, R. and Okido, M. (2010). Effect of anodizing potential on the surface morphology and corrosion property of AZ31 magnesium alloy. *Materials Transactions*, **51**(6), 1109–1113.

Sharma, A. K., Rani, R. U. and Giri, K. (1997). Studies on anodization of magnesium alloy for thermal control application. *Metal Finishing*, **95**(3), 43–51.

Sharma, A. K., Umarani, R., Bhojaraj, H. and Narayanamurthy, H. (1993). Galvanic black anodizing on Mg-Li alloys. *Journal of Applied Electrochemistry*, **23**, 500–507.

Shashikala, A. R., Umarani, R., Mayanna, S. M. and Sharma, A. K. (2008). Chemical conversion coatings on magnesium alloys—a comparative study. *International journal of Electrochemical Science*, **3**, 993–1004.

Shelley T. (2009). The continued development of a hard coating process is set to benefit mould toolmakers, Tom Shelley reports *http://www.eurekamagazine.co.uk/article/17500/Hard-work.aspx.*

Shi, Z., Song, G. and Atrens, A. (2005). Influence of the beta phase on the corrosion performance of anodised coatings on magnesium-aluminium alloys. *Corrosion Science*, **47**, 2760–2777.

Shi, Z., Song, G. and Atrens, A. (2006). Influence of anodising current on the corrosion resistance of anodised AZ91D magnesium alloy. *Corrosion Science*, **48**(8), 1939–1959.

Skar, J. I. and Albright, D. (2002). Emerging trends in corrosion protection of magnesium die-castings. *Magnesium Technology 2002, TMS*, 255–261.

Song, G. L. and Atrens, A. (1999). Corrosion mechanisms of magnesium alloys. *Advanced Engineering Materials*, **1**(1), 11–33.

Tagnite (2011). *Tagnite coating applications*. Retrieved 11.28.2011, from http://www.tagnite.com/applications/

Tomashov, N. D. (1966). *Theory of Corrosion and Protection of Metals*, New York: The Macmillan Company

Tomita, S. (1980). *Patent No. US4225399*.

Umehara, H., Terauchi, S. and Takaya, M. (2000). Structure and Corrosion Behaviour of Conversion Coatings on Magnesium Alloys. *Materials Science Forum*, **350–351**, 273–282.

Vujicic, V. and Lovrecek, B. (1985). A study of the influence of pH on the corrosion rate of aluminum. *Surface Technology*, **25**, 49–57.

Watarai, H. (2006). Trend of research and development for magnesium alloys—reducing the weight of structural materials in motor vehicles. *Science and Technology Trends*, **18**, 84–97.

Wu, C. S., Zhang, Z., Cao, F. H., Zhang, L. J., Zhang, J. Q. and Cao, C. N. (2007). Study on the anodizing of AZ31 magnesium alloys in alkaline borate solutions. *Applied Surface Science*, **253**(8), 3893–3898.

Yabuki, A. and Sakai, M. (2008). *Patent No. 013803*. Japan.

Yakovleva, N. M., Anicai, L., Yakovlev, A. N., Dima, L., Khanina, E. Y., Buda, M., *et al.* (2002). Structural study of anodic films formed on aluminum in nitric acid electrolyte. *Thin Solid Films*, **416**(1–2), 16–23.

Yerokhin, A. L., Nie, X., Leyland, A., Matthews, A. and Dowey, S. J. (1999). Plasma electrolysis for surface engineering. *Surface and Coatings Technology*, **122**, 73–93.

Zhang, R. F., Shan, D. Y., Chen, R. S., and Han, E. H. (2008). Effects of electric parameters on properties of anodic coatings formed on magnesium alloys. *Materials Chemistry and Physics*, **107**(2–3), 356–363.

Zhang, Y., Yan, C., Wang, F., Lou, H. and Cao, C. (2002). Study on the environmentally friendly anodizing of AZ91D magnesium alloy. *Surface and Coatings Technology*, **161**, 36–43.

Zhao, L., Cui, C., Wang, Q. and Bu, S. (2010). Growth characteristics and corrosion resistance of micro arc oxidation coating on pure magnesium for biomedical applications. *Corrosion Science*, **52**(7), 2228–2234.

9
Anodization and corrosion of magnesium (Mg) alloys

G.-L. SONG, General Motors Corporation, USA and
Z. SHI, The University of Queensland, Australia

DOI: 10.1533/9780857098962.2.232

Abstract: Anodization is one of the most important and effective surface pre-treatments for Mg alloys. This chapter systematically summarizes Mg alloy anodizing behavior, the compositions and microstructures of anodized films on Mg alloys and the anodization-influencing factors. Based on the anodizing voltage variation, gas evolution and sparking behavior in a typical anodizing process and the characteristic composition and microstructure of an anodized coating, a four-stage anodizing mechanism is postulated. Moreover, the corrosion performance of anodized Mg alloys is systematically reviewed and a corrosion model is proposed to explain the corrosion performance and electrochemical behavior. It is believed that some of the measured electrochemical features can be utilized to rapidly evaluate or compare the corrosion resistance of anodized Mg alloys.

Key words: Mg alloy, corrosion, anodizing, surface treatment, coating.

Note: This chapter has been adapted from Chapter 16 'Anodization and corrosion of magnesium (Mg) alloys' by G.-L. Song and Z. Shi, originally published in *Corrosion of magnesium alloys*, ed. Guang-Ling Song, Woodhead Publishing Limited, 2011, ISBN: 978-0-85709-41-3.

9.1 Overview of anodizing techniques

It is well known that magnesium (Mg) alloys are not corrosion resistant and require surface treatments or coatings in many applications. A variety of surface finishing processes are being used to protect magnesium alloys from corrosion, which include surface conversion treatments (e.g. chromating, phosphating), anodizing (e.g. DOW 17, HAE, Anomag, Keronite, Tagnite, Magoxid-Coat); galvanizing/plating (Zn, Cu, Ni, Cr). These processes can be used alone or in combination with application of organic coatings. In addition, there are also chemical vapor deposition (CVD), physical vapor deposition (PVD), flame or plasma spraying and laser/electron/ion beam surface treatments, etc., as options for corrosion protection of Mg alloys.

Among these techniques, anodizing appears to be one of the most effective corrosion protection techniques for Mg alloys. This is because a successful coating should meet the following requirements:

- be very stable in corrosive environments to protect the substrate;
- be resistant to scratching and flexible to follow the deformation of the substrate without cracking or delaminating;
- be a non-conductive layer to isolate the substrate and in case the layer is damaged, no significant galvanic effect of the coating to accelerate the corrosion of the substrate;
- have a relatively rough morphology for top-coating, if necessary;
- have high throwing power to treat components from all sides.

Anodizing is one of the most popular industrial processes (Blawert *et al.*, 2006), which applies an anodic current or voltage to a substrate metal to produce an oxide film. The nature of anodizing is an oxidation reaction in an aqueous environment driven by a current or voltage. Although many different terms, such as ‘arc’, ‘sparking’, ‘plasma’ and ‘anodic deposition’, are often used in various cases to describe this technique, they only refer to the phenomena exhibited in a certain stage of anodizing. The anodized coating has a porous morphology and ceramic-like composition, which is chemically stable. An anodized coating can be further sealed and/or top-coated for additional corrosion protection.

Anodizing is usually adopted as one of the important industrial surface treatment technologies for light metals, and it has been successfully used on aluminum (Al) alloys over many decades. It has also been further developed for use on Mg alloys. Although both Al and Mg alloys can be anodized, there are significant differences in their anodizing processes and anodized coatings. First, because of the high temperature caused by sparking during Mg alloy anodizing, there are a large number of species participating in the anodizing. The process on a Mg alloy is more complicated than that on an Al alloy. The latter normally does not have sparking. Second, the electrolytes used in the anodizing processes of an Al alloy and a Mg alloy are different. Third, the voltage used in anodizing a Mg alloy normally is much higher than in anodizing an Al alloy. Fourth, on an Al alloy, the formed anodized film is regular in structure and consists of a barrier layer adjacent to the metal surface as well as a layer containing uniform parallel pores normal to the surface, while the anodized film on a Mg alloy is irregularly porous in structure.

There are already several commercial anodized coatings that have been developed for magnesium alloys, such as Tagnite (Bartak *et al.,* 1993, 1995; Barton and Johnson, 1995), Anomag (Barton, 1998; Barton *et al.,* 2001), Magoxid (COMA, 2001; Schmeling *et al.,* 1990), Keronite (COMA, 2008;

Shatrov, 1999), HAE (Abulsain *et al.,* 2004; Evangelides, 1955), Dow 17 (Abulsain *et al.,* 2004; COMA, 1956), MGZ (Hillis and Murray, 1987; Kotler *et al.,* 1976a; Leyendecker, 2001), etc. The first anodizing coating for Mg alloys is Dow 17 developed by the Dow Chemical Company. It uses an AC or DC current at a voltage below 100 V in a solution comprising sodium dichromate, ammonium acid fluoride and phosphoric acid at pH ~5 above 70°F. The composition of the formed film is mainly MgF_2, $NaMgF_3$, $Mg_{x+y/2}O_x(OH)_y$ and smaller amounts of Cr_2O_3. However, the use of chromate in the electrolyte is a big concern limiting its applications. After that, the HAE anodic coating technique was developed for Mg alloys. The anodizing electrolyte contains potassium permanganate, potassium fluoride, trisodium phosphate, potassium hydroxide and aluminum hydroxide with a high pH value (approximately 14) and anodizing is operated between 20°C and 30°C with an AC current at a voltage below 125 V.

Anomag is a technique commercialized by Magnesium Technology Ltd in New Zealand. The anodizing bath consists of ammonia, sodium ammonium phosphate, etc. No chromium or other heavy metals are used. Anodizing is obtained without formation of high-energy plasma discharges (spark discharge), so it is a non-sparking process. The Magoxid-Coat is formed in an electrolyte containing borate or sulfate, phosphate and fluoride or chloride at pH 5~11 (preferably 8–9) buffered by amines at a DC current preferably with a voltage up to 400 V. The coating contains MgO, $Mg(OH)_2$, MgF_2 and $MgAl_2O_4$. Tagnite is another chromate-free commercial coating system. The coating has better corrosion and abrasion resistance than any chromate-based coating. The electrolyte is an alkaline solution containing hydroxide, fluoride and silicate species operated below room temperature (4–15°C). The anodizing current has a unique waveform optimized at voltages exceeding 300 V DC. The coating is white, consisting mostly of hard Mg oxide with minor deposition of hard fused silicates. Keronite process transforms Mg alloy surfaces into a hard, wear-resistant ceramic oxide coating at a temperature between 20°C and 50°C. This coating is a good corrosion and thermal barrier. Anodizing is performed with a bipolar (positive and negative) pulsed electrical current with a specific wave form in a proprietary, chrome and ammonia-free, low concentration alkaline (98% demineralized water) non-hazardous electrolyte. The coating is mainly composed of spinel ($MgAl_2O_4$) together with SiO_2 and SiP. Moreover, the high-voltage Cr-22 treatment is also commercially available, although the anodizing bath contains chromate, vanadate, phosphate and fluoride compounds (Hillis, 1995).

In addition to the commercially available anodizing processes, numerous new anodizing techniques are currently being developed (Blawert *et al.,* 2006). For example, a white glossy anodized oxide film formed in an aqueous solution was reported to contain silicate, carboxylate and alkali hydroxide (Kobayashi *et al.,* 1988). A film consisting of chemically stable and hard spinel compound of MgO–Al_2O_3 was found on Mg surfaces after anodizing (Kobayashi and Takahata, 1985). It was patented that magnesium substrate should be first pickled in an aqueous hydro-fluoric acid bath and then anodized in an aqueous bath composed of an alkali metal silicate and an alkali metal hydroxide to form an anodized film for magnesium (Kozak, 1980, 1986). Corrosion and wear-resistant protective layers on magnesium or Mg alloys can also be formed by anodic oxidation in an aqueous electrolyte bath containing borate or sulfate ions, phosphate ions fluoride ions, and alkali ions (Tanaka, 1993).

It should be noted that anodizing is a hot topic in the area of corrosion and protection of Mg alloys. It is expected that in near future, more corrosion-resistant, low cost, easily operated, environment-friendly and non-toxic anodizing techniques will become available for Mg alloys.

9.2 Characteristics of anodizing behavior

Although various anodizing techniques have been developed, which differ from one another for their different anodizing phenomena or coating performance, they do share some common characteristics.

9.2.1 Anodizing stages

Generally speaking, different stages can be observed in an anodizing process. For example, it was found that the DC anodizing of AM60 Mg alloy in 1.5M KOH + 0.5M KF + 0.25M $Na_2HPO_4 \bullet 12H_2O$ with addition of various concentrations of $NaAlO_2$ has three anodizing stages: traditional anodizing followed by micro-arc anodizing and finally arcing (Verdier *et al.,* 2004). Khaselev and co-worker studied pure Mg (Khaselev and Yahalom, 1998a, 1998b; Khaselev *et al.,* 1999), binary Mg–Al alloys (2 at.%, 5 at.% and 8 at.% of Al) and intermetallic $Mg_{17}Al_{12}$ in solution 3 M KOH + 0.6 M KF + 0.21 M Na_3PO_4, with and without addition of 0.4 and 1.1 M of aluminate ($Al(OH)_3$) under constant voltage and current conditions, and found that the anodizing behavior showed clearly different stages. It is also reported that the anodizing voltage of diecast AZ91D in a 40°C sodium aluminate + potassium fluoride electrolyte varied in the range of 240–600 V, the current density varied in the range of 0.5–5 A/dm^2, and micro-arcing was observed in the anodizing process. The process showed two stages. In the first stage, the cell voltage increased linearly at a very high rate of 80–300 V/min. With higher anodizing current densities and higher concentrations of the electrolyte components, the voltage increased more quickly. Approximately 3–20 min later, this process entered into the second stage. A steady state sparking was established on the surface and the cell voltage reached a relative stable value of 520–570 V. The formed ceramic coatings were composed of the spinel phase $MgAl_2O_4$. A few circular pores and micro-cracks formed on the ceramic coating surface; the number of the pores decreased, while the diameter of the pores increased as the anodizing treatment proceeded. Variation of treatment time in the range of 10–40 min caused no changes in the phase structure of the ceramic coatings.

An anodizing process may display different stages under potential control and current control modes. Generally speaking, a complete voltage-controlled anodic polarization curve of a Mg alloy under a potential control mode may be classified into four regions: (1) primary passivity; (2) breakdown of primary passivity and metal dissolution; (3) secondary passivity; and (4) breakdown of the secondary passivity. In practice, current-controlled anodizing is more popular. In this mode, an anodizing process may also be generally classified into four different stages: I-linear growing (the anodizing voltage increases quickly and linearly with time; there is no sparking, and gas evolution is not significant either); II-gas evolving (voltage keeps increasing, but the rate decreases with time; meanwhile, the gas evolution becomes much faster; sparking is insignificant); III-uniform sparking (anodizing voltage linearly increases with time again; sparking is clearly visible randomly over the specimen surfaces; the sparking activity together with gas evolution becomes more and more intense as the anodizing voltage increases with time); and IV-localized sparking (much more localized, intense vivid sparking arcs appear on the specimen surfaces; vigorous gas evolves from the sparking areas; the increasing of anodizing voltage with time slows down). In some cases, stages III and IV cannot be clearly separated and are combined together. Sometimes, stage I is too quick or short and may be overlooked. Therefore, two or three anodizing stages are more often reported in various studies. The four stages under current control mode do not necessarily correspond to the stages in potential control mode.

A schematic illustration of the anodizing stages is presented in Fig. 9.1. To obtain a steady micro-arcing and sparking process, the applied current density if under a current-controlled mode should be far higher than the active current peak in stage (2). Stage IV sometimes may correspond to stage (4). After stage (4) or IV, the surface film may collapse and cannot be anodized again; no more sparking can occur on a collapsed surface.

Anodizing voltage is a primary indicator of the formation of an anodized coating on Mg alloys under a current control mode. Initially, the voltage increases almost linearly with time until it reaches values of breakdown, at which a sharp deflection occurs and voltage oscillation and sparking start. The breakdown voltage increases with the Al content in the alloy and

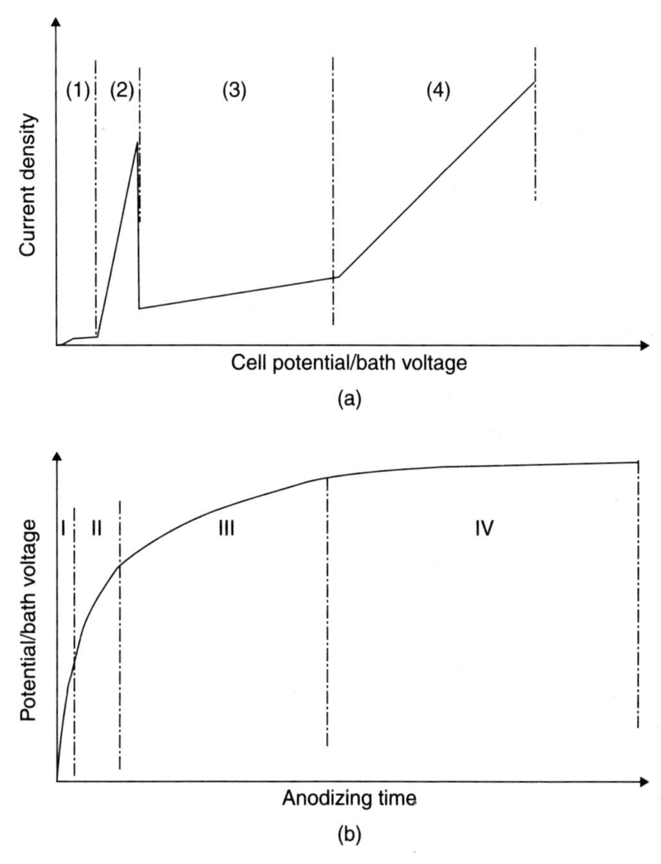

9.1 Schematic illustration of anodizing stages: (a) anodic polarization curve under a potential control and (b) dependence of cell potential (bath voltage) on anodizing time under a current control mode.

aluminate concentration in the bath solution. The highest breakdown voltage of 80 V was observed for intermetallic $Mg_{17}Al_{12}$ in a solution containing aluminate. This value is about 20 V higher than that observed for pure Mg (Khaselev and Yahalom, 1998a). In two-phase alloys, the anodizing is not uniform. Anodizing starts at α-phase while growth on β-phase requires voltages above 80 V. A uniform layer only forms above 120 V (Khaselev *et al.*, 1999).

Mg–Al alloys are passive at voltages up to 3 V (Takaya, 1988). Higher voltages lead to activation and dissolution of Mg–Al alloys accompanied by oxygen evolution. The dissolution is initiated in the form of separate pits. The number of pits gradually increases and the pits eventually join together to form large dissolution areas. The process continues with time until the Mg alloy surface is completely activated. When this happens, the surface of the alloy is fully black. Mg–Al alloys become passive again at an applied voltage of ~ 20 V. The secondary passivity is related to the formation of a MgO film. The presence of aluminate ion in the bath solution can promote the passivity. The aluminum content in the alloy is beneficial for the passivity of magnesium. Single phase Mg solid-solution alloy (Mg–0.8 at.% Cu and Mg–1.4 at.% Zn) was found to change appearance of its anodic film with applied voltage when anodized at up to 250 V at 10 mA/cm^{-2} in a 3 M ammonia + 0.05 M tri-ammonium orthophosphate electrolyte (pH 10.7) (Abulsain *et al.*, 2004). A relatively uniform film was developed at 26 V with dendritic-like and plate-like features; patches of porous material were formed at 160 V, which expanded as the voltage increased; and the final film formed at 240 V had typical sparking feature.

The anodizing stages vary from alloy to alloy, bath solution to bath solution and are influenced by applied potential and current density levels. It is well known that a film can be formed on a Mg alloy surface at a low voltage or current density. For example, for Mg in 1 M NaOH at voltages up to 3 V, when the current density remained low, a light grey protective film of $Mg(OH)_2$ is formed; a thick dark film of $Mg(OH)_2$ will be formed at higher voltages; above 20 V a protective coating will be produced (Huber, 1953). A barrier-type film or a semi-barrier film can be formed in an alkaline fluoride solution, which breaks down at around 5 V; above the breakdown potential, porous films were formed (Ono and Masuko, 2003; Ono *et al.*, 2003). Anodic films incorporating silicon can also be formed on AZ91D under a constant potential of 4 V in 3 M KOH solution with Na_2SiO_3. The films are uniform and thicker than the films formed in the bath solution without Na_2SiO_3. A few at % of silicon can be detected in the films, although the main component of the films was $Mg(OH)_2$ (Fukuda and Matsumoto, 2004).

9.2.2 Sparking

Sparking can be observed in many anodizing processes for Mg alloys at high voltages or current densities. For example, in 3 M NaOH at 24°C, anodizing voltage suddenly jumps to high levels when the applied anodic current density exceeds 20 mA/cm^2. After that, the potential oscillates in an irregular manner, and sparks appear and move around on the specimen surface (Carter *et al.*, 1999). Micro-arc, sparking or plasma oxidation of Mg alloys refers to the stage when applied voltage and current density are relatively high in an anodizing process. Normally, sparking process occurs when the applied voltage is over the dielectric breakdown potential of the anodized coating. The sparking voltage changes with bath composition and concentration. In most electrolytes, sparking does not occur on the surface of a magnesium alloy before the anodizing voltage exceeds 50 V.

Although an anodized film starts to form on Mg or a Mg alloy at a very low potential or current density, the film changes with potential or current density and breaks down at higher voltages. In other words, anodizing at high voltage or current density is a breaking down and repairing process of a film formed at a low current density or voltage, which accompanies a micro-arcing process. Initially the sparks are very small and are extinguished quickly. As the potential increases, the sparks become larger and begin to move over the specimen surface. During the whole sparking process, the specimen is actually being repeatedly activated and passivated. The passivation and activation correspond to the events that the film breaks down resulting in sparking and then it is rapidly repaired after sparking. As the rebuilt film at a sparked spot is thicker than the original film before sparking, the film keeps growing significantly during the sparking stage. The breakdown and repairing of the film can lead to a dramatic change in potential and current density. Thus, the current and potential fluctuations are always observed during sparking. If an anodized specimen remains at a constant potential for a long time, sparks may gradually diminish. If the potential was further increased slightly, the sparking will resume. A large positive potential step during an anodizing process can lead to suddenly intensified sparking. If sparking occurs continuously at a spot, not moving along, for example, at a sharp edge or a defective point on a specimen, there would be inevitably a large burned pit formed there. If this damaged area is too large, it may extinguish the sparking, resulting in failure of the anodizing process.

9.2.3 Gas evolution and anodizing efficiency

Apart from sparking, oxygen evolution is another important phenomenon in an anodizing process for Mg alloys. Oxygen evolution is closely related to sparking behavior. Before sparking, the oxygen evolution is insignificant. Vigorous oxygen evolution is normally observed when sparking occurs. For example, under a current-controlled mode, in the initial stage at a low voltage the evolved gas volume is very small; then the evolved gas volume increases gradually with voltage, and finally significantly at a high voltage.

The efficiency of film growth by anodizing is relatively low, only about 30% (Abulsain *et al.*, 2004). There is a close relationship between oxygen evolution and anodizing efficiency. It was found (Shi *et al.*, 2006b) that for AZ91D anodized in a silicate-containing bath solution, apart from film formation, the anodizing current was also used for electrochemical oxygen evolution. Consequently, a dramatic decrease in anodizing efficiency down to a negative value was measured after the anodizing getting into an intensive

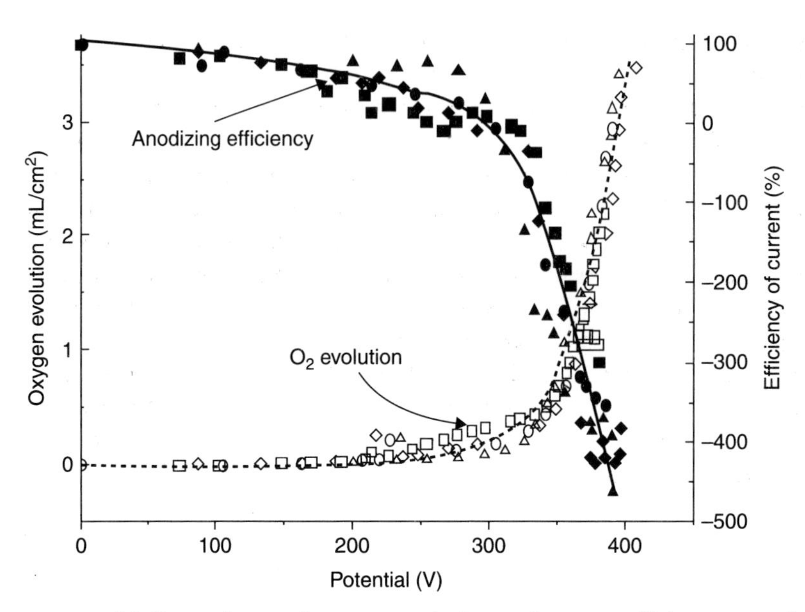

9.2 Dependence of oxygen evolution and current efficiency on cell voltage under constant electric charge (18 C/cm²) (Shi *et al.*, 2006b).

sparking stage (see Fig. 9.2). This result implies that a higher current density may produce a thicker coating quickly, but it is not economical in practice.

9.3 Anodized coating/film

An anodized coating/film can be characterized by its thickness, composition and microstructure. These parameters are important to its corrosion protection performance.

9.3.1 Coating thickness

During anodizing, the substrate magnesium is consumed by oxidation and the oxide film is deposited on the surface. The surface film layer grows about 50% into the original magnesium and 50% above the original surface level (Kurze and Banerjee, 1996). The thickness of the coating is dependent on the anodizing voltage or current density. Figure 9.3 shows the dependence of the thickness of an anodized coating under a current control mode on the final anodizing voltage. It is a linear relationship, that is, a higher cell voltage means a thicker anodized coating is formed. The increase in film thickness with final anodizing voltage can be easily understood if the resistance of the film is believed to be linearly dependent on the film thickness.

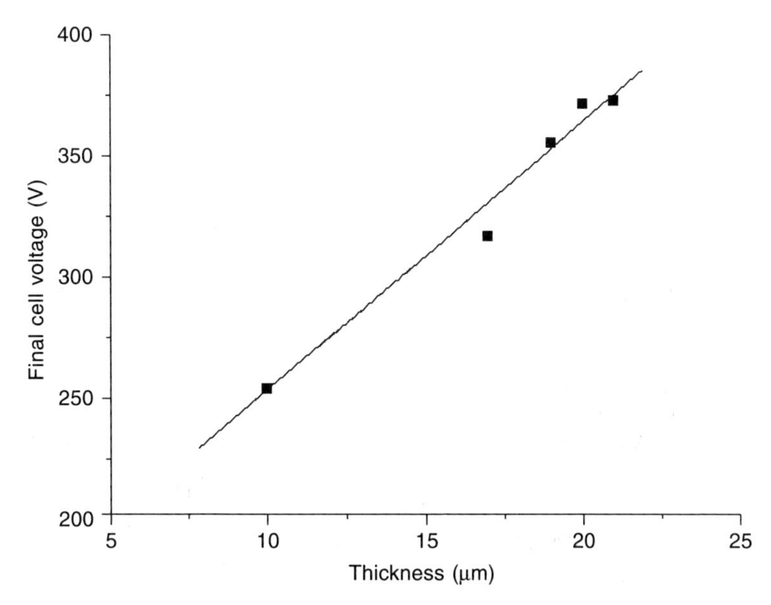

9.3 Dependence of the thickness of anodized coating on the final anodizing voltage (Shi *et al.*, 2006b).

9.3.2 Coating composition

The elements in a formed anodized film are mainly magnesium, oxygen and aluminum (if the substrate or the bath solution contains aluminum), which can usually be further specified as MgO and some $MgAl_2O_4$. Detailed composition varies widely, depending on many factors. For example, a coating formed through a Dow17 process on pure magnesium is basically composed of MgF_2, $Mg_{x+y/2}O_x(OH)_y$ and $NaMgF_3$. The first two components are formed at the metal/film interface and the latter formed in the middle of the film by a chemical reaction in presence of Na^+ in the electrolyte (Ono, 1998). Recently, it was found that a significant amount of Si from an SiO_3^{2-} -containing anodizing bath can be incorporated into the anodized coating after anodizing at a high voltage (Shi *et al.*, 2006b). The Si/Mg ratio of the anodized coating increased as the anodizing proceeded. In the first stage, the atomic Si/Mg ratio was only 0.5, indicating that the deposition rate of $Mg(OH)_2$ was the same as that of $MgSiO_3$. In the second stage, the atomic ratio of Si/Mg was higher (around 1), suggesting more silicate deposited into the coating from the bath solution. In stage III, the ratio increased to 2, implying more silicates are incorporated in the coating. In stage IV, the amount of silicate deposited was much greater than $Mg(OH)_2$ with a ratio of Si/Mg = 5.7, indicating that the silicate is the main composition of the coating. Figure 9.4 summarizes the chemical compositions of films formed at different anodizing voltages.

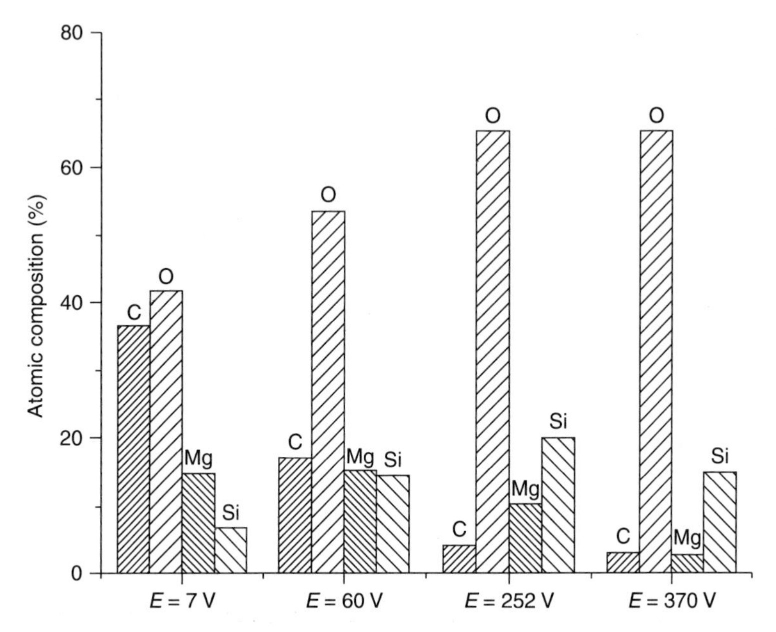

9.4 Chemical composition of anodized coatings formed at different anodizing voltages (based on Shi *et al.*, 2006b).

For an Al-containing Mg alloy, Al-enrichment can normally be detected in the anodized coating (Shi *et al.*, 2006b). Some other alloying elements may also be enriched in an anodized film after anodizing. For example, alloying element enrichment was found on sputter-deposited Mg–0.4 at.% W and Mg–1.0 at % W model alloys (Bonilla *et al.*, 2002a, 2002b). The alloys were anodized at 10 mA/cm^2 up to 150 V in a 3 M ammonium hydroxide + 0.05 M ammonium phosphate electrolyte at 293 K. During anodizing to about 50 V, a relatively smooth but finely porous film developed. With increasing voltage, the film transformed gradually to a coarse porous morphology due to dielectric breakdown. Enrichments of tungsten to at least 1.7×10^{15} and 2.9×10^{15} W atoms/cm^2 for Mg–0.4 at.% W and Mg–1.0 at.% W alloys, respectively were found. The enrichment mechanism may be analogous to that of dilute Al alloys, in which enrichment of Al oxides occurs in its surface film due to the Gibbs free energies of the alloying elements oxides being greater than that for the formation of alumina. Enrichment can also be observed in solid-solution Mg–0.8 at.% Cu and Mg–1.4 at.% Zn alloys anodized up to 250 V at 10 mA/cm^{-2} in 3 M ammonia + 0.05 M tri-ammonium orthophosphate electrolyte (pH 10.7) at 293 K (Abulsain *et al.*, 2004). Rutherford backscattering spectroscopy (RBS) revealed enrichments to about 4.1×10^{15} Cu atoms cm^{-2} and 5.2×10^{15} Zn atoms cm^{-2}, which can be correlated with the higher standard Gibbs free energies for formation of copper and zinc oxides than that for formation of MgO.

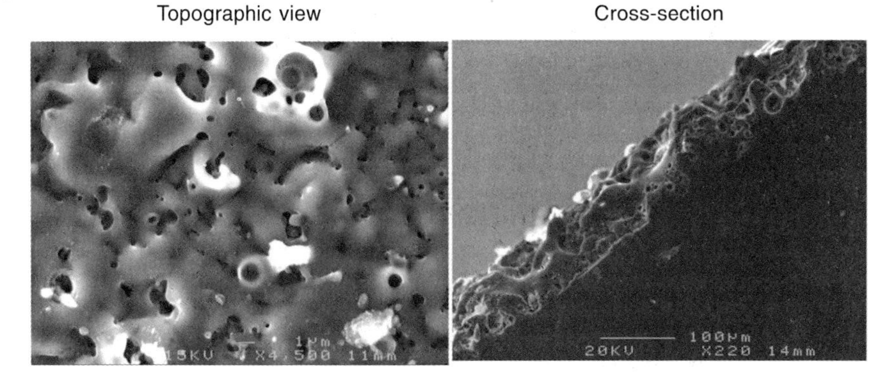

9.5 Typical microstructure of an anodized coating formed on AZ91D in a silicate-containing solution.

The enriched layers were about 1.5–4.0 nm thick as measured by medium energy ion scattering (MEIS). The anodized films, composed mainly of magnesium hydroxide, contained copper and zinc species throughout their thicknesses; the Cu:Mg and Zn:Mg atomic ratios were about 18% and 25% in those of the alloys, respectively. Phosphorus species were present in most of the film regions, with a P:Mg atomic ratio of about 0.16 (Abulsain *et al.*, 2004).

9.3.3 Coating microstructure

Generally speaking, in an anodizing process, a thin amorphous film can be formed during the first stage of anodizing. Prolonged anodizing causes local breakdown and crystallization of the initially formed anodic films. Anodized coatings formed at high voltages are porous. All the commercial anodized coatings that formed at high anodizing voltages, such as Dow17, HAE, Tagnite, Magoxide, Anomag and Keronite (most of them have got into the sparking stage) have been revealed to have a porous microstructure. A typical porous microstructure image of an anodized coating formed on AZ91D in a KOH and SiO_3^{2-} -containing solution is shown in Fig. 9.5.

In fact, anodized coatings are multi-layered. For example, using a Dow17 process (without sparking/electric breakdown), a cylindrical pore structure in the outer layer and barrier layer in the inner layer in contact with metal substrate can be formed on pure Mg (Ono, 1998). Figure 9.6 schematically illustrates the multi-layer structure of a Mg oxide coating.

It should be noted that the relatively compact inner barrier layer is normally very thin and can barely be revealed by scanning electron microscopy (SEM) in experiments. Compared with the barrier layer in the anodized film on an Al alloy, it is much less corrosion resistant and one cannot expect it to significantly retard the ingress of aggressive species.

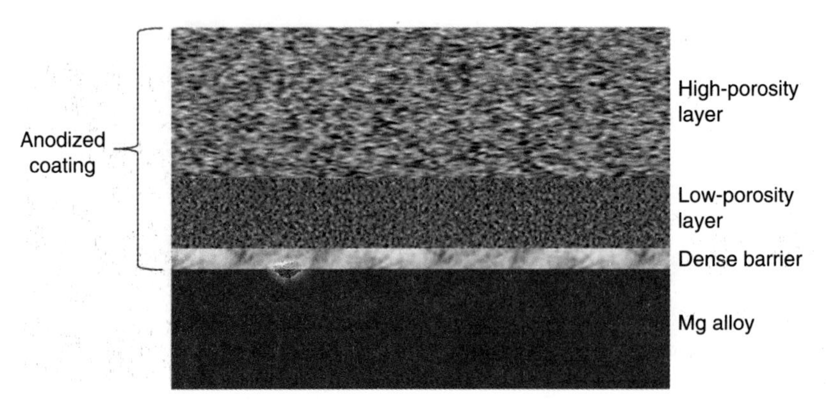

9.6 Schematic illustration of the multi-layer structure of Mg oxide coating (based on Blawert *et al.*, 2006).

The microstructure can vary with anodizing conditions. Mato *et al.* (2003) studied the anodizing behavior of sputter-deposited alloy, Mg–40% Ta, in ammonium pentaborate and sodium silicate electrolytes; in the pentaborate electrolyte, formation of an amorphous single layer composed of Ta_2O_5 and MgO was observed, while in the silicate electrolyte, a two-layered structure was obtained. It was found that faster outward diffusing Mg ions resulted in a magnesium-rich top layer on the inner Ta_2O_5/MgO layer.

A compact anodized coating with less defects, higher thickness and stable composition in aggressive environments would be expected to provide favorable corrosion protection to the Mg alloy substrate.

9.4 Influencing factors

Theoretically, the main factors influencing the anodizing of magnesium alloys include the composition and microstructure of the substrate magnesium alloy, the composition of the anodizing bath solution, and the operational parameters, such as anodizing current density, voltage and temperature. For example, the dimension and distribution of porosity are a function of the electrolyte composition, anodizing current density and voltage, which further determine the corrosion resistance of an anodized magnesium alloy (Blawert *et al.*, 2006; Shi, 2005; Shi *et al.*, 2006a, 2006b).

9.4.1 Anodizing current/charge and voltage

Anodizing current or charge and anodizing voltage can significantly affect the anodizing process. It has been found that the microstructure of an anodized coating is determined by the anodizing current charge used for the coating formation (Fig. 9.7a–9.7d).

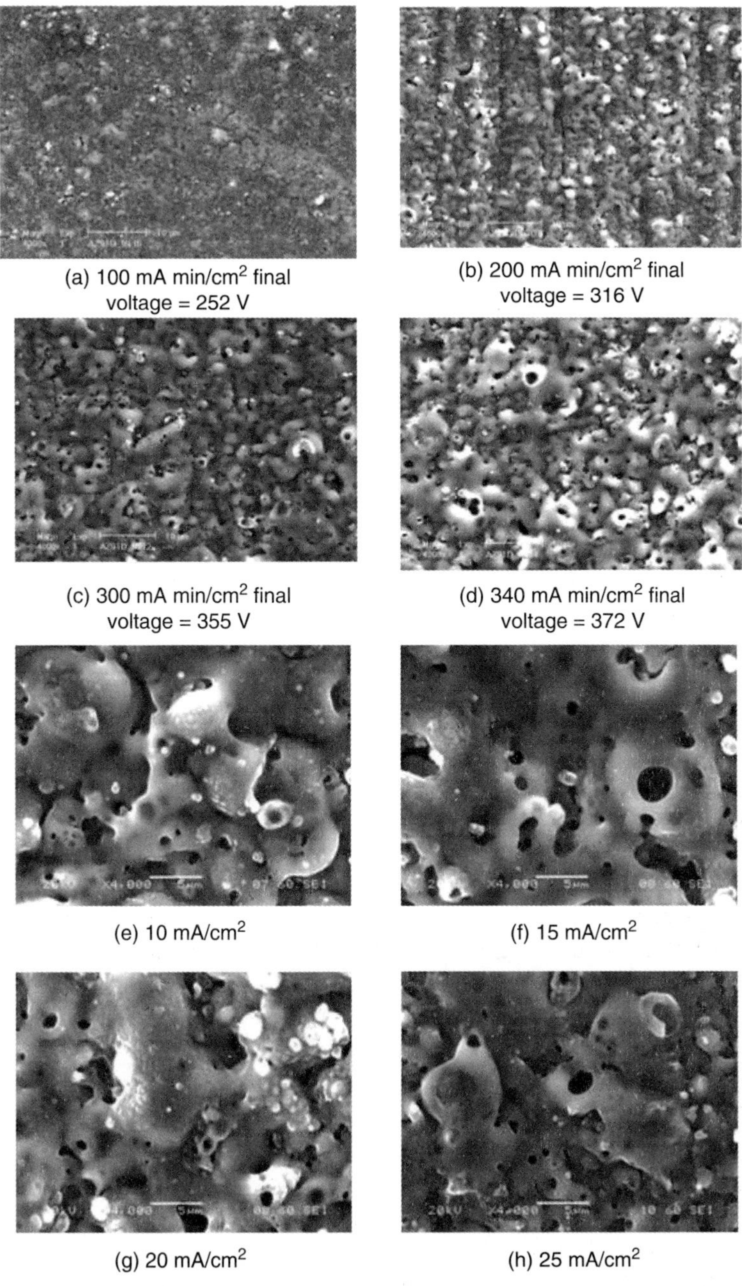

(a) 100 mA min/cm^2 final voltage = 252 V

(b) 200 mA min/cm^2 final voltage = 316 V

(c) 300 mA min/cm^2 final voltage = 355 V

(d) 340 mA min/cm^2 final voltage = 372 V

(e) 10 mA/cm^2

(f) 15 mA/cm^2

(g) 20 mA/cm^2

(h) 25 mA/cm^2

9.7 SEM images of anodized coatings formed in different stages (a), (b), (c) and (d) (with different anodizing current charges and final anodizing voltages) and under different current densities (e), (f), (g) and (h) with the same amount of total electric charge (18 C/cm^2) on AZ91D (Shi *et al.*, 2006b).

The pores in the coating formed in stage III are smaller in diameter (1–2 μm) than those in stage IV. With the same amount of total anodizing charge, the microstructure of a formed coating is relatively less significantly affected by anodizing current density (see Fig. 9.7e–9.7h).

However, if the current charge is not controlled, then a high current density can lead to a thicker and more porous anodic film. It has also been found that (Chai *et al.,* 2008) the corrosion resistance of an anodic film can be related to the applied current density. An anodized coating formed at a higher current density should be thicker and have higher corrosion resistance.

9.4.2 Anodizing bath solution

The composition of anodizing bath solution is the core part of many patents in this area, such as DOW17 (Hillis, 1994), HAE (Hillis, 1994), MGZ (Kotler *et al.,* 1976a, 1976b, 1977), Tagnite (Bartak *et al.,* 1995; Zozulin and Bartak, 1994), Magoxid (COMA, 2001), ANOMAG (Barton, 1998; Barton *et al.,* 2001) and Keronite (COMA, 2008), etc., because it can critically influence the coating composition, microstructure and performance. For example, although the Dow17, HAE and Tagnite anodized coatings are all porous, due to the different bath solutions used in these processes, their pore size and distribution characteristics are quite different. Figure 9.8 compares Anomag and Tagnite coatings on ZE44 Mg alloy, showing their different microstructures and thickness. Anomag bath normally contains phosphate and ammonia while Tagnite uses silicate and fluoride.

The effect of electrolyte on the characteristics of oxide coatings on Mg alloys has previously been studied (Barton, 1998; Fukuda and Matsumoto, 2004; Hsiao and Tsai, 2005; Khaselev *et al.,* 1999). In early studies, anodizing electrolytes mainly contained chromic acid or fluoride (Ono and Masuko, 2003; Zhang *et al.,* 2005). Recent research focused on chromate-free electrolytes, such as alkaline solution with additives such as phosphate, silicate, borate and organic substances (Cai *et al.,* 2006; Sharma *et al.,* 1997; Wu *et al.,* 2007). The effect of electrolyte on anodizing behavior aimed at improving coating quality is an important topic in investigations (Barton and Johnson, 1995; Hagans, 1984; Khaselev and Yahalom, 1998a; Khaselev *et al.,* 1999, 2001; Kotler *et al.,* 1976b, 1977; Takaya, 1987, 1988, 1989b; Takaya *et al.,* 1998; Zozulin and Bartak, 1994).

Although the influence of anodizing electrolyte on anodizing of Mg alloy is complicated, some common knowledge has been gained from existing studies. Generally speaking, some constituents from an anodizing electrolyte can be combined in the anodized film (Hillis, 2001; Ono *et al.,* 2003; Ono and Masuko, 2003; Takaya, 1989a). If an anodizing bath contains aluminates,

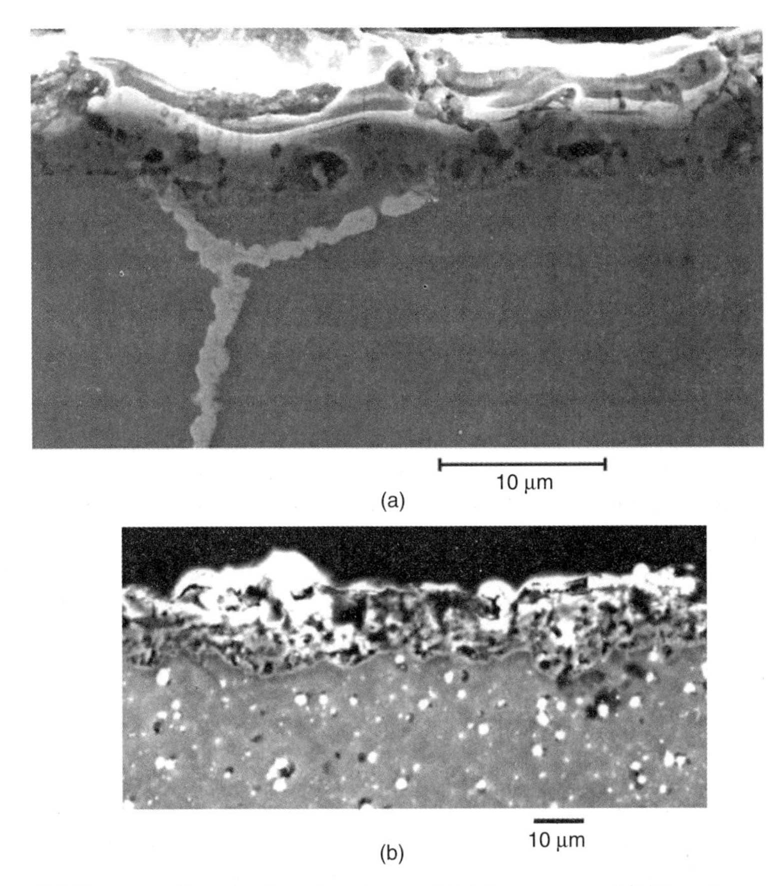

9.8 Cross-sectional microstructures of (a) Anomag and (b) Tagnite coatings.

then aluminum can normally get into the anodized coating (Khaselev and Yahalom, 1998b) as a significant component. The form of the aluminum in the coating is likely to be magnesium aluminate, combined with magnesium oxide and hydroxide. For example, when the electrolyte contains 1.1 M aluminate, the films on Mg are mainly composed of a single oxide phase – $MgAl_2O_4$ spinel (Khaselev *et al.*, 1999). Verdier further confirmed that the Al content in the film was a function of the aluminate concentration in the anodizing electrolyte (Verdier *et al.*, 2004). The addition of $Al(NO_3)_3$ into a 3 M KOH+0.21 M Na_3PO_4+0.6 M KF base electrolyte was found to contribute to a uniform sparking in anodizing AZ91 (Hsiao and Tsai, 2003). Without or with a low concentration of $Al(NO_3)_3$, a porous and non-uniform anodic film was formed. As the concentration of $Al(NO_3)_3$ was increased up to 0.15 M, uniform and compact anodic films were obtained. The addition of $Al(NO_3)_3$ into the base electrolyte resulted in the formation of Al_2O_3 and

$Al(OH)_3$ in the anodic film. The maximum amount of Al_2O_3 was found in the anodic film when the alloy was anodized in an electrolyte containing 0.15 M $Al(NO_3)_3$. However, the film thickness decreased as the concentration of $Al(NO_3)_3$ was increased.

Addition of fluoride, aluminate, phosphate or tetraborate to the electrolyte solution results in a mixed composition of various magnesium compounds in the anodized layer (Barton and Johnson, 1995). The presence of sodium fluoride in a bath solution leads to improved surface texture and opacity, but does not affect the thickness significantly (Shi *et al.,* 2003). Citrate in the electrolyte is helpful in controlling sparking process, and preventing pit formation during sparking. Iodide is a damaging electrolyte and can lead to uncontrolled pitting. Barton (Carter *et al.,* 1999) reported that certain electrolytes including aluminate and tetraborate could influence the coating thickness and microstructure, while other electrolytes, such as fluoride and phosphate, contributed to the color, opacity and uniformity of the coating. Tetraborate contributes both to coating thickness and colour, and lowers sparking voltage (Shi *et al.,* 2003). Kotler *et al.* (1976a, 1977) investigated an anodizing process on magnesium using an alternating current in an aqueous solution comprising chromate, vanadate, phosphate and fluoride and found that the coating was better than the earlier magnesium anodized coatings DOW17 (Hillis, 1994) and IIAE (Abulsain *et al.,* 2004) in terms of the corrosion resistance and wear resistance.

Obviously, in an anodizing bath, anions play a critical role in anodizing. The effects of SiO_3^{2-}, PO_4^{3-}, MoO_4^{2-} and AlO_2^- on anodizing behavior have been compared (Chai *et al.,* 2008). Among the selected oxysalts, sodium silicate could contribute to formation of an anodic film with the best anti-corrosion property, and the different concentrations of sodium silicate would cause the change of anodic film structure. Ma *et al.* (2004) showed that anodized coatings formed in Na_2HPO_4 solutions had better corrosion resistance than those produced from Na_2SiO_3 solutions, and the latter had slightly better erosion resistance than the former. They have different phase composition and corrosion failure processes (Cai *et al.,* 2006). Phosphate, fluoride and borate additives were reported to be beneficial to the improvement of corrosion resistance of anodized films on Mg alloy produced in silicate-based solutions (Duan *et al.,* 2007). Fluoride and phosphate were useful to enhance the corrosion resistance of inner layer of anodized films, while borate contributed to the growth of anodized film on Mg alloy.

Apart from inorganic chemicals in the anodizing bath, some organic additives can also influence the anodizing process and coating quality. For example, the addition of silica sol increased the thickness of the anodic film and improved the roughness of the film surface (Li *et al.,* 2006), and the anodic film formed in the electrolyte with addition of silicon–aluminum sol was more uniform and compact (Zhu and Liu, 2005). Titania sol can also lead to more uniform morphology with fewer structural imperfections and better corrosion resistance (Liang *et al.*, 2007).

Furthermore, anodizing of magnesium is even possible in organic electrolytes as demonstrated by Asoh and Ono (2003). In triethylamine/ethylene glycol solution, layers with transparent appearance and excellent corrosion resistance were obtained in the region between 10–40% of water content. However, the water content strongly affected the formation of the anodic film. Above 10% water content, a compact, transparent and enamel-like barrier-type film structure was observed. At higher water contents, the films became opaque and lost their good corrosion resistance. The properties were related to the incorporation of organic species (C, N, O) into the layer.

The solution temperature has a negative effect on the anti-corrosion property of an anodized film. This is probably because the large quantity of heat caused by sparking could not be released effectively (Chai *et al.*, 2008) when anodized at a higher temperature.

9.4.3 Influence of the substrate

Different Mg alloys have slightly different anodized films in the same bath solution. It has been reported (Hagans, 1984) that the coatings formed on Mg–Al alloys in a borate solution are enriched in aluminum compared with the base alloys and the increase in aluminum content in the alloy is beneficial. The enrichment of aluminum in the anodized coating becomes more evident with increasing aluminum content in the substrate alloy. It has also been found that the anodized coating formed on Mg–Al alloys in a bath containing aluminates is composed of $MgAl_2O_4$ spinel and that on intermetallic $Mg_{17}Al1_2$ (β-phase) comprised γ-Al_2O_3 and $MgAl_2O_4$ (Bonilla *et al.*, 2002a, 2002b; Clapp *et al.*, 2001; Verdier *et al.*, 2004; Wright *et al.*, 1999).

The difference in substrate can affect the anodizing behavior, which can be directly reflected by an anodizing voltage curve. Ono and co-workers (Ono and Masuko, 2003; Ono *et al.*, 2003) found a difference in anodizing behavior under a potential control mode between AZ31, AZ91 and pure Mg in 3 M KOH, 0.6 M KF and 0.2 M Na_3PO_4 (pH 13) at 29.8°C; the peak current at 5 V decreased with increasing Al content of the substrate; and the critical breakdown voltage was 60 V for 99.95% Mg, 99.6% Mg and AZ31B, but 70 V for AZ91D. Figure 9.9 shows that an Al-free alloy appears to have a lower anodizing voltage than an Al-containing alloy under a current control mode.

To some degree, even the microstructure of an anodized coating is dependent on the substrate, too. Figure 9.10 shows the different microstructures of the anodized coatings formed on α-matrix (Mg–1Al) and β-phase (Mg–41Al). It seems that the coating on the β-phase is more porous than the one on the α-phase.

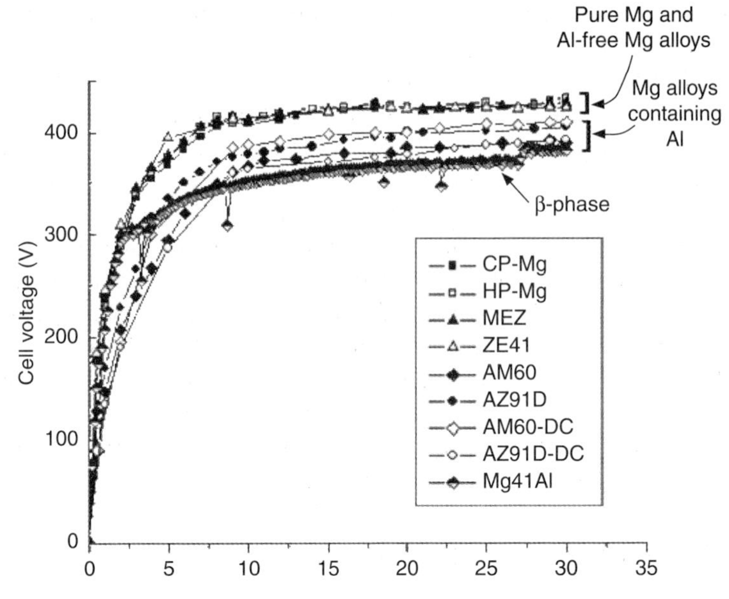

9.9 Anodizing voltage curves for different Mg alloys (Shi *et al.*, 2006c).

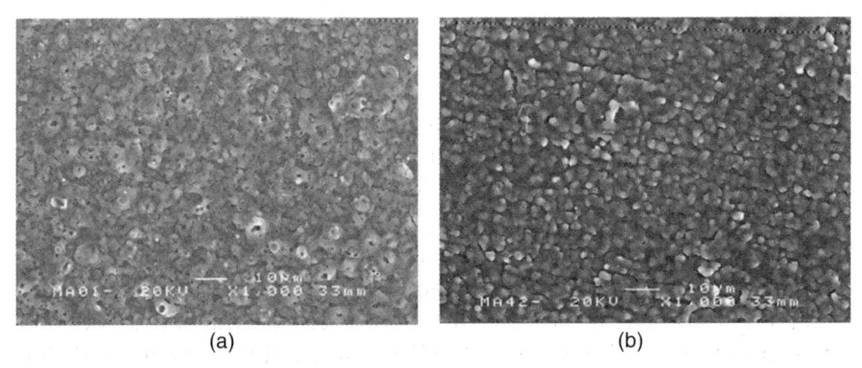

9.10 SEM images of the anodized coatings on (a) Mg–1Al (α-phase) and (b) Mg–41Al (β-phase) (Shi *et al.*, 2005).

9.5 Anodizing mechanism

9.5.1 Anodizing reactions

The formation of an anodized coating on a Mg alloy is a result of electrochemical and chemical reactions between the Mg alloy and an anodizing bath solution at an anodizing voltage (Barton and Johnson, 1995). The anodizing process of a Mg alloy includes complicated reactions between the alloy and the bath electrolyte.

Generally speaking, the electrochemical decomposition reaction of water occurs and oxygen from the anode and hydrogen from the cathode are generated before sparking. The intensity of the electrolysis reaction varies widely, depending on the electrolyte composition. Oxygen evolution from the anode is initially vigorous, then decreases with time as if the magnesium anode were somewhat passivated before sparking occurred (Barton and Johnson, 1995). When the voltage increases above a certain potential (also called initial sparking potential), sparking begins and a complex coating is generated with the arc travelling randomly over the surface. At the same time, oxygen evolution becomes more intense, because the evolution mechanism changes from electrochemical to chemical decomposition of water.

An anodizing process should include at least oxidation of the substrate Mg alloy, oxygen evolution and heat-decomposition of some electrolytes (e.g. aluminate, silicate and borate). The anodized coating on a Mg alloy is a mixture of the products from anodizing reactions, such as oxides and salt deposits of the parent metals and other high temperature decomposed oxide deposits from the solution (Avedesin and Baker, 1999). For example, during a Magoxid anodizing process (Kurze, 1998), at higher potentials (>100 V) the barrier layer is locally destroyed by injection of charged particles. The charge transfer and the diffusion of metal ions cause heat, and the release of energy is high enough to start the plasma chemical processes (gas discharges) in the metal surface/gas/electrolyte interfaces. Discharge channels are formed between the electrolyte/gas film (quasi cathode) interface and the metal surface (anode). This kind of arcing generates plasma similar to ionization of oxygen, melting and oxidizing the metal surface. Simultaneously, areas close to the discharge channels are thermally activated. Hence, on the entire surface a large number of partial anodes form, converting the whole surface into Mg oxide.

To further illustrate the anodizing mechanism and the reactions involved in Mg anodization, an anodizing process of a Mg alloy in a silicate-containing bath is analyzed as an example (Shi, 2005; Shi *et al.*, 2003, 2005, 2006a, 2006b, 2006c).

First stage (stage I)

The anodizing voltage increased linearly with time. Dissolution of Mg, formation and deposition of $Mg(OH)_2$ and $MgSiO_3$ and oxygen evolution occur. The possible detailed reactions include:

$$4OH^- \rightarrow 2H_2O + O_2 + 4e^- \quad [9.1]$$

$$Mg \rightarrow Mg^{2+} + 2e^- \quad [9.2]$$

$$Mg^{2+} + 2OH^- \rightarrow Mg(OH)_2 \downarrow \quad [9.3]$$

$$Mg^{2+} + SiO_3^{2-} \rightarrow MgSiO_3 \downarrow \quad [9.4]$$

The electrochemically evolved oxygen is the main component of the gas from the anode in the alkaline silicate solution. The evolution of oxygen from the anode in anodizing has been reported in previous work (Khaselev and Yahalom, 1998a, 1998b; Khaselev *et al.*, 1999). It is not surprising for oxygen to be electrochemically evolved at such a positive potential from an electrochemical point of view. In practice, the oxygen evolution was not significant in this stage because the overall voltage was mainly distributed in the bath solution and thus only a relatively small potential drop is distributed across the interface of electrode and solution where reaction [9.1] takes place.

Reactions [9.2] and [9.3] have been reported by Mizutani *et al.* (2003) to result in a film formed at a low anodizing voltage in alkaline solution consisting of $Mg(OH)_2$. Mg is first dissolved in solution and then deposited with hydroxyls on the electrode surface to form the primary coating. The most important evidence for reactions [9.3] and [9.4] is the chemical composition of the coating formed in this stage. This is determined by reaction [9.2]. The X-ray photoelectron spectroscopy (XPS) results of the coating formed in this stage show that the coating contains $Mg(OH)_2$ and $MgSiO_3$. The postulation of reaction [9.3] is inspired by Khaselev's work (Khaselev *et al.*, 2001) in which $MgAl_2O_4$ was found to be the composition of the anodized coating in an aluminate containing solution. Most of the anodizing current in this stage is consumed by the reaction [9.2] because reaction [9.1] is insignificant. The formed coating in this stage is thin because the anodizing time is only a few seconds.

Second stage (stage II)

The anodizing voltage further increases to about 190 V. Within this potential range, evolution of some tiny gas bubbles was observed. There was no sparking. It is postulated that in addition to anodizing reactions [9.1], [9.2], [9.3] and [9.4], another reaction starts to take place:

$$Mg + 2OH^- \rightarrow MgO \downarrow + H_2O + 2e^- \quad [9.5]$$

This is an electrochemical reaction resulting in coating formation directly. Mg is directly electrochemically oxidized into MgO.

Mizutani *et al.* (2003) reported that the anodized coating on Mg alloys formed at 60 V in an alkaline solution contained a mixture of magnesium hydroxide and magnesium oxide which has been confirmed by X-ray diffraction (XRD) analysis. Therefore, the involvement of reaction [9.5] in this stage is reasonable. The reasonability of reactions [9.3], [9.4] and [9.5] involved in this stage of anodizing can also be supported by experimental results. For reaction [9.4], magnesium silicate has been found by XPS analysis in the anodized coating. Moreover, the limited oxygen evolution observed in this stage indicates that reaction [9.1] is still slow in the second stage of the anodization. The high-resolution XPS results imply $Mg(OH)_2$, MgO and $MgSiO_3$ in this stage.

Third stage (stage III)

The anodizing voltage continues to increase to about 330 V. The most significant phenomena in this stage are uniform sparking and significant gas evolution from the anodized specimen surface. It is postulated that reactions [9.1], [9.2], [9.3], [9.4] and [9.5] continue in this stage and at the same time the following new reactions also occur:

$$Mg(OH)_2 \rightarrow MgO \downarrow + H_2O \qquad [9.6]$$

$$2Mg + O_2 \rightarrow 2MgO \downarrow \qquad [9.7]$$

In other words, the new reactions are mainly chemical precipitation processes. Reaction [9.6] is a dehydration process that can occur when temperature is higher than 350°C (Weast, 1976–1977). It has been reported (Yahalom and Zahavi, 1971) that the temperature of the sparking arc during the anodization of Mg alloys was far above 1000°C. Because of the sparking in this stage, some local areas where sparking occurs can easily exceed this dehydration temperature. However, not all the $Mg(OH)_2$ was changed into MgO. High-resolution XPS results show that there is still some $Mg(OH)_2$ in the coating. At the high temperature, direct oxidation of magnesium is also possible, particularly with the active oxygen freshly generated by reaction [9.1]. Therefore, reaction [9.7] is proposed in this stage of anodizing, and this is proved by the increased amount of MgO in the coating in this stage.

Fourth stage (stage IV)

After the cell voltage is over 330 V, oxygen evolution becomes vigorous, and sparking much more intense and localized. It is proposed that the following two reactions are also involved in the fourth stage of anodizing in addition to the reactions listed above.

$$MgO + yMgSiO_3 \rightarrow (MgO)_x \cdot (SiO_2)_y \qquad [9.8]$$

$$2H_2O \rightarrow 2H_2 \uparrow + O_2 \uparrow \qquad [9.9]$$

Reaction [9.8] is supposed to be a melting and solidification process of MgO and $MgSiO_3$ in the coating in the sparking spots. In fact, at high temperatures during sparking, the anodized coating could be locally melted by sparking spots and mixed together with some components or decomposed products of the bath solution, then rapidly solidified to reform coating. The melting points of MgO and $MgSiO_3$ are 2852°C and 1910°C, respectively (Weast, 1976–1977). Under a very intense sparking condition, some sparking arcs may generate sufficient heat to melt coating and the components of the bath solution in some particularly confined local sparking areas. The sparking cannot last long. It stops shortly, and thus the melted coating is rapidly cooled by the surrounding solution. In this way, the coating deposited in the first three stages is locally melted by sparking to form a thick and coarse coating. The significant increase in the ratio of Si/Mg in the coating formed in stage IV can be interpreted as the significant deposition of the silicates from the bath solution directly into the coating through the melting and solidification process.

The dramatically increased gas evolution in this stage cannot be simply ascribed to the electrochemical oxygen evolution. Thermal decomposition of water, reaction [9.9], should be considered. The normal thermal decomposition temperature of water was reported to be over 2000°C (Brown *et al.*, 2002). Since the temperature of the sparking arc during the anodization of magnesium alloys was reported to be far above 1000°C (Yahalom and Zahavi, 1971), the possibility exists that in some particular areas where the sparking is extremely intense, the local temperature becomes high enough to decompose water. This postulation is further verified by the anodizing current efficiency calculated below.

Based on the above model, the changes in the anodizing voltage, sparking, oxygen evolution, thickness, composition and microstructure of the coating will be explained as follows.

9.5.2 Anodizing voltage and coating growth

In the initial stage, the coating resistance increases with the formation and thickening of the anodized coating. The coating formation reactions [9.3] and [9.4] determined by reaction [9.2] should be proportional to the anodizing current density. Hence, at a constant anodizing current density, the film thickness increases linearly with time, and the increasing rate should be proportional to the anodizing current density.

In the second stage, the coating resistance continues to increase with the coating thickening. Therefore, the anodizing voltage also increases quickly in this stage. The thickness of the anodized coating continues to increase with the deposition of $Mg(OH)_2$, $MgSiO_3$ and MgO from reactions [9.3], [9.4] and [9.5]. As pores start to appear in the coating, the coating becomes less dense than in the first stage. Therefore, the increase of anodizing voltage starts to deviate from the linear increasing tendency and becomes relatively slower.

In the third stage, as the voltage is high enough to cause dielectric breakdown of the surface coating and sparking starts, more reactions occur than in the second stage. Due to the local high temperature, all the electrochemical reactions [9.1], [9.2] and [9.5] involved in this stage become relatively easy. Meanwhile, because of the sparking resulting in porosity of the formed coating, the coating resistance cannot increase significantly, even though the thickness of the coating keeps increasing. Therefore, an even slower increasing rate of the anodizing potential can be observed in this stage. Since sparking results in significantly increased deposition of Mg oxides, film thickening is evident in this stage. Dramatic potential oscillation always accompanies sparking behavior in this stage. For example, during sparking the potential oscillating randomly in the range of 60–90 V is often seen. This can be explained by the film breakdown and repair during sparking. Wright *et al.* (1999) anodized pure magnesium and some alloys of magnesium in alkaline solutions, using constant applied current, and found that the potential increased linearly with time to about 70 V as a thin barrier film grew on the metal surface. The electric field in the barrier film was found to be 9×10^8 V/m and there was no variation of field with the applied current. When the barrier film reached a critical thickness of around 80 nm, the observed potential dropped abruptly to a relatively low value of about 10 V. The potential then began to rise again linearly, until another sudden drop was observed. This pattern continued for a number of cycles, generating a saw-tooth pattern of potential as a function of time. The abrupt drop in potential is due to the rupture of the barrier film, forming a porous secondary layer. This rupture is caused by the tensile stresses in the barrier film, which has a molar volume much smaller than the metal from which it is formed.

When anodizing develops into the fourth stage, due to the intense sparking, relatively coarser pores are formed in the coating, which significantly decreases the coating resistance. So the anodizing voltage does not increase with time. The localized intensive sparking can cause rapid deposition of the coating in the local area. This significantly roughens the film and also leads to a thicker coating.

9.5.3 Sparking and microstructure

Although the bulk electrolyte temperature might be only 50°C, the local temperature in the plasma zone would probably be in excess of 1000°C, which can lead to formation of 'glassy' or 'ceramic' anodic coatings. The local high temperature can speed up reactions, such as deposition of the coating, gas evolution and oxidation of Mg alloy. It can also cause high-temperature decomposition of the components in the bath solution.

The high local temperature in an anodic film is a result of an extremely high current density passing through the pores or defects in the film. These elevated temperatures cause vaporization of the electrolyte and plasma formation in localized areas on the surface. The plasma or sparking in return produces even more heat and much higher temperatures locally. Nykyforchyn *et al.* (2003) investigated the plasma formation. Several hundred volts was applied to a metal–electrolyte system, resulting in electric discharges. Consequently, plasma develops in discharge channels and oxide deposition occurs on the metal surface. The density of the plasma electrons was determined to be $n_e \approx 10^{22}$ m^{-3} and the plasma electron temperature during the synthesis of oxides in the electrolyte plasma of spark discharges was $T_e \approx 10^4$ K. The calculation of temperature distribution in the spark zone suggests that the temperature in centre of the spark zone can be higher than the melting point of $MgAl_2O_4$; that should be a reason for the porous morphology of anodic films formed under continuous sparking (Khaselev *et al.*, 2001).

Sparking is normally involved in an anodizing process at high voltages. Magnesium is observed to exhibit spark discharges that move randomly over the surface. The sparking could start from some 'weak' sites of the film formed in the second stage, such as thin areas or areas with defects or pores. At a sparking spot, the sparking has two consequences. First, the dielectric breakdown during sparking results in pores. The direct evidence for this is the porous morphology of the formed coatings in this stage. As the sparking becomes more intense with increasing cell voltage, the pore size and the porosity of the coating increase. Second, the high local temperature resulting from sparking can to some extent consolidate the coating and seal the pores caused by dielectric breakdown of the coating. The heat generated by sparking could melt the coating surrounding the sparking spots. After the sparking stops at those spots, the melted coating rapidly solidifies there. This leads to the overlapped pearl-like morphology surrounding the pores and to some extent sealing the porosity of the coating. Therefore, it is difficult for sparking to reoccur in a sparked spot. It always travels or occurs randomly over a Mg surface. In some cases, after intense sparking causes too large a pore or damaged area in an anodized coating, sparking and anodizing will stop, because the large pore or damage area will allow the passage of all the anodizing current through (all the anodizing current will be leaking through there) and thus a sufficiently high potential drop across the interface between the specimen and solution cannot be sustained for sparking. This is why the anodizing voltage sometimes suddenly drops to a very low value and sparking stops after a couple of blinding flashes on the specimen surface in practical anodizing.

Since the porosity of an anodized coating formed at high voltage results from sparking, the intensity of the sparking process can to a great extent determine the coating porosity. If the intensity of sparking is low, the anodized coating is formed slowly. Correspondingly, the sparking area is small and the breakdown and melting reaction are not severe. As a result, the coating microstructure becomes fine and compact because the pores formed during the sparking process were small. On the other hand, more intense sparking can lead to a thicker, coarser and more porous anodized coating. All of these expectations have been proved by SEM observation of the microstructure.

9.5.4 Oxygen evolution and anodizing efficiency

Oxygen evolution is found to occur after the cell voltage exceeds 3 V (Shi, 2005; Shi *et al.*, 2005, 2006a, 2006b), and its rate increases with voltage. The trend of oxygen evolution dramatically accelerates after the anodizing develops into sparking stage, indicating that different reactions are responsible for the evolved gas before and after sparking. Before sparking, the evolved oxygen simply results from electrochemical decomposition of water Reaction [9.1]. However, in the sparking stage, thermal decomposition of water (Reaction [9.9]), is the major reaction responsible for the gas evolution. The gas evolution caused by the thermal decomposition of water is dependent on the heat or temperature. At a high temperature, the rate of the gas evolution can be very high.

The current efficiency can be defined by:

$$\eta = \frac{I_F}{I_T} \times 100\% \qquad [9.10]$$

where I_T is the total applied current and I_F is that part of the total current resulting in film formation. If it is assumed that the by-product forming reaction is simply the production of oxygen by the reaction [9.1], then I_F may be estimated from:

$$I_F = I_T - I_{O_2} \qquad [9.11]$$

where I_{O_2} is the rate of electrochemical evolution of oxygen (Reaction [9.1]). Hence,

$$\eta = \frac{I_T - I_{O_2}}{I_T} \times 100\% \qquad [9.12]$$

Theoretically, I_{O_2}, the electrochemical evolution of oxygen, cannot be greater than I_T. If the anodizing efficiency is calculated based on an assumption that the evolved gas is electrochemically evolved oxygen, a negative efficiency could be obtained. In other words, a negative η means a non-electrochemical evolution of gas involved in the anodizing process.

Figure 9.2 shows that the anodizing efficiency decreases with the increasing cell voltage, particularly in the fourth stage down to a negative value. It confirms that in the fourth stage (an intense sparking stage), a non-electrochemical gas evolution mechanism starts to operate. This non-electrochemical process is the thermal decomposition of water due to the intense sparking in this stage.

9.6 Corrosion of anodized magnesium (Mg) alloys

9.6.1 Corrosion performance

An anodized Mg alloy is much more corrosion resistant than an un-anodized one. A commercially anodized Mg alloy without any post-treatments or sealing can survive in standard salt spray (ASTM B117) for about 400–600 hours. Figure 9.11 presents the corrosion performance of a recently developed anodized coating (Song, 2004), which shows that the anodized AZ91D survives in 5 wt% NaCl environments for one month and no obvious corrosion damage can be visualized on the surfaces.

The corrosion of an anodized coating usually starts from tiny pitting corrosion and then progresses into localized or filiform corrosion. For example, in a 5% NaCl solution, specks and pits can be seen on the surface of an anodized AZ91D specimen after 16 h of immersion; after 88 hours, the

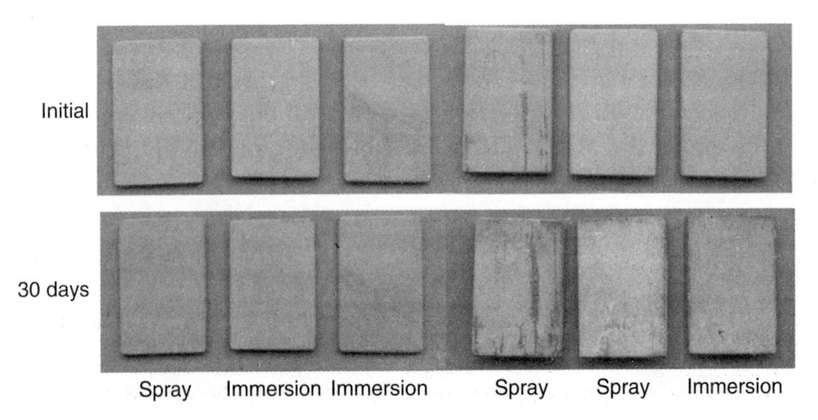

9.11 Corrosion damage of anodized AZ91D (Song, 2004) after 5 wt% NaCl immersion and salt spray.

specks become larger and develop into localized corrosion and filiform corrosion (Shi *et al.*, 2006b). According to the corrosion morphologies of anodized AZ91D in 5 wt% NaCl, the corrosion resistance of the anodized coating formed at stage IV is better than that formed at stage III, which can mainly be ascribed to the larger thickness of the coating formed at stage IV.

The corrosion performance of an anodized magnesium alloy depends on the corrosion performance of the substrate alloy. For example, the corroded areas of the anodized commercial alloys are measured and plotted versus the corrosion rates of these alloys in Fig. 9.12. There is a good correlationship in corrosion damage degree between the anodized specimens and the corresponding un-anodized alloy.

9.6.2 Protectiveness of an anodized coating

The penetration of corrosive species through the anodized coating is a critical step to the corrosion performance of an anodized Mg alloy if the coating is relatively compact. The corrosion resistance of a coated electrode should mainly be determined by the resistance of the coating to the corrosive species penetration. However, for a porous anodized coating on a magnesium alloy (Blawert *et al.*, 2006; Shi *et al.*, 2005, 2006a, 2006b, 2006c), the ingress of aggressive ions through the pores or defects in the coating to the substrate/coating interface would not be too difficult. In this case, both the resistance

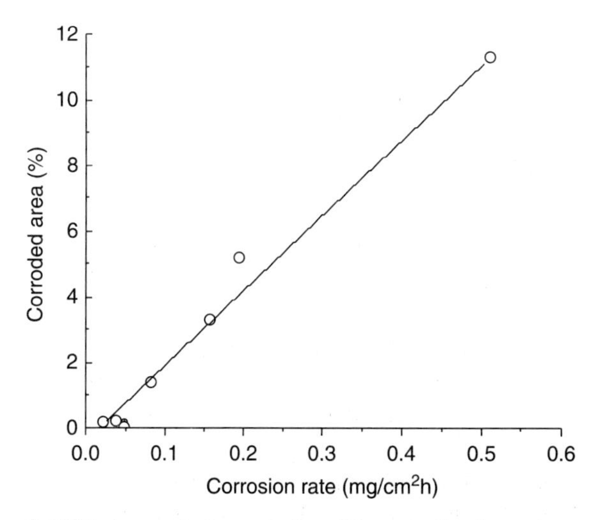

9.12 The corroded area ratio of the anodized specimens *vs.* the corrosion rate of the substrate alloy under 5 wt% NaCl immersion condition (Shi, 2005).

of the coating to the ingress of corrosive species and the resistance of the substrate/coating interface to corrosion would be responsible for the corrosion of the anodized alloy. The main role of an anodized coating is to retard the ingress of corrosive species and delay the corrosion of magnesium alloys, which is termed as 'retarding effect' in this study.

In an anodized coating, not all the pores are through from the top surface of the coating to the substrate/coating interface. The cross-sections of the anodized coatings (Figs 9.5 and 9.8) indicate that the possibility of through-pores is in effect not very high. Although the porosity of an anodized coating appears to be high, most of the pores are discontinuous and the number of the through-pores is very small. There may be a dense inner layer in an anodized coating next to the substrate (Blawert *et al.,* 2006), but this layer could be too thin to be observed in this study, and it could be in nature similar to the surface film spontaneously formed on a Mg alloy, which might be more conveniently considered as a surface layer of the substrate, rather than a layer in the anodized coating. In the event that a corrosive solution finally reaches the substrate/coating interface through the small number of the through-pores, the actual total area of the substrate exposed to the aggressive solution at the bottom of the through-pores is very limited. Therefore, an anodized coating can effectively restrict the area of the substrate Mg alloy exposed to corrosive solutions, which is called the 'blocking effect'.

Apart from retarding the ingress of corrosive species and blocking the exposed substrate of the alloy to solution, the most important effect of an anodized coating is suppressing the substrate surface defects or active points that are susceptible to corrosion. It is well known that the corrosion of metals normally starts from some active or defective sites. For Mg alloys, the active sites can be grain boundaries or grain central areas (Song, 2005a) depending on their alloying elements. During anodizing, these active areas are likely anodized preferentially. If there is an anodized coating on a Mg alloy, the chance of the active sites to be right at the bottom of the through-pores will be extremely low. In other words, even after corrosive solution has arrived at the substrate/coating interface underneath a through-pore, it is unlikely that the substrate at the bottom of the through-pore happens to be an active or defective site. By this mechanism, the corrosion resistance of the substrate alloy at the bottom of the through-pore should be much higher than the alloy without an anodized coating. Therefore, the presence of an anodized coating on a magnesium alloy will eliminate or suppress most active or defective sites. This is a kind of 'passivating effect'.

In summary, it is proposed that anodizing can improve the corrosion resistance of a magnesium alloy through three different mechanisms: (1) retarding effect, that is, slowing down the ingress of corrosive ions, (2) blocking effect, reducing the possibility or the area of magnesium alloys directly exposed to corrosive media, and (3) passivating effect, that is, suppressing the active sites for corrosion.

9.6.3 Corrosion model of an anodized Mg alloy

The ingress of corrosive ions into an anodized coating through its pores is not a slow process for a porous anodized coating. After corrosive ions reach a critical concentration threshold at the film/substrate interface and trigger the corrosion of the substrate Mg alloy, the corrosion of the anodized magnesium starts. Mg is dissolved into the solution in the pores, which gradually makes the solution saturated with $Mg(OH)_2$ due to the dissolved Mg^{2+}. Thus, the substrate magnesium alloy is actually mostly exposed to a $Mg(OH)_2$ saturated solution in the through-pores, and the corrosion of an anodized Mg alloy in this sense can be simply regarded as the corrosion of a very small surface area of the alloy in a $Mg(OH)_2$ saturated corrosive solution.

After corrosion starts at the bottom of the through-pores of an anodized coating, the corrosion will further penetrate and spread out from there, leading to collapse of the adjacent coating due to the expansion of the corrosion products. In this stage, the corrosion of an anodized coating covered substrate is in nature similar to the corrosion of the substrate alloy un-anodized.

Based on the above analyses, the typical microstructure and corrosion initiation of an anodized coating can be schematically illustrated as in Fig. 9.13.

9.6.4 Explanation of corrosion behavior

According to the corrosion mechanism, a porous anodized coating can only slow the ingress of aggressive species, reduce the areas of the substrate exposed to a corrosive solution and suppress the defective sites of the substrate for corrosion. All these are dependent on the porosity or more specifically the number of the through-pores in the anodized coating. It implies that a porous anodized coating itself may only slow down the corrosion and mitigate the corrosion damage, but not prevent the corrosion of the anodized Mg alloy. A decrease in porosity or the number of the through-pores of an anodized coating is the most effective way to improve its corrosion resistance. Sealing and top-coating with a primer or paint are effective ways of reducing the porosity or the number of the through-pores. It has also been found that the applied anodizing current density can significantly alter the porosity of an anodized coating and thus the corrosion resistance of an anodized Mg alloy (Shi *et al.*, 2006b).

The thickness also affects the penetration of corrosive solution due to its retarding effect and blocking effect. The ingress of corrosive solution into a

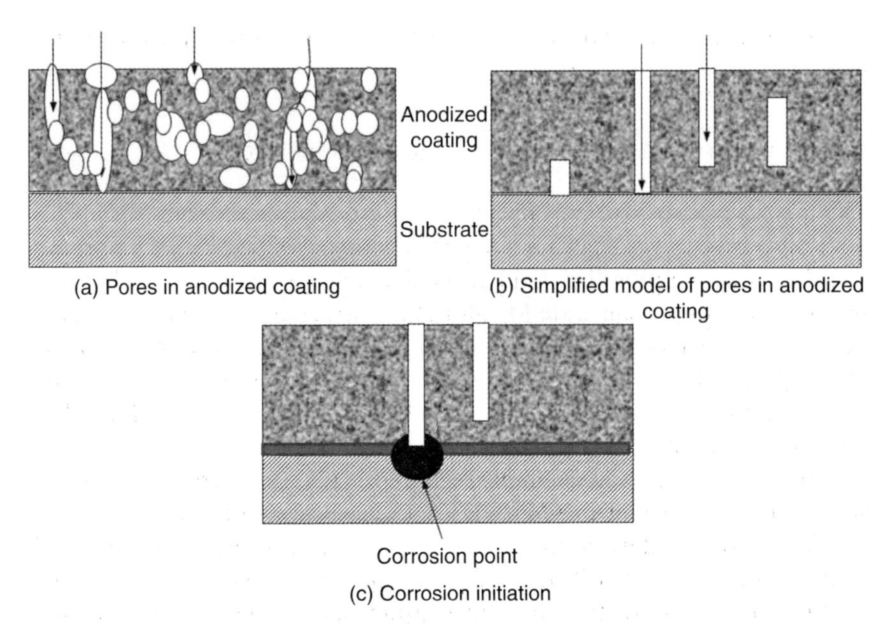

9.13 Schematic diagrams of (a) the microstructure of an anodized coating on the magnesium alloys (based on Shi *et al.,* 2005), (b) the simplified microstructure of the coating (based on Shi *et al.,* 2005) and (c) the corrosion of anodized magnesium alloy.

thicker anodized coating is more difficult or slower than into a thinner coating. Moreover, the thickness of an anodized coating can affect the number of the through-pores; a pore is more likely blocked or becomes discontinuous somewhere in a thick coating than in a thin one. Therefore, a thick anodized coating is usually more corrosion resistant than a thin one. Therefore, when the Anomag coating was found to be thinner than Tagnite under a certain anodizing condition (see Fig. 9.8), it displayed relatively lower corrosion resistance under atmospheric exposure conditions (Song *et al.,* 2006).

Another implication of this corrosion model is the influence of the substrate on the corrosion resistance of an anodized Mg alloy. According to the corrosion model, after the corrosive solution reaches the film/Mg interface, the corrosion resistance of the substrate will directly influence the corrosion behavior of the anodized Mg alloy. As mentioned earlier, the porous anodized coating cannot very effectively retard and block the ingress of corrosive species. The corrosion performance of an anodized Mg alloy should be to a great extent determined by the passivating effect, which is closely related to the intrinsic corrosion resistance of the Mg alloy. Therefore, the corrosion of anodized Mg alloys can be significantly affected by the corrosion resistance of these alloys before anodizing (see Fig. 9.12). It has been found that the corrosion resistance of anodized AZ91 was better than that of anodized

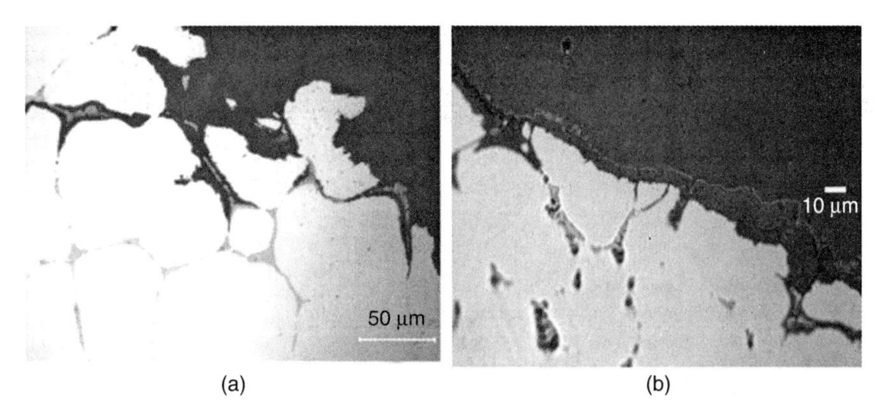

9.14 Optical photo of the cross-sections of (a) un-anodized ZE41 after 75 hours of immersion in 5 wt% NaCl solution and (b) Anomag anodized ZE41 after 20 hours in $Mg(OH)_2$ saturated 0.1 wt% NaCl solution.

AZ31 (Kotler *et al.,* 1977), because AZ91 alloy in general is more corrosion resistant than AZ31.

Strictly speaking, the surface of a magnesium alloy is not uniform. The anodized coating in different areas of the surface could be different in property. Therefore, the corrosion performance of an anodized Mg alloy can have different corrosion resistance in different areas. For ZE41, the grain boundaries are less corrosion resistant than the grain central areas before anodizing (Fig. 9.14a). In other worlds, the defective sites of the alloy are mainly distributed along the grain boundaries. After anodizing, if there happens to be some through-pores in the anodized coating formed over the grain boundaries, then the anodized alloy will be preferentially corroded there. That is why the anodized ZE41 still preferentially corroded along its grain boundaries and has corrosion morphology in microstructure similar to the un-anodized ZE41 (Fig. 9.14b). Along some of the grain boundaries or the secondary phases, corrosion penetrates deeper into the substrate.

As the coatings on α- and β-phases are porous (see Fig. 9.10), once the corrosive media penetrate the coating through the pores or the broken area and arrive at the substrate, the galvanic effect between the α- and β-phases will govern the corrosion of the anodized Mg alloy, although the galvanic effect is to a certain extent mitigated by the coating. In this sense, the corrosion mechanisms of the anodized and non-anodized Mg–Al alloys are similar, governed by the galvanic current density between the α- and β-phases, and the distribution of the β-phase in a substrate Mg alloy can affect the corrosion of the anodized alloy. The most likely or serious corrosion zone should be in the α matrix adjacent to β phase, which has been experimentally observed (see Fig. 9.15).

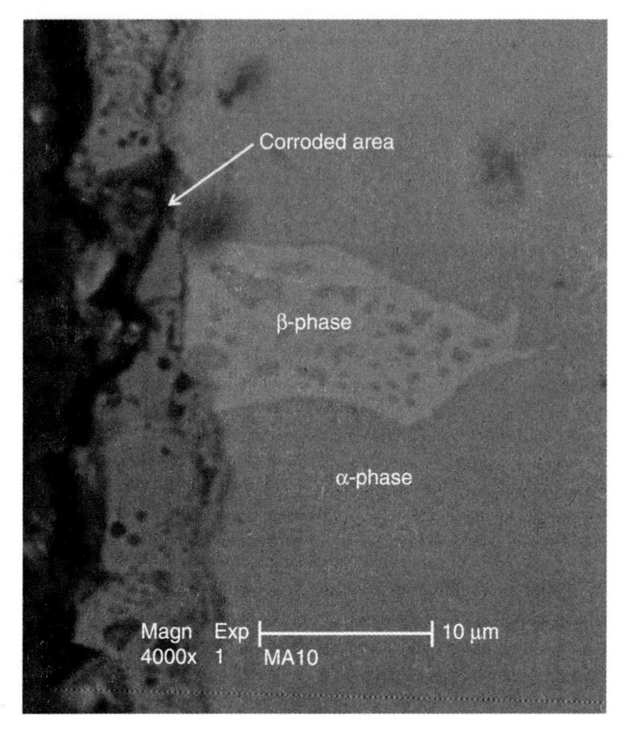

9.15 Corrosion damage of anodized Mg–10wt% Al two-phase alloy immersed in 5 wt% NaCl solution (Shi *et al.*, 2005).

9.6.5 Electrochemical behavior

An un-anodized Mg alloy and its anodized counterpart usually display similar polarization curves in the same (Cl^- containing) corrosive solution, except that the latter has lower polarization current densities and a more positive breakdown potential (Shi *et al.*, 2005) than the former. An example is provided in Fig. 9.16.

According to the corrosion model proposed above, the anodic polarization of the anodized ZE41 should be mainly the dissolution of the substrate at the bottom of the through-pores in the anodized coating in response to an applied potential. The different passive current densities and pitting corrosion potentials of anodized ZE41 from un-anodized ZE41 can be ascribed to the blocking effect and passivating effect of the anodized coating on ZE41. Because of the blocking effect, the anodic dissolution of an anodized Mg alloy can occur only in a very small area of the substrate at the bottom of the through-pores, so the passive current density is dramatically limited. At the same time, the passivating effect significantly reduces the number of active sites and thus leads to a lower passive current density. Therefore, the anodic

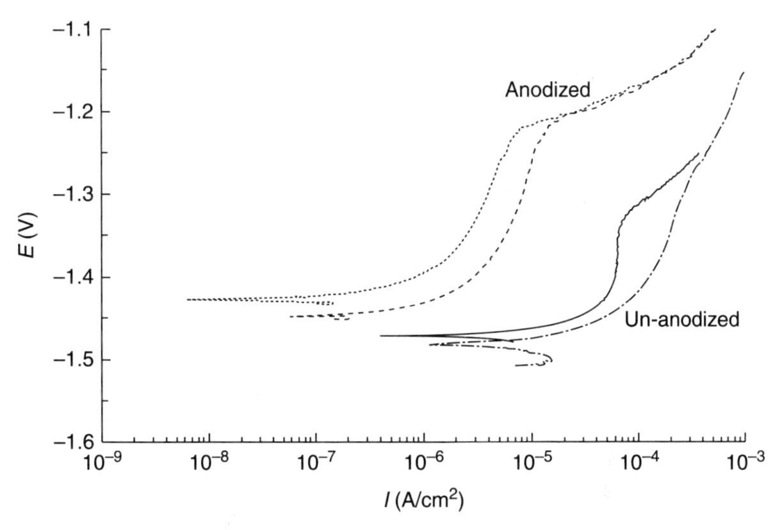

9.16 Polarization curves for un-anodized ZE41 and anodized ZE41 in $Mg(OH)_2$ saturated 0.1wt% NaCl solution (Song et al., 2006).

or passive current density of an anodized Mg alloy is much lower than that of an un-anodized Mg alloy. Since passive sites normally have more positive pitting potentials than defective sites, the onset of pitting corrosion of an anodized alloy would be more difficult. In addition, at the pitting potential, the ohm potential (IR) drop caused by the anodic dissolution current or passive current flowing through the anodized coating is significant, which is added to the measured pitting potential. Therefore, the breakdown potential of an anodized Mg alloy appears to be more positive than that of an un-anodized Mg alloy.

The presence of an anodized coating on a Mg alloy can significantly alter the AC impedance behavior of this alloy. For example, un-anodized ZE41 in $Mg(OH)_2$ saturated 0.1 wt% NaCl solution displays an AC impedance spectrum (EIS) containing two capacitive loops in the high- and intermediate-frequency ranges and an inductive loop in the low-frequency range (Fig. 9.17a). Mg and AZ91D have also similar AC impedance spectra in a chloride-containing solution (Song *et al.,* 1997, 2004). For an anodized ZE41 in the same corrosive solution basically, only one much larger capacitive loop in the high and intermediate-frequency ranges and some inductive characteristic in the low-frequency range can be detected.

The similarity of un-anodized ZE41 to Mg (Song *et al.,* 1997) and AZ91D (Song *et al.,* 2004) in EIS suggests that the corrosion of ZE41 and AZ91D follows the same corrosion mechanism: the Mg alloy surface has a spontaneously formed surface film, but it has some broken areas; Mg is oxidized into Mg^+ and dissolved into solution from the surface film broken

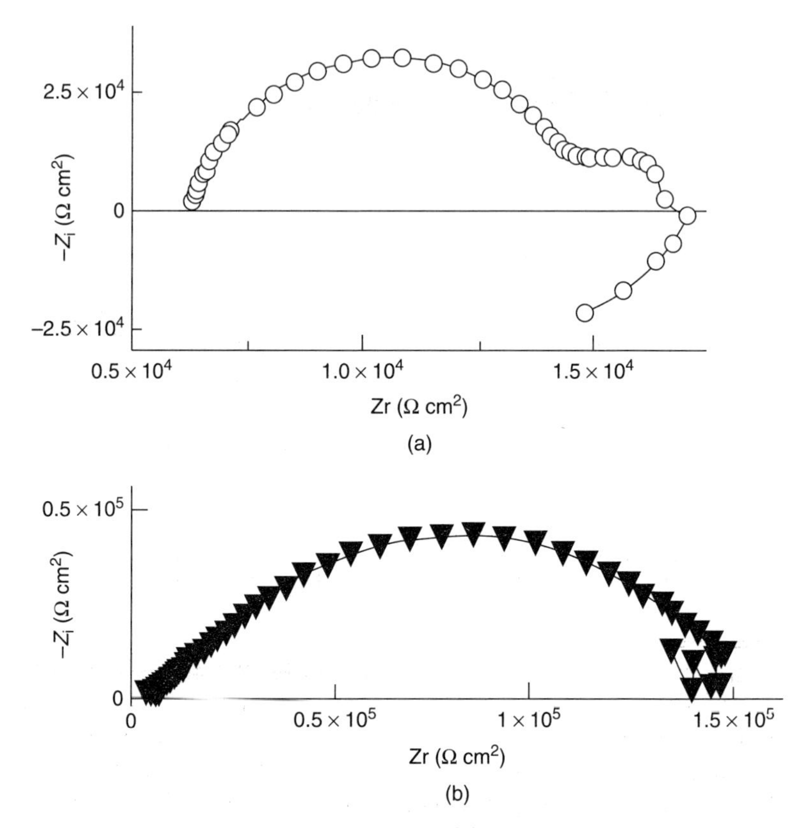

9.17 A typical AC impedance spectrum of (a) un-anodized and (b) anodized ZE41 immersed in $Mg(OH)_2$ saturated 0.1 wt% NaCl solution.

areas; Mg^+ further reacts with water to be oxidized into Mg^{2+}. Based on this understanding, the electrochemical interpretation of the EISs of Mg and its alloys was given by Song *et al.* (1997). The first capacitive loop in the high-frequency range can be ascribed to the charge transfer resistance R_t and the capacitance $C_{s/m}$ at the solution/Mg interface. The second capacitive loop in the mediate frequency range is attributed to the involvement of monovalent Mg^+ ions in the anodic dissolution at surface film broken areas. The inductive loop in the low-frequency range is a result of the response of broken areas of the surface film to the applied potential or Faraday current density.

For an anodized Mg alloy with a porous coating formed on its surface, according to the coating model (Fig. 9.13) proposed earlier, the anodized coating cannot change the electrochemical reactions involved in the corrosion of the alloy, but it can significantly reduce the rates of the reactions. The porosity of the anodized coating allows the penetration of aggressive

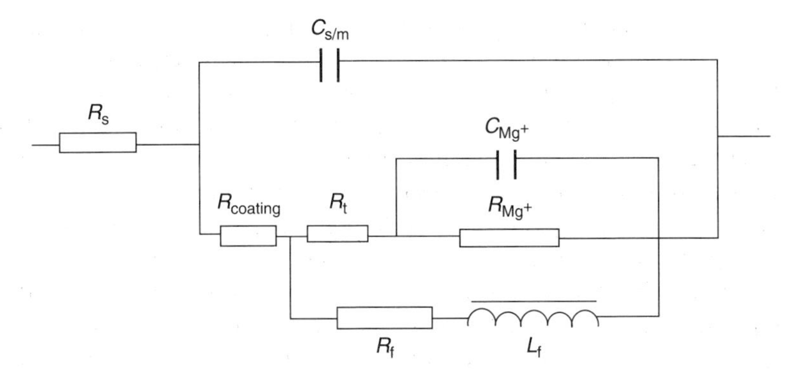

9.18 Equivalent circuit for corroding anodized Mg alloy.

solution through the coating and the arrival of the solution at the substrate surface. After this film breaks down by the corrosive solution in the through-pores, Mg is first oxidized into Mg^+ from the substrate Mg alloy and further reacts with water to form Mg^{2+}. All these electrochemical reactions are the same as those on an un-anodized Mg alloy. However, due to the blocking and passivating effects, the charge transfer step is significantly slowed down. Based on this, a corroding anodized Mg alloy can be depicted by an equivalent circuit as shown in Fig. 9.18. In the equivalent circuit, R_s is solution resistance between the reference electrode and Mg alloy specimen; $C_{s/m}$ is the capacitance between solution and magnesium substrate, which is equal to the coating capacitance for an anodized magnesium alloy; $R_{coating}$ is resistance of the anodized coating and its value is dependent on the coating thickness and porosity, particularly the number of through-pores; R_t is the charge transfer resistance at the bottom of the through-pores of the anodized coating; $C_{Mg}+$ and $R_{Mg}+$ are pseudo-capacitance and resistance caused by the involvement of Mg^+ in the corrosion (Song *et al.*, 1997); R_f and L_f are also pseudo-resistance and inductance resulting from breakdown of the surface film during corrosion (Song *et al.*, 1997).

Theoretically, a corroding anodized Mg alloy system with this equivalent circuit should have three time constants. On the Nyquist plane, there should be two capacitive loops in the high- and intermediate-frequency ranges and an inductive loop in the low-frequency range. In practice, the values of the components in the equivalent circuit may not be in suitable ranges. For example, when $R_{coating}$ is very large and $C_{s/m}$ is very small, the capacitive characteristic in the middle frequency range associated with the involvement of Mg^+ in corrosion can be overwhelmed by the coating related capacitive loop, and hence the displayed EIS on the Nyquist plane does not always have three clear loops, but simply one apparent capacitive loop in the high- and intermediate-frequency ranges and an inductive loop in the low-frequency range.

The reasonability of the equivalent circuit for a corroding anodized Mg alloy can be verified by regressing it into a simple one for an un-anodized Mg alloy. Simply let $R_{coating} = 0$ and slight modify the values of other equivalent components; the equivalent circuit will represent the electrochemical processes of a corroding Mg alloy. As there is no coating between the substrate and solution, $C_{s/m}$ becomes larger. For a film-free substrate with more defective sites, R_t should be considerably smaller. Certainly, the values of the other corresponding equivalent components C_{Mg^+}, R_{Mg^+}, R_f and L_f for the un-anodized alloy will be different from those of its anodized counterpart. After these modifications, the equivalent circuit produces an EIS spectrum with two capacitive loops in the high- and intermediate-frequency ranges and an inductive loop in the low-frequency range.

9.6.6 Evaluation of corrosion resistance

The corrosion resistance of most commercial anodized coatings is normally measured and compared according to salt spray test (ASTM B 117). In fundamental studies, the immersion corrosion test is also used to assess the corrosion performance of anodized coatings. The corrosion resistance of anodized Mg alloys is compared according to their corrosion damage degrees (corrosion morphologies) after the tests. For example, Fig. 9.19 shows the corrosion damage of anodized Mg and its alloys after immersion and salt spray, based on which it can be concluded that the anodized AZ91D is much better than the anodized Mg.

However, both salt spray and immersion tests are time consuming and only provide the final corrosion damage information. A rapid method of evaluating corrosion performance of anodized coatings for Mg alloys is needed and electrochemical technique can serve this purpose.

The electrochemical behavior of anodized Mg alloys has been measured using cyclic voltammogram (CV) and following the EIS techniques. Anodized AZ91 and AM50 after exposure in salt spray for 21 days displayed different corrosion performance on their two side surfaces. The anodized coating on side B of the specimen was better than side A. There were a few corroded pits on side B, while side A was intact; side B was still under the protection of the anodized coating while side A suffered from pitting or filiform corrosion. Figure 9.20 shows the CV polarization curves of anodized AM50 measured in 5% NaCl solution after 21 days of salt spray testing.

The forward scanned curves show that the open circuit potential of the coating on side A was more negative than that of side B. For side A, the pitting potential is equal to the open circuit potential on the forward scanned curve. For side B, the pitting potential is more positive than its open circuit potential before the coating breaks down. After forward

9.19 Appearance of anodized Mg and alloys after salt immersion and salt spray tests (based on Shi *et al.*, 2005). (a) Anodized Mg immersed in 5 wt% NaCl for 23 h; (b) anodized ZE41 immersed in 5 wt% NaCl for 48 h; (c) anodized AZ91D immersed in 5 wt% NaCl for 48 h; (d) anodized Mg exposed to 5 wt% NaCl salt spray for 23 h; (e) anodized ZE41 exposed to 5 wt% NaCl salt spray for 87 h; and (f) anodized AZ91D exposed to 5 wt% NaCl salt spray for 87 h.

scanning, the coating on side B was also broken, similar to the damage to side A. Therefore, the backward scanned curve of the polarization curve of side B is almost the same as that of side A. The anodic polarization current on the reversed curves is larger than that on the forward scanned

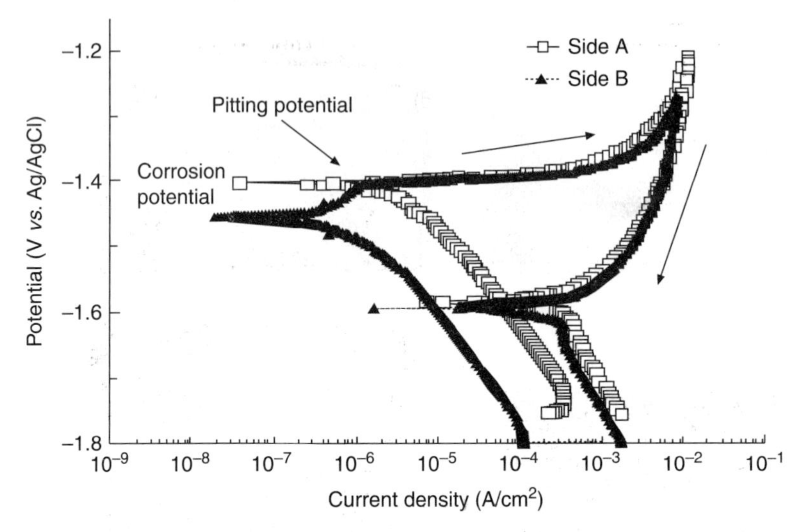

9.20 Polarization curves of anodized AM50 after 21 days salt spray testing. -□- side A, corrosion damaged; -▲- side B, no corrosion damage; OCP, open circuit potential; and PP, pitting potential (Song *et al.*, 2006).

curves due to the increased corrosion area of the anodized coating. The zero current potential of side A and side B is at the same position on the reversed curves.

Figure 9.21 presents typical EIS spectra of an anodized coating before and after corrosion damage. The EIS of side A consists of a capacitance arc and an inductance arc, but only one large capacitance arc for side B. A single large capacitive arc means that the coating can be treated as a good barrier layer on the electrode, and hence indicates that the coating should be undamaged. An inductive arc at low frequencies should be an indication that localized or pitting corrosion is taking place. According to the equivalent circuit and the corrosion mechanism of an anodized Mg alloy proposed earlier, the inductive characteristic is closely associated with the breakdown of surface film and the initiation of corrosion. This interpretation is also consistent with the corrosion models of other film-covered metals (Cao, 1990; Song, 2005b). Therefore, the inductive characteristic of an anodized coating should signify possible breakdown of surface film or initiation of the corrosion of an anodized magnesium alloy. Therefore, the EISs in Fig. 9.21 imply that the anodized side B was in good condition and side A suffered localized corrosion. Furthermore, it can be seen that the diameter of the capacitive loop for the anodized coating on side B is several orders larger than that of side A. This further confirms that the anodized coating on side A was corrosion damaged while that on side B was not.

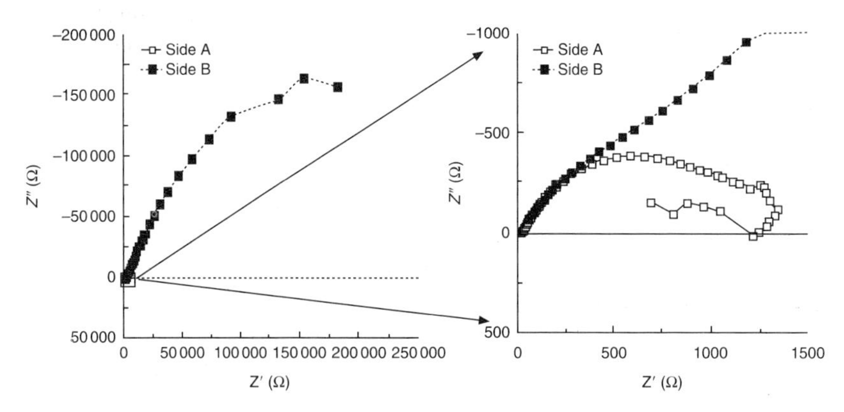

9.21 EIS of anodized AM50 in 5% NaCl solution after 21 days of salt spray testing (HF, high-frequency range; LF, low-frequency range; CP, capacitance loop; ID, inductance loop) (Song *et al.*, 2006).

9.7 Application examples

Commercial anodized coatings without any sealing or post-treatments can offer a few hundred hours of protection for the substrate Mg alloy in a standard salt spray testing environment. Although an anodized coating is much better than a conversion coating or phosphated coating, it is rarely used alone for corrosion protection. More often, an anodized coating is used as a surface treatment to provide a base for top-paint or coat. Sealing and post-treatments are strongly recommended for an anodized Mg alloy component for long service life when no organic coating is applied after anodizing.

9.7.1 Sealing

In principle, sealing is a process for species from a sealing solution to deposit on the anodized coating, particularly in the defective sites (such as the pores and cracks in the coating) to seal the active points in order to prevent preferential dissolution in these sites or aggressive solutions from penetrating the coating to attack the substrate. Existing sealing techniques normally include immersing an anodized Mg alloy specimen in a phosphate, silicate or borate-containing solution to allow these anions to react with Mg or the coating and form low-solubility salts in the defect pores or cracks, thereby blocking the corrosion paths (Blawert *et al.*, 2006). It is believed that sealing can significantly enhance the corrosion resistance of an anodized coating.

The above-mentioned sealing techniques are usually a reversible process. Salts can be deposited or formed in the defect pores or cracks when the coating is immersed in a sealing solution that contains a high concentration of those anions. The deposited salts may be dissolved or leached from the pores or cracks in service environments where there are no high concentrations of those particular anions to stabilize the deposited salts. Therefore, the improvement of the corrosion resistance of an anodized coating by this kind of sealing technique is not very significant and the durability of the sealing effect cannot last long.

Recently, an irreversible sealing process for anodized Mg alloys was reported (Song, 2009b), which was claimed to maintain a long-lasting corrosion resistant sealing effect. The deposition of sealant in the defect pores or cracks of an anodized coating is irreversible. The deposition mechanism has been systematically investigated and is believed to have chemical reactions similar to an E-coating process (Song, 2008, 2010; Song *et al.,* 2009). As long as the sealant is deposited in the defective sites, it becomes insoluble and very stable even if the environmental solution is significantly changed.

A comparison in corrosion morphology between sealed and unsealed anodized ZE41 specimens after corrosion test is presented in Fig. 9.22. The entire surface of both specimens (a) and (b) was originally anodized. The right side of specimen (a) and the left side of specimen (b) are as-anodized. The left side of specimen (a) and the right side of specimen (b) were sealed after anodizing, which appear to be darker than the as-anodized parts. The bottom parts of both specimens (a) and (b) have been immersed in 5 wt% NaCl solution for 2 days. It is clearly shown that corrosion only occurred in the as-anodized surface. Particularly on specimen (b) corrosion stopped exactly before the edge of the sealed surface. No corrosion damage can be detected in the sealed sections after immersion.

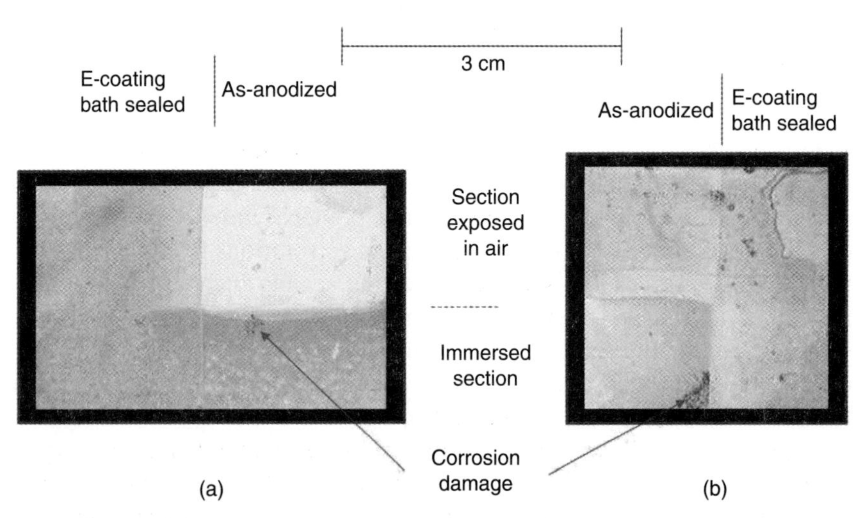

9.22 (a, b) Corrosion morphologies of anodized ZE41 half sealed after half-immersion in 5 wt% NaCl solution for 2 days (black and white, no color) (Song, 2009b).

These corrosion results suggest that the sealing treatment can significantly improve the corrosion resistance of an anodized Mg alloy. In fact, this sealing technique can also be applied to a phosphated coating (Song, 2009a) to obtain improved corrosion performance.

9.7.2 Organic top-coating

More commonly, an anodized coating is used together with an organic coating, such as an E-coating or a powder coating on its top. This is because its porous feature favors the formation of these organic coatings. In fact, these organic coatings can very well penetrate into the pores, defective points and cracks in an anodized coating and greatly seal the anodized coating, resulting in a slightly smoother surface and significantly higher corrosion resistance than the original anodized coating. Figure 9.23 shows the cross-section of an anodized coating + E-coating system as an example to illustrate the beneficial sealing and bonding effect of an E-coating on a commercial anodized coating.

A coating system consisting an anodized coating and an E-coating or a powder coating can usually survive in the standard salt spray test (ASTM B117) for a few thousand hours. In theory, this coating system should offer sufficient protection for Mg alloys. However, owing to the bad reproducibility and some unexpected defects in the coating system, premature failure

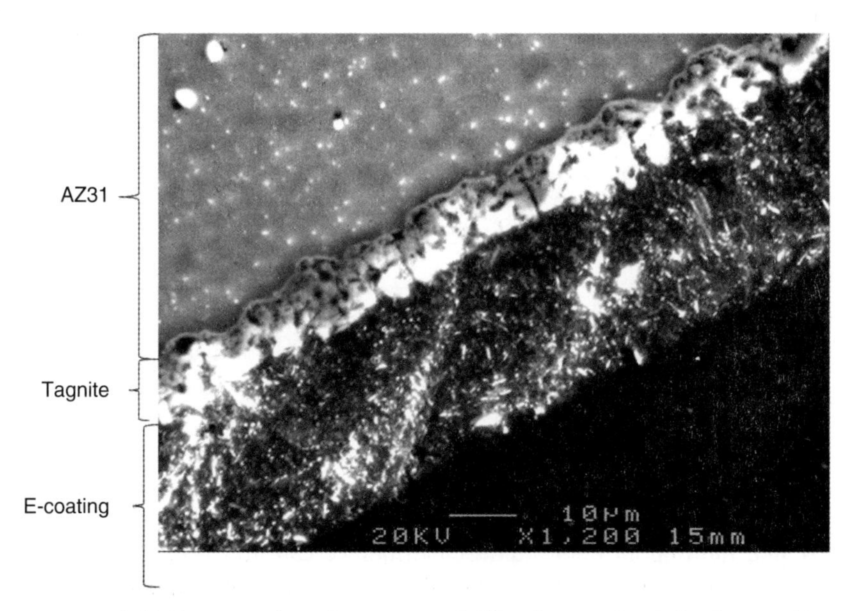

9.23 Cross-section of a commercial Tagnite anodized specimen with an E-coating on the top.

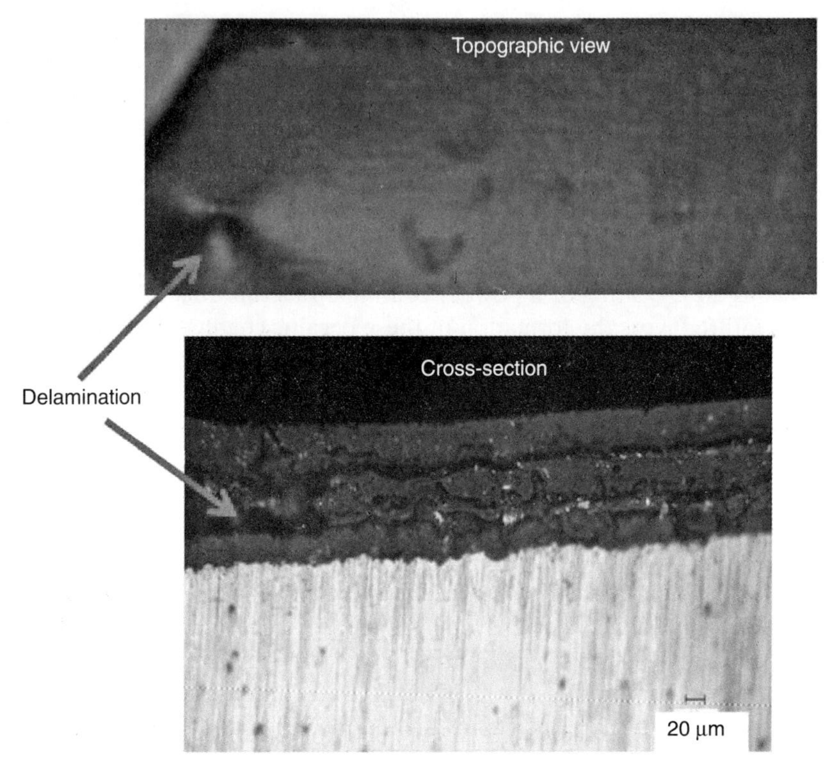

9.24 Delamination of E-coating on an anodized AZ31 alloy.

can occur. Similar to organic coatings on other materials, the most common damage of such a coating system on a Mg alloy is still delamination. An example is given in Fig. 9.24.

9.7.3 Corrosion rate control

In some particular cases, we need to control corrosion rate rather than completely stop corrosion. For those applications, an anodized coating may be used alone without sealing or organic coating as a top-paint. An example is the control of degradation rate of Mg alloy as a biodegradable material.

Magnesium is potentially a wonderful implant biomaterial because of its non-toxicity to the human body. Since rapid degradation is almost an intrinsic response of magnesium to a chloride-containing solution (Song and Atrens, 2003), like the human body fluid or plasma, we can utilize its poor corrosion performance to make magnesium into a biodegradable implant material (Song, 2007a, 2007b, Song and Song, 2007; Song *et al.,* 2007). Theoretically, a biodegradable material should have a controllable dissolution rate or

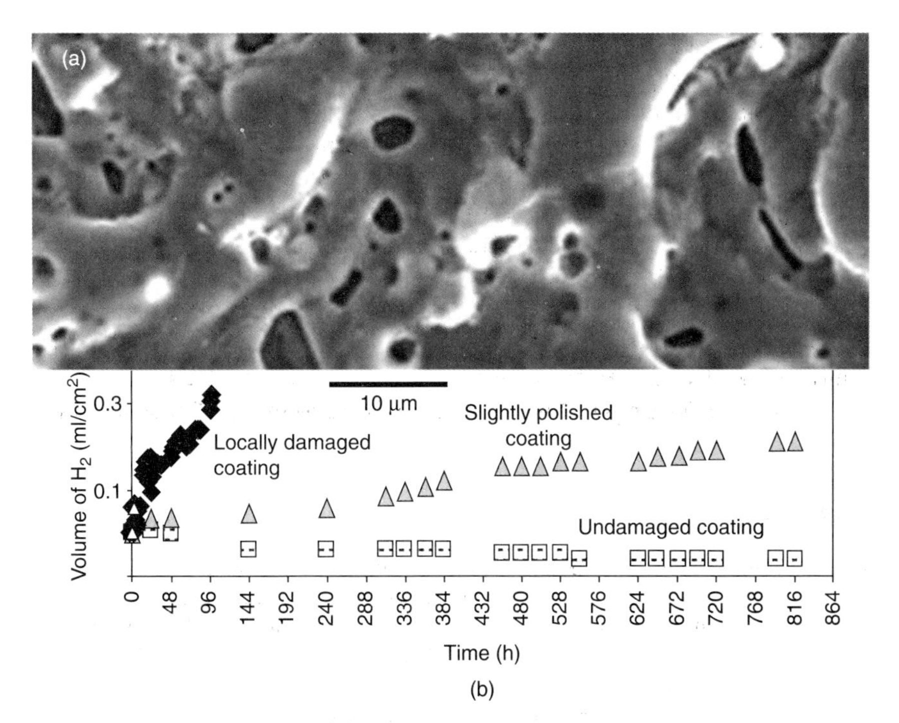

9.25 Anodized Mg: (a) microstructure of anodized coating and (b) hydrogen evolution from anodized coupons in the simulated body fluid (based on Song, 2007b).

a delayed degradation process (Hassel *et al.*, 2005). An implant made of such a material should fully function before the surgical region recovers or heals. After that, the implant should gradually dissolve, or be consumed or absorbed. Unfortunately, magnesium corrodes rapidly in simulated body fluids (Fonternier *et al.*, 1975; Kaesel *et al.*, 2003). The rapid biodegradation of magnesium can lead to a large amount of dissolved Mg^{2+}, a large volume of hydrogen gas and remarkable local alkalization of body fluid in the human body. Therefore, a straightforward strategy to solve these problems is to control the biodegradation rate of magnesium, so that the human body would have sufficient time to gradually consume, adsorb or release those degradation products (particularly hydrogen) of magnesium. The control of corrosion rate can be realized through a corrosion-resistant coating.

An example for such an application has been given by Song (2007b) who used an anodized coating containing less than 30% silicon oxides and hydroxides in addition to magnesium oxides and hydroxides which is hard, non-toxic, and in some sense similar to a ceramic layer. Although its microstructure is porous (Fig. 9.25a), it is corrosion resistant in the simulated body fluid (SBF) solution. From an undamaged anodized coupon (no scratch), no detectable hydrogen evolution was measured after one month testing (Fig. 9.25b). This means that the anodized coating can significantly delay the biodegradation process of Mg.

However, this anodized coating may be too protective, which may lead to insufficient degradation and a magnesium implant with such a coating may not biodegrade within a designated period. Fortunately, the slow biodegradation process can be accelerated by slightly polishing the anodized surface, making it thinner or less perfect. The slightly faster hydrogen evolution process shown in Fig. 9.25b verifies this idea.

Moreover, while implanting, the surface of an implant could be scratched. It was found that the degradation rate of an anodized coupon with a mechanically damaged corner was significantly increased (see Fig. 9.25b).

Nevertheless, if compared with an un-anodized coupon, its biodegradation rate is still evidently low.

9.8 References

Abulsain, M, Berkani, A, Bonilla, F A, Liu, Y, Arenas, M A, Skeldon, P, Thompson, G E, Bailey, P, Noakes, T C Q, Shimizu, K and Habazaki, H (2004), 'Anodic oxidation of Mg–Cu and Mg–Zn alloys'. *Electrochimica Acta*, **49**, 899–904.

Asoh, H and Ono, S (2003), 'Anodizing of Mg in amine-ethylene glycol electrolyte', *Materials Science Forum*, **419–422**, 957–962.

Avedesin, M and Baker, H (eds.) (1999), *Magnesium and Magnesium Alloys*, Materials Park, OH: ASM International.

Bartak, D E, Lemieux, B E and Woolsey, E R (1993), *Two-step chemical/electrochemical process for coating magnesium alloys*. US Patent: 5240589.

Bartak, D E, Lemieux, B E and Woolsey, E R (1995), *Hard anodic coating for magnesium alloys*. US Patent: 5470664.

Barton, T F (1998), *Anodization of magnesium and magnesium based alloys*. US Patent: 5792335.

Barton, T F and Johnson, C B (1995), 'Effect of electrolyte on the anodized finish of magnesium alloy'. *Plating and Surface Finishing*, **82**, 138–141.

Barton, T F, Macculloch, J A and Ross, P N (2001), *Anodization of magnesium and magnesium based alloys*. US Patent: 6280598 (B1).

Blawert, C, Dietzel, W, Ghali, E and Song, G (2006), 'Anodizing treatments for magnesium alloys and their effect on corrosion resistance in various environments'. *Advanced Engineering Materials*, **8**, 511–533.

Bonilla, F, Berkani, A, Skeldon, P, Thompson, GE, Habazaki, H, Shimizu, K, John, C and Stevens, K (2002a), 'Enrichment of alloying elements in anodized magnesium alloys'. *Corrosion Science*, **44**, 1941–1948.

Bonilla, F A, Berkani, A, Liu, Y, Skeldon, P, Thompson, G E, Habazaki, H, Shimizu, K, John, C and Stevens, K (2002b), 'Formation of anodic films on magnesium alloys in an alkaline phosphate electrolyte'. *Journal of the Electrochemical Society*, **149**, B4–B13.

Brown, L, Besenbruch, G, Schultz, K, Showalter, S, Marshall, A, Pickard, P and Funk, J (2002), 'High efficiency generation of hydrogen fuels using thermochemical cycles and nuclear power'. In: AIChE 2002 Spring National Meeting, New Orleans, LA. 139b.

Cai, Q Z, Wang, L S, Wei, B K and Liu, Q X (2006), 'Electrochemical performance of micro-arc oxidation films formed on AZ91D magnesium alloy in silicate and phosphate electrolytes'. *Surface & Coating Technology*, **200**, 3727–3733.

Cao, C N (1990), 'On the impedance plane displays for irreversible electrode reactions based on the stability conditions of the steady-state – I. One state variable besides electrode potential'. *Electrochimica Acta*, **35**, 831.

Carter, E A, Barton, T F and Wright, G A (1999), 'Anodisation of magnesium alloys at high voltages'. *Surface Treatment IV : 4th International conference on computer methods and experimental measurements for surface treatment effects*, Assisi, Italy, September 1999, WIT Press, Southampton, ROYAUME-UNI (Monographie), pp. 169–177.

Chai, L, Yu, X, Yang, Z, Wang, Y and Okido, M (2008), 'Anodizing of magnesium alloy AZ31 in alkaline solutions with silicate under continuous sparking'. *Corrosion Science*, **50**, 3274–3279.

Clapp, C, Kilmartin, P A and Wright, G A (2001). 'Anodisation of magnesium alloys'. *Proceedings – Corrosion and Prevention-2000. 40th Annual conference of the Australasian Corrosion Association*, Auckland, New Zealand/Australasian Corrosion Association, 12–22 November, 2000. pp. 297–301, Australasian Corrosion Association Inc, Auckland, New Zealand.

COMA (1956), *DOW17*. GB Patent: 762195.

COMA (2001), *A hard anodic coating for magnesium* [Online]. Luke Engineering & MFG Co. Available: http://www.lukeng.com/magoxid.htm [Accessed April 19 2001].

COMA (2008), *Keronite Plasma Electrolytic Oxidation (PEO) technology* [Online]. Keronite Advanced Surface Technology. Available: http://www.keronite.com/technology_centre.asp [Accessed 2008].

Duan, H, Yan, C and Wang, F (2007), 'Effect of electrolyte additives on performance of plasma electrolyte oxidation films formed on magnesium alloy AZ91D', *Electrochimica Acta*, **52**, 3785.

Evangelides, H A (1955), *Method of Electrochemically coating magnesium and electrolyte there-for*. US Patent: 2723952.

Fonternier, G, Freschard, R and Mourot, M (1975), 'Study of the corrosion *in vitro* and *in vivo* of magnesium anodes involved in an implantable bioelectric battery', *Medical and Biological Engineering*, **15**, 683.

Fukuda, H and Matsumoto, Y (2004), 'Effects of Na_2SiO_3 on anodization of Mg–Al–Zn alloy in 3 M KOH solution'. *Corrosion Science*, **46**, 2135–2142.

Hagans, P L (1984), Surface modification of magnesium for corrosion. In: 41st World Magnesium Conference, London, England. International Magnesium Association, 30–38.

Hassel, T, Bach, F W, Krause, C and Wilk, P (2005), Corrosion protection and repassivation after the deformation of magnesium alloys coated with a protective magnesium fluoride layer. In: Neelameggham, N R, Kaplan, H I and Powell, B R (eds.) *Magnesium Technology 2005*, TMS. 485–490.

Hillis, J E (1994), 'Surface engineering of magnesium alloys'. *ASM Handbook, Surface Engineering.* ASM International. 819.

Hillis, J E (1995), Corrosion testing and standards: application and interpretation In: Baboian, R (ed.) *ASTM Manual Series: MNL 20, Magnesium, Chapter 45.* Philadelphia, ASTM. 438–446.

Hillis, J E (2001). In: Proc. of 40th Annual Conf. of Metallurgists of CIM. 3–26.

Hillis, J E and Murray, R W (1987), 'Finishing alternatives for high purity magnesium alloys'. In: SDCE 14th International. Society of Die Casting Engineers, Inc., #G–T87-003.

Hsiao, H Y and Tsai, W T (2003), In: Corrosion 2003. NACE Int., Paper No: 03212.

Hsiao, H Y and Tsai, W T (2005), 'Characterization of anodic films formed on AZ91D magnesium alloy'. *Surface and Coatings Technology*, **190**, 299–308.

Huber, K (1953), 'Anodic formation of coatings on magnesium, zinc, and cadmium'. *Journal of the Electrochemical Society*, **100**, 376–382.

Kaesel, V, Tai, P T, Bach, F W, Haferkamp, H, Witte, F and Windhagen, H (2003), Approach to control the corrosion of magnesium by alloying. In: Kainer, K U (ed.) Magnesium, Proceedings of the 6th International Conference on Magnesium Alloys and Their Applications. 534–539.

Khaselev, O and Yahalom, J (1998a), 'The anodic behavior of binary Mg–Al alloys in KOH–aluminate solutions'. *Corrosion Science*, **40**, 1149–1160.

Khaselev, O and Yahalom, J (1998b), 'Constant voltage anodizing of Mg–Al alloys in KOH–$Al(OH)_3$ solutions'. *Journal of the Electrochemical Society*, **145**, 190–193.

Khaselev, O, Weiss, D and Yahalom, J (1999), 'Anodizing of pure magnesium in KOH-aluminate solutions under sparking'. *Journal of the Electrochemical Society*, **146**, 1757–1761.

Khaselev, O, Weiss, D and Yahalom, J (2001), 'Structure and composition of anodic films formed on binary Mg–Al alloys in KOH–aluminate solutions under continuous spaking'. *Corrosion Science*, **43**, 1295–1307.

Kobayashi, W and Takahata, S (1985), *Aqueous anodizing solution and process for colouring article of magnesium or magnesium–base alloy*. US Patent: 4551211.

Kobayashi, W, Uehori, K and Furuta, M (1988), *Anodizing solution for anodic oxidation of magnesium or its alloys*. US Patent: 4744872.

Kotler, G L, Hawke, D L and Aqua, E N (1976a), 'MGZ coating process for magnesium and its alloys'. *Light Metal Age*, **34**, 20–21.

Kotler, G R, Hawke, D L, Aqua, E N and Kotler, G L (1976b), MGZ coating process for magnesium and its alloys. In: 33rd Annual Meeting- International Magnesium Association Proc., Montreal, Que, Canada, 45–48.

Kotler, G R, Hawke, D L and Aqua, E N. (1977), A new anodic coating process for magnesium and its alloys. In: The Society of Die Casting Engineers (ed.) SDEC-77: 9th SDEC International Die Casting Exposition & Congress, MECCA Convention Center, Milwaukee, Wisconsin. G-T77–022.

Kozak, O (1980), *Anti-corrosive coating on magnesium and its alloys*. US Patent: 4184926.

Kozak, O (1986), *Method of coating articles of magnesium and an electrolytic bath therefor*. US Patent: 4620904.

Kurze, P (1998), 'Ceramic coatings on light metals by plasma-chemical treatment'. *Materialwiss. Werkstofftech.*, **29**, 85–89.

Kurze, P and Banerjee, D (1996), 'Eine neve anodische Beschichtung zur Verbesserung der korrosions-und VerschleiBbeständigkeit von Magneslumwerkstoffen', *Gießerei-Praxis*, **11**/12, 211–217.

Leyendecker, F (2001), Neue Entwicklungen Leichtmetall-Anwendungen. *In:* der Oberflächentechnik, Conf., Dt. Forschungsges. f. Oberflächenbehandlung, Dt. Ges. f. Galvano- u. Oberflächentech., DGO, Münster. Berichtsband DFO, 132–142.

Li, W, Zhu, L and Li, Y (2006), 'Electrochemical oxidation characteristic of AZ91D magnesium alloy under the action of silica sol'. *Surface and Coatings Technology*, **201**, 1085–1092.

Liang, J, Hu, L and Hao, J (2007), 'Preparation and characterization of oxide films containing crystalline TiO_2 on magnesium alloy by plasma electrolytic oxidation'. *Electrochim. Acta*, **52**, 4836–4840.
Ma, Y, Nie, X, Northwood, D O and Hu, H (2004), 'Corrosion and erosion properties of silicate and phosphate coatings on magnesium'. *Thin Solid Films*, **469–470**, 472.
Mato, S, Alcala, G, Skeldon, P, Thompsen, G E, Masheder, D, Habazaki, H and Shimizu, K (2003), 'High resistivity magnesium-rich layers and current instability in anodizing a Mg/Ta alloy'. *Corrosion Science*, **45**, 1779–1792.
Mizutani, Y, Kim, S J, Ichino, R and Okido, M (2003), 'Anodizing of Mg alloys in alkaline solutions'. *Surface and Coatings Technology*, **169–170**, 143–146.
Nykyforchyn, H M, Dietzel, W, Klapkiv, M D and Blawert, C (2003), Corrosion properties of Conversion Plasma Coated Magnesium Alloys. In: Kainer, K U (ed.) *Magnesium, 6th International Conference Magnesium Alloys and their Applications.* Wolfsburg (D).
Ono, S (1998), 'Anodic film growth on magnesium die-cast AZ91D (12)'. *Metallurgical Science and Technology*, **16**, 91–104.
Ono, S and Masuko, N (2003), 'Anodic films grown on magnesium and magnesium alloys in fluoride solutions'. *Materials Science Forum*, **419–422**, 897–902.
Ono, S, Kijima, H and Masuko, N (2003), 'Microstructure and voltage-current characteristics of anodic films formed on magnesium in electrolytes containing fluoride'. *Materials Transactions, JIM*, **44**, 539–545.
Schmeling, E L, Roschenbleck, B and Weidemann, M H (1990), *Method of producing protective coatings that are resistant to corrosion and wear on magnesium and magnesium alloys*. US & EP Patent: EP: 333049 US: 4978432.
Sharma, A K, Rnai, R U and Giri, K (1997), 'Studies on anodization of magnesium alloy for thermal control applications'. *Metal Finishing*, **95**, 43–51.
Shatrov, A S (1999), *Method for Producing Hard Protection Coatings on Articles made of Aluminium Alloys.* World Intellectual Property Organization (WO) Patent: WO 9931303.
Shi, Z (2005), *The corrosion performance of anodized magnesium alloys.* PhD Thesis, University of Queensland.
Shi, Z, Song, G and Atrens, A (2003), Effects of zinc on the corrosion resistance of anodized coating on Mg–Zn alloys. In: Dahle, A (ed.) Proceedings of the 1st International Light Metals Technology Conference, Brisbane, Australia. CAST, 393–396.
Shi, Z, Song, G and Atrens, A (2005), 'Influence of the β phase on the corrosion performance of anodized coatings on magnesium–aluminium alloys'. *Corrosion Science*, **47**, 2760–2777.
Shi, Z, Song, G and Atrens, A (2006a), 'Corrosion resistance of anodised single phase Mg alloys'. *Surface & Coatings Technology*, **201**, 492.
Shi, Z, Song, G and Atrens, A (2006b), 'Influence of anodizing current on the corrosion resistance of anodized AZ91D magnesium alloy'. *Corrosion Science*, **48**, 1939–1959.
Shi, Z, Song, G and Atrens, A (2006c), 'The corrosion performance of anodised magnesium alloys'. *Corrosion Science*, **48**, 3531–3546.
Song, G (2004), *Magnesium Anodisation*. Australia Patent: 2004904949.
Song, G (2005a), 'Recent progress in corrosion and protection of magnesium alloy'. *Advanced Engineering Materials*, **7**, 563–586.
Song, G (2005b), 'Transpassivation of Fe–Cr–Ni stainless steels'. *Corrosion Science*, **47**, 1953.

Song, G (2007a), 'Control of biodegradation of biocompatible magnesium alloys'. *Corrosion Science*, **49**, 1696–1701.

Song, G (2007b), 'Control of degradation of biocompatible magnesium in a pseudo-physiological environment by a ceramic like anodized coating'. *Advanced Materials Research*, **29–30**, 95–98.

Song, G (2008), *Self-deposited coatings on magnesium alloys*. US Patent: 61/047766.

Song, G (2009a), 'A prelimary quantitative – XPS study of the surface films formed on pure magnesium and on magnesium-aluminium intermetallics by exposure to high purity water'. *Materials Science Forum*, **618–619**, 269–271.

Song, G (2009b), 'An irreversible dipping sealing technique for anodized ZE41 Mg alloy'. *Surface & Coatings Technology*, **203**, 3618–3625.

Song, G (2010), '"Electroless" deposition of pre-film of electrophoresis coating anits corrosion resistance on a Mg alloy'. *Electrochimica Acta*, **55**, 2258–2268.

Song, G and Atrens, A (2003), 'Understanding magnesium corrosion – a framework for improved alloy performance'. *Advanced Engineering Materials*, **5**, 837.

Song, G and Song, S (2007), 'A possible biodegradable magnesium implant material'. *Advanced Engineering Materials*, **9**, 298–302.

Song, G, Atrens, A, St John, D, Wu, X and Nairn, J (1997), 'The anodic dissolution of magnesium in chloride and sulphate solutions'. *Corrosion Science*, **39**, 1981.

Song, G, Bowles, A L and St John, D (2004), 'Corrosion resistance of aged die cast magnesium alloy AZ91D'. *Materials Science & Engineering – A*, **366**, 74–86.

Song, G, Shi, Z, Hinton, B, McAdam, G, Talevski, J and Gerrard, D. (2006), Electrochemical evaluation of the corrosion performance of anodized magnesium alloys. In: 14th Asian-Pacific Corrosion Control Conference, Shanghai, China. Keynote-11.

Song, G, Song, S and Li, Z (2007), Corrosion control of magnesium as an implant biomaterial in simulated body fluid. In: Ultralight2007 – 2nd International Symposium on Ultralight Materials and Structures, Beijing (Invited Lecture).

Song, G, Atrens, A, StJohn, D, Wu, X and Nairn, J (2009), '"Electroless" E-coating – an innovative surface treatment for magnesium alloys'. *Electrochemical and Solid-State Letters*, **12**, D77–D79.

Takaya, M (1987), 'Anodizing of magnesium alloy in potassium hydroxide–aluminum hydroxide solutions'. *Keikinzoku*, **37**, 581–586.

Takaya, M (1988), Anodizing film of Magnesium, *Journal of the Metal Finishing Society*, **35**, 290.

Takaya, M (1989a), 'Anodizing of magnesium alloys in KOH-Al$(OH)_3$ solutions', *Aluminium*, **65**, 1244–1248.

Takaya, M (1989b), Anodizing film and its composition of magnesium alloys, *Gypsum & Lime*, **223**, 40.

Takaya, M, Inoue, T, Nakazato, D and Sugano, K (1998), *Surface treatment of magnesium-based metal material for forming anticorrosive and rustproofing coating*. Japan Patent: 10219496.

Tanaka, K (1993), *Anodization with the Colouring of magnesium and magnesium alloys.* European Patent: 0333048 BI.

Verdier, S, Boinet, M, Maximovitch, S and Dalard, F (2004), 'Formation structure and composition of anodic films on AM60 magnesium alloy obtained by DC plasma anodising'. *Corrosion Science*, **47**, 1429–1444.

Weast, R (ed.) (1976–1977), *Handbook of Chemistry and Physics*. CRC Press Inc., Cleveland, Ohio.

Wu, C S, Zhang, Z, Cao, F H, Zhang, L J, Zhang, J Q and Cao, C N (2007), 'Study on the anodizing of AZ31 magnesium alloys in alkaline borate solutions'. *Applied Surface Science*, **253**, 3893–3898.

Yahalom, J and Zahavi, J (1971), 'Experimental evaluation of some electrolytic breakdown hypotheses'. *Electrochimica Acta*, **16**, 603.

Zhang, Y, Yan, C, Wang, F and Li, W (2005), 'Electrochemical behavior of anodized Mg alloy AZ91D in chloride containing aqueous solution'. *Corrosion Science*, **47**, 2816–2831.

Zhu, L and Liu, H (2005), 'The effect of sol ingredients to oxidation film on magnesium alloys'. *Function Materials*, **36**, 923–926.

Zozulin, A J and Bartak, D E (1994), 'Anodized coatings for magnesium alloys'. *Metal Finishing*, **92**, 39–44.

10
Corrosion-resistant coatings for magnesium (Mg) alloys

X.-B. CHEN, M. A. EASTON and N. BIRBILIS,
Monash University, Australia,
H.-Y. YANG, Shanghai Jiao-tong University, China and
T. B. ABBOTT, Magontec Pty Ltd, Australia

DOI: 10.1533/9780857098962.2.282

Abstract: The escalation of ongoing research on the topic of conversion coatings for magnesium alloys confirms that the search continues for appropriate replacements to toxic chromate-based coatings. This chapter reviews the performance of the range of typical conversion coating technologies presently available. As a means to rationalise coating performance and characteristics more generally, thermodynamic analysis has been used to compare coating processes and coating pretreatment, the latter of which appears to be at least as significant as the choice of coating technology itself.

Key words: magnesium, corrosion protection, surface engineering, conversion coating.

10.1 Introduction

Magnesium (Mg) alloys present a unique corrosion protection challenge. To date, one of the most effective ways to impart corrosion protection to Mg alloys is to form a barrier-coating on the surface, to isolate the base material from the environment. This is done by either forming, or adding, some type of coating(s). Coating is not a single step process, and this is particularly true for Mg, possibly more so than any other alloy system. Coatings may additionally form a good base for subsequent organic coatings (Avedesian and Baker, 1999) or paints. Several surface coating techniques, including electrochemical plating (electroplating), conversion coatings, anodising, hybrid coatings, microarc oxidation and vapour-phase processes, have been developed and are applicable to Mg (Gray and Luan, 2002; Chen *et al.*, 2011b). Of these, chemical conversion treatment is an effective, comparatively low-cost and easily implemented method (Elsentriecy *et al.*, 2007b) that has been widely adopted in industrial processes for a range of commodity applications which require moderate levels of corrosion protection.

In the conversion coating processes, the substrates to be protected are immersed in a solution that reacts with the surface altering the metal ion concentration and the pH at the metal–solution interface. For Mg, this is nominally at neutral to acidic pH, where Mg dissolves and cannot passivate. The localised change in composition causes precipitation from the solution (which contains species deliberately added to result in such precipitation) onto the surface of the substrate, forming a coating. Consequently the coating is bonded to the metal surface. Many factors influence the quality of conversion coatings on Mg alloys, including composition of the Mg alloy substrates, chemical pretreatments, composition of the conversion electrolytes/formulations, post-treatments and operating parameters, such as temperature, pH (Lin *et al.*, 2006b), immersion time and degree of agitation (Song *et al.*, 2009a). These can all influence the structure, composition and performance of conversion coatings. One of the biggest challenges in the field of conversion coatings is to produce defect-free coatings with uniform coverage. The protective role of defective coatings will be compromised as the corrosive medium may easily penetrate to the metal surface through such defects. From this simple description alone, one can appreciate the numbers of variables, and possible coating exploration matrix, are not straightforward. It is for this reason that research in this topic remains highly active.

This chapter is structured such that the classes of conversion coatings are presented and reviewed individually (this appears to be the most logical approach as the field has studied a number of different alloys assessed by widely varying means), followed by a discussion of relevant pretreatments and lastly a general discussion. The alternative would have been to present the sections according to the substrate alloy type, which would have been too restrictive given that the majority of research to date has been on the AZ (Mg–Al–Zn) class of alloys. Solution phase inhibitors for corrosion protection and sol-gel coatings are not covered, since they strictly represent a different class of corrosion protection that does not require the fundamental step of 'chemical conversion'. This chapter is intended to give an up-to-date overview of the field, and allow general aspects to be conveyed to the reader in a holistic manner. For more specific and critical appraisal, the reader is directed to the specific papers cited herein.

10.2 Conversion coatings for Mg alloys: chromate and phosphate coatings

Development and corrosion resistance of chromate and phosphate conversion coatings.

10.2.1 Chromate coatings

Chromate conversion coatings are an effective means of protecting reactive metals such as Mg, Al and Zinc (Zn). The chromating solution contains hexavalent chromium (VI) (nominally from $HCrO_4^-$) and is nominally acidic (i.e. buffered aqueous solution). The substrate (M) is oxidised and Cr (VI) is reduced to trivalent Cr (III) as follows:

$$2HCrO_4^- + 14H^+ + (6/n)M \rightarrow 2Cr^{3+} + (6/n)M^{n+} + 8H_2O(Mg, Zn: n = 2, Al: n = 3) \quad [10.1]$$

The Cr (III) precipitates to form a coating of hydrated Cr and substrate oxides. This example represents a classical conversion coating formation. It is also possible that some Cr (VI) becomes incorporated into the coating, giving the coating a self-healing capability (Gray and Luan, 2002) – however this has not been explicitly shown for Mg alloys. Cr-conversion coatings can provide high levels of corrosion protection and would form part of a suitable protection strategy for Mg, if not for the toxicity issue (meaning that whilst reviewed here, Cr-conversion coatings are not industrially feasible). Cr (VI) is one of six substances listed in the European Union's Restriction of Hazardous Substances Directive (RoHS). Consequently, products requiring RoHS certification cannot involve a chromating process and as such are not discussed in greater detail herein. Figure 10.1 depicts the predominant phases for Cr (VI) (Fig. 10.1a) and Cr (III) (Fig. 10.1b) states. Cr (VI) is soluble over the full range of the concentrations and pH levels shown. When reduced to the trivalent state Cr becomes insoluble except at low pH levels.

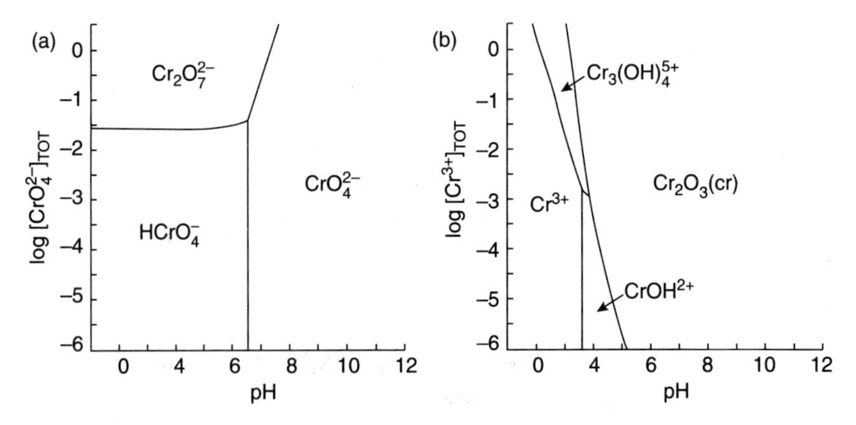

10.1 Predominance diagrams for chromium in the (a) hexavalent, and (b) trivalent oxidation states. These diagrams, and all others herein, were calculated using the MEDUSA software package.

10.2.2 Phosphate coatings

Metal phosphates are generally insoluble in aqueous solution at neutral pH, but increasingly soluble in acidified solutions. The overwhelming majority of the conversion coatings for Mg alloys reported in the past decade have been related to the phosphate system, and hence this section is gives proportionate coverage to this area. Figure 10.2 illustrates how the predominance of Mg, manganese (Mn), calcium (Ca) and Zn in the presence of phosphate varies with pH and metal ion concentrations.† These diagrams were calculated for a solution containing 0.2 M total phosphate (equivalent to 20 g/L H_3PO_4). For each of the metals presented there is a boundary between an insoluble phosphate and soluble metal ions that extends diagonally from

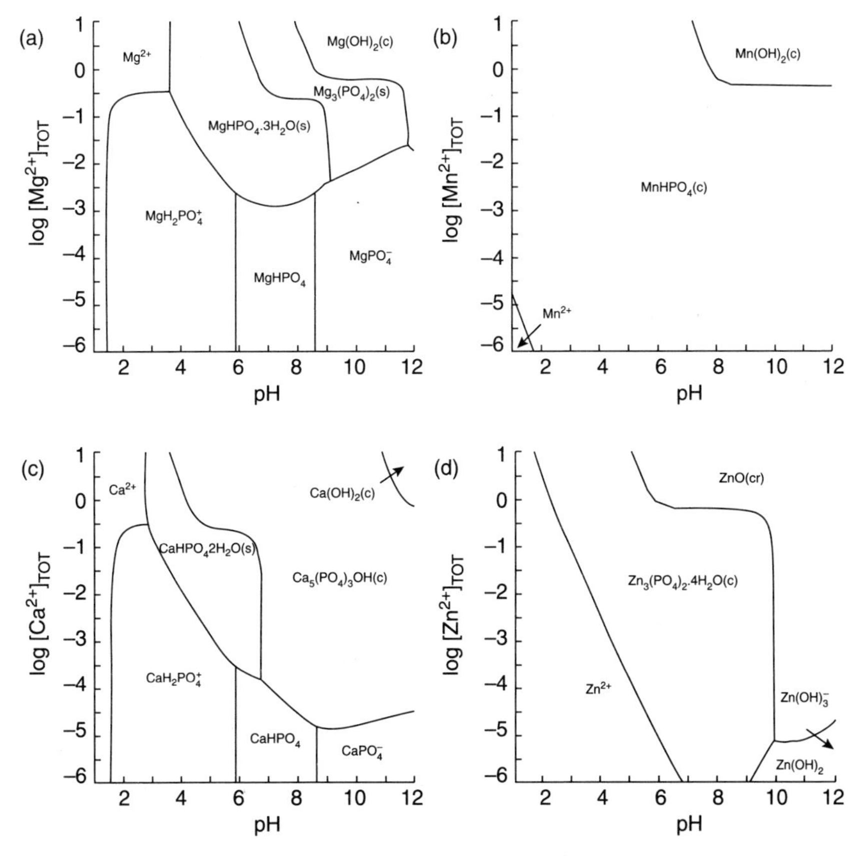

10.2 (a–d) Predominance area diagrams for solutions containing 0.2 M phosphate at concentrations of Mg, Mn, Ca and Zn ions and pH.

† *Thermodynamic diagrams were calculated using MEDUSA software.*

low pH, high metal ion concentration to high pH, low metal ion concentration. The location of this boundary varies according to metal species. Placing a metallic (M) item in an appropriately acidified phosphate-containing solution causes the following reaction at the interface:

$$M + nH^+ \rightarrow M^{n+} + (n/2)H_2 \uparrow \quad [10.2]$$

This results a local rise in both pH and metal ion concentration, leading to the precipitation of metal phosphate on the surface. In the case of Mg, the phosphate species ($MgHPO_4 \cdot 3H_2O$) is poorly protective because it is insoluble only over a narrow range of conditions (Fig. 10.2a), particularly when environmental exposure or in-service conditions are nominally neutral and tending to acidic.

In order to form a suitable phosphate conversion coating on Mg, other species must be present. These additional species serve several roles:

- enable formation of coatings with lower solubilities in acid (and alkaline) solutions.
- act as oxidising agents to prevent or limit hydrogen evolution.
- modify the coating formation process to create dense and uniform coatings (where coating defects are typical in instances where the coating formation reaction occurs too rapidly).

Several different species have been reported in the literature as additions to phosphate based coatings for Mg and its alloys. These include species containing Zn, Mn and Ca. Figure 10.2b reveals that theoretically Mn-based phosphate coatings are stable over a wide range of conditions. As an example of such a system, a bath of Mn dihydrophosphate ($Mn(H_2PO_4)_2$) produced a 10 μm thick phosphate conversion coating on Mg alloy AZ91D (Zhou *et al.*, 2008). In that instance, hydrogen evolution upon the β phase ($Mg_{17}Al_{12}$) increased the bath pH by consuming H^+ ions and drove the hydrolysing equilibrium for the $Mn(H_2PO_4)_2$ to generate insoluble phosphate precipitation. The conversion film was amorphous film also containing Mg and Al from the substrate. However, both wide and narrow crack networks were produced in the coating and this was attributed to the substrate microstructure of the β phase, which served to compromise the protective nature of the coating since it made it heterogeneous. Microstructural effects are discussed rather generally in this chapter, however if the reader wishes to gain a greater appreciation of the microstructure and metallurgy of Mg alloys, they are referred to the monograph of Polmear (1995).

Figure 10.2d reveals that Zn phosphate is less soluble than Mg phosphate, but more soluble than Mn phosphate. However a $Zn_3(PO_4)_2$ layer with conditional insolubility and protectiveness was grown upon Mg alloy AM60 (Kouisni *et al.*, 2004). The $Zn_3(PO_4)_2$ coating formed after 10 min immersion in the bath provided optimum protection, though that was still not satisfactory enough for functional corrosion protection, dissolving in mildly acidic or alkaline solutions (Kouisni *et al.*, 2005). Fine crystalline Zn particles surrounded by phosphate crystals can fill the interstices of the insoluble phosphate and ultimately improve the corrosion resistance of the $Zn_3(PO_4)_2$ coating (Niu *et al.*, 2006). In addition, the microporous $Zn_3(PO_4)_2$ coating can adsorb paint molecules tightly and create a better bonding of paint film than that of paint on Cr-based coating.

Addition of polyvalent layer-forming cations of, for example, Mn, Cobalt (Co) and Nickel (Ni) into the $Zn_3(PO_4)_2$ conversion bath leads to an improvement of corrosion resistance of the resultant Zn phosphate coatings (Cape, 2008). Conversion baths containing Ni contents greater than 6 g/L can produce phosphate coatings with fine crystals but darker appearance. The latter feature is not satisfactory when coatings with high reflection are required but it can be avoided by partially or completely replacing the Ni with Co and Mn (Cape, 2008).

Complexing agents, such as sodium tartrate and film-formation assistant, meta-vanadate, have also been extensively explored in conversion coating. The sodium tartrate plays a critical role in stabilising the conversion coating bath and preventing the coating surface from dusting (Li *et al.*, 2005). Meta-vanadate (NH_4VO_3, $NaVO_3$ or KVO_3) improves the coating process to produce a dense, adhesive, protective and crack-free Zn phosphate coating (Chang *et al.*, 2005; Wei *et al.*, 2009).

Permanganate (MnO_4^-) salt, a strong oxidising agent analogous to $HCrO_4^-$, may be used as an accelerator to produce phosphate conversion coatings. The generated phosphate-permanganate conversion coating can protect Mg-Li alloys to some extent (Zhang *et al.*, 2008), however, a network of cracks is a general feature (Chong and Shih, 2003; Umehara *et al.*, 2003; Zucchi *et al.*, 2007), ascribed to hydrogen evolution or conversion coating dehydration effects (Chong and Shih, 2003; Lin *et al.*, 2006a). A suitable accelerator must not contain any anions that can be reduced to an elemental metallic state by Mg (which would induce galvanic corrosion on the surface). MnO_4^- ions are an ideal choice because, like $HCrO_4^-$, when reduced they form low valency insoluble oxides (based on Mn^{3+} or Mn^{4+}), which can co-precipitate with the phosphate conversion coating layer on Mg substrates, rather than being reduced to metallic Mn (Umehara *et al.*, 2001; Chong and Shih, 2003) as:

$$MnO_4^- + 4H^+ + 3e^- \rightarrow MnO_2 \downarrow + 2H_2O \qquad [10.3]$$

$$2MnO_2 + 2H^+ + 2e^- \rightarrow Mn_2O_3 \downarrow + H_2O \qquad [10.4]$$

$$MnO_4^- + 2H_2O + 3e^- \rightarrow MnO_2 \downarrow + 4OH^- \quad [10.5]$$

Vanadate and $HCrO_4^-$ also share these advantages. Ions of more noble metals such as Co and Ni may be detrimental if reduction to the metallic state occurs. Table 10.1 presents a comparison of oxidation–reduction potentials of the metals discussed in this section.

Adding HF to a permanganate-phosphate bath is beneficial for a thin film with fewer defects (Umehara *et al.*, 2003) because Mg reacts with F^- ions in the solution to give rise to insoluble MgF_2 that acts as a passive surface layer, and efficiently retards Mg matrix dissolution. A continuous film could be produced only if the $Mg_{17}Al_{12}$ precipitates partially dissolved together with the α Mg matrix (Chong and Shih, 2003). H_3PO_4 pickling and HF activating are commonly needed prior to coating growth (Zucchi *et al.*, 2007). To simplify the whole coating procedure, acid pickling can sometimes be eliminated (Zhao *et al.*, 2006b). Conversely, in some cases, post-treatment is also necessary to seal the open micro-pores in the permanganate–phosphate conversion coating (Zhou *et al.*, 2004).

Whilst molybdates have been applied extensively on steel and Zn metals for corrosion protection for many years, they are not a good choice for Mg alloys because of lower oxidising capability and their susceptibility to reduction by Mg (Hawke and Albright, 1995). They can be used to form an optimum crack-free composite coating with phosphate in a molar ratio of 1:2 on AM60 Mg alloy (Yong *et al.*, 2008).

Ca^{2+} cations are another option for the formation of corrosion-resistant phosphates on Mg since they cannot be reduced by Mg (see Table 10.1). This property is postulated to make Ca^{2+} superior to other divalent competitors, like Co^{2+}, Ni^{2+} and Zn^{2+}. Ca and zirconium (Zr) combined hydrophilic and corrosion-resistant phosphate coating was applied to Mg substrates under weak acidic conditions (pH 2.6–3.1) (Tomlinson, 1995a). A novel and simple pure calcium phosphate conversion film was developed on Mg-8.8Li (Song *et al.*, 2009b). Both the cathodic hydrogen evolution and anodic corrosion reaction were concurrently restrained by the coating thus slowing the dissolution rate.

Chen *et al.* (2011a, 2011b, 2012a, 2012b). have also explored Mg and its alloy surfaces with pretreatment and calcium phosphate conversion coatings for improved corrosion resistance and controlled biodegradation Thermodynamic equilibrium calculations were exploited to design suitable solutions and operating conditions to form a highly corrosion-resistant hydroxyapatite (HA) coating on Mg and its alloys. In addition, to protect Mg from rapid corrosion in simulated body fluid, HA coatings are also bioactive and thus make Mg alloys promising implant candidates for tissue engineering (Chen *et al.*, 2011a, 2012a). Waterman *et al.* (2012) found that

Table 10.1 Oxidation–reduction potentials of reactions considered for Mg conversion coatings

Metal	Reaction	Standard potential (V)*
Mn	(1) $MnO_4^- + 4H^+ + 3e^- = MnO_2 + H_2O$	−1.68
	(2) $Mn^{2+} + 2e^- = Mn$	+1.17
V	(1) $V^{5+} + e^- = V^{4+}$	−1.21
	(2) $V^{2+} + 2e^- = V$	+1.20
Cr	(1) $Cr_2O_7^{2-} + 14H^+ + 6e^- = 2Cr^{3+} + 7H_2O$	−1.33
	(2) $Cr^{3+} + 3e^- = Cr$	+0.74
Ca	(1) $Ca^{2+} + 2e^- = Ca$	+2.87
Co	(1) $Co^{2+} + 2e^- = Co$	+0.28
Ni	(1) $Ni^{2+} + 2e^- = Ni$	+0.25
Zn	(1) $Zn^{2+} + 2e^- = Zn$	+0.76
Mo	(1) $Mo^{6+} + e^- = Mo^{5+}$	−0.53
	(2) $Mo^{3+} + 3e^- = Mo$	+0.20
Sn	(1) $Sn^{4+} + 2e^- = Sn^{2+}$	+0.154
	$Sn^{2+} + 2e^- = Sn$	+0.136
Mg	(1) $Mg^{2+} + 2e^- = Mg$	+2.37

* The standard potential of reactions of $2H^+ + 2e^- \rightarrow H_2$ and $H_2 \rightarrow 2H^+ + 2e^-$ is 0.00 V.

pre-coating Mg with CaOH prevented defects in the subsequent calcium phosphate coating and led to longer lasting corrosion protection.

10.3 Conversion coatings for Mg alloys: fluoride (F) and stannate coatings

Development and corrosion resistance of fluoride and stannate conversion coatings are discussed below.

10.3.1 Fluoride conversion coatings

Alkaline earth elements (except beryllium) readily form insoluble F-species that are stable over a wide range of conditions. This is in contrast to other common active metals such as Zn and Al that are attacked by F-ions. Figure 10.3 compares the predominant phases of Mg, Ca, Al and Zn in the presence of 0.2 M F^- solutions (equivalent to 3.8 g/L). Figure 10.3a and 10.3b demonstrate MgF_2 and CaF_2 are stable over a wide range while Fig. 10.3c and 10.3d reveals that for Al and Zn the most common phases are the ionic species AlF_4^- and ZnF^+.

The F-passivation layer formed on Mg has been reported to be $Mg(OH)_{2-x}F_x$ film (Gulbrandsen *et al.*, 1993) that becomes closer to MgF_2 with increasing F concentration in the film (Verdier *et al.*, 2004). Its promising protective property attracts many researchers to generate coatings onto Mg alloys in F solutions rather than regarding F solutions only as a means of surface pretreatment. The anti-corrosive effect in aqueous solutions seems to depend on solution pH and F concentration (Gulbrandsen *et al.*, 1993).

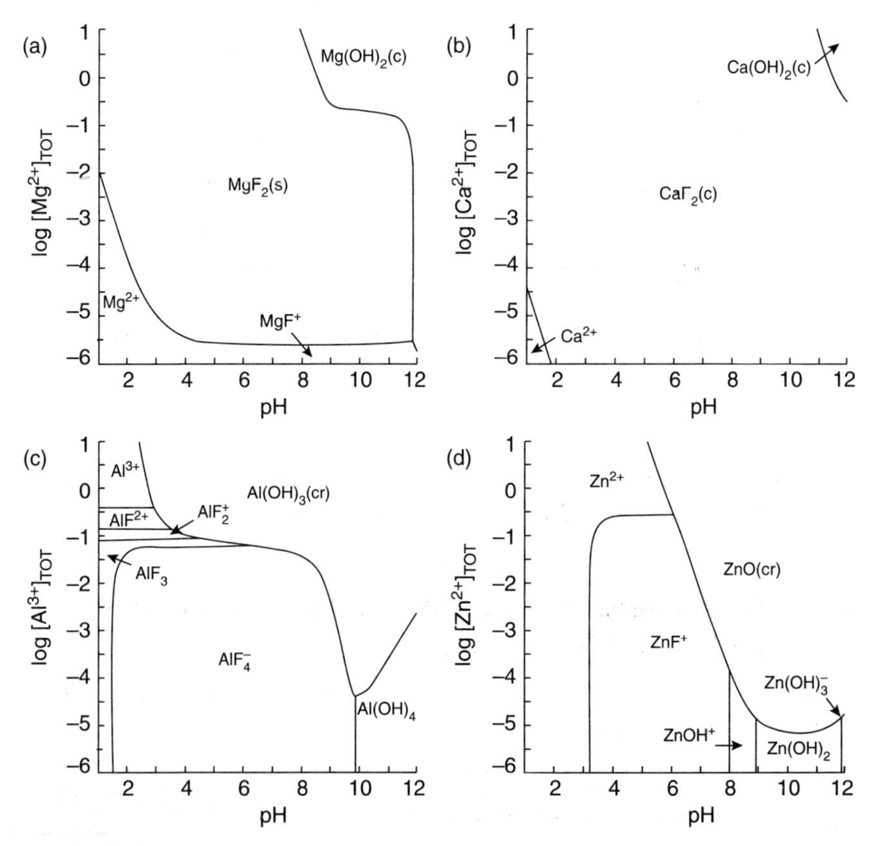

10.3 Predominance area diagrams for Mg, Ca, Al and Zn in solutions containing 0.2 M F^-. The diagrams were calculated using the MEDUSA software package.

F-conversion coating can be formed in both basic (Gulbrandsen *et al.*, 1993) and acidic (Verdier *et al.*, 2004) baths. The native porous $Mg(OH)_2$ film on Mg surface was transformed into a continuous, more compact and thicker form by the incorporation of F-ions in an alkaline bath (Gulbrandsen *et al.*, 1993). Increasing F^- concentration to a certain level (~ 0.5 M) transformed the surface film from hydroxide to fluoride with an extraordinary corrosive protection performance $\left(i_{corr} = 10^{-9} - 10^{-8}\ \text{Acm}^{-2}\right)$, though the composition of this film could not be identified. The native $Mg(OH)_2$ film on an AM60 Mg alloy dissolved and hydrogen gas evolved upon exposure to HF (Verdier *et al.*, 2004). However, after a few seconds, a passive surface formed on the AM60 with an elevated corrosion potential. The exact composition of such F-containing films is not fully determined yet. Even the concentrated 48 wt.% HF has little impact on the alteration of surface roughness of Mg (Chiu *et al.*, 2007). The high concentration of F^- ions, low temperature and long treatment time all favoured the formation of a uniform, dense and smooth coating. A big concern arising from the application of this research however is the large volumes of HF required.

The group IVB metal fluoride conversion coatings, such as flourozirconate and -titanate (Dolan and Carlson, 2007; Greene and Vonk, 2008) are also promising options for corrosion protection of Mg alloys since they can form three-dimensional polymeric metal or metalloid-oxide matrices, from aqueous solutions as Cr (III) does (Gray and Luan, 2002), but they are environmentally innocuous. The commercial availability and lower cost characteristics nowadays make Zr popular in many applications (Tomlinson, 1995b). Zr coatings may contain the dioxide form (ZrO_2), hydroxyl-oxide or hydroxyl-fluoride, while, the dioxide TiO_2 was the only chemical state present in the Ti films. F is selected to maintain the group IVB and other metals in working solution as complex fluorides and also keep dissolved substrate metal ions in solution (Reghi, 1982). The local increase in pH resulting from the reduction of water at the surface of Mg alloy led the Zr or Ti complexes to precipitate in the solutions (Verdier *et al.*, 2005). The film-formation process was determined by the pH, F and Zr or Ti ion concentrations in the conversion solution (Verdier *et al.*, 2006). Increasing the pH from 2.2 to 5.3 gave rise to a more passive film with higher corrosion potential. Higher F^- ion concentration increased coating roughness with lower surface coverage and had a subsequent detrimental effect on protective film-formation.

10.3.2 Stannate coatings

Stannate based conversion coatings have been proposed as environmentally acceptable alternatives to Cr-coatings. Generally, resistance to pitting and crevice corrosion is enhanced by such coatings due to the formation of a continuous protective $Mg(OH)_2$ layer doped with tin oxide (SnO_2) that can also act as a barrier to prevent oxygen diffusion to the metal surface. A layer of crystalline $MgSnO_3{\cdot}3H_2O$ (Gonzalez-Nunez *et al.*, 1999), $MgSnO_3{\cdot}H_2O$ (Huo *et al.*, 2004), Mg-Sn oxides (Lin *et al.*, 2006b), or mixed $MgSn(OH)_6$ hydroxides (Anicai *et al.*, 2005) have also been identified. The bath pH, stannate ion content and temperature had a critical impact on the coating property (Lin *et al.*, 2006b). An incubation time was detected prior to the noticeable growth of nuclei on the metal plate in the stannate bath. Finer particles that were preferably formed in less alkaline conditions with higher stannate ion concentration, led to fewer and thinner defects and better anti-corrosion characteristics, though some contrary results existed (Hamdy, 2008). Elevated bath temperature had a positive influence on the mass gain rate of the substrate, though little effect on the maximum mass gain. The existence of the metallic Sn ions, potentially reduced from stannate ions via Mg, could be detrimental to the corrosion protection of the coating (Lin *et al.*, 2006b).

A new pickling pretreatment for the formation of stannate conversion coatings on AZ91D Mg alloy was studied (Elsentriecy *et al.*, 2007a); the optimal condition reported for removal of the surface oxide layer without severely damaging the surface was 0.25 wt.% HCl + 0.25 wt.% HF for 20 s. After 1 h immersion in a stannate bath, the coating on the pickled sample was denser than that on the non-pickled sample, ascribed to the higher Mg^{2+} ion concentration on the pickled sample during the coating process. Hence the pickled sample dissolved immediately upon exposure to the coating solution and continuous metal dissolution provided sufficient Mg^{2+} ion concentration for the stannate coating film precipitation; presumably the native oxide film on the non-pickled sample lowered the dissolution rate of the metal, causing a longer coating initiation and decreased coating density.

Meanwhile, porous stannate conversion coating can be used as a base for the subsequent electroless Ni-P plating to achieve a hybrid coating system with optimal corrosion resistance and sometimes good appearance (Huo *et al.*, 2004). The porous structure provided good adhesion to the Ni-P plating to eliminate cavities or crevices. Additionally the potential difference between the Ni-P film and the substrate was reduced by the stannate deposition (Huo *et al.*, 2004).

10.4 Conversion coatings for Mg alloys: rare earth (RE) and other coatings

Development and corrosion resistance of rare earth and other conversion coatings are discussed below.

10.4.1 Rare earth coatings

Conversion films made from rare earth (RE) salts, especially those based on cerium (Ce) and lanthanum (La), have been identified to be environmentally friendly corrosion inhibitors on steel and Al alloys. Protection is attributed to a hydrated RE oxide blocking film over the cathodic sites on the metal surface. Recently, several studies (Dabala *et al.*, 2003; Brunelli *et al.*, 2005; Morris, 2008; Phelps *et al.*, 2008) have successfully exploited acidic aqueous solutions containing RE salts on Mg and its alloys to deliver conversion coatings. It is noted that weak acidic, neutral or alkaline baths are not suitable for Ce salts because of their high hydrolysis tendency (for instance, when $[Ce^{3+}]$ is at 0.05 M concentration, pH must be lower than 3.6) (Rudd *et al.*, 2000). However, the unsatisfactory long term stability of such RE coatings upon Mg substrates is still one of the largest concerns (Rudd *et al.*, 2000).

Selecting an appropriate pretreatment prior to the RE conversion coating processes is also critical. Acid pretreatment is of benefit for adhesion and corrosion resistance, attributed to the high amount of subsequent cathodic sites, high Al content and surface roughness induced by the acid pretreatment (Brunelli *et al.*, 2005). A citric acid ethanol solution was used in an attempt to improve corrosion resistance and adhesion of a RE coating on AZ91D Mg alloy (Wang *et al.*, 2009), where α-Mg dissolved slightly and formed insoluble Mg_3Cit_2 in ethanol. Thus, the Ce conversion coating was mainly formed on α phase, differing from the coatings formed in the aqueous conversion solutions.

Some stable compounds, such as, Niobium (Nb) and Zr oxides, were added into the CeO_2 conversion film on AZ91D and AM50 (Ardelean and Marcus, 2004, 2007; Ardelean *et al.*, 2008). The corrosion current density of the Ce-Zr-Nb coating was reported as two decades lower than that of pure Ce or Ce-Zr coatings and Dow 22 chromate coatings. But a 24 h treatment is necessary to seal the open pores in the initially formed film.

Though different theories have been proposed to explain the formation of Ce and La conversion films on iron (Fe), Zn (Böhm *et al.*, 2000), Ni (Kobayashi and Fujiwara, 2006), copper (Cu) (Aldykiewicz *et al.*, 1996) and Al alloys (Campestrini *et al.*, 2004; Yu and Li, 2004), little is really known about the coating mechanism on Mg alloys to date. Some observations suggest that a phenomenology is: upon immersion in an acidic $Ce(NO_3)_3$ bath, where pH is about 5.2 at 30°C, air-formed $Mg(OH)_2$ surface film on AZ31 dissolves first, followed by metal dissolution. Al^{3+} ions precipitate back to the surface in the form of $Al(OH)_3$ or Al_2O_3 due to its low solubility when pH is above 4.0. Then Mg^{2+} and OH^- ions deposit on the surface as a porous layer when pH rises over 8.5. Meanwhile, Ce^{3+} ions, in sufficient amounts in the bath, also locate on the top of the $Mg(OH)_2$ layer and a compact layer forms

(Lin and Fang, 2005). The process of the growth of Ce conversion films was investigated and a three-step mechanism was postulated (Montemor *et al.*, 2007): (1) dissolution of air-formed oxide, formation of OH^- ions and hence pH increase; (2) growth of an initial mixed layer of $Ce(OH)_4$ and $Ce(OH)_3$ upon immersion; (3) thickening of the surface film and milder pH changes with preferential deposition of $Ce(OH)_4$ and conversion into CeO_2, forming a top layer, containing sufficient Ce^{4+} species. The corrosion behaviour of AZ31 Mg alloy converted in the baths with different Ce salts, that is, $CeCl_3$, $Ce(NO_3)_3$, $Ce_2(SO_4)_3$ and $CePO_4$, was investigated by the same authors (Montemor *et al.*, 2008). Results revealed that the anions in the coating bath play a critical role in the thickness and corrosion protection efficiency on the conversion coating. The $CeCl_3$ bath generated the thickest coating while the thinnest one was formed in the $CePO_4$ solution.

In terms of La-based coatings, lanthanum nitrate ($La(NO_3)_3$) is commonly used as La source to form $La(OH)_3$ on Mg alloys (Yang *et al.*, 2008). The $La(OH)_3$ coating displayed passivation in a wide potential region and the corrosion current density (i_{corr}) decreased by about two orders of magnitude compared to that of the bare substrate due to its stability and cathodic inhibiting characteristics. On the contrary, after converting a series of Mg alloys, say AZ31, AZ61, AZ91 and AM60, in a $Mg(NO_3)_2$ and/or $La(NO_3)_3$ bath, it was observed that a coating having good corrosion protection was produced only in a bath containing both $Mg(NO_3)_2$ and $La(NO_3)_3$ (Takenaka *et al.*, 2007, 2008). Again, a drawback of such coatings is the minimum 1 h immersion time required to produce a protective coating.

10.4.2 Other conversion coatings

Ionic liquid (IL) and molten salt conversion coatings

Ionic liquid (IL) and molten salt are not only a good medium to electroplate Mg alloys with metallic films but can be used as conversion coating. A bis(trifluoromethanesulfonyl) amide (TFSA) based IL coating was produced upon pure Mg (Birbilis *et al.*, 2007). The properties of the resultant coating were significantly influenced by coating time. A continuous, adhesive and defect-free film owing the optimum corrosion protective performance was achieved when the IL treatment time was less than 2 h. An attempt was made to use aluminium chloride ($AlCl_3$)-NaCl molten salt to deposit a dense and protective coating on AZ91D alloy (He *et al.*, 2008). The hypothesis behind this work was that the corrosion rate of the Mg-Al alloys could be significantly reduced by enhancing the volume fraction of the β phase ($Mg_{17}Al_{12}$) in the skin to a high level. There was a linear correlation between the coating thickness and the treatment temperature. High operating temperature (above 300°C) generated cracks in the deposition due to the different thermal expansion co-efficiency between the coating and the AZ91D substrate, which compromised the corrosion resistance of the coating.

Multi-elements complex coatings (MECC)

Zhao *et al.* have presented an environmentally acceptable process to develop a 'multi-elements complex coating' (MECC) on AZ91D (Zhao *et al.*, 2006a), where HF pickling was avoided for the sake of cost and pollution control. $Ca_{0.965}Mg_2Al_{16}O_{27}$, $Mn_{5.64}P_3$, $ZnAl_2O_4$ and $(Mg_{0.66}Al_{0.34})(Al_{0.83}Mg_{0.17})_2O_4$ were the major components in the MECC. The less than 10 min coating time is promising for industrial adoption. Though networks of tiny cracks in the MECC were created by the hydrogen released during the coating process, they did not approach the substrate. The coating formation process was divided into three stages according to the thickness, mass loss, and phase and element composition (Zhao *et al.*, 2007). In the first 3 min, the coating thickness steadily increased but the mass loss was positive due to metal matrix corrosion. During the metaphase stage (3 ~ 5 min), the value of the thickness decreased distinctly and mass loss was still positive, but the coating was a mixture of amorphous and crystalline phases. In the last 2 min, the final stage, the coating grew thicker, smoother and more compact. No amorphous phase was detected. The results revealed that the MECC could provide effective protection better than that of the conventional Dow No.1 coating.

Vanadate coatings

Vanadate is a good corrosion inhibitor used widely in paint and pigment systems, but there has been little work done to convert Mg alloys with vanadates. One reason might be the hazardous effect of the vanadate ions in the waste solution. V_2O_5, $NaVO_3$ and $Na_2V_2O_4$ act as the nominal vanadate source in conversion coating systems (Ger *et al.*, 2004; Buchheit *et al.*, 2006). The influence of coating conditions, such as vanadate concentration, immersion time and bath temperature, on corrosion resistance was investigated (Yang *et al.*, 2007). The coating thickness grew with increased immersion time and temperature. An optical crack-free coating was realised in 30 g/L $NaVO_3$ at 80°C for 10 min.

Titanate coatings

Yang and colleagues (Yang *et al.*, 2012b) proposed a titanate conversion coating on AZ31 Mg alloy for corrosion resistance. It was found that the formation of the titanate conversion coating proceeded via the formation of a porous $Mg(OH)_2$ and MgF_2 layer. Compact $Si(OH)_4$ and $Ti(OH)_4$ were then dissolved locally on top of the porous layer. Continuing immersion up to 6 min thickened the porous layer and extended the coverage of the compact $Si(OH)_4$ and $Ti(OH)_4$ to a larger extent, which, in turn, reduced the corrosion area fraction and enhanced the polarisation resistance of the AZ31. However, prolonged immersion up to 10 min resulted in more cracks with larger openings, reducing the corrosion resistance of the titanate conversion coating. As a result, the titanate conversion coating formed at immediate immersion times (3–6 min) offered the best corrosion protection on the AZ31.

Organic coatings

Though organic conversion coatings are rarely applied upon Mg alloys, some organic compounds, such as tannic acid ($C_{76}H_{52}O_{46}$) (Chen *et al.*, 2009) and organic phytic acid ($C_6O_{18}O_{24}P_6$) (Liu *et al.*, 2006), due to their structural uniqueness, are capable of chelating with metal ions to make stable complexes. A tannic acid based conversion coating comprising pentahydroxy benzamide-Mg complex, Al_2O_3 and MgF_2, was performed on AZ91 Mg alloy (Chen *et al.*, 2009). A protective coating, without coarseness and uneven coverage, was obtained when the samples were subjected to the coating process for 300–600 s. Besides the ability to form a complex with metal ions, organic phytic acids are also non-toxic and provide Mg alloys comparative corrosion protection to that of the conventional chromate coating (Liu *et al.*, 2006).

10.5 Coating pretreatment processes

The field of conversion coating technology for Mg alloys is significantly less advanced than that of Al, principally owing to Al being the staple of the aerospace industry for many decades. The aerospace related research has allowed technological advances in Al coatings, and concomitantly, the fact that Mg is presently not routinely used in high-value applications means that aspects such as surface pretreatment have not received the dedicated attention they desperately deserve. This chapter has revealed that for the same conversion coating, the pretreatment selected can have a major (if not the major) impact on the ultimate coating efficacy. Consequently, a brief summary of the scope and role of pretreatment is given here – and is considered very important to the topic. However, it should be borne in mind, that very few publications dealing with coating pretreatments alone exist, with the exception of Yang *et al.* (2012a), which reveals just how important pretreatment is.

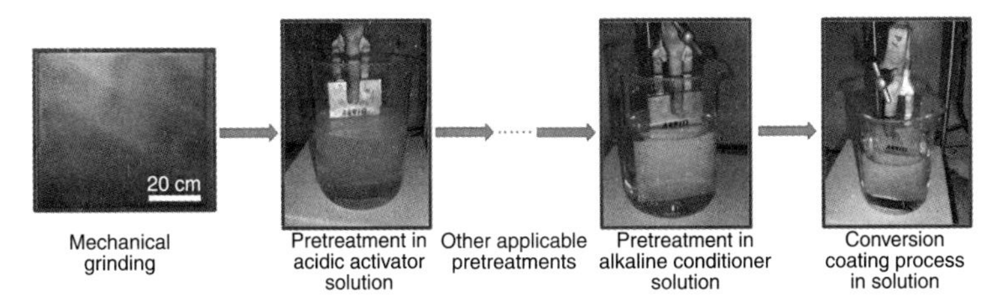

10.4 Conventional conversion coating process of Mg alloy AZ91D.

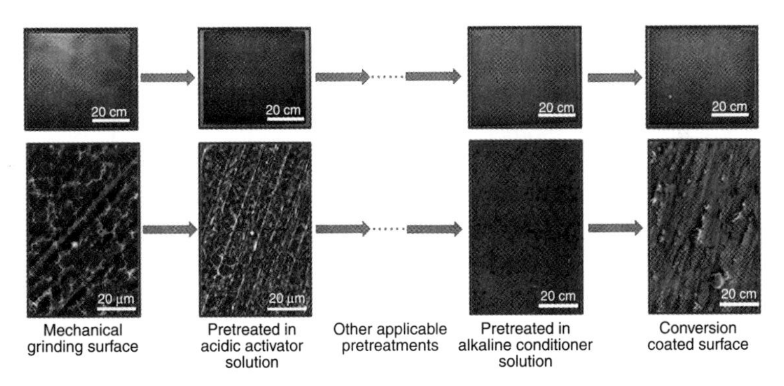

10.5 Photographs (top) and SEM images (bottom) of the surface of AZ91D plates after different treatments.

According to Yang *et al.* (2012a) using AZ91D as the principal example, the typical conversion coating process including pretreatment can be described and depicted as per Fig. 10.4. The flow chart in Fig. 10.4 reveals rather obviously that the conversion coating is as simple as immersion in appropriate chemicals, and can be preceded by any number of pretreatment steps to aid in the optimisation of the coating (final step). The pretreatments must therefore necessarily impart some sort of functionality to, by modification of, the surface of the alloy.

The macro and microstructural impact from the most common pretreatment procedures is depicted in Fig. 10.5 for AZ91D. It is clear that pretreatments change the macroscopic colour of the surface (Fig. 10.5 top), and the concomitant changes to the underlying microstructure are captured in Fig. 10.5 (bottom). What is seen is that following mechanical polishing, the AZ91D has a surface consisting of the Mg matrix populated with β-phase intermetallic particles. Yang *et al.* (2012a) discovered that the β-phase of AZ91D Mg alloy can be preferentially dissolved/removed in conditioner (concentrated NaOH) since the Al-containing phase dissolves at high pH, and the Mg matrix is essentially passive and unaffected. But the concomitant formation of $Mg(OH)_2$ film upon the Mg matrix can eventually inhibit

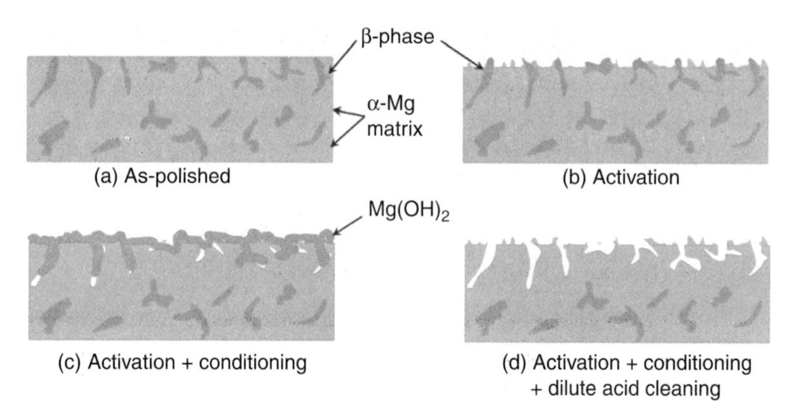

10.6 (a–d) Schematic representation of surface following the various pretreatment processes.

dissolution of the β-phase. As such, dissolution efficiency was dramatically improved by prior activation in acid solution, which was shown to attack the Mg matrix and thus was an effective way to generate protruding phase upon the surface that is then readily dissolved by conditioning through increasing contact region. Phenomenologically for the case of AZ91D, this is presented in Fig. 10.6 which reveals the ability of pretreatments to create a surface of uniform composition (*viz.* electrochemically homogenous) and the reader could foresee how it may be possible to customise the pretreatment process for either: (a) alloys of different composition and microstructure, or (b) performance requirement of the conversion coatings.

One thing that is not appreciated in a large portion of, if not the majority of papers reviewed herein, is that surface pretreatment is as important, if not more important, than the ultimate 'coating' step (since the latter is dependent on the quality and preparedness of the incoming surface). In order to provide the reader with a more general overview of the morphology of conversion coatings, a selection of micrographs is included in Fig. 10.7. The SEM images reveal that a uniform conversion coating conforms to the substrate topography (Fig. 10.7a and 10.7b) that is typical of ground surfaces. The higher magnification imaging (Fig. 10.7c) displays a very distinct and dense coating that varies in thickness between ~ 2 and ~ 3 μm (Fig. 10.7d). Such coatings enhanced the corrosion resistance of the AZ91D substrate by one order of magnitude.

10.6 Evaluation of the coating (corrosion) performance

Details about the conversion coating procedures from the literature reviewed herein and the protection afforded by the conversion coatings was extracted

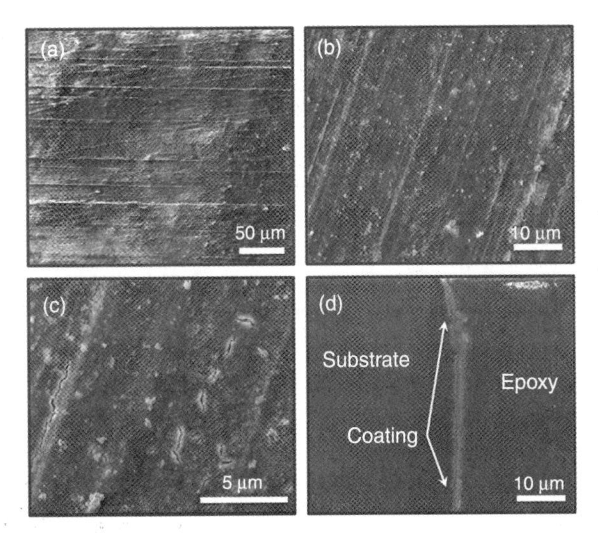

10.7 Representative SEM images of calcium phosphate conversion coating on Mg alloy AZ91D: (a), (b) and (c) plain, and (d) cross-section view.

and summarised in Table 10.2. It lists the key features of the coatings, including any pretreatment, the coating chemicals, the coating processes (times, temperatures, etc.) and the ultimate corrosion performance. It is clear that there is no common method of assessment being used, with performance metrics varying widely from study to study with respect to techniques used (i.e. immersion tests, EIS, salt spray, etc.).

In order to further abridge the useful information in Table 10.2, the authors' assessment of the anti-corrosive performance of the various conversion coatings can be seen in the literature (adapted and updated from the original form in Chen *et al.*, 2011b). By interpreting and weighing the relative merits of the coatings, such an assessment could be made, however as the reader will note, rather than a detailed scrutiny, the performance has been categorised in groups. In interpreting performance, conventional chromate-based conversion coatings were used as a benchmark to evaluate the coatings.

The classifications are given as: Grade A means the resulting coating can provide a better corrosion protection than that of chromate coatings. The coatings with grade B either performed comparatively to chromate coatings, or reduced the i_{corr} dramatically by inducing a form of passivation. A coating was classified as C grade when it decreased the i_{corr} value within one order of magnitude. Grade D provided little or minimal protection (in some cases barely enough to justify the coating cost or effort) and grade E had no sufficient corrosion protective ability that could persist for in-service

Table 10.2 Summary of coating procedures and corrosion performance of the resulting coatings

	Case No.	Pretreatment*	Formulation composition	Coating methodology	Postulated phases in coating	Film properties	Corrosion test medium	Corrosion performance test employed	Corrosion protection enhancement
PO_4/PO_4-MnO_4,	1 (Kouisni *et al.*, 2005)	None	Na_2HPO_4 20 g/L, H_3PO_4 7.4 mL/L, $NaNO_3$ 3 g/L, $ZnNO_3$ 5 g/L and NaF 1 g/L	47°C, 10 min	$Zn_3(PO_4)_2$	Porosity decreased with treatment time	Borate buffer solution of pH 9.2	Potentiodynamic polarisation (PDP), open circuit potential (OCP) and impedance spectroscopy measurements (EIS)	Not sufficiently high in the field of corrosion protection
	2 (Niu *et al.*, 2006)	Conditioning	H_3PO_4 17.5 g/L, ZnO 3.2 g/L, NaF 1.7 g/L, organic amine 0.18 g/L, tartaric acid 2.2 g/L, $NaNO_2$ 0.83 g/L and $NaNO_3$ 0.17 g/L	40–45°C, less than 40 min, pH 1.8–2.5	$Zn_3(PO_4)_2$, $MgZn_2(PO_4)_2$	6 μm thick	5 wt.% NaCl (0.86 M)	Salt spray ASTM 117 standard (SS)	Similar to that of Cr-based conversion coating, but with better adhesion property
	3 (Song *et al.*, 2009b)	None	$Ca(NO_3)_2$ 25 g/L, $NH_4H_2PO_4$ 25 g/L and H_3PO_4 or HNO_3 to adjust pH	40°C, 5 min, pH 3.0	$CaHPO_4{\cdot}2H_2O$, $Ca_3(PO_4)_2$ and $Mg_3(PO_4)_2$	Lamellar structure with leaf-like particles	0.6 M NaCl, 0.6 M Na_2SO_4, 0.6 M NaOH	PDP and EIS	Better than uncoated Mg-8.8Li in NaCl and Na_2SO_4
	4 (Chong and Shih, 2003)	Alkaline decreasing, acid pickling and activating	$KMnO_4$ 20 g/L and $MnHPO_4$ 60 g/L	50°C, 10 min	(Mg, Al and Mn) oxides, $MgAl_2O_4$, hydroxides and phosphates	compact, 10 μm thick with cracks	5 wt.% NaCl	PDP	Equivalent or slightly better than conventional Cr-based coatings

	5 (Zhao *et al.*, 2006b)	Solvent degreasing and alkaline cleaning	$KMnO_4$ 40 g/L, K_2HPO_4 150 g/L, and H_3PO_4 to adjust pH	40–70°C, 10 min, pH 3–6	Mg, O, P, K, Al, Mn	Porous, 7–10 μm thick	5 wt.% NaCl	PDP and SS	Equivalent to the conventional Cr-based coatings (Dow 1)
	6 (Zhou *et al.*, 2004)	Alkaline degreasing, acid pickling and activating	$Mn(NO_3)_2$ 30 g/L, NaH_2PO_4 40 g/L	85°C, 40 min, pH 2	P, Mn, O	Amorphous	3.5 wt.% NaCl	Mass Loss (ML)	Equivalent to that of the DOW 7 Cr-based coating
	7 (Chen *et al.*, 2011a, 2012a)	None	$Ca(NO_3)$ 0.01 M and 0.017 M, Na_3PO_4 0.01 M, pH 3–4.5 adjusted by HNO_3	pH 3.0, 65°C	HA, $Mg(OH)_2$, and $CaH(PO_4)\cdot 2H_2O$	Flake-like structure, compact and uniform	Minimum Essential Medium	PDP, ML, Hydrogen evolution measurement	i_{corr} of coated Mg and Mg alloys 2 orders of magnitude lower than uncoated ones
			NaOH 10 g/L	80°C, 4 h					
	8 (Waterman *et al.*, 2012)	$Ca(OH)_2$ Electrostatic coating	$CaCl_2$ 1.65 g/L, KH_2PO_4 0.3 g/L $Na_2HPO_4\cdot 4H_2O$	pH 6, 24 h, 37°C	$CaH(PO_4)\cdot 2H_2O$	Flake-like structure	Hank's solution	PDP and OCP	i_{corr} of coated Mg 1 order of magnitude lower than uncoated one
Stannate	9 (Huo *et al.*, 2004)	None	NaOH 10 g/L, $Na_2SO_3\cdot 3H_2O$ 50 g/L and $NaC_2H_3O_2\cdot 3H_2O$ 10 g/L	90°C, 60 min moderate stirring	$MgSnO_3\cdot H_2O$	3–5 μm thick porous coating composed of spherical grains	0.6 M NaCl,	Hydrogen evolution (HE) and PP	Passivation occurred during anodic polarisation in 3.5 wt.% NaCl solution

(*Continued*)

Table 10.2 Continued

	Case No.	Pretreatment*	Formulation composition	Coating methodology	Postulated phases in coating	Film properties	Corrosion test medium	Corrosion performance test employed	Corrosion protection enhancement
	10 (Elsentriecy *et al.*, 2008)	Acid pickling	0.25M $Na_2SO_3 \cdot 3H_2O$, 0.073 M $NaC_2H_3O_2 \cdot 3H_2O$, 0.13 M $Na_3PO_4 \cdot 12H_2O$, and 0.125 M NaOH	80°C, 60 min, pH 13.1	$MgSnO_3 \cdot 3H_2O$	Compact coating with crevices blocked by small particles	0.15 M H_3BO_3 + 0.05 M $Na_2B_4O_7$	PDP	i_{corr} of coated AZ91D 2 orders of magnitude lower than uncoated one
	11 (Liu *et al.*, 2009)	None	0.25 M $Na_2SO_3 \cdot 3H_2O$, 0.073 M $NaC_2H_3O_2 \cdot 3H_2O$, 0.13 M $Na_3PO_4 \cdot 12H_2O$ and 0.05 M NaOH	60°C, 60 min	Mg/Sn hydroxide and oxide	Continuous and compact dual layer, 1 μm thick	0.05 M NaCl + 0.01 M NaOH	PDP and EIS	Compact structure in the coating blocked the penetration of aggressive media (Cl^-), which led to great improvement of the corrosion resistance of resulting coating
Fluoride	12 (Chiu *et al.*, 2007)	None	HF 48 wt.%	RT, 1440 min	MgF_2	1.5 μm thick, dense, crack-free	Hank's solution	PDP, OCP and EIS	30 or 40 times improvement of polarisation resistance from uncoated sample

RE	13 (Montemor *et al.*, 2007)	None	10^{-3} M $Ce(NO_3)_3$	22°C, 60 min, pH 6.0	Mg $(OH)_2$, Ce_2O_3, CeO_2	Coating thickened with the increase in treatment time	0.005 M NaCl	PDP, OCP and scanning vibrating electrode technique (SVET)	10 times decrease in i_{corr}
	14 (Montemor *et al.*, 2007)	None	$La(NO_3)_3$ 10^{-3} M		$La(OH)_3$				
	15 (Ardelean *et al.*, 2008)	Alkaline degreasing and acid pickling	(0.03–0.1 M) $Ce(NO_3)_3$, (0.03–0.1 M) $ZrO(NO_3)_2$, (0.02–0.05 M) $Nb_xO_yF_x$	RT, 3–1440 min, pH 4.0	MgO, MgF_2, $MgCO_3$, Ce_2O_3, CeO_2, ZrO_2, Nb_2O_5	Compact coating	0.5 M Na_2SO_4	PDP, OCP, EIS and SS	Better than that of Cr-based coatings (Dow 22)
	16 (Wang *et al.*, 2009)	None	$Ce(NO_3)_3 \cdot 6H_2O$ 5 g/L, Citric acid 1 g/L	RT, 120 min	CeO_2, Ce_2O_3, $CePO_4$, $Mg_3(PO_4)_2$, Al_2O_3, MgO	Adhesive, cracked and 0.5 μm thick	3.5 wt.% NaCl	ML and PDP	Better corrosion resistance than that of Dow 1 coating
	17 (Yang et al., 2008)	None	$La(NO_3)_3$ 5 g/L	RT, 20 min, pH 5.0	$La(OH)_3$	10 μm thick, homogeneous uniform with aciculate structure	3.5 wt.% NaCl	PDP and OCP	Corrosion current density of coating decreased 2 orders of magnitude compared to bare metal
	18 (Cui *et al.*, 2012)	None	$Nd(NO_3)_3$ 5 g/L Phytic acid 5 g/L (Post-treatment)	50°C, 20 min 40°C, 15 min, pH 8.0	Nd_2O_3, Al_2O_3, $Mg(OH)_2$, PO_4^{3-} and HPO_4^{2-} ions	Cracked and compact	3.5 wt.% NaCl	OCP and EIS	Post-treatment gave the coating better corrosion resistance

(*Continued*)

Table 10.2 Continued

	Case No.	Pretreatment*	Formulation composition	Coating methodology	Postulated phases in coating	Film properties	Corrosion test medium	Corrosion performance test employed	Corrosion protection enhancement
Others	19 (Birbilis *et al.*, 2007)	None	$P_{6,6,6,14}$TFSA	RT, 60 min	Likely MgF_2	87 nm, uniform	0.1 M NaCl	ML, PDP and EIS	Defect-free resulted in an increase in the impedance of pure Mg of as much as 3 orders of magnitude
	20 (He *et al.*, 2008)	None	AlCl3-NaCl molten salt	250–400°C, 480 min	$Mg_{17}Al_{12}$ and Mg_2Al_3	Dense, 10.8 μm thick	0.6 M NaCl	PDP, OCP and EIS	Sample with 300°C treatment had the best corrosion resistance
	21 (Zhao *et al.*, 2006a)	Alkaline degreasing	H_3PO_4, $MnCO_3$, $Mn(NO_3)_2$	55°C, 7 min	$Ca_{0.965}Mg_2Al_{16}O_{27}$, $Mn_{5.64}P_3$, $ZnAl_2O_4$, $(Mg_{0.66}Al_{0.34})$ $(Al_{0.83}Mg_{0.17})_2O_4$	7–15 μm thick, tiny cracks, continuous	5 wt.% NaCl	PDP and SS	Corrosion resistance 5 times better than Dow 1 Cr-based coatings
	22 (Yang *et al.*, 2007)	None	$NaVO_3$ 30 g/L	80°C, 10 min pH 8.0	Vanadium oxide	0.8 μm thick, cracked	0.1 wt.% NaCl	PDP and OCP	Superior to chromate coating

23 (Chen *et al.*, 2009)	Alkaline degreasing	HNO_3 0.1 g/L, Na_3PO_4 1.0 g/L, $Na_2B_7O_4$ 5.0 g/L, $C_{76}H_{52}O_{64}$ 0.8 g/L, NH_4VO_3 1.0 g/L, H_2ZrF_6 1.1 g/L	RT, 5–10 min,	Penta-hydroxy benzamide-Mg complex, Al_2O_3 and MgF_2	Dense, 1.6 μm thick	3 wt.% NaCl	PDP, OCP and EIS	Corrosion resistance R_f value 23 000 Ω far higher than that (57.9 Ω) of uncoated one
24 (Ng *et al.*, 2010)	Hydrothermal treatment in Teflon container	Stearic acid $(CH_3(CH_2)_{16}COOH)$	100°C, 2 min	$Mg(OH)_2$, stearic acid and magnesium stearate	230 μm thick, smooth and compact	Hank's solution	PDP, OCP and EIS	Corrosion resistance improved as high as 4 orders of magnitude in the initial period, and gradually dropped to about 40 times in the long run
25 (Yang *et al.*, 2012b)	None	0.01 M $TiCl_4$, 0.01M H_2SiF_6, HNO_3 5 ml/L	40°C, pH 4.0, up to 10 min	$Mg(OH)_2$, MgF_2, $Si(OH)_4$ and $Ti(OH)_4$	Crack	0.05 M NaCl, 0.1 M Na_2SO_4 and 5 wt.% NaCl	PDP, OCP, EIS and SS	The optimal titanate coating had a corrosion area fraction less 5% after 24 h salt spray test

*Excluding mechanical polishing, normal water cleaning and solvent degreasing procedures.

application. A brief overview relating to the top four performing coatings is given below.

The assessment in the literature (Chen *et al.*, 2011b) suggests that phosphate/phosphate-permanganate conversion coating technology produces protective films on Mg alloys. With respect to case number 2, the performance observed appears to be excellent, which is in line with emerging manuscripts covering phosphate coatings. The quality of coating 2 in particular appears to arise from the combination of using a conditioner pretreatment (presumably homogenising the surface) and the coating formulation that includes several functional components, including NaF. The NaF presumably decomposes to give free F in solution (of low pH) and allowing high rates of Mg reaction and coating formation.

The past two years have seen activity in the area of conversion coating of Mg alloys intended for biomaterial/bio-implant applications. Such coatings are nominally based on Ca and phosphate-containing species that are insoluble in physiological fluids. Ca and phosphate compounds are a major component of bone, therefore are used to improve the biocompatibility of metallic implants. Ca-P compounds, such as coatings 7 and 8, have been studied to improve the corrosion properties for biomedical applications, including Mg and its alloys.

The fluoride coating (coating 12) performed well, however the use of 48 wt.%HF is unfeasible in an industrial setting, with the Occupational Health & Safety (OHS) restrictions in most nations preventing industrialisation of this process. Additionally, the lengthy time required for coating is not entirely practical. None the less, the improvement in corrosion resistance observed is not surprising if the surface could be coated with MgF_2. The MgF_2 would be insoluble in most aqueous media and form a suitable barrier-coating if defect-free.

Coating 15 showed good performance, and emphasis was given in the role of pretreatment with this technology, indicating that the alkaline and acid pretreatments play an important role in delivering a good final coating. Again, the long coating times make this coating a little impractical, however the ultimate performance of the coating would be presumed to be high if indeed MgF_2 and non-Mg oxides could be well bonded to the Mg surface.

Coating 21 also showed some promise. The pretreatment in this case included alkaline degreasing, and the chemicals used were considered to be OHS friendly compared with F-containing solutions. The coating time was short, and performance in a concentrated chloride seems excellent. The characterisation of surface products makes the mechanism of protection elusive, but again points to future work.

This chapter has demonstrated that a large number of conversion coating processes can be used for corrosion protection of Mg alloys. Of the best performing coatings, pretreatment was used in all cases (with the exception of the very concentrated HF immersion example). To be a real substitute for chromate coatings, there is still significant research to be done which requires at a minimum: strategic deployment of pretreatment processes and the minimisation of coating time down to a couple of minutes for the sake of practical applications. Each technique reviewed has its own merits and limitations. In regard to the novel coatings reviewed, IL and molten salt conversion coatings are not applicable for practical use at this stage (and may never be owing to technological and cost issues); vanadium based coatings still have toxicity concerns; and stearic acid coatings require high operational temperatures. There is no doubt however, that researchers are inspired by the urgent industrial demand to develop better, simpler, cleaner, safer and more cost-effective conversion coatings to realise a growth in the applications of Mg alloys.

10.7 Conclusions

Chemical conversion coatings provide a cost-effective method for corrosion protection of Mg and its alloys. Generally speaking, the protective performance of a conversion coating can be determined by several coating characteristics, such as thickness, number and type of defects, adhesion to the metallic substrate, resistance to the chemical attack of the test environment/electrolyte, etc. Thicker conversion coatings seem to be more effective in retarding corrosion activity, but only when no trans-cracks are present in the coating. Chemical composition of the conversion coating also plays a critical role in the corrosion protectiveness, however it was found that the coating performance also relies heavily on appropriate pretreatments required to functionalise the surface.

Though a large number of chromium-free coating technologies are nowadays available (or at least reported) for Mg alloys, the widespread use of Mg alloys in industry (e.g. automotive industry) is still limited due to the lack of appropriate protection for resisting auto service conditions. Every coating system reviewed in this chapter has relative merits and limitations.

It is apparent that coating formation mechanisms, from a fundamental science point-of-view, are still lacking in the literature; this will most likely change as workers seek to secure technology rights or patents for their coatings. Some recent exceptions to this are emerging in the context of Ca-phosphate and hydroxyapatite coatings in the past few years, on the quest to control the inherent rapid biodegradation/corrosion process of Mg biomaterials in simulated body fluid. Standard methods to quantitatively and qualitatively evaluate real corrosion resistance of conversion coated Mg and its alloys is however also lacking – with some papers using one or more of: immersion tests, salt spray, impedance, potentiodynamic polarisation or hydrogen collection (but not necessarily the same methods as others). This makes cross correlations and comparisons of efficiency very subjective, a challenge that was presented herein via the construction of Table 10.2.

Developing effective and environmental friendly coating technologies is, and will remain, a hot topic needing further investigation. This is essential such that the mechanical physical properties of Mg alloys can be fully exploited in practical applications.

10.8 References

Aldykiewicz, J. A. J., Davenport, A. J. and Isaacs, H. S. (1996). Studies of the formation of cerium-rich protective films using X-ray absorption near-edge spectroscopy and rotating disk electrode methods. *Journal of the Electrochemical Society*, **147**, 147–154.

Anicai, L., Masi, R., Santamaria, M. and Quarto, F. D. (2005). A photoelectrochemical investigation of conversion coatings on Mg substrates. *Corrosion Science*, **47**, 2883–2900.

Ardelean, H., Frateur, I. and Marcus, P. (2008). Corrosion protection of magnesium alloys by cerium, zirconium and niobium-based conversion coatings. *Corrosion Science*, **50**, 1907–1918.

Ardelean, H. and Marcus, P. (2004). *Composition and method for treating magnesium alloys*. US patent application WO2002FR01843 20020531.

Ardelean, H. and Marcus, P. (2007). *Composition and method for treating magnesium alloys*. US patent application 10/478861.

Avedesian, M. M. and Baker, H. (1999). *ASM Specialty Handbook: Magnesium and Magnesium Alloys*. International Metals Park, OH, ASM.

Birbilis, N., Howlett, P. C., Macfarlane, D. R. and Forsyth, M. (2007). Exploring corrosion protection of Mg via ionic liquid pretreatment. *Surface and Coatings Technology*, **201**, 4496–4504.

Böhm, S., Greef, R., McMurray, H. N., Powell, S. M. and Worsley, D. A. (2000). Kinetic and mechanistic studies of rare earth-rich protective film formation using in situ ellipsometry. *Journal of the Electrochemical Society*, **147**, 3286–3293.

Brunelli, K., Dabalà, M., Calliari, I. and Magrini, M. (2005). Effect of HCl pre-treatment on corrosion resistance of cerium-based conversion coatings on magnesium and magnesium alloys. *Corrosion Science*, **47**, 989–1000.

Buchheit, R. G., Guan, H. and Laget, V. N. (2006). *Corrosion resistant coating with self-healing characteristics.* US patent application 7,135, 075.

Campestrini, P., Terryn, H., Hovestad, A. and Wit, J. H. W. D. (2004). Formation of a cerium-based conversion coating on AA2024: relationship with the microstructure. *Surface and Coatings Technology*, **176**, 365–381.

Cape, T. W. (2008). *High manganese cobalt-modified zinc phosphate conversion coating*. US patent application 12/135,520.

Chang, M. Z., Chen, L. and Gong, Q. (2005). *Method for processing Mg alloy surface.* China patent application CN20051115372 20051116.

Chen, X.-B., Abbott, T. and Birbilis, N. (2011a). A simple route towards a hydroxyapatite-$Mg(OH)_2$ conversion coating for magnesium. *Corrosion Science*, **53**, 2263–2268.

Chen, X. -B., Birbilis, N. and Abbott, T. B. (2011b). A review of corrosion resistant conversion coatings for magnesium and its alloys. *Corrosion*, **67**, 035005.

Chen, X.-B., Kirkland, N. T., Krebs, H., Thiriat, M.-A., Virtanen, S., Nisbet, D. and Birbilis, N. (2012a). In vitro corrosion survey of Mg-xCa and Mg-3Zn-yCa alloys with and without calciumphosphate conversion coatings. *Corrosion Engineering, Science and Technology*, **47**, 365–373.

Chen, X. M., Li, G. Y., Lian, J. S. and Jiang, Q. (2009). Study of the formation and growth of tannic acid based conversion coating on AZ91D magnesium alloy. *Surface and Coatings Technology*, **204**, 736–747.

Chen, X.-B., Yang, H.-Y., Abbott, T. B., Easton, M. A. and Birbilis, N. (2012b). Magnesium: engineering the surface. *Journal of the Minerals, Metals and Materials Society*, **64**, 650–656.

Chiu, K. Y., Wong, M. H., Cheng, F. T. and Man, H. C. (2007). Characterization and corrosion studies of fluoride conversion coating on degradable Mg implants. *Surface and Coatings Technology*, **202**, 590–598.

Chong, K. Z. and Shih, T. S. (2003). Conversion-coating treatment for magnesium alloys by a permanganate-phosphate solution. *Materials Chemistry and Physics*, **80**, 191–200.

Cui, X. F., Jin, G., Yang, Y. Y., Liu, E. B., Lin, L. L. and Zhong, J. G. (2012). The formation of neodymium conversion coating and the influence of post-treatment. *Applied Surface Science*, **258**, 3249–3254.

Dabala, M., Brunelli, K., Napolitani, E. and Magrini, M. (2003). Cerium-based chemical conversion coating on AZ63 magnesium alloy. *Surface and Coatings Technology*, **172**, 227–232.

Dolan, S. E. and Carlson, L. R. (2007). *Process for coating metal surfaces.* US patent application 7,175,882.

Elsentriecy, H. H., Azumi, K. and Konno, H. (2007a). Effect of surface pretreatment by acid pickling on the density of stannate conversion coatings formed on AZ91 D magnesium alloy. *Surface and Coatings Technology*, **202**, 532–537.

Elsentriecy, H. H., Azumi, K. and Konno, H. (2007b). Improvement in stannate chemical conversion coatings on AZ91D magnesium alloy using the potentiostatic technique. *Electrochimica Acta*, **53**, 1006–1012.

Elsentriecy, H. H., Azumi, K. and Konno, H. (2008). Effects of pH and temperature on the deposition properties of stannate chemical conversion coatings formed by the potentiostatic technique on AZ91 D magnesium alloy. *Electrochimica Acta*, **53**, 4267–4275.

Ger, M.-D., Yang, K.-H., Sung, Y., Hwu, W.-H. and Liu, Y.-C. (2004). *Method for treating magnesium alloy by chemical conversion.* US patent application 10/167,479.

Gonzalez-Nunez, M. A., Skeldon, P., Thompson, G. E. and Karimzadeh, H. (1999). Kinetics of the development of a nonchromate conversion coating for magnesium alloys and magnesium-based metal matrix composites. *Corrosion*, **55**, 1136–1143.

Gray, J. E. and Luan, B. (2002). Protective coatings on magnesium and its alloys – a critical review. *Journal of Alloys and Compounds*, **336**, 88–113.

Greene, J. A. and Vonk, D. R. (2008). *Conversion coatings including alkaline earth metal fluoride complexes*. US patent application 20030230364.

Gulbrandsen, E., Taftø, J. and Olsen, A. (1993). The passive behaviour of Mg in alkaline fluoride solutions. Electrochemical and electron microscopical investigations. *Corrosion Science*, **34**, 1423–1440.

Hamdy, A. S. (2008). The effect of surface modification and stannate concentration on the corrosion protection performance of magnesium alloys. *Surface and Coatings Technology*, **203**, 240–249.

Hawke, D. and Albright, D. L. (1995). A phosphate-permanganate conversion coating for magnesium. *Metal Finishing*, **93**, 34–38.

He, M. F., Liu, L., Wu, Y. T., Tang, Z. X. and Hu, W. B. (2008). Corrosion properties of surface-modified AZ91D magnesium alloy. *Corrosion Science*, **50**, 3267–3273.

Huo, H., Li, Y. and Wang, F. (2004). Corrosion of AZ91D magnesium alloy with a chemical conversion coating and electroless nickel layer. *Corrosion Science*, **46**, 1467–1477.

Kobayashi, Y. and Fujiwara, Y. (2006). Chemical deposition of cerium oxide thin films on nickel substrate from aqueous solution. *Journal of Alloys and Compounds*, **408**–412, 1157–1160.

Kouisni, L., Azzi, M., Zertoubi, M., Dalard, F. and Maximovitch, S. (2004). Phosphate coatings on magnesium alloy AM60 part 1: study of the formation and the growth of zinc phosphate films. *Surface and Coatings Technology*, **185**, 58–67.

Kouisni, L., Azzi, M., Zertoubi, M., Dalard, F. and Maximovitch, S. (2005). Phosphate coatings on magnesium alloy AM60 part 2: electrochemical behaviour in borate buffer solution. *Surface and Coatings Technology*, **192**, 239–246.

Li, G. Y., Niu, L. Y., Lian, J., Jiang, Z. H., Liu, X. L. and Chen, J. W. (2005). *Magnesium alloy phosphorization solution and its phosphorized technology.* CN200410011015 20040802.

Lin, C. S. and Fang, S. K. (2005). Formation of cerium conversion coatings on AZ31 magnesium alloys. *Journal of the Electrochemical Society*, **152**, B54–B59.

Lin, C. S., Lee, C. Y., Li, W. C., Chen, Y. S. and Fang, G. N. (2006a). Formation of phosphate/permanganate conversion coating on AZ31 magnesium alloy. *Journal of the Electrochemical Society*, **153**, B90–B96.

Lin, C. S., Lin, H. C., Lin, K. M. and Lai, W. C. (2006b). Formation and properties of stannate conversion coatings on AZ61 magnesium alloys. *Corrosion Science*, **48**, 93–109.

Liu, J. R., Guo, Y. N. and Huang, W. D. (2006). Study on the corrosion resistance of phytic acid conversion coating for magnesium alloys. *Surface and Coatings Technology*, **201**, 1536–1541.

Liu, X. L., Zhang, T., Shao, Y. W., Meng, G. Z. and Wang, F. H. (2009). Effect of alternating voltage treatment on the microstructure and corrosion resistance of stannate conversion coating on AZ91D alloy. *Corrosion Science*, **51**, 2685–2693.

Montemor, M. F., Simões, A. M. and Carmezim, M. J. (2007). Characterization of rare-earth conversion films formed on the AZ31 magnesium alloy and its relation with corrosion protection. *Applied Surface Science*, **253**, 6922–6931.

Montemor, M. F., Simões, A. M., Ferreira, M. G. S. and Carmezim, M. J. (2008). Composition and corrosion resistance of cerium conversion films on the AZ31 magnesium alloy and its relation to the salt anion. *Applied Surface Science*, **254**, 1806–1814.

Morris, E. L. (2008). *Corrosion resistant conversion coatings*. US patent application 11/002,741.

Ng, W. F., Wong, M. H. and Cheng, F. T. (2010). Stearic acid coating on magnesium for enhancing corrosion resistance in Hanks' solution. *Surface and Coatings Technology*, **204**, 1823–1830.

Niu, L. Y., Jiang, Z. H., Li, G. Y., Gu, C. D. and Lian, J. S. (2006). A study and application of zinc phosphate coating on AZ91D magnesium alloy. *Corrosion Science*, **200**, 3021–3026.

Phelps, A. W., Sturgill, J. A. and Swartzbaugh, J. T. (2008). *Non-toxic corrosion-protection conversion coats based on rare earth elements.* US patent application US20030625915 20030723

Polmear, I. J. (1995). *Light alloys: metallurgy of the light metals*. London, E . Arnold.

Reghi, G. A. (1982). *Coating composition and method.* US patent application 168811.

Rudd, A. L., Breslin, C. B. and Mansfeld, F. (2000). The corrosion protection afforded by rare earth conversion coatings applied to magnesium. *Corrosion Science*, **42**, 275–288.

Song, Y., Shan, D., Chen, R., Zhang, F. and Han, E. -H. (2009a). Formation mechanism of phosphate conversion film on Mg-8.8Li alloy. *Corrosion Science*, **51**, 62–69.

Song, Y., Shan, D., Chen, R., Zhang, F. and Han, E. -H. (2009b). A novel phosphate conversion film on Mg-8.8Li alloy. *Surface and Coatings Technology*, **203**, 1107–1113.

Takenaka, T., Narazaki, Y., Uesaka, N. and Kawakami, M. (2008). Improvement of corrosion resistance of magnesium alloys by surface film with rare earth element. *Materials Transactions*, **49**, 1071–1076.

Takenaka, T., Ono, T., Narazaki, Y., Naka, Y. and Kawakami, M. (2007). Improvement of corrosion resistance of magnesium metal by rare earth elements. *Electrochimica Acta*, **53**, 117–121.

Tomlinson, C. E. (1995a). *Conversion coatings for metal surfaces*. US patent application 5,380,374.

Tomlinson, C. E. (1995b). *Hydrophilic coatings for aluminum*. US patent application 5,441,580.

Umehara, H., Takaya, M. and Kojima, Y. (2001). An investigation of the structure and corrosion resistance of permanganate conversion coatings on AZ91D magnesium alloy. *Materials Transactions*, **42**, 1691–1699.

Umehara, H., Takaya, M. and Terauchi, S. (2003). Chrome-free surface treatments for magnesium alloy. *Surface and Coatings Technology*, **169–170**, 666–669.

Verdier, S., Delalande, S., Laak, N. V. D., Metson, J. and Dalard, F. (2005). Monochromatized x-ray photoelectron spectroscopy of the AM60 magnesium alloy surface after treatments in fluoride-based Ti and Zr solutions. *Surface and Interface Analysis*, **37**, 509–516.

Verdier, S., Laak, N. V. D., Dalard, F., Metson, J. and Delalande, S. (2006). An electrochemical and SEM study of the mechanism of formation, morphology, and composition of titanium or zirconium fluoride-based coatings. *Surface and Coatings Technology*, **200**, 2955–2964.

Verdier, S., Laak, N. V. D., Delalande, S., Metson, J. and Dalard, F. (2004). The surface reactivity of a magnesium-aluminium alloy in acidic fluoride solutions studied by electrochemical techniques and XPS. *Applied Surface Science*, **235**, 513–524.

Wang, C., Zhu, S. L., Jiang, F. and Wang, F. H. (2009). Cerium conversion coatings for AZ91D magnesium alloy in ethanol solution and its corrosion resistance. *Corrosion Science*, **51**, 2916–2923.

Waterman, J., Birbilis, N., Dias, G. J., Woodfield, T. B. F. and Staiger, M. P. (2012). Improving in vitro corrosion resistance of biomimetic calcium phosphate coatings for Mg substrates using calcium hydroxide layer. *Corrosion Engineering, Science and Technology*, DOI 10.1179/1743278212Y.0000000018.

Wei, Z. L., Shen, Y., Chen, Q. R., Li, C. M. and Huang, Y. W. (2009). *P-Ca-V composite phosphating solution on magnesium alloy surface and chemical conversion processing method.* China patent application CN20081201854 20081028.

Yang, H.-Y., Chen, X.-B., Guo, X.-W., Wu, G.-H., Ding, W.-J. and Birbilis, N. (2012a). Coating pretreatment for Mg alloy AZ91D. *Applied Surface Science*, **258**, 5472–5481.

Yang, K. H., Ger, M. D., Hwu, W. H., Sung, Y. and Liu, Y. C. (2007). Study of vanadium-based chemical conversion coating on the corrosion resistance of magnesium alloy. *Materials Chemistry and Physics*, **101**, 480–485.

Yang, L., Li, J., Yu, X., Zhang, M. and Huang, X. (2008). Lanthanum-based conversion coating on Mg-8Li alloy. *Applied Surface Science*, **255**, 2338–2341.

Yang, Y. C., Tsai, C. Y., Huang, Y. H. and Lin, C. S. (2012b). Formation mechanism and properties of titanate conversion coating on AZ31 magnesium alloy. *Journal of the Electrochemical Society*, **159**, C226-C232.

Yong, Z. Y., Zhun, J., Qiu, C. and Liu, Y. L. (2008). Molybdate/phosphate composite conversion coating on magnesium alloy surface for corrosion protection. *Applied Surface Science*, **255**, 1672–1680.

Yu, X. W. and Li, G. Q. (2004). XPS study of cerium conversion coating on the anodized 2024 aluminum alloy. *Journal of Alloys and Compounds*, **364**, 193–198.

Zhang, H., Yao, G. C., Wang, S. L., Liu, Y. H. and Luo, H. J. (2008). A chrome-free conversion coating for magnesium-lithium alloy by a phosphate-permanganate solution. *Surface and Coatings Technology*, **202**, 1825–1830.

Zhao, M., Wu, S. S., An, P., Fukuda, Y. and Nakae, H. (2007). Growth of multi-elements complex coating on AZ91D magnesium alloy through conversion treatment. *Journal of Alloys and Compounds*, **427**, 310–315.

Zhao, M., Wu, S. S., An, P., Luo, J. R., Fukuda, Y. and Nakae, H. (2006a). Microstructure and corrosion resistance of a chromium-free multi-elements complex coating on AZ91D magnesium alloy. *Materials Chemistry and Physics*, **99**, 54–60.

Zhao, M., Wu, S. S., Luo, J. R., Fukuda, Y. and Nakae, H. (2006b). A chromium-free conversion coating of magnesium alloy by a phosphate–permanganate solution. *Surface and Coatings Technology*, **200**, 5407–5412.

Zhou, W. Q., Shan, D. Y., Han, E. H. and Ke, W. (2004). *Preparation method of magnesium alloy chromeless chemical conversion film and its used film forming solution*. China patent application CN1475602A.

Zhou, W. Q., Shan, D. Y., Han, E. H. and Ke, W. (2008). Structure and formation mechanism of phosphate conversion coating on die-cast AZ91D magnesium alloy. *Corrosion Science*, **50**, 329–337.

Zucchi, F., Frignani, A., Grassi, V., Trabanelli, G. and Monticelli, C. (2007). Stannate and permanganate conversion coatings on AZ31 magnesium alloy. *Corrosion Science*, **49**, 4542–4552.

Part III

Coatings

11
Corrosion-resistant electrochemical plating of magnesium (Mg) alloys

X.-B. CHEN, M. A. EASTON and N. BIRBILIS,
Monash University, Australia,
H.-Y. YANG, Shanghai Jiao-tong University, China and
T. B. ABBOTT, Magontec Pty Ltd, Australia

DOI: 10.1533/9780857098962.3.315

Abstract: With the exception of a very few applications, the industrial deployment of magnesium alloys requires anti-corrosion coatings. This chapter looks at aqueous electrochemical plating systems (including pretreatment, undercoating, electroplating and electroless plating) and non-aqueous plating systems (including high temperature molten salts and ionic liquids). The performance assessment of various plated coatings upon magnesium alloys is discussed, along with the general pros and cons of plating techniques.

Key words: magnesium, electroplating, electroless plating, corrosion, coatings.

11.1 Introduction

Electrochemical plating is of interest to protect Mg and its alloys against corrosion due to its unique advantages. For instance, the resultant metallic coatings have good solderability, electrical conductivity and decorative appearance, in addition to hardness, and resistance to corrosion and wear, combined with convenient operation and relative cost-efficiency. They can be subdivided into two groups: electroplating and electroless plating. Electroplating (often termed as electrodeposition) is a process that employs an electrical signal provided by an external power source to reduce cations of a desired metal in solution and produce a metallic coating. Electroless plating is a 'self-reduction' process, which relies upon the auto-catalytic reduction process of metal ions in an aqueous solution containing a chemical reducing agent (typically sodium hypophosphite, NaH_2PO_2). At present, electroless Ni-P plating coating is the primary process. As the name implies, no external power source is required. Owing to their different natures, electroplating can yield a thicker and more

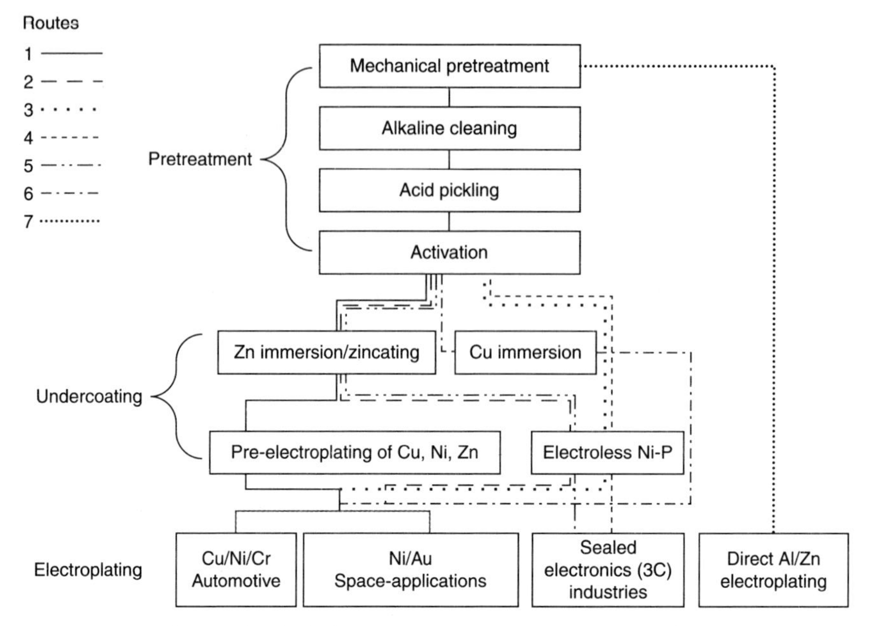

11.1 Generalised plating process current applied onto Mg alloys.

compact metallic coating compared to electroless plating. Thinner electroless coatings have micro-pinholes that are beneficial for sealing or subsequent electroplated coatings. However, uneven current distribution during electroplating is a challenge for coating components with complex shapes, while electroless plating is a good solution to this issue. Depending on the service conditions and shape of the components, these two plating methods can be applied separately or in combination upon Mg alloys. A generalised plating process currently applied to Mg alloys is summarised in Fig. 11.1. It should be noted that in all of the processes discussed there are water rinse stages between the steps shown. These have been omitted from the diagrams for clarity.

The metal-coating/Mg system is a typical example of 'a cathodic coating on an anodic substrate' as most metals are nobler than Mg alloys with respect to electrochemical potential. Therefore, the metal coatings are only physical barriers against corrosion attack for Mg substrates. They must be uniform, adhesive and pore-free, otherwise rapid corrosion occurs. It is not straightforward to plate such protective coatings onto Mg alloys because of: (1) rapid formation of $MgO/Mg(OH)_2/MgCO_3$ films upon Mg exposed to air or water, which prevents adhesion of the plating coatings; (2) aggressive attacks upon Mg substrates by conventional acidic aqueous plating electrolytes; (3) loose immersion layers formed on the surface by replacement, which inhibits successive electrodeposition and is detrimental to adhesion and uniformity; (4) the electrochemically heterogeneous surface of substrates, in which each component behaves differently in the plating bath, leading to non-uniform coating growth; and (5) intense cathodic hydrogen evolution, resulting in micro-pinholes within a coating and low current efficiency (metal Cr $< 25\%$). Hence, in aqueous plating electrolytes, pretreatment is the key step to coating Mg. Pretreatment mainly includes four steps: mechanical pretreatment, alkaline cleaning, acid pickling and activation. Undercoating is also an essential step prior to final plating to improve adhesion and reduce galvanic corrosion between the metal coating and Mg substrates. So far, two types of undercoating are used for subsequent electroplating: zinc (Zn) immersion followed by pre-electroplating (Cu, Ni, Ni/Cu or Zn/Cu) and electroless Ni plating (see Fig. 11.1) (Gray and Luan, 2002). Currently, multilayered metal coatings, rather than single galvanic layers, are typically employed to protect Mg alloys. Conventional multilayered Cu/Ni/Cr coatings are deposited on pre-plated Mg samples for outdoor use (e.g. automotive applications). Noble metals like Ni/Au multilayer coatings have been explored for Mg alloys in the aerospace field. Electroless Ni-P plating after sealing is mainly used for mild use conditions, such as computer and electronic (3C) industries.

Recently, plating metals onto Mg substrates in non-aqueous electrolytes, especially ionic liquids (ILs), has been increasingly studied to avoid the detrimental effects of aqueous plating bath. ILs demonstrate some advantages over aqueous electrolytes in plating (Abbott *et al.*, 2008). Firstly, a wider range of metals and alloys can be plated in ILs owing to their larger electrochemical windows and the closer electrodeposition potential between metals in ILs, compared to water. Hence, metals such as Al, Ti and their alloys, which are difficult to plate in aqueous electrolyte but have high corrosion resistance, are prepared in ILs with improved quality. Secondly, pretreatment can be significantly simplified. For instance, Al was electrodeposited on Mg after just mechanical grinding followed by cleaning with alcohol (Chang *et al.*, 2007, 2008; Chuang *et al.*, 2010). Thirdly, high current efficiency can be realised in ILs (close to 100%) since cathodic hydrogen evolution is eliminated in such non-aqueous systems (Yang *et al.*, 2011a). Finally, ILs are environmentally safe solvents because of their low vapour pressure. Thus IL systems appear as one option to widen the plating field in the future.

In the following sections, the chapter will focus upon conventional aqueous plating (including pretreatments, undercoatings and electroplating), electroless plating (which is used as undercoating in some cases but discussed in separate section since it is an important coating technology for Mg), and non-aqueous plating (including high temperature molten salts and ionic liquids), followed by a general discussion.

11.2 Aqueous plating: pretreatment, undercoating and electroplating

Development of plating in aqueous solution: pretreatment, undercoating and electroplating are discussed below.

11.2.1 Pretreatment

A review of the specific literature, along with general coating practices, indicates that an appropriate pretreatment process is essential for successful plating upon Mg alloys. It is still a significant challenge to obtain a protective deposit through a direct (one-step) electro- or electroless plating method in an aqueous plating bath (Mahallawy *et al.*, 2008). To achieve satisfactory plated coatings either complex pretreatments plus undercoating on the surface of Mg alloys or special bath formulations (e.g. non-aqueous electrolytes) are required. The former is discussed herein and the latter is discussed in the section on non-aqueous plating.

The general pretreatment process for plating Mg alloys is (Gray and Luan, 2002): mechanical pretreatment → alkaline cleaning → acid pickling/etching → activation (details in Table 11.1).

Mechanical pretreatment

Mechanical pretreatment (Kurze, 2003, 2006), including grinding and polishing, abrasive-finishing, particle-blasting and brushing, is mainly used to remove impurities such as oxides or dirt and burred edges from Mg parts and/or to produce special decorative effects. Selection of these methods depends on both the shape and surface quality of the candidate components, and the specific characteristics of the final surface state. Mechanical treatment should be carried out immediately prior to the other surface treatment(s). However, mechanical treatment is often not needed when the Mg parts are in shapes involving great precision, for example, die-cast (details in Table 11.1).

Alkaline precleaning

Alkaline cleaning removes oil, grease and organic matter, but not oxides/hydroxides from Mg alloy surfaces. Some cleaners are listed in Table 11.1. Alkaline cleaner components fall into three categories, known as builders, surfactants and sequestering agents.

Builders including Na/K hydroxides, carbonates (CO_3^{2-}) and silicates (SiO_3^{2-}) provide mild and high alkalinity because metals such as Al and Zn are attacked/dissolve while Mg stays stable at high pH (Pourbaix, 1974). Phosphates are often added to increase the sequestering power of the baths

Table 11.1 Pretreatment baths for Mg alloys

Types		Composition and conditions			Application
Mechanical pretreatment (Mahallawy *et al.*, 2008; Kurze, 2003)	1	Grinding and polishing: wet, rate: 900 and 1800 m/min, agent: SiC and Al_2O_3 (60–400 mesh)			Removal of oxides and dirt
	2	Abrasive-finishing: neutral or alkaline reagent			Removal of burred edges, final-machining of the surface
	3	Particle-blasting: purified sand (SiO_2), corundum (Al_2O_3) and glass balls			Producing special decorative effects Removal of oxide and dirt
	4	Brushing			
Alkaline cleaning	1(Innes, 1971, Smyth, 1985)	NaOH	50 g/L	10–15 min 65°C	Removal of oils, grease or organic matter, all alloys
		$Na_3PO_4 \cdot 12H_2O$	10 g/L		
		OP emulgator	3 mL/L		
	2(Pearson and Cordero-Rando, 2007)	NaOH	25 g/L	3 min, 65°C, 4–6V	
		$NaHOCH_2(CH(OH))_4CO_2$	25 g/L		
Acid pickling (Kurze, 2003)	1	CrO_3	180 g/L	1–15 min, 20–100°C	Removal of oxide, flux, corrosion products, all alloys
	2	CrO_3	125 g/L	1–3 min	Removal of rolling scale, oxides and burnt on graphite lubricant, wrought alloys and cast alloys
		HNO_3	110 mL/L	20–30°C	
	3	CH_3COOH	200 mL/L	3–5 min	Removal of rolling scale, wrought alloys
		$NaNO_3$	50 g/L	20–30°C	
	4	HNO_3 (65%)	90 mL/L	10–15 s	Removal of impurities after sandblasting Cast alloys
		H_2SO_4 (98%)	20 mL/L	20–30°C	
Activation (Lei *et al.*, 2010b)	1	HF(40%)	385 mL/L	5–10 min, 20–30°C	For direct electroless plating
	2	$K_4P_4O_7 \cdot 3H_2O$	76 g/L	75°C, 2–5 min	For Zn immersion
		Na_2CO_3	15 g/L		
		$KF_2 \cdot H_2O$	7 g/L		

and as a water softener. The surfactants act as detergents, assisting the cleaning by lowering surface tension. The ASM Handbook (Cormier, 1994) gives examples of the make-up of cleaners for Al, Fe and Zn with typical surfactants or wetting agents including sodium lauryl sulphonate (used with Al and Zn), and naphthalene sulphonate (used with Al). No standardised process for Mg presently exists. Tetraethylammonium perfluorooctane sulphonate (FT248) is an example that has been used with Mg (Olsen and Halvorsen, 1982). Additives are typically 0.5 g/L sequestering agents, also called chelating agents, which form metal complexes and so increase the capacity (lifetime) of the cleaners.

The surface cleaning process in an alkaline bath can be aided by electrolytic means, for example, treating the Mg work piece as a cathode (i = 1~4 A/dm^2; E = 4~6 V; t = 1~3 min). The hydrogen generated at the cathode has a positive effect on impurity removal (Pearson and Cordero-Rando, 2007).

Acid etching/pickling

Following alkaline cleaning, Mg alloys should be etched/pickled in acidic solution to remove oxide or hydroxide layers, which cannot dissolve in alkaline cleaning. Removal of Mg-oxides necessitates neutral or acidic pH (Pourbaix, 1974) and hence typical acidic baths comprise chromic, nitric or sulphuric acid component(s). Some typical acid pickling baths for Mg alloys are also listed in Table 11.1. The extreme toxicity of the chromium (VI) containing compounds is a matter of concern. Furthermore, surface roughness of the substrates can be noticeably increased by the acidic pretreatment (Liu and Gao, 2006b, 2006c). Surface roughness plays a critical role in determining the resultant coating qualities, for example, coating porosity, deposition rate and adhesion. It is proposed that the rougher the surface, the higher the deposition rate, which is attributed to more nucleation sites on the rougher surface (Liu and Gao, 2006c). Beyond a certain level of roughness, the porosity within the coating is not acceptable (Tomlinson and Mayor, 1988). Suitable surface roughness is also of benefit for coating adhesion, owing to enhanced interlocking (Brunelli *et al.*, 2005; Liu and Gao, 2006b, 2006c).

Activating

Fluoride (F) activation of Mg alloy surfaces is both a common and (presently) important step in the electro/electroless plating industry. For the case of the AZ series of Mg alloys whose α (matrix or eutectic) and β (intermetallic) phases support different reaction rates, the F activation treatment is posited to create an equipotentialised film (MgF_2) on the surface, a good (temporary) base for subsequent uniform coating deposition (Turhan *et al.*,

2009). At present, it is an indispensable pretreatment step in the processing of electro-/electroless plating on Mg alloys. Typical baths however, which operate at pH = 3~5, suggest that HF acid will be present and this is an increasingly unacceptable occupational hazard.

11.2.2 Undercoating

The undercoating process is one of the most difficult, yet important, steps in Mg plating. Once a suitable undercoating is formed upon Mg surfaces, many desirable metals can subsequently be plated. To date, two types of undercoating are common: Zn immersion followed by pre-electroplating (Cu, Ni or Zn/Cu) (route 1 in Fig. 11.1) and electroless plating Ni-P coating (routes 2 and 3 in Fig. 11.1), prior to subsequent electroplating. The Zn immersion route is discussed below and details of the electroless Ni-P coatings are given individually in the next section.

Zn immersion

Zn immersion, also termed zincating, is a common pretreatment for both electro- and electroless plating. In the Zn immersion process, electrons for metal reduction are supplied by substrate oxidation. Zn deposits on the surface by displacement of Mg: Mg metal is oxidised to form Mg^{2+} ions while Zn^{2+} ions are reduced to form metallic Zn.

A zincated surface is relatively easy to achieve, and whilst it does not provide a means to final corrosion protection (since Zn is an active metal itself though more noble than Mg), it does catalyse the subsequent plating. Plating upon Mg necessitates that Mg be the cathode (i.e. site of reduction). The physical characteristics of Mg, namely that it has the lowest exchange current density amongst all metals (Bockris and Reddy, 2000), demonstrates its inherent ability to sustain low reduction reactions. As a result, plating for Mg would be a sluggish, albeit possible process, in the absence of zincating. Zn supports cathodic reactions – the basis for electroplating – at rates much higher than Mg.

One typical zincating procedure is the Dow process depicted in Table 11.2 (DeLong, 1957; DeLong, 1961; Hillis, 1994). Precise control in treating duration is indispensable to deposit a compact Zn film with high coverage and adhesion (Zhang *et al.*, 2009c) since short duration results in incomplete Zn coating, whilst a long immersion incurs a loose Zn film with poor adhesion. In many cases, the surface of the base metals is unevenly covered by spongy and loose Zn deposits, specifically on the intermetallic phases. Such non-uniform and non-adhesive interlayers compromise subsequent platings and their corrosion protection. Doubling the zincating treatment can refine Zn crystal grains and improve adhesion and compactness, leading to a better

Table 11.2 Dow zincating processes

Stage	Component	Value		Component	Value		Component	Value
Surface	CrO_3 $Fe(NO_3)_3 \cdot 9H_2O$ KF	180 g/L 40 g/L 3.5–7 g/L	or	H_3PO_4 85 wt.%	Undiluted	or	CH_3COOH	280 g/L
condition	*T*	20–30°C		*T*	20–30°C		$NaNO_3$	20–30°C
(Pickle)	*t*	15 s–2 min		*t*	15 s–5 min		*T*	30 s–2 min
							t	
↓								
Activation	Acid	200 mL/L	or	Alkaline	40 g/L		*Note: Acid solution	
(Acid or	H_3PO_4 85 wt.%	100 g/L		$Na_4P_2O_7$	70 g/L		attacks Fe in rack	
alkaline	NH_4HF_2	20–30°C		$Na_2B_4O_7 \cdot 10H_2O$	20 g/L		which redeposits onto	
floride)*	*T*	15 s–2 min		NaF	75°C		specimen. Alkaline	
	t			*T*	2–5 min		solution cannot be used	
				t			if chromate present.	
↓								
Zinc	$ZnSO_4 \cdot H_2O$	30 g/L	**Note: Generally used without applied voltage except in US					
immersion	$Na_4P_2O_7$	120 g/L	2 811 484 where 1 A/dm^2 is recommended					
(Zincate) **	NaF or LiF	5 or 30 g/L						
	Na_2CO_3	5 g/L						
	pH	10.2–10.4						
	T	80–85°C						
	t	1–10 min						

Sources: De Long, 1961; Hillis, 1994; De Long, 1957.

coverage of the Mg surface (Armyanov *et al.*, 1982). Adding a small amount of $FeCl_3$ into the zincating bath alters the Zn crystal structure and thereby coverage and adhesion of the resulting Zn layer (Chen *et al.*, 2006). Some additional external conditions, for example, ultra-sonication, also improve Zn immersion process (Tang and Azumi, 2011a).

The Cu immersion process has also been studied aiming to substitute Zn immersion for subsequent electro-/electroless plating (Yang *et al.*, 2005). Little research work on Cu immersion exists; however, similarly to Zn, the Cu surface can sustain faster cathodic reaction rates than Mg and Zn, which could benefit subsequent electroplating. An orthogonal experimental design (Yang *et al.*, 2005) was used to optimise the process outputs, given the complex effect of different parameters. After analysing the effect of the tested parameters on the Cu immersion coatings, the optimal experimental procedure was identified. The HF acid concentration in the immersion bath was identified as possibly the most critical factor. But use of concentrated HF raises serious occupational health and safety concerns, indicating why the process remains non-commercial.

Pre-plating

In a number of instances, Zn is too reactive to be a suitable base for direct plating of other metals (Gray and Luan, 2002). Thus, an inert Cu deposition upon Zn coatings produced in a cyanide bath usually acts as underlayer to enable subsequent metal plating. Popular use of cyanide-containing baths is however decreasing owing to (severe) toxicity. Nowadays, one alternative attempted is pre-plating of Cu onto AZ31, AZ61 and AZ91 substrates in a pyrophosphate bath (Huang *et al.*, 2010a, 2010b). Ni undercoatings are another choice for subsequent Mg plating. In such instances, a pure Ni deposition on Mg alloys provides good corrosion resistance, especially when the Ni grain structure is fine (nanocrystalline) and compact (Gu *et al.*, 2006). The hardness value of the nanocrystalline Ni coating is about 580 VHN, which is well in excess of uncoated AZ91D (~100 VHN) (Gu *et al.*, 2006). The chemical stability of Ni is also excellent across the pH range of atmospheric/environmental exposures (Pourbaix, 1974). In some cases, Cu/Ni (Tang and Azumi, 2011b) and Zn/Cu (Jiang *et al.*, 2005) dual pre-platings are used as undercoating to benefit subsequent (standard) multilayer coatings, for example Cu/Ni/Cr.

11.2.3 Electroplating

Cu/Ni/Cr multilayer coating system

Cu/Ni/Cr is an effective coating system for many indoor and mild outdoor applications owing to its corrosion and wear resistance, combined with excellent appearance. The elasticity of the Cu bottom layer balances out tension between the components and the coatings. It ensures good adhesion and acts as a leveller. The type of Cu-plating finish determines the final gloss effect of the subsequent coatings. Regardless of the selected coating thickness, Ni is always highly corrosion resistant. Cr is a high-strength and tarnish-resistant material, hence these three materials are combined to form a coating 'system' for Mg alloys.

Cu/Ni/Cr multilayer coatings generated from cyanide-free solutions have been achieved on Mg components industrially, for example, AZ91D engine shell and AM60B wheel hub (Lei *et al.*, 2010b). The appearance of the plated products is presented in Fig. 11.2. Three pre-plating processes were used as undercoating: direct electroless Ni-P coating (route 5 in Fig. 11.1); Zn immersion followed by pre-electroplating Ni; and Zn immersion followed by pre-electroplating Cu (route 1 in Fig. 11.1) as depicted in Fig. 11.2. Performance indicates that the Cu/Ni/Cr multilayer coatings on the wheel hub and motor engine shell provide good adhesion through thermal shock testing and scribe grid testing. The corrosion resistance to salt spray testing of the electroplated products reached to Grade 9 (Chinese Standard GB/T6461-2002) where Grade 10 is the highest score. From a practical perspective, note that it is necessary to increase the auxiliary anode (in electroplating) in the cavity of the shell to improve uniform coverage. Bubbles, which might assemble in the corners of component cavities, cause serious plating faults. Plating work pieces should be aligned in the correct orientation so that the gas bubbles can easily escape from the cavity of the work pieces.

Ni/Au dual-layer coating system

Plating Ni/Au dual-layer films on Mg has been proven to be useful for aerospace applications. Successful gold plating was achieved on Mg-Li alloy with electro- and electroless plated Ni undercoating. The electroplated Ni film is porous but catalyses subsequent electroless Ni-P deposition. The resultant uniform electroless Ni-P coating, though porous with micro-cracks, is desirable for the final crack-free gold plating (Sharma, 1988, 1991), which has good mechanical, thermal and optical properties, along with environmental stability (route 1 in Fig. 11.1). Ni/Au dual-coating has been attempted on Mg alloy RZ5 with a variety of undercoatings (Rajagopal *et al.*, 1990), such a Ni undercoating produced in a non-aqueous bath, and Zn immersion followed by electroless Ni-P (route 4 in Fig. 11.1). The former is not cost-effective, while the latter demonstrates poor support to enable subsequent gold plating as a result of the chemically heterogeneous microstructure on the surface of the Mg substrates. Such poor support is mediated by an AC electrolytic treatment (5 V) activation step and subsequently reaps a final, uniform, gold

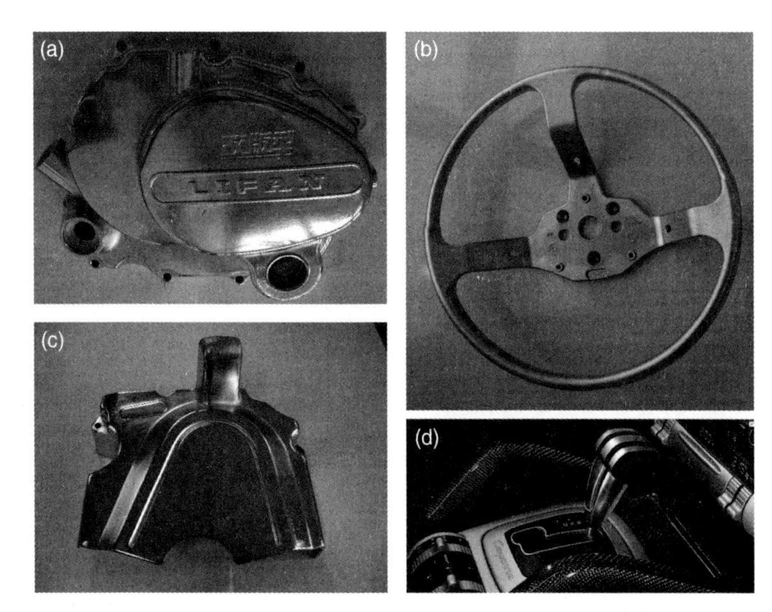

11.2 Photos of Cu/Ni/Cr plated components of Mg alloy: (a) top cover of engine using direct electroless Ni-P undercoating, (b) motor steer wheel using Zn immersion followed by pre-electroplated Ni undercoating, (c) top cover of engine using Zn immersion followed by pre-electroplated Ni undercoating and (d) interior panel decoration of Porsche Cayenne using Zn immersion followed by Cu/Ni/Cr plating. (a and c are adapted by kind permission of Maney Publishing).

coating. The expected life span of satellite components was prolonged by two years with this technique.

Electroplating techniques are relatively complicated as many variables can alter the properties of the final coatings including process factors, such as power control and bath solution longevity (not covered herein), but from an industrial perspective this is significant as Mg dissolution can have an enormous impact on bath pH and chemistry. Also, since the external current or voltage is essential during the electroplating process, there is a cost concern in certain industries.

11.3 Electroless Ni-P plating

Unlike conventional electroplating, electroless deposition of metals is an auto-catalytic reaction, therefore no external electrical voltage or current is required. Another advantage is that electroless baths provide uniform deposition upon parts with irregular shape. A sharp edge or a blind hole nominally achieves the same thickness of deposit. As such, electroless deposits are also adopted in many industrial fields owing to the ease of obtaining coating uniformity on either conductive or non-conductive components. Compared to that of bare Mg substrates, electroless coatings can provide improved corrosion resistance, appearance, solderability, etc.

Ni-P alloy coating is the most common form of electroless Ni plating and is used in applications where the combination of wear resistance, hardness and corrosion protection is required. The P content in the electroless Ni coatings ranges from 1 to 5 wt.% (low P content) and 10 to 13 wt.% (high P content), produced at high and low pH, respectively. The former is crystalline and poorly corrosion resistant, whilst the latter is amorphous but displays high corrosion resistance.

Combining electroless Ni-P layers with use of either multilayered Ni-P coating with higher P content, sol-gel multilayer or direct current (DC) electroplating can achieve desirable appearance, adhesion and resistance to wear and corrosion (Gu *et al.*, 2005a, 2006; Tan *et al.*, 2005). Of the commercially available electroless methods commonly used for industrial Mg components, Ni-P coatings have the greatest market share.

Zn immersion followed by electroless Ni deposition is a standard plating procedure for Mg alloys nowadays (route 4 in Fig. 11.1). In addition, electroless plating has been developed to simplify the complex pretreatment process and reduce the cost (route 5 in Fig. 11.1). Direct electroless Ni plating procedures are suitable for the Mg alloys with high Al content. However use of HF and/or cyanide in the pretreatments (Ambat and Zhou, 2004; Li *et al.*, 2006b) increases the risk of the operation.

Alkaline Ni plating baths (Ambat and Zhou, 2004; Li *et al.*, 2006b), though expensive, of low efficiency and impractical in many applications, are still more commonly used because Mg alloys corrode severely in acidic solutions (Pourbaix, 1974). Commercial baths still contain HF, as the dissolution of Mg in HF is predictable and uniform.

11.3.1 Standard/direct electroless plating

Standard plating processes refer to using the conventional Zn immersion followed by pre-plating Cu as undercoating for the subsequent electroless Ni-P plating (route 4 in Fig. 11.1). The Ni-P coating exhibits satisfactory adhesion, compact morphology and brightness. Zn immersion can effectively eliminate residual oxides and hydroxides to produce a thin Zn layer, which prevents re-oxidation of the Mg surface (Gray and Luan, 2002). The Cu undercoating greatly improves the appearance of the resultant coatings. The substrate becomes catalytically active after the surface oxides dissolve into the plating bath, which makes the replacement reaction between the substrate and Ni ions occur. Appropriate Zn immersion enables electroless Ni-P deposit applicable on some Mg alloys, namely, AZ31B, AE42 and ZE11 (which was named as ZRE1 by Mahallawy *et al.* (2008)). The final coating systems are adhesive, compact, uniform, corrosion protective and crack-free. However, Zn immersion usually does not achieve satisfactory deposits on Mg alloys with high Al content (e.g. AZ91) (Gray and Luan, 2002).

To address the difficulties in plating Mg alloys with high Al content, direct electroless plating was developed. After alkaline cleaning, Mg alloy is immersed in a chromic acid pickling bath and then activated by HF. The chromic acid pickling significantly corrodes the surface and forms a reduced chromium film on the surface. The fluoride activation can remove such reduced chromium film and control the deposition rate and thus the quality of the coating. One concern about this process is the use of toxic chemicals. AHC (Germany) (Kurze, 2003) developed an environmentally friendly pretreatment process. The first step is surface cleaning in heated mild alkaline solution in which ultrasound is needed to supplement the degreasing. In the second step, mild alkaline stain media is exploited to attack the surface in a one-step process. Activation is the third step and makes a suitable surface for direct Ni-P electroless plating. To avoid the rapid corrosion of Mg substrate in the acidic plating bath, an alkaline electroless process was developed as pretreatment to produce a Ni-P base containing low P content (Zuleta *et al.*, 2012). On top of such base, a corrosion resistant, a Ni-P coating with high P content was then grown in a low-pH bath.

Some notable variants to the conventional Ni-P baths are summarised below. These processes, whilst noteworthy, are yet to be deployed in any industrial practices. The basic Ni carbonate baths containing HF give rise to major issues in terms of environmental and cost control. In one instance (Li *et al.*, 2006a), a F-free alkaline electroless plating bath using Ni sulphate was developed to provide Ni ions and Na carbonate as a buffering agent. The existence of the Na carbonate promotes the deposition rate of the Ni-P plating coating and enhances the plating adhesion. The pH value of the plating bath is maintained between 8.5 and 11.5, the optimum range for plating. The obtained plating coatings are crystalline, and of low-phosphorus content, high density and low porosity. Currently, all technical Mg alloys, such as AZ91, AZ61, AZ31, AM50 and AM20, can be electroless plated directly with Ni-P deposits. Some applications in electronics are given in Fig. 11.3.

11.3.2 Effect of the substrate characteristics

Substrate characteristics, especially microstructure and roughness, have a significant influence on the nucleation and growth of the direct electroless Ni-P deposits (Ambat and Zhou, 2004; Gu *et al.*, 2005b; Liu and Gao, 2006b, 2006c). Study of the effect of microstructure and roughness of Mg alloy AZ91 on the properties of the electroless plated Ni-P reveals that finer microstructure (Liu and Gao, 2006b) and rougher surface (Liu and

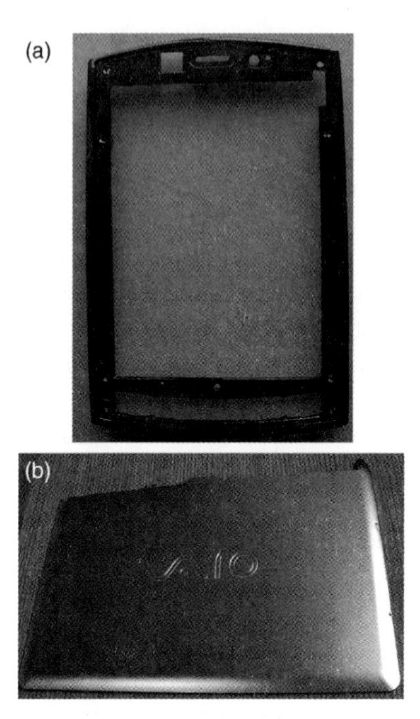

11.3 Electroless plating Ni-P coated Mg components: (a) mobile-phone case and (b) laptop case.

Gao, 2006c) of the substrates leads to higher deposition rates. While coating adhesion rises with an increase in roughness, wear tests reveal that the Ni coatings on the roughened substrate have higher friction coefficients than those on the polished surfaces.

Investigation of the effect of the AZ91D substrate microstructure on the formation mechanism of electroless Ni-P platings discovered that the platings preferentially deposit on the β-$Mg_{17}Al_{12}$ intermetallic particles in the early stages (Ambat and Zhou, 2004), and then extend to eutectic α Mg and primary α Mg phases. The authors proposed the electrons produced by the anodic dissolution of Mg in the plating bath from the α phase are consumed by the cathodic Ni deposition on the β phase. No experimental evidence in the work of Ambat and Zhou (2004) or Gu *et al.* (2005b) supports the claim that the Al-rich β phase is more cathodic than the α Mg phase in the coating bath, since a considerable amount of HF in the coating bath makes the situation complicated. However, direct evidence was provided by Anik and Körpe (2007) – confirming what is expected from conventional electrolytes. This was done by using various alloys as substrates that were controls to the individual behaviour of each microstructural constituents of AZ91 alloy in the coating baths. It was also suggested that coupling of the β phase and α matrix phase in the coating bath decelerates the deposition rate on the α phase, and the initial sites for Ni nucleating preferentially attract further Ni deposition growth.

In general, the electroless Ni-P deposits preferentially form at sites with high cathodic activity, which is in contrast to conversion coatings (which preferentially form at sites of high anodic activity). This distinction between coating mechanisms can lead to better selection/deployment of specific pretreatments to enhance the cathodic activity upon the surfaces.

11.3.3 Alternatives to chromate and fluoride pretreatments

Equalising the surface potential of Mg alloys via Cr and F pretreatments is increasingly restricted because of safety and environmental concerns, thus other coating technologies, including conversion coatings, organic coatings and anodising have been studied as an effective alternative.

Conversion coatings have been used as barrier layers to minimise the electrode potential difference over the surface of AZ91D Mg alloys (Huo *et al.*, 2004; Lian *et al.*, 2006; Zhang *et al.*, 2007a). A porous stannate ($MgSnO_3{\cdot}H_2O$) conversion coating (Huo *et al.*, 2004) renders excellent corrosion resistance to the final electroless Ni-P plating. Phosphate-manganese (Zhang *et al.*, 2007a) and molybdate (Yang *et al.*, 2009) conversion coatings, playing as interlayer between the electroless Ni-P deposition and Mg alloys, effectively reduce both the potential difference between α and β phases and the corrosion current density of substrates. The subsequent Ni-P plating on the interlayers is more compact and defect-free compared to the direct electroless platings. Zn phosphate conversion coatings were also studied as an interlayer for electroless Ni-P platings (Lian *et al.*, 2006). The numerous dispersed Zn particles in the interlayer serve as catalyst nuclei for subsequent dense and fine Ni-P deposition. The addition of Na_2MoO_4 in the phosphating bath results in the increase of Zn content in the coating. Salt spray tests reveal that the electroless Ni-P coating withstands 150 h without corrosion.

Organic coatings are a popular, simple and cost-effective layer to protect metallic components against corrosion. They are usually used as the final process owing to their corrosion and wear resistance arising from the barrier effect. However herein, organic coatings have been introduced as an interlayer in the pretreatment of Mg surfaces prior to electroless plating. The surface of the organic coatings must be sensitised and activated before plating. A commercial product, organosilicon heat-resisting varnish (Zhao *et al.*, 2007) was used as a barrier between AZ31 substrate and the top, uniform and adhesive Ni-P coating. The organic interlayer can also protect the substrate against corrosion when the Ni-P film is broken (serving as a barrier layer). In an extension of that work, it was also demonstrated that silver

(Ag) can be deposited upon Mg. Compact and uniform Ag deposits were plated onto AZ31 with the aid of a commercial organosilicon heat-resisting varnish product (Zhao and Cui, 2007; Zhao *et al.*, 2007).

Anodising/plasma electrolytic oxidation (PEO) coatings are capable of producing a stable oxide film for Mg alloys. These films can improve corrosion resistance as well as provide a good base for subsequent painting and interlayer between Mg substrates and the electroless metallic layers to further improve corrosion resistance and adhesion. Note that the anodising film is usually activated in a palladium (Pd) salt solution before plating but TiB_2 powder can also be used as catalyst (Sun *et al.*, 2008). The presence of a PEO film as the interlayer offers high density of nucleation sites for electroless plating Ni plating as a catalyst (Liu and Gao, 2006a; Zeng *et al.*, 2010). As a result, porosity of the Ni-P coating is decreased and adhesion strength is improved. The salt spray test and polarisation test results illustrate that the pitting corrosion resistance has been improved greatly.

11.3.4 Ternary/composite electroless plating

Co-deposition of metals, ceramic, polymer, or other particles, such as tungsten (W), Sn, ZrO_2, B_4C and SiC, dispersed within a metal matrix can enhance properties such as wear and corrosion resistance, hardness, and self-lubricating behaviour. Nevertheless, it is well known that the reinforced solid particles, especially nanoparticles, are prone to agglomerate in electroless solutions due to high surface energy, which results in poor mechanical properties and corrosion resistance being attained. Hence it is a big challenge to achieve good dispersion of the particles in solutions and improve their distribution in subsequent coatings. Various methods including vigorous agitation, air injection, ultrasonic vibration and addition of surfactants, have been examined to help the dispersion of reinforced particles in the plating baths (Guo *et al.*, 2011).

W was introduced into electroless plated Ni-P coating due to its high hardness, excellent corrosion and wear resistance. Zhang *et al.* (2007b) fabricated a Ni-W-P ternary alloy on AZ91D alloy in a Na carbonate buffered alkaline plating bath (pH 9.0). Co-deposition of the tungsten leads to a ternary Ni-W-P coating with P content of 4.9 wt.% and tungsten content of 4.5 wt.%. The hardness of the ternary Ni-W-P coating is about 660 VHN, which is higher than that of the binary Ni-P coating with similar P (5.6 wt.%) content (approximately 580 HV). Porosity test combined with immersion test and electrochemical measurements reveals the presence of W leads to a denser coating, which would be suitable for corrosion resistance applications. An amorphous ternary electroless Ni-Sn-P coating, with P content of 8.51 wt.% and Sn content of 2.48 wt.% was developed on AZ91D Mg alloy by the same authors (Zhang *et al.*, 2008). Hardness of the as-deposited Ni-Sn-P coating is about 670 HV. Better corrosion protection compared to the conventional binary Ni-P coating was achieved due to the low porosity and high compactness with the addition of Sn into the Ni-P coating system.

Nano-sized ZrO_2 particles (diameter 20 nm) were added into the conventional Ni-P plating bath via ultrasonic stirring and the electroless composite Ni-P-ZrO_2 plating technique was optimised (Song *et al.*, 2008). It was found the corrosion resistance of the resulting composite coating is superior to that of Ni-P coating due to its higher adhesion and lower porosity, and also the ZrO_2 content is beneficial to enhance coating hardness, corrosion resistance and wear resistance.

B_4C particles (1.5 μm in diameter) were dispersed into a conventional electroless plating bath to deposit Ni-P-B_4C composite coating onto AZ91 alloy (Araghi and Paydar, 2010). Hardness of the Ni-P-B_4C composite coatings was ca. 1200 MPa, higher than that of Ni-P coatings (ca. 700 MPa) with the same thickness (50 μm). Pin-on-disc wear testing, demonstrates that ternary Ni-P-B_4C composite coatings have a good wear resistance superior to Ni-P coating. However, the presence of B_4C particles decreases the corrosion resistance, attributed to the presence of micro-cracks in the composite coatings.

Ni-P-TiO_2 composite coatings with dispersed TiO_2 nanoparticles (15 nm) were prepared on AZ31 substrate by adding transparent TiO_2 sol into an alkaline based electroless Ni solution (Zhang *et al.*, 2010b). The microhardness of the composite coatings were significantly increased to ca. 1025 HV_{200} compared to 710 HV_{200} when produced by the solid particle mixing method. A Ni-P-TiO_2 composite coating was successfully extended to NZ30K Mg-RE alloy and the similar improvement in wear and corrosion resistant was observed (Guo *et al.*, 2011).

In summary, however, any such variation to the conventional Ni-P coatings tends to increase cost and hence such variants are rarely (if ever) relevant to commercial practice.

11.3.5 Post pretreatment

Thermal treatment can harden the electroless Ni-P platings by transforming the crystal structure from amorphous to crystalline. Generally, the heat treatment regime is 400°C for 1 h as it yields maximal hardness (Apachitei *et al.*, 1998). However, the adhesion may be damaged by the heat treatment due to the formation of Al-Ni intermetallic sub-layers in the regions where $Al_{17}Mg_{12}$ phase is present in the substrate (Novák *et al.*, 2011). Usually, temperatures above 400°C are therefore not recommended. Meanwhile, some reports claim that coating wear resistance does not significantly depend on the heat treatment (Novák *et al.*, 2011) and heating at 200°C is ideal for dehydrogenation (Zhang *et al.*, 2010a).

Thin electroless Ni-P coatings may still have micro-pinholes, which can cause severe galvanic corrosion between the coatings (cathode) and Mg alloy substrates (anode). Therefore, it is often necessary to carry out sealing, electroplating or subsequent coating. Chromate is the widely used passivation process. Its excellent self-healing function provides a desirable protective effect. However, because Cr (VI) is toxic there is a growing need for an alternative process. Zou *et al.* (2005) presented a way to seal the micropores in a Ni-P coating on Mg by means of a post treatment with fluoro-polymers. The coating hardness approaches 600 HV, and the coefficient of friction is less than 0.15, which is suitable for dry lubrication. The corrosion resistance is also excellent in enduring 5 wt.% neutral salt spray test for 500 h, however sealing treatments are, and will remain, an important research topic.

11.4 Platings from non-aqueous electrolytes

One of the challenges posed in electroplating Mg alloys in conventional aqueous electrolytes is their susceptibility to surface corrosion/dissolution and formation of a homogenous surface immersion layer. Poor adhesion of the surface layers affects the quality of successive electro-/electroless deposition. Replacing water with non-aqueous electrolytes not only addresses the above issues, but allows for larger electrochemical windows in the case of electrodeposition (meaning Mg can be plated at low cathodic potentials, $<1\ V_{SCE}$, given the low potential of Mg).

11.4.1 High temperature molten salt systems

Ionic materials usually melt at high temperature to form liquids that consist of only ions, that is, high temperature molten salts. High temperature molten salts have been extensively used for the electro-winning of metals such as Al, Ti and Li at temperatures up to 1000°C. They provide wide potential window, high conductivity and high solubility for metal salts. They inherit most of the advantages whilst being free of most of the demerits of the aqueous solutions. One limitation however, is that the operating conditions are difficult to achieve and limit the selection of substrates that can be used for deposition. Of the high temperature molten salts, $AlCl_3$-NaCl-KCl is the most popular due to the relative low melting point (ca. 190°C) in the context of Mg (Abbott *et al.*, 2008).

The use of undercoating such as Zn, though itself not protective, is still the key support for the subsequent Al–Mn alloy deposition with good adhesive strength at the interface and satisfactory short- and long-term corrosion resistance (Zhang *et al.*, 2009a; 2009b). Without undercoating, AZ31B substrates corrode severely in the molten salt during the deposition process, generating poor adhesion of the Al–Mn alloy coating (Zhang *et al.*, 2009a). Other molten salts, for instance LiCl-KCl-$YbCl_3$ (2 wt.%) can be applied to electroplate a thin Mg–Yb alloy film (ca. 200–400 nm thick) on Mg, but relatively extreme operating conditions are needed, say 500°C and under argon atmosphere (Chen *et al.*, 2010), which is not practical for industrial production.

11.4.2 Ionic liquid systems

The ideal alternative to high temperature molten salts is an ionic substance that is liquid at room temperature. Since 'molten salts' is the specific term for high temperature ones, for clarity, the 'room temperature IL' is used and 'IL' for short (Simka *et al.*, 2009). ILs are nominally organic ions that may be present in various structural orders. Undergoing electroplating in non-aqueous ILs for Mg alloys is a promising avenue. Other merits of ILs are non-volatility, inflammability, low toxicity, good solubility for many metal halides, high thermal stability, wide electrochemical windows, high polarity and high purity. However, some shortcomings include low conductivity, high viscosity, high density, water sensitivity and low cost-efficiency (Simka *et al.*, 2009). Nevertheless, ILs enables a significant widening of the electroplating field, which is critical to the Mg industry.

Coating Mg alloys with Al has long been considered desirable yet exclusive from aqueous baths. Pure Al electroplated on AZ91D in aluminium chloride ($AlCl_3$)-1-ethyl-3-methylimidazolium chloride (EMIC) IL (Bakkar and Neubert, 2007; Chang *et al.*, 2007, 2008; Chuang *et al.*, 2010; Yang *et al.*, 2011b) is however now being intensely investigated for the following reasons: (1) the overall alloy density is still close to AZ91D; (2) continuous Mg-Al intermetallic compounds with somewhat improved corrosion resistance are generated near the surface; (3) already established anodisation and electrolysis coloration techniques for Al can be applied directly to Mg alloys; and (4) the Al-coated material is readily recyclable.

Applying a low potential of −0.2 V *vs.* Al wire can succeed a crack-free and conductive Al coating upon AZ91D Mg with improved corrosion resistance in an 1-ethyl-3-methylimidazolium chloride (EMIC) IL bath containing $AlCl_3$ (Chang *et al.*, 2008). Higher deposition potentials lead to an increased deposition rate, which incurs poor adhesion and cracks within the Al coatings. Al-Zn alloy coatings have also been attempted in a $ZnCl_2$-containing $AlCl_3$–EMIC IL under a constant applied potential (Pan *et al.*, 2010). The potential for Zn reduction is more positive than that for Al. At −0.1 V, the deposit primarily comprises Zn dendrites with a rough surface. When the applied potential is below −0.2 V *vs.* Al wire, the chemical compositional homogeneity of the Al-Zn coating is lost and the average Zn content in the deposit declines.

However, achieving adhesive Al alloy coatings to Mg is still a challenge, particularly after simple mechanical pretreatment. In some of the above-mentioned reports, coatings detached from substrates in their cross-section profiles. A further 10 min post heat treatment at 450°C improved the adhesion and corrosion resistance due to inter-diffusion at the Al/Mg interface and healing of internal defects (Chuang *et al.*, 2010). An ageing treatment of Al-coated AZ91D at 200°C produced a triple layer structure (α-Al(Mg)/β/γ), improving adhesion, corrosion resistance and surface hardness of the resultant coating (Yang *et al.*, 2011b). Dilute phosphoric acid pickling enhances coating adhesion and density as well as corrosion resistance (Qian *et al.*, 2009). It ought to be emphasised, however, that some of the abovementioned coating procedures were executed under a purified nitrogen atmosphere (glove box), which is impractical for upscaled industrial applications. None the less, the ability to coat Mg with Al is very promising, and any means to do this really represents the state of the art.

Recently, a cheaper family of ILs was developed, formed by mixing a quaternary ammonium salt, choline chloride (2-hydroxyethyl-trimethylammonium), with a hydrogen-bond donor species such as urea (but also including other amides, glycols or carboxylic acids) (Abbott *et al.*, 2000). They have been accepted as a promising solvent in plating conventional metallic coatings, including Zn, Ni, Cr, Sn and their alloys. These solvents (often called 'deep eutectics') are considered as eutectic-based ILs since they have the relevant properties of ILs. In particular, a mixture of ChCl and urea with a molar ratio of 1:2 (1:2 ChCl-urea) displays attractive features with respect to electrodeposition, say, being stable in air and water, easy to prepare, low-cost, concerning environmental compatibility and toxicology. At this point in time, this type of IL is the most promising solvent for industrial applications of surface finishing of Mg alloys. Abbott *et al.* (2000) claimed that electroplating Cr, Zn and its alloys via those ILs is currently in the beta-test scale using 50 l baths to coat hydraulic systems after 4 years work which is based on a €13 million integrated project involving 33 companies and universities.

Trials have been made to electroplate pure Zn in this type of IL on Mg alloys including pure Mg, AZ31, AZ61, AZ91, AS41, AE42, WE43-T6, QE22, MgGd5Sc1 and MgY4Sc1 (Bakkar and Neubert, 2007). However, it only succeeded upon the Al-free Mg-RE alloys, whilst powdery deposits occurred on Mg–Al alloys. The supplied power type plays an important role in the coating properties of the successfully plated cases. Pulsed cathodic current is superior to constant current. The former produced uniform, shiny, dense and crack-free Zn deposits exhibiting corrosion behaviour similar to pure Zn. Recently, researchers from Monash University and Shanghai Jiao-tong University found an easy way to achieve Zn plating upon Mg–Al alloy AZ91D in 1:2 ChCl-urea IL following the standard Zn immersion

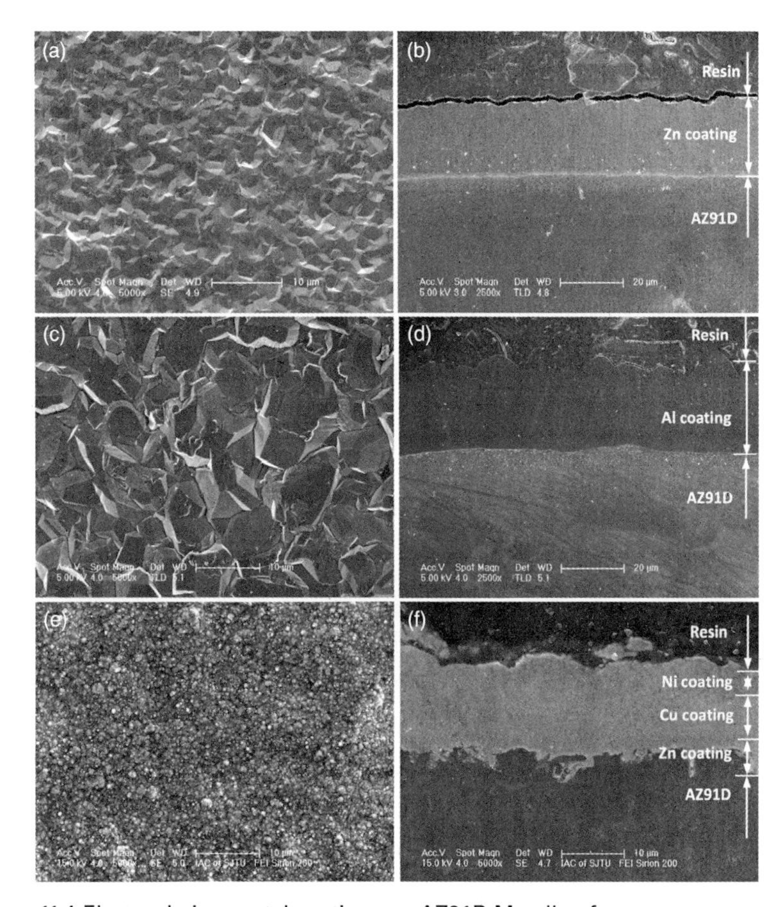

11.4 Electroplating metal coatings on AZ91D Mg alloy from eutectic-base ionic liquid: (a) and (b) Zn coating, (c) and (d) Al coating, (e) and (f) Ni coating laptop case.

pretreatment (Yang *et al.*, 2010). Furthermore, Ni and Zn-Ni alloy (Zn less than 50 at.%) were also electroplated from the same IL upon on a Cu substrate (Yang *et al.*, 2011a, 2012), and also these coatings can extend to Mg substrates after Cu pre-plating. Corrosion resistance of the Ni/Ni-Zn coatings generated from IL is much better than that from aqueous solution due to its porosity-free nature and good adhesion. Some typical Zn, Al and Ni coatings on AZ91D are depicted in Fig. 11.4.

11.5 Evaluation of coatings

Electro-/electroless plating methods have become well established with further promising innovations for coating Mg alloys in development leading to production of corrosion resistant coatings. Whilst nowadays electroplated Mg alloys are widely used in decorative and non-load bearing applications (mobile phone chassis, auto door handles, electronic parts in routers, etc.), there is no generalised protocol for Mg-alloy coatings as yet. Descriptions of the methods used in electro-/electroless plating vary with a lack of cohesiveness and numerous researchers are studying different methods. As a result, the successful plating (in a broader sense) of Mg alloy components in industry is still relatively low. This, in turn, increases the production cost and limits the spread of plating techniques being trialled commercially (unfortunately leaving the current industrial best practice using toxic and dangerous chemicals). The exception is routine, large volume production, where the process has worked and in such cases uptake (more specifically volume) is high.

Many factors such as treatment time, temperature, pH and chemical concentration in the plating bath, can influence coating properties. The fact that the explosion in research into electro-/electroless plating for Mg alloys in the last decade is in its early stage is confirmed by the notion that, in the vast majority of cases, the pretreatment is more influential on the final properties of the plating than the plating itself. Such variation in quality is not industrially acceptable, and serves to indicate that the final plating itself is only half of the issue. This chapter reveals that there are many one-off studies, many of which are not amenable to upscaling, with all studies suggesting an improvement in the corrosion resistance afforded compared to the substrates (even if minor). This situation is actually stifling to some extent, in that there is no report of increased corrosion rates, even with the realistically high chance of adverse galvanic couples being formed.

In addition, it is difficult to compare the corrosion protection afforded by different coatings since in the majority of studies the methods used to determine corrosion rate vary (i.e. between EIS, DC polarisation, salt spray, immersion, etc). In order to get the reader up to speed on the state of the art, abridged information about the electro-/electroless plating procedures from the papers reviewed herein, key features of the coatings and the protection afforded by the plated coatings was extracted and summarised in Table 11.3.

This chapter demonstrates that a large number of plating processes can be used for corrosion protection of Mg alloys. One standout feature of Table 11.3 is that in all cases an improvement in corrosion resistance is reported. Each technique reviewed has its own merits and limitations and the authors' assessment of the anti-corrosive performance of these techniques can be seen in the literature (adapted and updated from the original form (Chen *et al.*, 2012)). Many metals can be electroplated onto pre-plated Mg alloys for corrosion protection, including Ni, Cu, Cr and Au, etc. These metallic coatings are highly inert and corrosion resistant. However the corrosion

Table 11.3 Summary of plating procedures and corrosion performance of the resulting coatings

Material plated	Substrate	Reference	Electroplated (EP) / electroless plated (EL)	Pretreatment process	Comments on application and methods	Corrosion resistance performance (and method of evaluation)
Al	AZ91D	Yang *et al.*, 2011b	EP	Mechanical grinding (MG) only	Used ionic liquid plating bath, post heat treatment promoted adhesion and strength	$Mg_{17}Al_{12}$ intermetallic layers formed whose corrosion rate is lower than base AZ91D (EIS, salt spray and hydrogen evolution in 3.5% NaCl)
Zn	AZ91D	Li *et al.*, 2007		None	Alkaline plating bath with SnO_2 addition to improve coating adhesion	Better than pure Zn plated without SnO_2 addition (Polarisation in 5% NaCl)
Ni	AZ91D	Gu *et al.*, 2006		MG, alkaline cleaning, acid pickling, fluoride activating, electroless plating	Nanocrystalline Ni deposition, high wear resistance, toxic CrO_3 and 40 vol% HF used	More noble E_{corr} and lower corrosion current than EL Ni film (Polarisation in 3% NaCl and immersion test in 10% HCl)
Zn/Sn/Zn-Sn	AZ91D	Zhu *et al.*, 2006		Alkaline cleaning and Zn immersion	Adhesive, compact Zn single coating, while low adhesion for Sn single with defects. Thermal treatment causes diffusion to form double layer	Corrosion resistance: Double layer >> single Zn layer > single Sn layer (Polarisation in 5% NaCl)
Zn-Ni	AZ91	Jiang *et al.*, 2005		MG, alkaline cleaning, acid pickling, activating, Zn immersion	Pulse potential plating produced finer grain structure and higher nucleation rate	Intermetallic phase, γ-$Zn_{21}Ni_5$ offered high corrosion resistance (Polarisation in 5% NaCl and salt spray)

(*Continued*)

Table 11.3 Continued

Material plated	Substrate	Reference	Electroplated (EP) / electroless plated (EL)	Pretreatment process	Comments on application and methods	Corrosion resistance performance (and method of evaluation)
Al-Mn	AZ31B	Zhang *et al.*, 2009a, 2009b		MG, alkaline cleaning, activating, pre-Zn plating	Plating amorphous Al-Mn composite coating in molten salt, pre-plated Zn is important for Al-Mn plating	Passive $Al(OH)_3$ film on Al-Mn coating offered satisfactory short-term and long-term corrosion resistance (Polarisation and EIS in 3.5% NaCl)
Cu (then Cr)	AZ91D	Huang *et al.*, 2010b		MG and rinse in water	No HF or CrO_3 used, Cr was plated onto Cu as Cu is prone to oxidise in air	As good as pure Cr (Polarisation in 0.1 M H_2SO_4/ 0.1 M H_2SO_4 with different NaCl)
Cu	AZ91D	Tang and Azumi, 2011a		MG, pickling in HF+HCl, zincating	HF used in pretreatment, adhesion of Cu was improved	Not mentioned
MgO	Mg-6Zn-Ca	Cai *et al.*, 2009; Lei *et al.*, 2010a		Polishing only	Highly alkaline solution electrolyte	MgO coating enhanced corrosion resistance of Mg-6Zn-Ca alloy (Polarisation in 3.5% NaCl)
Ni-P	AZ91, A4 and A42	Anik and Körpe, 2007	Direct EL	MG, alkaline cleaning, acid pickling and HF activating	Toxic chemical used (40 vol% HF) in the plating bath, weak acidic electrolyte used	Better performance than the bare alloy by up to 8 times (Polarisation)
	AZ91	Li *et al.*, 2006a		MG, alkaline cleaning, acid pickling and HF activating,	Ni sulphate plating bath without HF, but HF used in pretreatment	The free corrosion potential of the coated sample was improved to –0.4 V compared to –1.5 V of the bare alloy (Free corrosion potential)

Ni-W-P	AZ91D	Zhang *et al.*, 2007b		MG, alkaline cleaning and conversion coating	Alkaline-citrate-based baths used. No toxic chemicals	Improved corrosion resistance by 50 times than the bare substrate (Polarisation and immersion in 10% HCl)
Ni-Sn-P	AZ91D	Zhang *et al.*, 2008		MG, alkaline cleaning and conversion coating	Alkaline-citrate-based baths used. No toxic chemicals. Produced a conversion film prior to EL	Better performance than Ni-P coating by 20 times (Polarisation and immersion in 10% HCl)
Ni-P-ZrO_2	AZ91D	Song *et al.*, 2008		MG, alkaline cleaning, activating and Ni-P EL plating	Formed nano ZnO_2 particles in the coating	Claims to be superior to Ni-P and better than untreated substrate by 3 orders of magnitude (Polarisation and EIS in 0.6 M NaCl/Na_2SO_4/NaOH; Salt spray in 5%NaCl; and immersion in 0.6 M NaCl)
Ni-P	AZ31B, AE42 and ZRE1	Mahallawy *et al.*, 2008	Zn immersion /interlayer EL	MG, alkaline cleaning, acid pickling, acid activating, alkaline activating, zincating	Complicated process, long plating duration (35 min) not applicable for industrial application	Significant improvement in corrosion resistance after Ni-P plated (by 15 times) (Polarisation in 3% NaCl)
	AZ91D	Huo *et al.*, 2004		MG and stannate conversion coating	A stannate conversion coating used as interlayer, its porous feature benefit for Ni-P plating	Satisfactory corrosion resistance (Polarisation and immersion in 3.5% NaCl)

(*Continued*)

Table 11.3 Continued

Material plated	Substrate	Reference	Electroplated (EP) / electroless plated (EL)	Pretreatment process	Comments on application and methods	Corrosion resistance performance (and method of evaluation)
	AZ91D	Zhang *et al.*, 2007a		MG, alkaline cleaning, acid pickling and conversion coating	Phosphate-manganese conversion film as interlayer, no hazardous Cr or HF used	Corrosion resistance improved by 20 times compared to bare metal (Immersion in 10% HCl and polarisation in 3% NaCl)
	AZ31	Zhao *et al.*, 2007		Acid cleaning, interlayer coating and alkaline surface roughening	Organic interlayer used, simple operation, basic bath, no hazardous chemicals	Better corrosion resistance compared to bare substrate by 2 orders of magnitude (Polarisation in 3% NaCl and immersion test in 10% HCl)
	AZ91D	Chen *et al.*, 2006		MG, acid pickling, activating, zincating and electroplating Zn	Zn transition interlayer applied, complicated but environmental-friendly process	Zn interlayer protected substrate from corrosion, subsequent Ni-P EL further improved the corrosion resistance by 10 times (Polarisation and immersion in 5% NaCl)

resistance is often not as good as the metals in their bulk forms, since the heterogeneous nature of the Mg substrates presented for coating in most instances generates some defect sites in the plating coatings and eventually compromises the corrosion protection.

Factors to consider which will require future work are that the corrosion test results reported are nominally from short-term tests, and hence the long-term performance of most of the coatings described herein are unknown, particularly in regards to environmental exposure. Galvanic corrosion is often not reported, even though the theoretical tendency is high, reflecting that short-term tests may offer limited prediction of in-service conditions. Additionally, the corrosion protection afforded is somewhat qualitative in many instances, with no benchmarking or standard practice in the field. To date, no effective and sustainable anodic coatings for Mg have been reported in the scientific literature and patents.

11.6 Conclusions

Electrochemical coatings are a cost-effective method to assist in corrosion protection of Mg and its alloys – as evidenced by recent industrial uptake. Electrochemical plating produces a beautiful surface finish, which is well adhered and presents corrosion protection. The plating process is indeed complicated, made more difficult by an apparent lack of systematic investigation of the topic. This may be related to the fact that the vast majority of relevant papers in the area are post-2000, suggesting that the field is maturing. However, at present there is no best practice across the Mg community. In this chapter we attempted to summarise the primary methods used to produce such electrochemical coatings on Mg, and abridged them for a consolidated presentation.

It is apparent that pretreatment is a complex multi-step in its own right, and it is critical when considered in the context of alloy microstructure effects. Hence, any effort to advance plating upon Mg alloys must not simply focus solely upon the coating science (i.e. the 'wet' side) but also consider the metal side of the interface and how to tailor it for the coating proposed. In addition, post heat treatments also offer engineers some latitude in tuning the coating metallurgy and properties.

It is also apparent that the electrochemical plating of Mg alloys employs a host of extremely toxic chemicals. For instance, concentrated HF, chromate compounds and, in some cases, cyanide are still common ingredients in both the pretreatment and plating procedures. This poses a risk not only to the environment but also to the operators. Alternative chemicals have not been significantly explored in the case of Mg, with the exception of some very recent and promising work employing IL solvents, which presents an attractive alternative (particularly where the IL is not water sensitive).

As such, whilst the industry is still trying to optimise protective coatings, a simultaneous (and not trivial) effort to develop better, simpler, cheaper and 'greener' plating technologies is and will remain a hot topic needing further investigation likely to be fuelled by the advantages of the weight reduction Mg can provide when fully exploited in practical applications.

11.7 References

http://www2.le.ac.uk/departments/chemistry/people/academic-staff/professor-andrew-p.-abbott [Online]. [Accessed].

Abbott, A. P., Capper, G., Davies, D. L., Rasheed, R. and Tambyrajah, V. (2000). *Ionic liquids and their use*. U.K. patent application, PCT/GB01/04300.

Abbott, A. P., Dalrymple, I., Endres, F. and Macforlane, D. R. (2008). Why use ionic liquids for electrodeposition? In: Endres, F., MacFarlane, D. R. and Abbott, A. P. (eds.) *Electrodeposition from Ionic Liquids*. Weinheim: WILEY-VCH Verlag GmbH & Co. KgaA.

Ambat, R. and Zhou, W. (2004). Electroless nickel-plating on AZ91D magnesium alloy: effect of substrate microstructure and plating parameters. *Surface and Coatings Technology*, **179**, 124–134.

Anik, M. and Körpe, E. (2007). Effect of alloy microstructure on electroless NiP deposition behavior on Alloy AZ91. *Surface and Coatings Technology*, **201**, 4702–4710.

Apachitei, I., Duszczyk, J., Katgerman, L. and Overkamp, P. J. B. (1998). Electroless Ni-P composite coatings: the effect of heat treatment on the microhardness of substrate and coating. *Scripta Materials*, **38**, 1347–1353.

Araghi, A. and Paydar, M. H. (2010). Electroless deposition of Ni-P-B_4C composite coating on AZ91D magnesium alloy and investigation on its wear and corrosion resistance. *Materials and Design*, **31**, 3095–3099.

Armyanov, S., Vangelova, T. and Stoyanchev, R. (1982). Pretreatment of Al-Mg alloys for electrodeposition by immersion zinc and electroless nickel. *Surface Technology*, **17**, 89–100.

Bakkar, A. and Neubert, V. (2007). Electrodeposition onto magnesium in air and water stable ionic liquids: From corrosion to successful plating. *Electrochemistry Communications*, **9**, 2428–2435.

Bockris, J. O. M. and Reddy, A. K. N. (2000). *Modern Electrochemistry*. New York: Springer.

Brunelli, K., Dabala, M., Calliari, I. and Magrini, M. (2005). Effect of HCl pre-treatment on corrosion resistance of cerium-based conversion coatings on magnesium and magnesium alloys. *Corrosion Science*, **47**, 989–1000.

Cai, Z.-P., Lu, D.-S., Li, W.-S., Liang, Y. and Zhou, H.-B. (2009). Study on anodic oxidation of magnesium in 6 M KOH solution by alternative current impedance. *International Journal of Hydrogen Energy*, **34**, 467–472.

Chang, J.-K., Chen, S.-Y., Tsai, W.-T., Deng, M.-J. and Sun, I.-W. (2007). Electrodeposition of aluminum on magnesium alloy in aluminum chloride ($AlCl_3$)-1-ethyl-3-methylimidazolium chloride (EMIC) ionic liquid and its corrosion behavior. *Electrochemistry Communications*, **9**, 1602–1606.

Chang, J. -K., Chen, S. -Y., Tsai, W. -T., Deng, M. -J. and Sun, I. -W. (2008). Improved corrosion resistance of magnesium alloy with a surface aluminum coating electrodeposited in ionic liquid. *Journal of The Electrochemical Society*, **155**, C112–C116.

Chen, J. -L., Yu, G., Hu, B. -N., Liu, Z., Ye, L. -Y. and Wang, Z. -F. (2006). A zinc transition layer in electroless nickel plating. *Surface and Coatings Technology*, **201**, 686–690.

Chen, X. -B., Yang, H. -Y., Abbott, T. B., Easton, M. A. and Birbilis, N. (2012). Corrosion resistant electrochemical platings on magnesium alloys: A state-of-the-art review. *Corrosion*, **68**, 518–535.

Chen, Y., Ye, K. and Zhang, M. (2010). Preparation of Mg-Yb alloy film by electrolysis in the molten LiCl-KCl-$YbCl_3$ system at low temperature. *Journal of Rare Earths*, **29**, 128–133.

Chuang, M. -H., Chang, J. -K., Tsai, P. -J., Tsai, W. -T., Deng, M. -J. and Sun, I. -W. (2010). Heat-treatment induced material property variations of Al-coated Mg alloy prepared in aluminum chloride/1-ethyl-3-methylimidazolium chloride ionic liquid. *Surface and Coatings Technology*, **205**, 200–204.

Cormier, G. J. (1994). Alkaline cleaning. *ASM Handbook Volume 5: Surface Engineering*. 10 ed. Materials Park, OH: ASM International.

Delong, H. K. (1950). *Method of producing a metallic coating on magnesium and its alloys*. U.S. patent application, 2,526,544.

Delong, H. K. (1957). *Electrodeposition of zinc on magnesium and its alloys*. U.S. patent application, 2,811,484.

Delong, H. K. (1961). Plating on magnesium by electrodeposition and chemical reduction methods. In: 48th Annual Proceedings of the American Electroplater's Society 1961, American Electroplaters' Society, Newark, NJ.

Gray, J. E. and Luan, B. (2002). Protective coatings on magnesium and its alloys – a critical review. *Journal of Alloys and Compounds*, **336**, 88–113.

Gu, C. -D., Lian, J. -S., He, J. -G., Jiang, Z. -H. and Jiang, Q. (2006). High corrosion-resistance nanocrystalline Ni coating on AZ91D magnesium alloy. *Surface and Coatings Technology*, **200**, 5413–5418.

Gu, C. -D., Lian, J. -S. and Jiang, Z. -H. (2005a). Multilayer Ni-P coating for improving the corrosion resistance of AZ91D magnesium alloy. *Advanced Engineering Materials*, **7**, 1032–1036.

Gu, C. -D., Lian, J. -S., Li, G. -Y., Niu, L. -Y. and Jiang, Z. -H. (2005b). Electroless Ni-P plating on AZ91D magnesium alloy from a sulfate solution. *Journal of Alloys and Compounds*, **391**, 104–109.

Guo, X. W., Wang, S. H., Yang, H. Y., Peng, L. M. and Ding, W. J. (2011). Characterization of Ni-P/TiO_2 MMC coatings prepared by electroless plating process on Mg-Nd-Zn-Zr magnesium alloys. *Materials Science Forum*, **690**, 422–425.

Hillis, J. E. (1994). Surface engineering of magnesium alloys. *ASM Handbook Vol 5: Surface Engineering*. 10 ed. Materials Park, OH: ASM International.

Huang, C.-A., Lin, C.-K. and Yeh, Y.-H. (2010a). Corrosion behavior of Cr/Cu-coated Mg alloy (AZ91D) in 0.1 M H_2SO_4 with different concentrations of NaCl. *Corrosion Science*, **52**, 1326–1332.

Huang, C.-A., Lin, C.-K. and Yeh, Y.-H. (2010b). Increasing the wear and corrosion resistance of magnesium alloy (AZ91D) with electrodeposition from eco-friendly copper- and trivalent chromium-plating baths. *Surface and Coatings Technology*, **205**, 139–145.

Huo, H.-W., Li, Y. and Wang, F.-H. (2004). Corrosion of AZ91D magnesium alloy with a chemical conversion coating and electroless nickel layer. *Corrosion Science*, **46**, 1467–1477.

Innes, W. P. (1971). Surface Treatments for Magnesium. *Metal Finishing Guidebook and Directory*. Hackensack, N.J.: Metal and Plastic Publications.

Jiang, Y.-F., Zhai, C.-Q., Liu, L.-F., Zhu, Y.-P. and Ding, W.-J. (2005). Zn-Ni alloy coatings pulse-plated on magnesium alloy. *Surface and Coatings Technology*, **191**, 393–399.

Kurze, P. (2003). Corrosion and corrosion protection of magnesium. In: Kainer, K. U. (ed.) *Magnesium Alloys and Technology.* Weinheim: WILEY-VCH Verlag GmbH & Co. KG aA.

Kurze, P. (2006). Corrosion and surface protection. In: Friedrich, H. E. and Mordike, B. L. (eds.) *Magnesium Technology, Metallurgy, Design Data, Applications*. Berlin Heidelberg: Springer-Verlag.

Lei, T., Ouyang, C., Tang, W., Li, L.-F. and Zhou, L.-S. (2010a). Preparation of MgO coatings on magnesium alloys for corrosion protection. *Surface and Coatings Technology*, **204**, 3798–3803.

Lei, X. P., Yu, G., Zhu, Y. P., Zhang, Z. P., He, X. M., Hu, B. N. and Chen, Y. (2010b). Successful cyanide free plating protocols on magnesium alloys. *Transactions of the Institute of Metal Finishing*, **88**, 75–80.

Li, J.-Z., Shao, Z.-C., Zhang, X. and Tian, Y.-W. (2006a). The electroless nickel-plating on magnesium alloy using NiSO46H2O as the main salt. *Surface and Coatings Technology*, **200**, 3010–3015.

Li, J.-Z., Tian, Y.-W., Huang, Z.-Q. and Zhang, X. (2006b). Studies of the porosity in electroless nickel deposits on magnesium alloy. *Applied Surface Science*, **252**, 2839–2846.

Li, W.-P., Zhu, L.-Q. and Li, M. (2007). Zn coatings on AZ91D magnesium alloy prepared by electroplating from the electrolyte containing SnO_2 gel. *Materials Science Forum*, **546–549**, 593–596.

Lian, J.-S., Li, G.-Y., Niu, L.-Y., Gu, C.-D., Jiang, Z.-H. and Jiang, Q. (2006). Electroless Ni-P deposition plus zinc phosphate coating on AZ91D magnesium alloy. *Surface and Coatings Technology*, **200**, 5956–5962.

Liu, Z. and Gao, W. (2006a). A novel process of electroless Ni-P plating with plasma electrolytic oxidation pretreatment. *Applied Surface Science*, **253**, 2988–2991.

Liu, Z.-M. and Gao, W. (2006b). The effect of substrate on the electroless nickel plating of Mg and Mg alloys. *Surface and Coatings Technology*, **200**, 3553–3560.

Liu, Z.-M. and Gao, W. (2006c). Electroless nickel plating on AZ91 Mg alloy substrate. *Surface and Coatings Technology*, **200**, 5087–5093.

Mahallawy, N. E., Bakkar, A., Shoeib, M., Palkowski, H. and Neubert, V. (2008). Electroless Ni-P coating of different magnesium alloys. *Surface and Coatings Technology*, **202**, 5151–5157.

Novák, M., Vojtech, D., Novák, P. and Vítu, T. (2011). Tribological properties of heat-treated electroless Ni-P coatings on AZ91 alloy. *Applied Surface Science*, **257**, 9982–9985.

Olsen, A. L. and Halvorsen, S. T. (1982). *Method for the electrolytical metal coating of magnesium articles*. US patent application, 4,349,390.

Pan, S.-J., Tsai, W.-T., Chang, J.-K. and Sun, I.-W. (2010). Co-deposition of Al-Zn on AZ91D magnesium alloy in $AlCl_3$-1-ethyl-3-methylimidazolium chloride ionic liquid. *Electrochimica Acta*, **55**, 2158–2162.

Pearson, T. and Cordero-Rando, M. D. M. (2007). *Pretreatment of magnesium substrates for electroplating*. US patent application, 7,704,366.

Pourbaix, M. (1974). *Atlas of Electrochemical Equilibria in Aqueous Solutions*. Houston, Tex.: National Association of Corrosion Engineers (NACE).

Qian, H. M., Li, Y. and Ling, G. P. (2009). Influence of acid pickling on electrodeposition of aluminum on magnesium alloy in room temperature molten salts. *Chinese Journal of Nonferrous Metals*, **19**, 854–860.

Rajagopal, I., Rajam, K. S. and Rajagopalan, S. R. (1990). Plating on magnesium alloy. *Metal Finishing*, **88**, 43–47.

Sharma, A. K. (1988). Electrodeposition of gold on magnesium-lithium alloys. *Metal Finishing*, **86**, 33–34.

Sharma, A. K. (1991). Gold pating on magnesium-lithium alloys. *Metal Finishing*, **89**, 16–17.

Simka, W., Puszczyk, D. and Nawrat, G. (2009). Electrodeposition of metals from non-aqueous solutions. *Electrochimica Acta*, **54**, 5307–5319.

Smyth, M. D. (1985). Magnesium Surface Finishing. *Metal Finishing Guidebook and Directory*. Hackensack, N.J.: Metal and Plastic Publications.

Song, Y.-W., Shan, D.-Y. and Han, E.-H. (2008). High corrosion resistance of electroless composite plating coatings on AZ91D magnesium alloys. *Electrochimica Acta*, **53**, 2135–2143.

Sun, S., Liu, J.-G., Yan, C.-W. and Wang, F.-H. (2008). A novel process for eletroless nickel plating on anodized magnesium alloy. *Applied Surface Science*, **254**, 5016–5022.

Tan, A. L. K., Soutar, A. M., Annergren, I. F. and Liu, Y.-N. (2005). Multilayer sol-gel coatings for corrosion protection of magnesium. *Surface and Coatings Technology*, **198**, 478–482.

Tang, J.-W. and Azumi, K. (2011a). Influence of zincate pretreatment on adhesion strength of a copper electroplating layer on AZ91 D magnesium alloy. *Surface and Coatings Technology*, **205**, 3050–3057.

Tang, J. W. and Azumi, K. (2011b). Effect of copper pretreatment on the zincate process and subsequent electroplating of a protective copper/nickel deposit on the AZ91D magnesium alloy. *Electrochimica Acta* **56**, 8776–8782.

Tomlinson, W. J. and Mayor, J. P. (1988). Formation, microstructure, surface roughness and porosity of electroless nickel coatings. *Surface Engineering*, **43**, 235–238.

Turhan, M. C., Lynch, R., Killian, M. S. and Virtanen, S. (2009). Effect of acidic etching and fluoride treatment on corrosion performance in Mg alloy AZ91D (MgAlZn). *Electrochimica Acta*, **55**, 250–257.

Yang, H.-Y., Guo, X.-W., Birbilis, N., Wu, G.-H. and Ding, W.-J. (2011a). Tailoring nickel coatings via electrodeposition from a eutectic-based ionic liquid doped with nicotinic acid. *Applied Surface Science*, **257**, 9094–9102.

Yang, H.-Y., Guo, X.-W., Chen, X.-B., Wang, S.-H., Wu, G.-H., Ding, W.-J. and Birbilis, N. (2012). Electrodeposition behavior of nickel-zinc alloys from a eutectic-based ionic liquid. *Electrochimica Acta*, **63**, 131–138.

Yang, H.-Y., Guo, X.-W., Wu, G.-H. and Ding, W.-J. (2010). Electrodeposition of Zn on AZ91D in choline chloride-urea ionic liquids. *Journal of Chinese Society for Corrosion and Protection*, **30**, 155–160.

Yang, H.-Y., Guo, X.-W., Wu, G.-H., Ding, W.-J. and Birbilis, N. (2011b). Electrodeposition of chemically and mechanically protective Al-coatings on AZ91D Mg alloy. *Corrosion Science*, **53**, 381–387.

Yang L., Li, J., Zheng, Y., Jiang, W. and Zhang, M. (2009). Electroless Ni-P plating with molybdate pretreatment on Mg-8Li alloy. *Journal of Alloys and Compounds*, **467**, 562–566.

Yang, L.-X., Luan, B., Cheong, W.-J. and Jiang, J.-R. (2005). Optimization and performance of copper immersion coating on AZ91 magnesium alloy. *Journal of Coatings Technology and Research*, **2**, 493–498.

Zeng, L., Yang, S., Zhang, W., Guo, Y. and Yan, C. (2010). Preparation and characterization of a double-layer coating on magnesium alloy AZ91D. *Electrochimica Acta*, **55**, 3376–3383.

Zhang, J.-F., Yan, C.-W. and Wang, F.-H. (2009a). Electrodeposition of Al-Mn alloy on AZ31B magnesium alloy in molten salts. *Applied Surface Science*, **255**, 4926–4932.

Zhang, J.-F., Zhang, W., Yan, C.-W., Du, K.-Q. and Wang, F.-H. (2009b). Corrosion behaviors of Zn/Al-Mn alloy composite coatings deposited on magnesium alloy AZ31B (Mg-Al-Zn). *Electrochimica Acta*, **55**, 560–571.

Zhang, J.-W., Hou, L.-F. and Xu, B.-S. (2010a). Effect of heat treatment on microstructure and properties of electroless Ni-P deposits on AZ91D magnesium alloy *Transactions of Materials and Heat Treatment*, **31**, 137–140.

Zhang, S., Li, Q., Zhong, X., Dai, Y. and Luo, F. (2010b). Corrosion resistance of AZ91D magnesium alloy with electroless plating pretreatment and Ni-TiO_2 composite coating. *Materials Characterization*, **61**, 269–276.

Zhang, W.-X., He, J.-G., Jiang, Z.-H., Jiang, Q. and Lian, J.-S. (2007a). Electroless Ni-P layer with a chromium-free pretreatment on AZ91D magnesium alloy. *Surface and Coatings Technology*, **201**, 4594–4600.

Zhang, W.-X., Huang, N., He, J.-G., Jiang, Z.-H., Jiang, Q. and Lian, J.-S. (2007b). Electroless deposition of Ni-W-P coating on AZ91D magnesium alloy. *Applied Surface Science*, **253**, 5116–5121.

Zhang, W.-X., Jiang, Z.-H., Li, G.-Y., Jiang, Q. and Lian, J.-S. (2008). Electroless Ni-Sn-P coating on AZ91D magnesium alloy and its corrosion resistance. *Surface and Coatings Technology*, **202**, 2570–2576.

Zhang, Z.-P., Yu, G., Ouyang, Y.-J., He, X.-M., Hu, B.-N., Zhang, J. and Wu, Z.-J. (2009c). Studies on influence of zinc immersion and fluoride on nickel electroplating on magnesium alloy AZ91D. *Applied Surface Science*, **255**, 7773–7779.

Zhao, H. and Cui, J.-Z. (2007). Electroless plating of silver on AZ31 magnesium alloy substrate. *Surface and Coatings Technology*, **201**, 4512–4517.

Zhao, H., Huang, Z.-H. and Cui, J.-Z. (2007). A new method for electroless Ni-P plating on AZ31 magnesium alloy. *Surface and Coatings Technology*, **202**, 133–139.

Zhu, L.-Q., Li, W.-P. and Shan, D.-D. (2006). Effects of low temperature thermal treatment on zinc and/or tin plated coatings of AZ91D magnesium alloy. *Surface and Coatings Technology*, **201**, 2768–2775.

Zou, H. Q., Lu, J. C., Zhang, J., Wu, H. C. and Fang, M. (2005). Synergistic fluoropolymer coating on magnesium alloys. *Materials Science Forum*, **488–489**, 673–676.

Zuleta, A. A., Correa, E., Sepúlveda, M., Guerra, L., Castaño, J. G., Echeverría, F., Skeldon, P. and Thompson, G. E. (2012). Effect of NH4HF2 on deposition of alkaline electroless Ni-P coatings as a chromium-free pre-treatment for magnesium. *Corrosion Science*, **55**, 194–200.

12
Plating techniques to protect magnesium (Mg) alloys from corrosion

K. AZUMI, Hokkaido University, Japan,
H. H. ELSENTRIECY, Central Metallurgical Research and Development Institute, Egypt and J. TANG,
Hokkaido University, Japan

DOI: 10.1533/9780857098962.3.347

Abstract: Plating on Mg alloys has proved difficult due to their high susceptibility to the degradation reactions such as galvanic corrosion in the plating processes. In this chapter, surface conditioning, activation, electroless- or electroplating methods are discussed from viewpoints of corrosion engineering. Degradation reactions of the Mg substrate in each plating stage are considered to figure out how to suppress degradation of substrate and to achieve uniform deposition of a plating layer. Building up an interface between the substrate and the plating layer is also an important subject to establish enough adhesion strength of the plating layer and low defect density to prohibit corrosion progress of the substrate. Al coating plating on Mg alloys electrodeposited from an ionic liquid bath is also introduced to provide coatings with less-galvanic coupling property with a less-noble Mg substrate.

Key words: galvanic corrosion, uniform activation, adhesion strength, ionic bath.

12.1 Introduction

Mg alloys are expected to play an important role in future engineering because of their light weight, large strength–weight ratio, natural abundance and non-toxicity to the human body and environment. On the other hand, poor corrosion resistance has limited their use in practical applications, especially for outdoor structures. Various corrosion protection technologies have been developed for Mg alloys, including thermal oxide coating, anodized oxide coating, silicate coating, chemical conversion coating, and metal plating. In the application of traditional electroplating (EP) or electroless plating (EL) to Mg alloys, formation of a defect-less metal coating and establishment of good adhesion of the coating layer to the substrate are major issues. For example, pinholes in the plating layer can result in galvanic coupling between the less-noble Mg substrate and relatively noble plating layer and allow penetration of water or moisture that leads to severe localized corrosion of the substrate beneath the coating layer. Achievement of sufficient adhesion of the plating layer to the substrate has also been an important issue. A Mg substrate is easily corroded in the plating process, and the resultant corrosion layer inhibits strong binding of the plating layer to the substrate metal surface. To overcome these problems, considerable efforts have been made to establish optimized surface pretreatment methods, suitable plating bath composition and condition and combination of metal plating with other coating methods, such as anodized coating and organic coating. These studies have been introduced and summarized in review articles.[1–3] Research on surface coating of Mg alloys has recently increased dramatically in China.[4] To achieve an excellent corrosion-protective coating on a Mg alloy using the EP or EL method, the following issues must be considered.

12.1.1 Substrate property

As seen in the Pourbaix diagram, Mg corrodes in the wide potential range and pH range lower than pH 11.5.[5] This means that a Mg substrate dissolves in a neutral or acidic bath in the coating process. Mg is categorized as a 'passivation' metal on which a thick corrosion layer with low corrosion protection forms easily and may prevent metallic binding of the plating layer with the metal substrate. On the other hand, the fact that Mg substrate is easy to dissolve and easy to precipitate as a hydroxide are used in chemical conversion coating.[6–9] It has been reported that the corrosion resistance of pure Mg to atmospheric exposure was high but the contribution of impurities or alloying elements of Mg to corrosion susceptibility is crucial to protect Mg in some corrodible conditions.[10,11] Mg is generally used in the form of alloys to improve mechanical properties, workability and corrosion resistivity. The AZ series of alloys is amongst the popular Mg alloys composed with Al, Zn and other elements. Depending on composition, concentration and dispersion of these elements, the substrate surface reveals non-uniform activity for reactions in the coating process. For example, AZ91D has a binary structure composed of a Mg-rich α-phase matrix and an Al-rich β-phase of $Mg_{17}Al_{12}$ eutectic dispersion; the electrochemical activities of these phases are slightly different because of the difference in their compositions.[12–15] The Mg-rich α-phase dissolves more easily than the β-phase surface in a bath at pH lower than 10.5, while the α-phase passivates at pH higher than 11. Such a non-uniform property may induce acceleration of electrochemical reactions, such as rapid and non-uniform metal deposition and gas evolution, and result in not only a non-uniform appearance of the coating but also in a weak binding area of the coating to the substrate. Suitable coating conditions for substrate materials depending on their alloy composition and impurities must therefore be considered.[16–21]

12.1.2 Preparation of the substrate and coating bath

Generally, the EP process is conducted by the sequence of rinsing, pickling, activating and EP. Various kinds of additional processes such as alkaline degreasing, acidic etching, zincating, pre-coating and post heating are also applied.[3] A Mg substrate is easily dissolved in traditional Ni plating baths of acidic or neutral pH. Direct EP on Mg alloys is therefore difficult and tends to result in a plating layer with weak adhesion. Protection of the substrate from excess dissolution by zincating and fluoride layer formation can therefore be applied.[2,18,20] Another strategy is the use of an alkaline baths in which Cu, Ni and Ni/Zn alloys are electrodeposited without dissolution of the Mg substrate.[22,23] Elimination of chromium from the bath and thus from the plating layer must also be considered to respond to regulations being introduced to restrict chromium use.[24,25] The use of a non-aqueous system such as molten salt and ionic liquid (IL) has been opening up new possibilities for EP because these baths are not restricted by the narrow potential window of water decomposition reactions and thus enable deposition of less-noble metals such as Al and their alloys.[26–32] Some auxiliary methods such as ultrasound irradiation[33–35] and pulse polarization[22,36,37] have also been used to improve the quality of coatings and performance of the plating process. These methods agitate or mitigate the depletion layer of Ni^{2+} ions near the substrate surface to accelerate rapid and uniform deposition.

12.1.3 Type of coating

The coating layer deposited on a Mg alloy after pre-coating of other metals provides a multilayer structure such as Ni-P/Cu/Zn.[38,39] Duplex coating of, for example, Ni-P and NiB,[40] and composite coating with alumina[37] and TiO_2 particles[41] have also been proposed to improve mechanical strength of the coating layer. Electrodeposition of various alloys such as Zn-Ni, Co-Zn, Ni-Sn-P and Ni-W-P is also possible by controlling the deposition polarization and bath composition.[22,42–44] Alloys and composites including Al have also been electrodeposited on Mg alloys in a molten salt bath or IL bath.[29,45–47] The combination of EP or EL with the oxide layer formed by anodizing or plasma electrolytic oxidation (PEO) has also been investigated because these oxides adhere strongly to the substrate.[33,48–50] Organic coatings[51,52] and conversion coatings[53] as undercoating layers have also been examined.

12.1.4 Evaluation of coatings

Structure, morphology, porosity, adhesion strength, mechanical property and corrosion resistance of coatings formed on Mg alloys have been evaluated. For corrosion protection, formation of pinholes in the coating should be prevented because they cause severe corrosion of the substrate Mg as mentioned above. Porosity and defect density in the coating have therefore been evaluated.[54,55] One of the origins of defect structure in EP coatings is non-uniformity or low density of the deposition nucleation at the initial stage of the EP process. The Mg surface is not sufficiently active for metal deposition reaction because the surface is easily covered with a thick corrosion layer, resulting in inhibition of electron transfer to metal ions in the bath. Small amounts of impurities and eutectic crystals therefore provide relatively active sites for the deposition reaction, resulting in non-uniform and less-dense nucleation of the EP coating, as seen in the case of AZ91D alloy.[56–58] Suitable activation to provide uniform and dense activation sites to the surface is therefore necessary. Adhesion strength is also important, especially for coating on Mg alloys because partial detachment of the coating from the substrate allows induction of defects in the coating and penetration of a corrodible solution or moisture to the coating–substrate interface. Due to its less-noble property, protection of a Mg substrate by a coating layer with a sacrificial anodic function is impossible. The coating should maintain its adherence to the substrate not only in the production line but also in use of products over an extended period of time. However, most reports have shown the adhesive property of coatings evaluated only from a convenient tape test in which adhesive tape was peeled off from the crosscut coatings. Quantitative measurements of adhesion strength have revealed that heat treatment provides improvement in adhesion strength to the substrate.[30,31,55,59]

12.2 Coating processes

Attempts to improve the EP coating formed on typical Mg alloys of AZ91D and AZ31 in aqueous and non-aqueous baths from the viewpoint of corrosion control are described in this section. The plating process of Mg alloys used in this study is shown in Fig. 12.1. Since Mg is susceptible to corrosion in the aqueous bath used in the coating sequence, suppression of the corrosion or uniform progress of the corrosion of the Mg alloy substrate in each process is important. Bath composition and conditions of individual processes used in this study are shown in Table 12.1. In the following subsections, conditioning of the substrate surface for uniform activation, suppression of corrosion of the substrate surface in the bath and formation of less-defective and better adhesive metal coatings are introduced.

Table 12.1 Bath compositions and condition of processes used for surface coating of Mg alloys[6,56–58]

Process	Bath chemicals	Concentration (mol/dm^3)	Condition
(a) Pickling	HF	0–0.25 wt.%	RT, 20 s
	HCl	0–0.25 wt.%	
(b) Activation	$K_4P_2O_7$	0.26	pH 10.5, RT, 120 s
	Na_2CO_3	0.14	
	KF	0.077	
	$CuSO_4{\cdot}5H_2O$	0 or 0.32 m	
(c) Zincating	$ZnSO_4{\cdot}7H_2O$	0.16	pH 10, RT, 1 ks
	$K_4P_2O_7$	0.42	
	Na_2CO_3	0.047	
	KF	0.10	
(d) Cu plating or undercoating	$CuSO_4{\cdot}5H_2O$	0.26	pH13.5, RT
	$KNaC_4H_4O_6{\cdot}4H_2O$	1.3	St. 600 rpm
	NaOH	3.0	Ic = –8 mA/cm^2, 1 ks
	H_3BO_3	0.32	
(e) Ni plating	$NiSO_4{\cdot}6H_2O$	1.14	pH 4.2, RT, 600 s
	$NiCl_2{\cdot}6H_2O$	0.2	I_C = –50 mA/cm^2
	H_3BO_3	0.5	
(f) Al electrodeposition	EMIC	EMIC/$AlCl_3$ = 0.5	RT, Dry N_2 atmosphere
	$AlCl_3$	(molar ratio)	Galvanostatic or pulse polarization
(g) Corrosion test	H_3BO_3	0.15	pH 8.5, RT
	$Na_2B_4O_7$	0.05	Cyclic voltammogram

RT: Room temperature, St: Stirring.

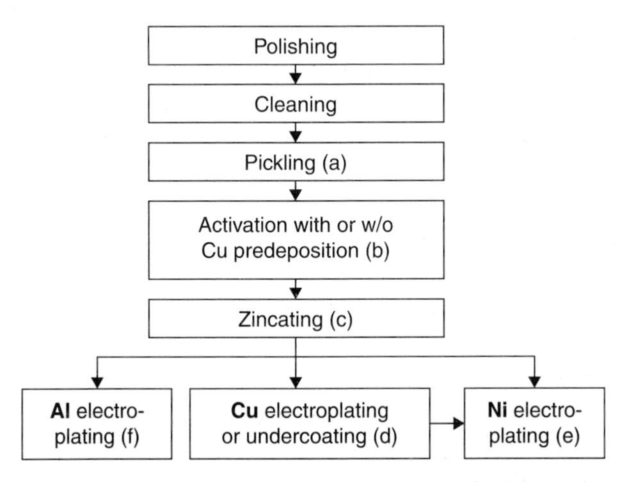

12.1 Coating procedure for Mg alloys used in this study.

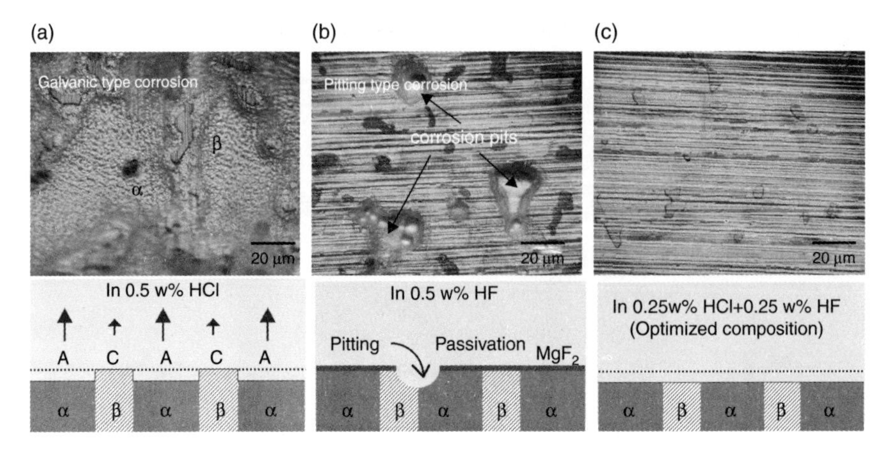

12.2 Results of pickling treatment of AZ91D Mg die-cast alloy immersed in (a) 0.5 wt.% HCl, (b) 0.5 wt.% HF, and (c) 0.25 wt.% HCl + 0.25 wt.% HF solutions.

12.2.1 Suppression of non-uniform dissolution in the pickling process

Pickling treatment is used to remove corrosion films from the substrate surface prior to the following processes by rinsing the specimen in an acidic bath. In this treatment, an appropriate pickling bath suitable for the target substrate must be chosen.[6,60] Figure 12.2 shows optical microscope images of the AZ91D surface after pickling treatment in three acidic solutions. In the HCl solution shown in Fig. 12.2a, typical galvanic dissolution was observed, that is, galvanic coupling between the anode of the Mg-rich α-phase and cathode of the Al-rich β-phase caused preferential dissolution of the α-phase surface. On the other hand, in the HF solution shown in Fig. 12.2b, pitting-type corrosion was observed. This indicated that the substrate surface was covered with a corrosion-protective fluoride film such as MgF_2 and that local dissolution of the substrate occurred at the defects of the film. These non-uniform substrates are not desirable for the following coating process because defective coating is easily formed on these substrates. Figure 12.2c, shows the results of pickling treatment in the mixed acid solution with optimized concentration. A flat surface was obtained due to the moderate protection property of the fluoride film and weakened dissolution ability of chloride ions. In the following processes, specimens pickled in this solution were used.

12.2.2 Uniform seeding for Zn deposition in the zincate process using Cu predeposition

Mg is hard to plate directly with other metals because its surface is easily covered with a thick oxide or hydroxide film acting as an electric insulator.

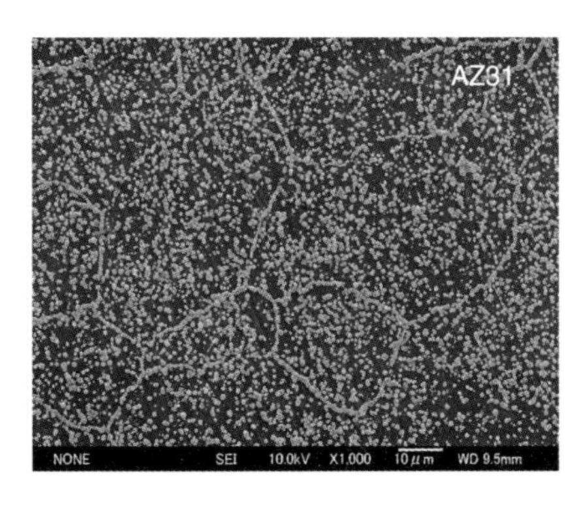

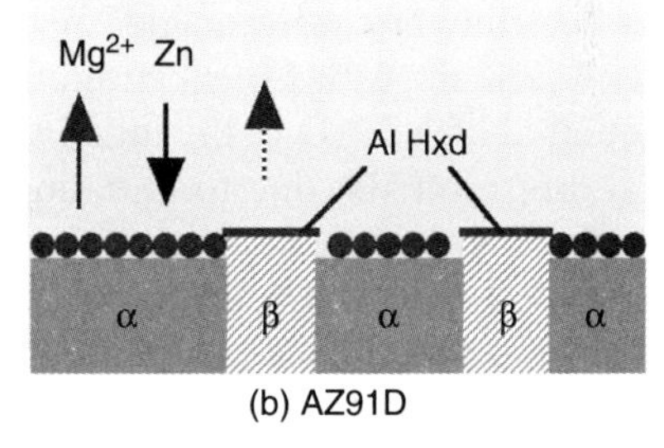

12.3 SEM images of (a) AZ31 and (b) AZ91 D alloys after zincate treatment for 1000 s.

Zincating pretreatment has therefore been used, in which a specimen is immersed in a zincate bath containing Zn(II) ions. First, the oxide (or hydroxide) film on the substrate is chemically dissolved and then the Mg substrate begins to dissolve. This electrochemical oxidation reaction leaves electrons in the substrate and causes the electrode potential to drop in a less-noble direction thus promoting deposition of Zn on the substrate. The Zn deposits not only protect the substrate from further dissolution but also provide seeds for the following plating process.

Taking into consideration the potential pH diagram of Mg and the requirement for substrate dissolution in the zincate process, the pH of the bath should be lower than pH 11.5 but not too low, so as to avoid the undesirable excess dissolution. As shown in Table 12.1, the typical pH value is around 10. Figure 12.3 shows SEM images of Mg alloys after zincate pretreatment. In the case of AZ31, preferential Zn deposition was observed at the grain boundary, probably due to the slightly active property of the grain boundary for metal deposition reaction. In the case of AZ91D, Zn was deposited only on the α-phase surface because the Al-rich β-phase might be covered with Al hydroxide formed by preferential dissolution of Mg and thus by enrichment of Al on the β-phase surface at the initial stage of the zincating process. This non-uniform Zn deposition causes the formation of a non-uniform plating layer with low adhesion strength in the subsequent EP process.

To overcome the problems of the non-uniformity, Cu pretreatment prior to the zincate process has been proposed[56–58,61] in which the specimen is immersed in an activation bath containing a low concentration of Cu(II) ions. In this process, fine Cu particles are densely deposited on the substrate in a substitution reaction with Mg substrate dissolution because the low concentration of Cu(II) ions disables continuous growth of Cu deposits. Cu deposition occurs not only on the grain boundary but also on the grain surface in the case of AZ31, and on both the α-phase surface and β-phase surface in the case of AZ91D. These fine Cu deposits act as an effective seed for Zn deposition in the zincate process which follows because charge transfer between the substrate and Zn(II) ions in the bath is more efficient on Cu deposits than on the Mg substrate. Figure 12.4 shows the change in immersion potential of AZ91D at the initial stage of the zincating process. The potential drops initially due to initiation of substrate dissolution. The potential then changes to the noble direction due to coverage of the substrate surface with Zn deposits, which suppress the dissolution reaction of the substrate. The duration of the less-noble potential shift for the Cu-predeposited specimen is shorter than that of the non-predeposited sample, indicating rapid coverage of the substrate with Zn deposits. The negative peak potential of the Cu-predeposited sample is also more noble than that of the non-predeposited sample. Such a difference in electrode potential is explained by the illustration in Fig. 12.4b. The surface of the Mg substrate may be covered with a corrosion film even after the activation process, and less-effective charge

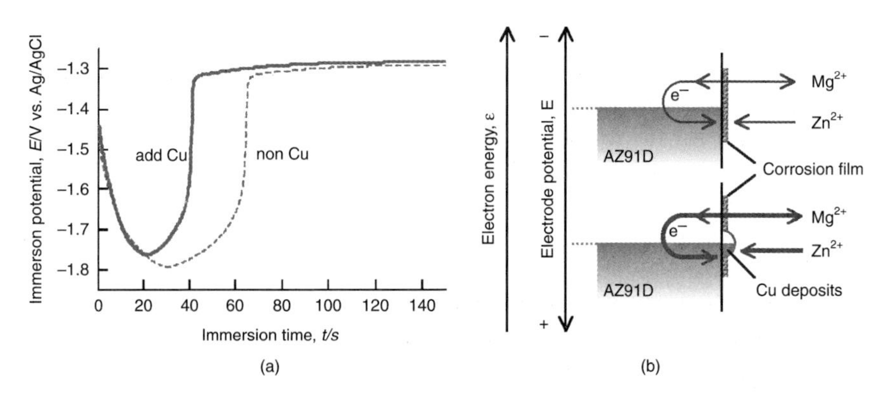

12.4 (a) Changes in the immersion potential of AZ91D Mg alloys with or without Cu predeposition during the zincate process. (b) Schematic representation of immersion potential difference in the Mg alloy with or without Cu predeposition.

transfer through the film causes the larger overpotential for Zn deposition. On the other hand, Cu deposits with a high density provide effective charge transfer sites for rapid Zn deposition.

SEM images of Cu-predeposited AZ31 and AZ91D after zincate pretreatment are shown in Fig. 12.5. In the case of AZ31, the entire surface was covered with Zn deposits, although slightly preferential deposition at the grain boundary can be recognized. In the case of AZ91D, Zn was deposited not only on the α-phase surface but also on the β-phase surface, as confirmed by EDS measurement.[56] In both specimens, uniformity of Zn deposition was drastically improved due to Cu predeposition as schematically illustrated in the figure. High coverage also suppresses the undesirable dissolution of the substrate surface.

Figure 12.6 shows surface and cross-sectional SEM images of AZ91D after electrodeposition of Ni on zincated substrates prepared without and with

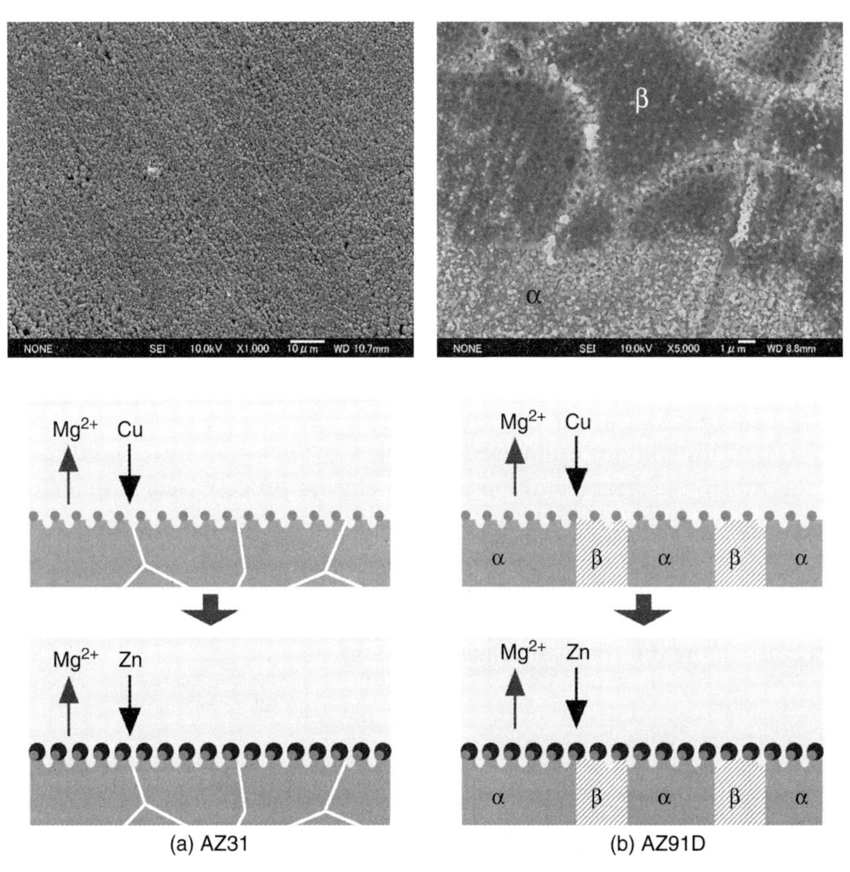

12.5 SEM images of (a) AZ31 and (b) AZ91 D alloys prepared with Cu pretreatment after zincate treatment for 1000 s.

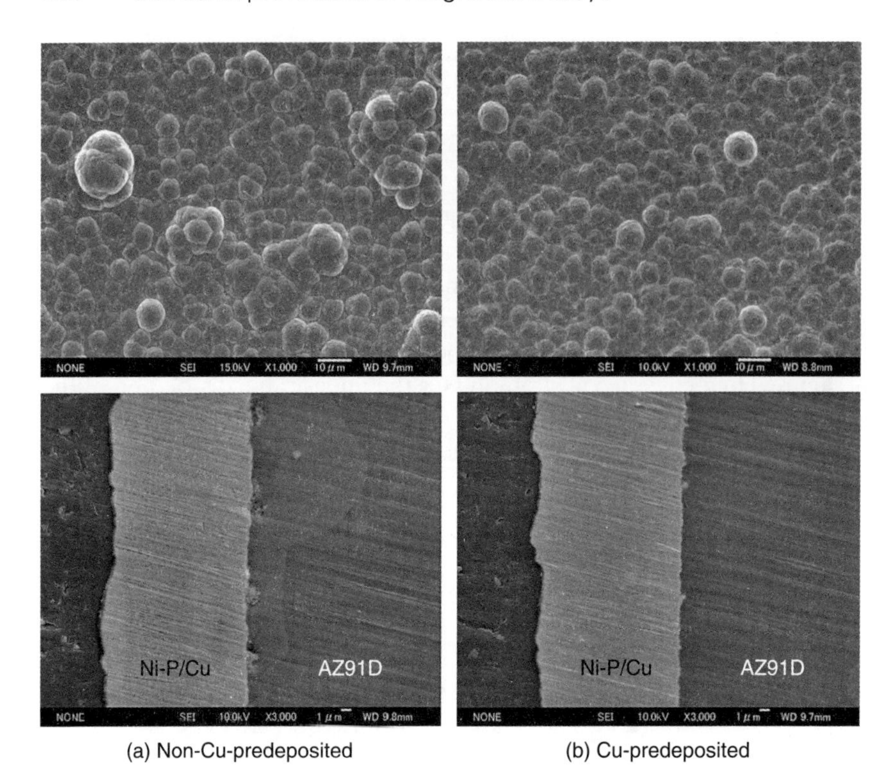

(a) Non-Cu-predeposited (b) Cu-predeposited

12.6 Surface and cross-sectional SEM images of AZ91D after Ni electrodeposition on specimens zincated with (a) or without (b) Cu predeposition.

Cu predeposition. The sample without Cu predeposition shows irregular surface morphology and a defective interface between the plating layer and substrate. This non-uniformity was caused by the non-uniform conditioning of the substrate surface in the preceding zincate process. Such a defective interface also resulted in relatively low adhesivity of the plating layer to the substrate, as seen in Table 12.2 in which results of pull-off tests of the plating films are summarized. On the other hand, the Cu-predeposited sample shows improvements in uniformity of the surface morphology and coherent structure of the interface as well as adhesion strength.

12.2.3 Protection of the Mg substrate in the electroplating process by using undercoating

After the zincating process, the specimen was subjected to EP of Ni. Since the traditional Ni plating bath is acidic or neutral, the Mg substrate may be damaged during the plating process even if the surface is covered with Zn

Table 12.2 Adhesion strength of the plating layer deposited on Mg alloys measured by a pull-off test. Values are averaged data of three runs for each condition

#	Substrate	Cu predeposition	Zincate	Undercoat	Coating	Polarization condition[a]	Adhesion/ MPa
1	AZ31		Yes		Cu	GP	9
2	AZ31		Yes		Cu	GP	9
3	AZ31	Yes	Yes		Cu	GP	14.5
4	AZ91D					Im	10
5	AZ91D		Yes			Im	9
6	AZ91D		Yes		Cu	GP	9
7	AZ91D		Yes		Cu	GP	7
8	AZ91D	Yes		Cu	Ni	GP	7
9	AZ91D	Yes	Yes	Cu	Ni	GP	13
10	AZ91D	Yes	Yes		Al	GP	2–3
11	AZ91D	Yes	Yes		Al	BCP[b]	>11

[a] Im: immersion, GP: galvanostatic polarization, BCP: bipolar current pulse polarization.
[b] $I_C = -25$ mA/cm^2, $I_A = 0.5$ mA/cm^2, $t_C = 100$ ms, $t_A = 400$ ms.

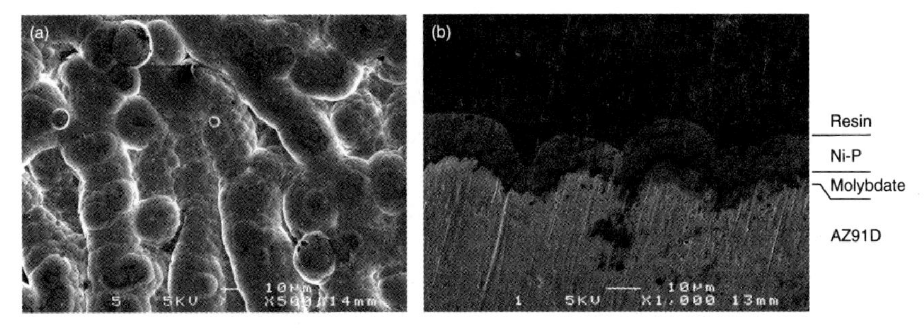

12.7 (a) SEM images of AZ91D after electroless Ni-P plating for 3.6 ks on the substrate covered with a molybdate coating. (b) Cross-sectional SEM image.

deposits. For protection of the substrate, a Cu precoat was electrodeposited in an alkaline bath at pH 13.5 prior to the Ni coating.[56] Other kinds of undercoats such as an oxide film[33,48–50] and chemical conversion films[53,62] were also examined. Figure 12.7 shows an example of a molybdate chemical conversion film used as an undercoat for EP of the Ni-P layer.[53] The advantages of this undercoating are that (a) strong adhesion between the conversion coating and substrate is provided, (b) the porous surface of the conversion film enables strong adhesion between the EP layer and conversion undercoat and (c) the formation of a convexo-concave substrate induced by non-uniform dissolution of the substrate in the conversion process is suitable for further coating such as coating with paint. These attempts indicate the possibility of new functions of Mg coatings using multilayer coatings.

12.3 Electroplating of Al from a non-aqueous plating bath

As mentioned above, Ni or Cu could be successively plated on the Mg alloy by appropriate preparation of the substrate and optimization of each process. From the viewpoint of corrosion engineering, however, corrosion-protective coating using noble metals such as Ni and Cu has a potential risk of galvanic coupling corrosion of the substrate, that is, the coating film of a noble metal over a large area may couple with the less-noble substrate in a small area at a pinhole in the coating, resulting in severe local corrosion of the substrate under the coating. To avoid such a risk, corrosion-protective coating with less-noble metals such as Al and Ti is desirable. Despite the less-noble properties of Al and Ti, chemically stable and mechanically strong oxide films form on them, providing very good corrosion protection. Since the redox potential of these metals is less noble than the hydrogen evolution potential that causes water decomposition, a non-aqueous bath is required for electrodeposition of these metals. Ionic liquid has suitable properties for a plating bath for these metals including a wide potential window, stability at a high temperature, high electric conductivity and the ability to dissolve metal ions at high concentrations. There have been many studies in which an IL bath such as 1-ethyl-3-methylimidazolium chloride (EMIC) containing aluminum chloride ($AlCl_3$) was used for Al EP on Mg alloys.[26–32] Although it was not clearly stated in those reports, the adhesion strength of an Al coating directly deposited on the Mg substrate is not usually good because of degradation of the substrate surface in the coating processes. Zincate pretreatment combined with Cu predeposition described in the previous subsection was therefore used for substrate conditioning of Mg alloys for Al EP in an EMIC IL bath.

The polarization method for EP in an IL bath was optimized as follows. Figure 12.8 shows typical polarization methods used in this study: (a) galvanostatic polarization (GP), (b) monopolar current pulse polarization (MCP) and bipolar current pulse polarization (BCP). The pulse polarization method is as effective for EP in an IL bath as in an aqueous bath.[63–67] In Fig. 12.8, a Pt working electrode (WE) was used instead of the Mg alloy in order to clarify the basic electrochemical response of the Al electrodeposition reaction.[66] Potential response of the WE in each method was also measured against an Al wire reference electrode (RE).

In the case of GP shown in Fig. 12.8a, initiation of polarization caused a sudden potential drop, short recovery in a noble direction, gradual shift in a less-noble direction and almost stable potential. These potential responses corresponded to cathodic polarization of the WE, nucleation and growth of Al deposits, continuous deposition of Al and expansion of the diffusion layer of Al^{3+} ions, respectively. Growth of the diffusion layer can lead to the

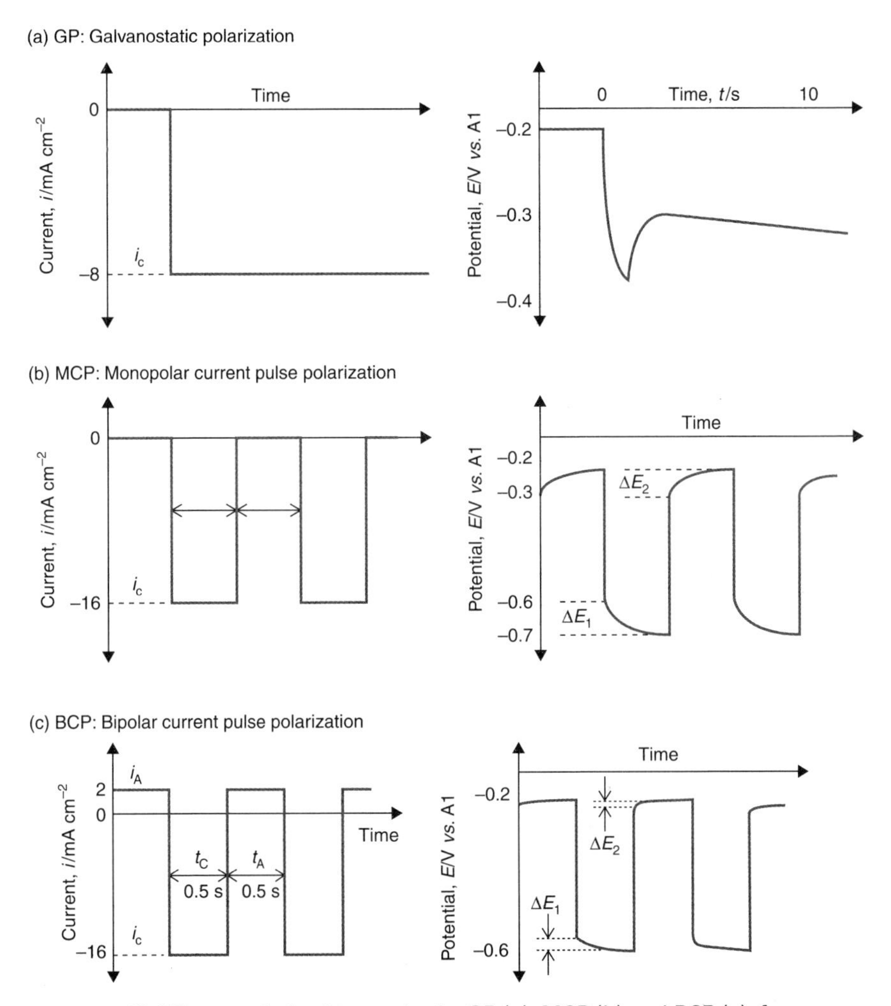

12.8 Three polarization methods, GP (a), MCP (b) and BCP (c), for electrodeposition of Al, and examples of their current waveform and potential response. Pt working electrode, Al counter electrode and Al reference electrode were used in these experiments.

following phenomena: one is formation of porous or dendrite deposits due to insufficient supply of Al(III) ions from the bulk of the bath, as shown in Fig. 12.9a; another is increase of nucleation density due to an increase of overpotential for Al deposition caused by depletion of Al(III) ions. This means that the morphology of the deposits changes with current density. Suitable conditions for Al deposition using the GP method were therefore investigated experimentally.

Current pulse polarization is used to moderate the depression of Al(III) ions near the substrate surface by superimposing the rest period to the GP,

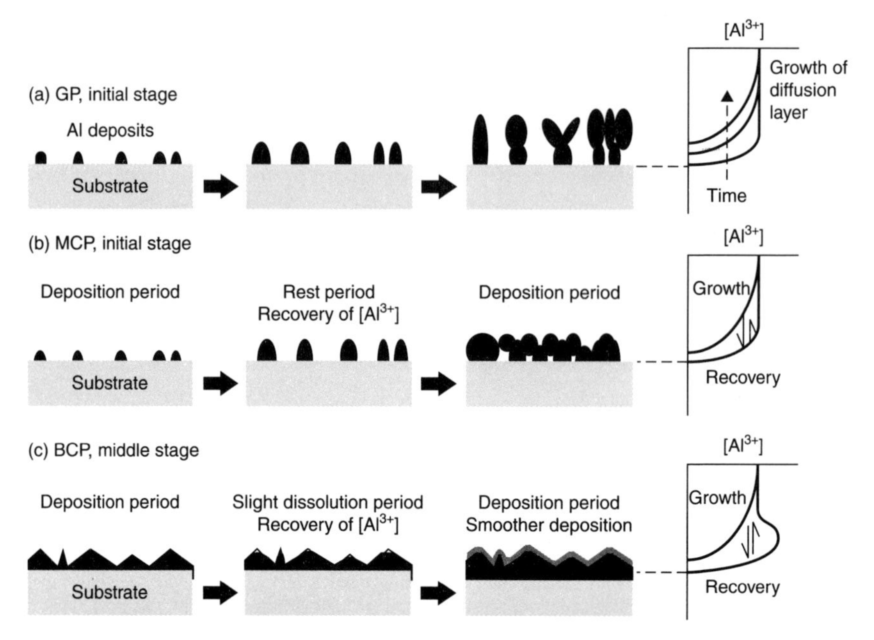

12.9 Relationship between the growth of Al electrodeposits and distribution of the diffusion layer of Al^{3+} ions near the surface during GP (a), MCP (b) and BCP (c) in an [EMIm]Cl/$AlCl_3$ bath.

as shown in Fig. 12.8b. The potential response in the cathodic polarization period, t_C, shows a sudden shift in a less-noble direction at the initial stage and a subsequent gradual increase in overpotential due to growth of the diffusion layer of Al(III) ions. The potential response in the rest period, t_{Off}, shows a sudden shift and a subsequent gradual shift in a noble direction corresponding to recovery of the surface concentration of Al(III) ions by supply from the bulk of the bath. The potential difference, ΔE_1 and ΔE_2, shown in the figure therefore reflects the growth and decrease of the diffusion layer of Al(III) ions. Recovery of Al(III) ion concentration near the surface aids the uniform deposition of Al in the polarization period of the next pulse, as shown in Fig. 12.9b. Improvement in uniformity of deposits depends considerably on pulse conditions such as deposition current density, i_C, and the periods of t_C and t_{Off} (or frequency and duty ratio of them). The potential of Al deposits in the rest period was less-noble than that of the Al RE by ~ 0.2 V, indicating a less stable (or more active) property of Al deposits than that of the Al plate because of their small particle form.

Bipolar current pulse (BCP) is used to achieve further improvement in uniform deposition. In this pulse form, a slight anodic polarization period, t_A, is periodically superimposed on the cathodic polarization, instead of the

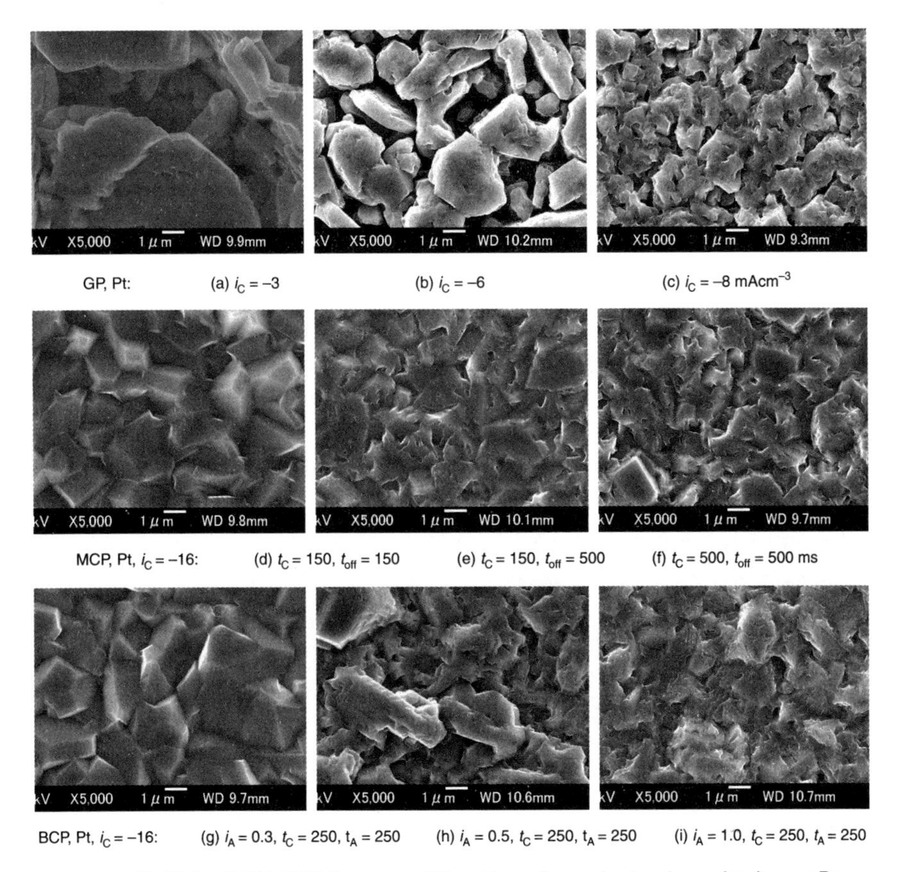

12.10 (a–i) FE-SEM images of the Al surface electrodeposited on a Pt plate from an [EMIm]Cl/$AlCl_3$ bath under various conditions. i_C and i_A are cathodic and anodic current densities [mA/cm^2], respectively, and t_C and t_A are cathodic and anodic polarization periods [ms], respectively.

rest period in the MCP. This anodic oxidation causes slight dissolution of the Al deposits mainly at the active sites such as the edges or prominent parts of deposits and increases the concentration of Al(III) ions near the surface, as shown in Fig. 12.9c. The increase of Al(III) ions is expected to suppress preferential deposition on the active sites and supply sufficient Al(III) ions to enable uniform deposition during the following deposition period. The small potential shift of ΔE_1 in the deposition period, t_C, in the BCP shown in Fig. 12.8c confirms suppression of the growth of the diffusion layer.

SEM images of the Al deposits on Pt using three current forms are shown in Fig. 12.10. In the case of GP, Al deposits were composed of large planar particles at a low cathodic current density due to aging of the crystal, and the size of particles decreased with an increase in current density as shown in Fig. 12.10a–12.10c due to increase in nucleation density. In the case of

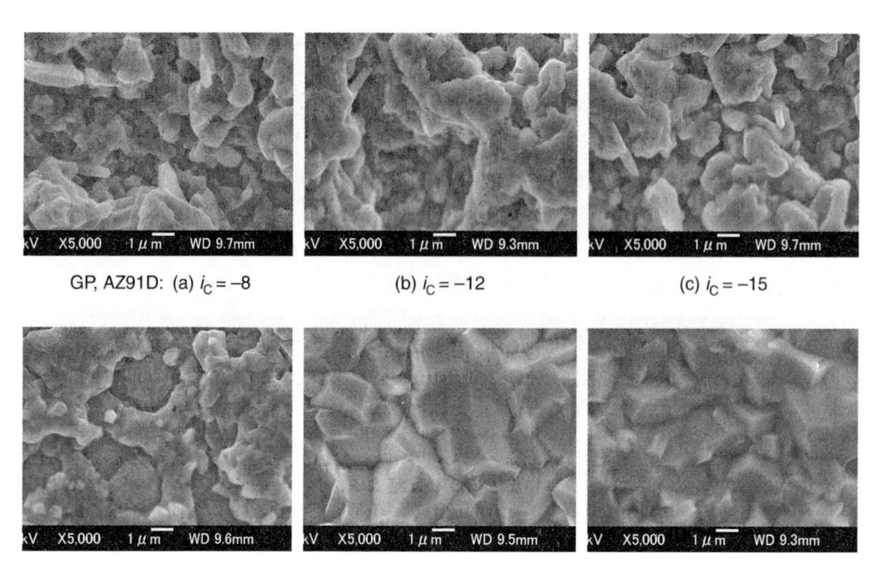

12.11 (a–f) FE-SEM images of the Al surface electrodeposited on zincated AZ91D from the [EMIm]Cl/$AlCl_3$ bath under various conditions. i_C and i_A are cathodic and anodic current densities [mA/cm^2], respectively, and t_C and t_A are cathodic and anodic polarization periods [ms], respectively.

MCP, Al deposits became denser compared with GP deposits. Flatness of the deposits was improved when the rest period was increased, as seen in Fig. 12.10e. When the deposition period was increased, however, the flatness was lessened probably due to growth of the diffusion layer of Al(III) ions as seen in Fig. 12.10f. Al deposits formed with BCP show morphology similar to that of MCP deposits at a low anodic current density, i_A, as seen in Fig. 12.10g. When i_A increased, the crystalline form of Al particles was distorted due to anodic dissolution and the whole surface was flattened due to uniform deposition, as shown in Fig. 12.10h and 12.10i. These results indicate that pulse polarization methods can reduce porosity and improve density and flatness of the electrodeposits using a suitable current waveform.

Using the information on Al electrodeposition obtained for Pt electrodes, Al was then electrodeposited on AZ91D prepared with Cu predeposition and zincate pretreatment using GP and BCP methods in an IL bath. SEM images of Al deposits using the GP method shown in Fig. 12.11a–12.11c indicate that the deposits are more porous than those on the Pt substrate shown in Fig. 12.10c even at a higher deposition current density. The discrepancy of porosity of Al deposits on AZ91D and Pt substrates is probably due to lower nucleation density of Al electrodeposition on AZ91D alloy than on Pt. The BCP coating deposited on AZ91D shown in Fig. 12.11d also shows a porous structure compared with the deposits on Pt under the same

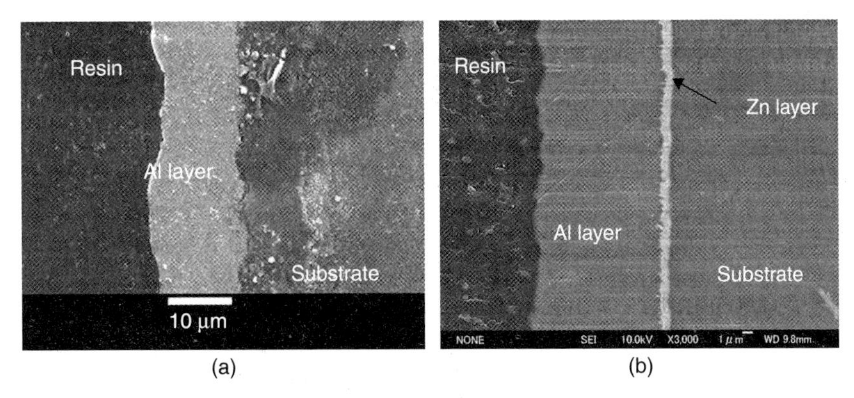

(a) (b)

12.12 Cross-sectional SEM images of the coatings shown in (a) Fig. 12.11(c) and (b) Fig. 12.11(f).

deposition conditions shown in Fig. 12.10h. In this case, increase in deposition current density or increase in anodic polarization period provided higher density as seen in Fig. 12.11e and 12.11f.

Cross-sectional SEM images of some samples are shown in Fig. 12.12. The interface between the GP coating and substrate is uneven and the coating contains defects or voids, as seen in Fig. 12.12a. For the BCP coating, a thin and uniform Zn layer is clearly observed between the AZ91D substrate as seen in Fig. 12.12b and the Al coating layer and no defect is seen. Adhesion strength of BCP coatings was considerably higher than that of GP coatings as shown in Table 12.2. Since detachment of the dolly from the substrate used in the pull-off test always occurred at the interface between the BCP coating surface and adhesive epoxy resin, the actual adhesion strength of the BCP coating to the substrate is fairly high. Such strong adhesion is clearly provided by the low defective interface structure formed by the combination of Cu predeposition, appropriate zincating and optimized BCP methods. Although the adhesion strength of electroplated Al-coatings shown here or Ni coatings shown in the previous subsection looks sufficiently high, it is still lower however than, for example, Ni-P coatings plated on Al alloys. Ni or Al coating can peel off from the edge of the specimen or from the scratches of the coatings, while a well-plated Ni-P coating on Al alloy never peels off. As suggested by previous studies[30,31,55,59] post-treatment such as low temperature heat treatment seems to be promising for achieving further improvement of adhesion strength.

One of the criteria for evaluation of coating quality is corrosion protection performance. An Al coating electrodeposited on the AZ91D alloy was therefore evaluated by potentiodynamic polarization in 3.5 wt.% NaCl aqueous solution as shown in Fig. 12.13. The large anodic polarization current of bare AZ91D confirms its susceptibility to corrosion. The GP coating deposited on zincated AZ91D shows a low corrosion resistance due to the

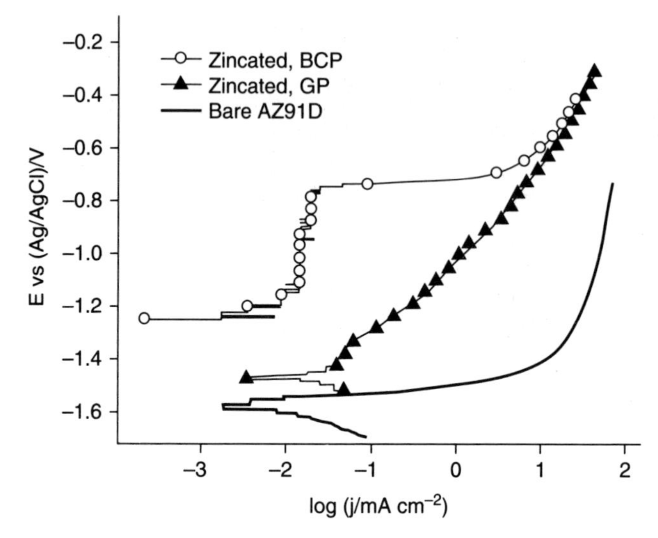

12.13 Potentiodynamic polarization curves of Al-coated AZ91D prepared with and without zincate pretreatment by using BCP and GP methods in 3.5 wt.% NaCl solution. BCP: $I_C = -25$ mA/cm², $I_A = 0.5$ mA/cm², $i_C = -25$, $t_C = 100$ ms, $t_A = 400$ ms; GP: $I_C = -8$ mA/cm².

rather porous structure of the coating as seen in Fig. 12.11a. The BCP coating shows a wide passive potential region as found for pure Al, indicating good corrosion resistance. Improvement in corrosion resistance of the BCP coating is clearly provided by the dense structure of the BCP coating as seen in Fig. 12.11f. Electrochemical voltammogram has been generally used to evaluate the corrosion protection performance of the coatings, although a Mg alloy is generally not used in an aqueous environment containing NaCl. Corrosion tests suitable for the actual usage of coatings on Mg alloys such as wet and dry cycling tests and exposure tests to the actual environment seem to be better for evaluation of their corrosion protection ability.

12.4 Conclusions

Development of superior coatings for Mg alloys is important for the practical use of Mg alloys. Despite the difficulty in obtaining high-quality coatings by using EP or EL methods on Mg alloys, persistent efforts in fundamental research in this field have been clarifying suitable surface preparation, plating bath composition and optimum operating conditions. In all individual processes for coating including picking, activation, pre-coating and EP or EL, not only must the intended reactions proceed smoothly but also suppression of any corrosion reaction or uniformization of the surface condition of the Mg alloy substrate must always be considered. Achievement of defect-less coating seems to be important for corrosion protection of the Mg alloy substrate. This approach may not, however, be effective in some practical situations because partial damage to the coating can cause propagation of severe corrosion from this damage due to formation of galvanic coupling between the substrate anode and plating layer cathode. From this viewpoint, coatings of less-noble and corrosion-protective metals such as Al and Ti from a non-aqueous bath seem to be useful. However, the use of expensive IL chemicals may cause economical problems in a practical production line. From this standpoint, a molten salt bath is more attractive because of its low cost and easiness in operation. In this bath, however, high corrosion activity of the Mg substrate at a high temperature should be controlled. Other approaches such as electro- or electroless metal coatings on an oxide layer or organic undercoating layer and a composite with functional elements have also been attempted. Development of a self-healing coating or undercoating by using, for example, conducting polymers such as polypyrrole is one of the interesting trial, although it is still at the fundamental stage of research.[68,69] In conclusion development of the surface protection methods for Mg alloys using various coating technologies is on the road to their establishment and many possibilities are waiting for further investigation.

12.5 References

1. Gray J.E. and Luan B., 'Protective coatings on magnesium and its alloys — a critical review', *Journal of Alloys and Compounds* (2002), **336**, 88–113.
2. MahAllawy N. El., 'Surface treatment of magnesium alloys by electroless Ni –P plating technique with emphasis on zinc pre-treatment: a review', *Key Engineering Materials* (2008), **384**, 241–262.
3. Czerwinski F., *Magnesium Alloys – Corrosion and Surface Treatments*. 1st ed. India: InTech (2011).
4. Zhang J. and Wu C., 'Corrosion and protection of magnesium alloys – a review of the patent literature', *Recent Patents on Corrosion Science* (2010), **2**, 55–68.
5. Pourbiax M., *Atlas of Electrochemical Equilibria in Aqueous Solutions*. 2nd ed. Houston: NACE (1974).
6. Elsentriecy H.H., Azumi K. and Konno H., 'Effect of surface pretreatment by acid pickling on the density of stannate conversion coatings formed on AZ91 D magnesium alloy', *Surface and Coatings Technology* (2007), **202**, 532–537.
7. Elsentriecy H.H., Azumi K. and Konno H., 'Improvement in stannate chemical conversion coatings on AZ91 D magnesium alloy using the potentiostatic technique', *Electrochimica Acta* (2007), **53**, 1006–1012.
8. Elsentriecy H.H., Azumi K. and Konno H., 'Effects of pH and temperature on the deposition properties of stannate chemical conversion coatings formed by the potentiostatic technique on AZ91 D magnesium alloy', *Electrochimica Acta* (2008), **53**, 4267–4275.
9. Chen X.B., Birbilis N. and Abbott T.B., 'Review of corrosion-resistant conversion coatings for magnesium and its alloys', *Corrosion (NACE)* (2011), **67**, 035005-1-035005-16.

10. Takaya M., 'Recent tendency of surface treatment technology for magnesium', *Journal of the Japan Institute of Light Metals* (2000), **50**, 567–576.
11. Takatani Y., Tamai R., Yamamoto A. and Tsubakino H., 'Corrosion behavior of an AZ31 magnesium alloy with deposited high purity magnesium film', *Journal of the Japan Institute of Light Metals* (2005), **55**, 357–362.
12. Song G., Atrens A., Wu X. and Zhang B., 'Corrosion Behaviour of AZ21, AZ501 and AZ91 in Sodium Chloride', *Corrosion Science* (1998), **40**, 1769–1791.
13. Baril G., Blanc C., and Pebere N., 'AC impedance spectroscopy in characterizing time-dependent corrosion of AZ91 and AM50 magnesium alloys characterization with respect to their microstructures', *Journal of the Electrochemical Society* (2001), **148**, B489–B493.
14. Ambat R. and Zhou W., 'Electroless nickel-plating on AZ91D magnesium alloy: effect of substrate microstructure and plating parameters', *Surface and Coatings Technology* (2004), **179**, 124–134.
15. Ballerini G.,. Bardi U, Bignucolo R. and Ceraolo G., 'About some corrosion mechanisms of AZ91D magnesium alloy', *Corrosion Science* (2005), **47**, 2173–2184.
16. Armyanov S., Vangelova T. and Stoyanchev R., 'Pretreatment of Al-Mg alloys for electrodeposition by immersion zinc and electroless nickel', *Surface Technology* (1982), **17**, 89–100.
17. Dennis J. K., Wan M. K. Y. Y. and Wake S. J., 'Plating on magnesium alloy diecastings', *Transactions of the Institute of Metal Finishing* (1985), **63**, 74–80.
18. Dennis J. K., Wan M. K. Y. Y. and Wake S., 'Plating on magnesium alloy die castings – the role of fluoride and pH', *Transactions of the Institute of Metal Finishing* (1985), **63**, 81–84.
19. Zhang H., Wang S., Yao G. and Hua Z., 'Electroless Ni–P plating on Mg–10Li–1Zn alloy', *Journal of Alloys and Compounds* (2009), **474**, 306–310.
20. Zhang Z., Yu G., Ouyang Y., He X., Hu B., Zhang J., Wu Z., 'Studies on influence of zinc immersion and fluoride on nickel electroplating on magnesium alloy AZ91D', *Applied Surface Science* (2009), **255**, 7773–7779.
21. Jin J., Liu C., Fu S., Gao Y. and Shu X., 'Electroless Ni-P plating on Mg-10Gd-4.8Y-0.6Zr magnesium alloy with a new pretreatment process', *Surface and Coatings Technology* (2011), **206**, 348–353.
22. Jiang Y.F., Liu L.F., Zhai C.Q., Zhu Y.P. and Ding W.J., 'Corrosion behavior of pulse-plated Zn–Ni alloy coatings on AZ91 magnesium alloy in alkaline solutions', *Thin Solid Films* (2005), **484**, 232–237.
23. Huang C.A., Wang T.H., Weirich T. and Neubert V., 'A pretreatment with galvanostatic etching for copper electrodeposition on pure magnesium and magnesium alloys in an alkaline copper-sulfate bath', *Electrochimica Acta* (2008), **53**, 7235–7241.
24. Zhang W.X., He J.G., Jiang Z.H., Jiang Q. and Lian J.S., 'Electroless Ni–P layer with a chromium-free pretreatment on AZ91D magnesium alloy', *Surface and Coatings Technology* (2007), **201**, 4594–4600.
25. Lei X., Yu G., Gao X., Ye L., Zhang J. and Hu B., 'A study of chromium-free pickling process before electroless Ni–P plating on magnesium alloys', *Surface and Coatings Technology* (2011), **205**, 4058–4063.
26. Bakkar A. and Neubert V., 'Electrodeposition onto magnesium in air and water stable ionic liquids: From corrosion to successful plating', *Electrochemistry Communications* (2007), **9**, 2428–2435.

27. Chang J.-K., Chen S.-Y., Tsai W.-T., Deng M.-J and Sun I-W., 'Electrodeposition of aluminum on magnesium alloy in aluminum chloride (AlCl3)-1-ethyl-3-m ethylimidazolium chloride (EMIC) ionic liquid and its corrosion behavior', *Electrochemistry Communications* (2007), **9**, 1602–1606.
28. Chang J.K., Chen S.Y., Tsai W.T., Deng M.J. and Sun I.W., 'Improved corrosion resistance of magnesium alloy with a surface aluminum coating electrodeposited in ionic liquid', *Journal of the Electrochemical Society* (2008), **155**, C112–C116.
29. Pan S.-J., Tsaia W.T., Changa J.-K. and Sun I.-W., 'Co-deposition of Al–Zn on AZ91D magnesium alloy in $AlCl_3$-1-ethyl-3-methylimidazolium chloride ionic liquid', *Electrochimica Acta* (2010), **55**, 2158–2162.
30. Chuang M., Chang J., Tsai P., Tsai W., Deng M. and Sun I., 'Heat-treatment induced material property variations of Al-coated Mg alloy prepared in aluminum chloride/1-ethyl-3-methylimidazolium chloride ionic liquid', *Surface and Coatings Technology* (2010), **205**, 200–204.
31. Yang H.Y., Guo X.W., Wu G.H., Ding W.J. and Birbilis N., 'Electrodeposition of chemically and mechanically protective Al-coatings on AZ91D Mg alloy', *Corrosion Science* (2011), **53**, 381–387.
32. Kui-ren L., Quan L., Qing H. and Gan-feng T., 'Electrodeposition of Al on AZ31 magnesium alloy in TMPAC-$AlCl_3$ ionic liquids', *Transactions of Nonferrous Metals Society China* (2011), **21**, 2104–2110.
33. Wang X.M., Zhu L.Q., Liu H.C. and Li, W.P. 'Influence of surface pretreatment on the anodizing film of Mg alloy and the mechanism of the ultrasound during the pretreatment', *Surface and Coatings Technology* (2008), **202**, 4210–4217.
34. Sheng M., Wang C., Zhong Q., Wei Y. and Wang Y., 'Ultrasonic irradiation and its application for improving the corrosion resistance of phosphate coatings on aluminum alloys', *Ultrasonics Sonochemistry* (2010), **17**, 21–25.
35. Yang L., Luan B., Cheong W.-J, Shoesmith D., 'Sono-immersion deposition on magnesium alloy', *Journal of the Electrochemical Society* (2005), **152**, C131-C136.
36. Jiang Y.F., Zhai C.Q., Liu L.F., Zhu Y.P. and Ding W.J., 'Zn-Ni alloy coatings pulse-plated on magnesium alloy', *Surface and Coatings Technology* (2005), **191**, 393–399.
37. Bahrololoom M.E. and Sani R., 'The influence of pulse plating parameters on the hardness and wear resistance of nickel–alumina composite coatings', *Surface and Coatings Technology* (2005), **192**, 154–163.
38. Huang C. A., Wang T.H., Weirich T. and Neubert V., 'Electrodeposition of a protective copper/nickel deposit on the magnesium alloy (AZ31)', *Corrosion Science* (2008), **50**, 1385–1390.
39. Zhang S., Cao F., Chang L., Zheng J., Zhang Z., Zhang J. and Cao C., 'Electrodeposition of high corrosion resistance Cu/Ni–P coating on AZ91D magnesium alloy', *Applied Surface Science* (2011), **257**, 9213–9220.
40. Zhang W.X., Jiang Z.H., Li G.Y., Jiang Q. and Lian J.S., 'Electroless Ni-P/Ni-B duplex coatings for improving the hardness and the corrosion resistance of AZ91D magnesium alloy', *Applied Surface Science* (2008), **254**, 4949–4955.
41. Zhang S., Li Q., Yang X., Zhong X., Dai Y. and Luo F., 'Corrosion resistance of AZ91D magnesium alloy with electroless plating pretreatment and Ni–TiO_2 composite coating', *Materials Characterization* (2010), **61**, 269–276.

42. Karwas C. and Hepel T., 'Morphology and composition of electrodeposited cobalt-zinc alloys and the influence of boric acid', *Journal of the Electrochemical Society* (1989), **136**, 1672–1678.
43. Zhang W.X., Huang N., He J.G., Jiang Z.H., Jiang Q. and Lian J.S., Electroless deposition of Ni–W–P coating on AZ91D magnesium alloy', *Applied Surface Science* (2007), **253**, 5116–5121.
44. Zhang W.X., Jiang Z.H., Li G.Y., Jiang Q. and Lian J.S., 'Electroless Ni–Sn–P coating on AZ91D magnesium alloy and its corrosion resistance', *Surface & Coatings Technology* (2008), **202**, 2570–2576.
45. Tsuda T., Arimoto S., Kuwabata S. and Hussey C.L., 'Electrodeposition of Al–Mo–Ti ternary alloys in the Lewis acidic aluminum chloride–1-ethyl-3-methylimidazolium chloride room-temperature ionic liquid', *Journal of the Electrochemical Society* (2008), **155**, D256–D262.
46. Tsuda T. and Hussey C.L., 'Electrodeposition of photocatalytic AlInSb semi-conductor alloys in the Lewis acidic aluminum chloride–1-ethyl-3-methylimid azolium chloride room-temperature ionic liquid', *Thin Solid Films* (2008), **516**, 6220–6225.
47. Zhang J., Zhang W., Yan C., Du K. and Wang F., 'Corrosion behaviors of Zn/Al–Mn alloy composite coatings deposited on magnesium alloy AZ31B (Mg–Al–Zn)', *Electrochimica Acta* (2009), **55**, 560–571.
48. Liu Z. and Gao W., 'A novel process of electroless Ni-P plating with plasma electrolytic oxidation pretreatment', *Applied Surface Science* (2006), **253**, 2988–2991.
49. Sun S., Liu J., Yan C. and Wang F., 'A novel process for electroless nickel plating on anodized magnesium alloy', *Applied Surface Science*, (2008), **254**, 5016–5022.
50. Zeng L., Yang S., Zhang W., Guo Y. and Yan C., 'Preparation and characterization of a double-layer coating on magnesium alloy AZ91D', *Electrochimica Acta* (2010), **55**, 3376–3383.
51. Zhao H., Huang Z. and Cui J., 'A novel method of electroless plating on AZ31 magnesium alloy sheet', *Journal of Materials Processing Technology* (2008), **203**, 310–314.
52. Zhao H., Huang Z. and Cui J., 'A new method for electroless Ni–P plating on AZ31 magnesium alloy', *Surface and Coatings Technology* (2007), **202**, 133–139.
53. Elsentriecy H.H. and Azumi K., 'Electroless Ni–P deposition on AZ91 D magnesium alloy prepared by molybdate chemical conversion coatings', *Journal of the Electrochemical Society* (2009), **156**, D70–D77.
54. Li J., Tian Y., Huang Z. and Zhang X., 'Studies of the porosity in electroless nickel deposits on magnesium alloy', *Applied Surface Science* (2006), **252**, 2839–2846.
55. Lian J.S., Li G.Y., Niu L.Y., Gu C.D., Jiang Z.H. and Jiang Q., 'Electroless Ni–P deposition plus zinc phosphate coating on AZ91D magnesium alloy', *Surface and Coatings Technology* (2006), **200**, 5956–5962.
56. Tang J. and Azumi K., 'Effect of copper pretreatment on the zincate process and subsequent electroplating of a protective copper/nickel deposit on the AZ91D magnesium alloy', *Electrochimica Acta* (2011), **56**, 8776–8782.
57. Tang J. and Azumi K., 'Influence of zincate pretreatment on adhesion strength of a copper electroplating layer on AZ91 D magnesium alloy', *Surface & Coatings Technology* (2011), **205**, 3050–3057.

58. Tang J. and Azumi K., 'Effect of copper pretreatment on the zincate process and subsequent copper electrodeposition of AZ31 magnesium alloy', *Journal of the Electrochemical Society* (2011), **158**, D535–D540.
59. Zhu L.Q., Li W.P. and Shan D.D., 'Effects of low temperature thermal treatment on zinc and/or tin plated coatings of AZ91D magnesium alloy', *Surface & Coatings Technology* (2006), **201**, 2768–2775.
60. Zhu Y.P., Yu G., Hu B. N., Lei X.P., Yi H.B. and Zhang J., 'Electrochemical behaviors of the magnesium alloy substrates in various pretreatment solutions', *Applied Surface Science* (2010), **256**, 2988–2994.
61. Azumi K., Egoshi S., Kawashima S. and Koyama Y., 'Effect of copper pretreatment on the double zincate process of aluminum alloy films', *Journal of the Electrochemical Society* (2007), **154**, D220–D226.
62. Cui X., Jin G., Li Q., Yang Y., Li Y. and Wang F., 'Electroless Ni–P plating with a phytic acid pretreatment on AZ91D magnesium alloy', *Materials Chemistry and Physics* (2010), **121**, 308–313.
63. Manoli G., Chryssoulakis Y. and Poignet J.C., 'Study of pulsed electrolytic deposition of aluminum onto aluminum, platinum and iron electrodes', *Plating and Surface Finishing* (1991), **78**, 64–69.
64. Yang C.C., 'Electrodeposition of aluminum in molten $AlCl_3$-n-butylpyridinium chloride electrolyte', *Materials Chemistry and Physics*, **37** (1994), 355–361.
65. Li B., Chen Y., Yan L.G. and Ma J.D., 'Pulse current electrodeposition of Al from an AlCl3-EMIC ionic liquid containing $NdCl_3$', *Electrochemistry* (2010), **78**, 523–525.
66. Tang J. and Azumi K., 'Optimization of pulsed electrodeposition of aluminum from $AlCl_3$-1-ethyl-3-methylimidazolium chloride ionic liquid', *Electrochimica Acta* (2011), **56**, 1130–1137.
67. Li B., Fan C.H., Chen Y., Lou J.W. and Yan L.G., 'Pulse current electrodeposition of Al from an $AlCl_3$-EMIC ionic liquid', *Electrochimica Acta* (2011), **56**, 5478–5482.
68. Jiang Y.F., Guo X.W., Wei Y.H., Zhai C.Q. and Ding W.J., 'Corrosion protection of polypyrrole electrodeposited on AZ91 magnesium alloys in alkaline solutions', *Synthetic Metals* (2003), **139**, 335–339.
69. Sheng N. and Ohtsuka T., 'Preparation of conducting poly-pyrrole layer on zinc coated Mg alloy of AZ91D for corrosion protection', *Progress in Organic Coatings* (2012), **75**, 59–64.

13
Electroless nickel-boron plating to improve the corrosion resistance of magnesium (Mg) alloys

Z.-C. WANG, L. YU and Z.-B. QI, Xiamen University, China
and G.-L. SONG, General Motors Corporation, USA

DOI: 10.1533/9780857098962.3.370

Abstract: This chapter briefly reviews current electroless plating techniques for Mg alloys. Among them, a recently developed Ni-B plating is particularly selected for further discussion. Following that, the Ni-B plating methodology is presented and its electroless plating mechanism is proposed. The deposited Ni-B layer on AZ91D is then systematically characterized for its microstructure, porosity, thickness, adhesion and corrosion resistance. Based on the plating performance and characteristics, some possible applications and future developments of this new plating are predicted.

Key words: magnesium alloy, electroless plating, Ni-B coating, corrosion.

13.1 Introduction

13.1.1 Current electroless plating techniques for Mg alloys

Magnesium and its alloys have many useful physical and mechanical properties, such as low density, high strength/weight ratio, wonderful castability and low heat capacity.[1–5] These properties make magnesium and its alloys ideal materials in many applications, including automotive and aerospace components, portable electronics and household equipment.[6–8] However, poor corrosion resistance has limited the application of magnesium and its alloys in more severe environments.[9,10] A practical approach to solving the corrosion problem is to coat Mg alloys with a barrier layer so that corrosion media will not be in direct contact with their surfaces.

To date various coating techniques have been developed for the corrosion protection of magnesium alloys, such as organic coatings,[8] chemical conversion coatings,[11] anodizing,[12] electroplating (EP) and electroless plating (EL).[13,14] Among these, EL has the unique advantage of forming a coating of uniform thickness even on an irregular surface simply by immersing the Mg alloy in an aqueous solution.[15–17] Therefore, it is a highly desirable metal coating and has been extensively studied on different Mg alloys.[5,18,19]

The electroless nickel plating is one of the most important such coatings on Mg alloys. It can be classified according to the reducing agent used in plating. Currently, the most widely studied is electroless nickel-phosphorous (Ni-P) plating which uses hypophosphates as reducing agents. The Ni-P plating can improve the wear- and corrosion resistance of Mg alloys due to the good corrosion resistance and high hardness of this Ni-P layer. Apart from the most popular electroless Ni-P plating, Ni-B plating has gradually gained some research interest recently. Currently, the electroless nickel-boron (Ni-B) plating with borohydrides as reducing agents is still far from as popular as Ni-P plating in practice, especially on magnesium alloys.[20] This is mainly because the highly efficient reducing agent borohydride is unstable in practical EL baths.[21–23] In addition, the high chemical reactivity of Mg alloys is also a problem with this plating process.[24–26] In operation, a uniform and protective film has to be deposited on magnesium alloy substrates to insulate the magnesium surface before EL. This often involves the use of hydrofluoric acid (HF),[17] palladium[19] or chromic compounds,[8] resulting in complex steps and highly toxic baths for this plating process. Nevertheless, the potential application of this new plating on Mg alloys is extremely attractive due to the unique properties offered by this coating.

13.1.2 Advantages of electroless Ni-B plating

Compared with the widely studied Ni-P plating, a Ni-B coating has some notable characteristics, such as superior bonding, low porosity, very good solderability, good conductivity, superior electromagnetic performance and hardness, as well as good wear- and corrosion resistance.[27–29] In addition, electroless Ni-B plating also provides a favorable alkaline bath environment for Mg alloys. For example, in our previous study,[5] the electroless Ni-B coating was found to exhibit a uniform surface, an amorphous structure, good corrosion resistance and good adhesion to the AZ91D substrate.

Previous investigations have also revealed that the traditional chromium oxide and HF pretreatment can be replaced with acetic acid in the pickling bath; the plating rate is affected by pH, temperature, and bath composition; different deposition reactions dominate in different stages of the plating process; the coating thickness, porosity and boron content of Ni-B coating are dependent on the EL bath composition; heat-treatment can lead to a transition of the plating from amorphous to crystalline and the precipitation of Ni, Ni_3B, Ni_2B and Ni_4B_3; and the corrosion resistance of Ni-B coating is influenced by both the coating thickness and boron content given the similar porosity. In this chapter, the electroless Ni-B plating process and performance

Table 13.1 Chemical composition of the AZ91D magnesium alloy (in wt.%)

Al	Mn	Zn	Si	Cu	Ni	Fe	Mg
9.1	0.33	0.91	0.04	0.018	0.0006	0.003	Bal

will be systematically summarized in the following sections. It is expected that the discussion regarding such an innovative plating technique will provide a basis for a better understanding of EL techniques for Mg alloys.

13.2 Electroless Ni-B plating process and deposition mechanisms

13.2.1 Plating process

In experiments, nickel-boron coatings were EL deposited on AZ91D (chemical composition shown in Table 13.1) in a round area with an average diameter of 10 mm. The deposited layer had an average thickness of 1mm. Before EL, the substrate AZ91D was treated with the following steps: mechanically polishing → alkaline cleaning → acid pickling.[5] The mechanical polishing was conducted using SiC papers (320# → 800# → 1500#). After being rinsed with distilled water, the samples were immersed in a 75°C alkaline solution containing Na_2CO_3, $Na_3PO_4 \cdot 12H_2O$ and polyoxyethylene octylphenyl ether for 10 min to remove the grease and dirt. Rinsed with distilled water again, the samples were dipped in an acid solution with CH_3COOH and $NaNO_3$ as the main components for less than 1 min. The oxides should all be dissolved during such an acid pickling process. Finally, the samples were immersed in a 85°C alkaline solution containing $Ni(CH_3COO)_2 \cdot 4H_2O$, $NaBH_4$, H_3BO_3, $H_2NCH_2CH_2NH_2$ for electroless Ni-B plating.

The detailed bath compositions, procedures and the operation conditions used in the cleaning, pickling and plating process are shown in Table 13.2. It should be noted that no surface activation step was involved in the above processes. In alkaline cleaning bath, polyoxyethylene octylphenyl ether was used as emulsifier for better degreasing of the samples. Sodium carbonate and sodium phosphate were used as pH adjusters to control the bath pH to a value higher than 12. In acid pickling solution, acetic acid was selected to pickle the sample surface. It is a relatively weak organic acid, but it can remove the impurities and the oxidation film on the magnesium alloy surface, and also etches the surface to improve coating adhesion.[30] To prevent excessive corrosion of the substrate, sodium nitrate was added as an inorganic corrosion inhibitor.

In the EL bath, nickel acetate was selected as the source of nickel ions as it can prevent over-corrosion of magnesium alloy. Sodium borohydride was

Table 13.2 Bath compositions and operating conditions for direct electroless Ni-B plating on AZ91D magnesium alloy

Procedures	Bath compositions		Conditions
1. Alkaline cleaning	Na_2CO_3	20 g/L	75°C
	$Na_3PO_4 \cdot 12H_2O$	20 g/L	10 min
	Polyoxyethylene octylphenyl ether	10 mL/L	
(Rinse with distilled water)			
2. Acid pickling	CH_3COOH	20 mL/L	25°C
	$NaNO_3$	40 g/L	30–60 s
(Rinse with distilled water)			
3. Direct electroless Ni-B plating	$Ni(CH_3COO)_2 \cdot 4H_2O$	38 g/L	85°C
	$H_2NCH_2CH_2NH_2$	50 mL/L	
	NaOH	36 g/L	
	$NaBH_4$	0.6 g/L	
	H_3BO_3	10 g/L	
	Toluene-p-sulphonic acid sodium salt	5 g/L	
	1,2-propanediol	2 g/L	
	5-Sulphosalicylic acid dihydrate	50 mg/L	
	pH	13.4	

used as a reducing agent for reduction of nickel ions on the Mg alloy surface. Sodium hydroxide created a strong alkali environment for stabilizing the reducing agent. Ethylenediamine, which is a complexing agent, controlled the content of free nickel ions in the bath. Boric acid can reduce the pH difference between the solution adjacent to the electrode surface and the bulk plating bath, and therefore suppressed hydrogen evolution, which is beneficial to the corrosion resistance of the deposited coating.[31] To improve the brightness of the coating, toluene-p-sulphonic acid sodium salt, propandioic acid and 5-sulphosalicylic acid dehydrate were added together in the bath. All the solutions were prepared using analytically pure chemical reagents and distilled water. The samples were thoroughly washed with distilled water as quickly as possible between any two steps in the technical flow.

13.2.2 Deposition mechanisms

According to the literature,[32] during electroless nickel-boron plating, the following reaction occurs:

$$4Ni^{2+} + 8BH_4^- + 18H_2O \rightarrow 2Ni_2B + 6H_3BO_3 + 25H_2 \uparrow \qquad [13.1]$$

Reaction [13.1] takes place at a high nickel ion concentration. At the same time, a side-reaction occurs, also releasing a small amount of hydrogen:

$$4Ni^{2+} + 2BH_4^- + 6OH^- \rightarrow 2Ni_2B + 6H_2O + H_2 \uparrow \quad [13.2]$$

Moreover, an electroless Ni-B plating mechanism[20,33] on mild steel has been proposed, which may also be applicable on a Mg alloy. The plating reaction mainly forms pure nickel at high pH:

$$4Ni^{2+} + BH_4^- + 8OH^- \rightarrow 4Ni + BO_2^- + 6H_2O \quad [13.3]$$

In addition, the hydrolysis of borohydride compounds should also be considered, which can be exacerbated if solution alkalinity declines:

$$2BH_4^- + 2H_2O \rightarrow 2B + 2OH^- + 5H_2 \uparrow \quad [13.4]$$

Therefore, an electroless Ni-B plating process may involve the above 4 possible reactions.

According to reactions [13.1] and [13.2], Ni_2B was the main product, where the mass ratio of Ni to B should be 9.2:1. However, the experimental results give a ratio of 5:1, much less than the expected value. The boron content near the substrate is higher than that in the plating, which has not been reported previously.[20,33] This is perhaps because different substrate materials were used in the studies. In fact, the substrate can strongly affect the deposition process.[34] After pretreatment, magnesium alloy substrate has a high activity. In the initial stage of plating, the replacement reaction [13.3] consumes a great deal of nickel ions, and the Mg surface is covered by Ni nuclei to some degree. Hereafter, the EL reaction [13.3] proceeds and leads to a reduced pH value. As a result, hydrolysis reaction [13.4] is triggered to form boron atoms in large numbers. The consumption of borohydride ions will gradually retard reaction [13.4]. The ion diffusion and the scouring effect of hydrogen evolution partially restore the concentration of nickel ions and the pH. In summary, the hydrolysis reaction of borohydride ions (Reaction [13.4]) dominates in the early stage. Subsequently, reactions [13.3] and [13.4] proceed at steady rates, leading to a stable ratio of Ni to B in the EL.

As far as the EL process is concerned, there are three simultaneous crystallization processes: nucleation, growth and coalescence of the three-dimensional crystallites.[30] Generally speaking, the exposed Mg surface possesses a large number of active sites for rapid replacement and nucleation reactions, forming small nickel nuclei from edge to centre of the plating pores. Meanwhile, deposited boron atoms can also impede the growth of nickel particles. Boron can be adsorbed to the Ni surface and promote the formation of new nuclei, leading to a microcrystal plating structure.[35]

On a Mg alloy surface, apart from the Ni-B covered areas, there is Mg hydroxide covered area and film-free fresh metal surface area. The former is relatively protruding over the latter. The fresh metallic magnesium along the magnesia film pore edges is favorable for nucleation. The growth of existing Ni nuclei, rather than nucleation, occurs predominantly to exhibit relative large nodular nickel. Finally, nickel particles cover the magnesia completely. Therefore, in effect the EL tends to form a flat surface in the initial stage.

During plating, corrosion of the substrate can also take place, but the corrosion attack to the substrate is relatively localized,[36,37] resulting in some recessed areas. The complicated interaction between the coating and substrate, as well as the existence of many unknown or not well understood influencing factors[38–43] may also help roughen the surface. This kind of surface roughening actually helps form mechanical interlocking between the Ni-B coating and magnesium alloy. The dissolution of the substrate stops when magnesia film and the exposed substrate areas are fully covered by deposited particles. Nodular nickel is a catalyst for continuous Ni deposition reaction. The growth mode during deposition is strongly affected by the diffusion layer in the plating solution. The diffusion of nickel determines the plating morphology.[20,33] On the relative protruding areas, to which reactants from the plating solution can easily diffuse, large spherical nickel particles appear because of their similar growth rates in various directions. Within the recessed areas, because the growth in the lateral direction is confined by the adjacent particles and diffusion layer, the faster growth in the vertical direction results in a columnar structure. Subsequently, deposited nickel particles accumulate layer upon layer to form a smooth and dense Ni-B coating. A schematic illustration of the deposition mechanism is shown in Fig. 13.1.

13.3 Ni-B plating characterization

An obtained Ni-B plating may vary significantly in physical and chemical properties, such as the coating adhesion and corrosion resistance, depending the coating chemical composition, microstructure, thickness and morphology. It is important to characterize and evaluate these coating parameters and coating performance.

13.3.1 Microstructure

A scanning electron microscope (SEM, LEO-1530, Germany) was employed to examine the micro morphologies of AZ91D after pickling and the electroless Ni-B plating. Due to the high anodic reactivity of the matrix phase and the galvanic effect between the alloy constituent phases,[44] it is essential to generate a uniform and less anodically active surface that is catalytically active for plating ions before plating. In this study, pickling is particularly critical as it is the last step before EL.

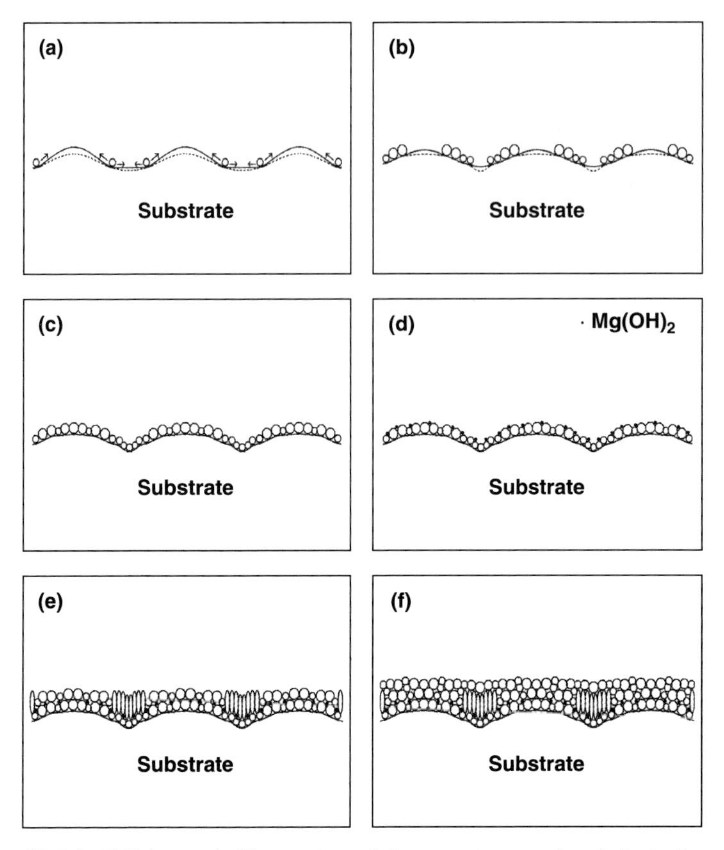

13.1 (a–f) Schematic illustration of the growing mode of electroless Ni-B coating on AZ91D magnesium alloy.

Acetic acid was the main component of the pickling solution. In the acid, sodium nitrate was also added as corrosion inhibitor to slow down the dissolution of AZ91D during picking. To optimize the pickling treatment, different bath compositions, including concentrations of acetic acid and sodium nitrate, and operational parameters, such as temperature and pickling time, were carefully controlled, adjusted and compared. The solution composition and operation conditions for acid pickling listed in Table 13.2 are actually optimized results based on the experimental observations and weight loss measurements of pickled AZ91D samples.

The micro morphology of the surface after acid pickling is shown in Fig. 13.2. It appears that the AZ91D surface is covered partially by a uniform honeycomb-like film with many voids. The adhesion of subsequent deposited coating on the substrate is expected to be increased by such a honeycomb structure. It should be noted that in some areas, metallic Mg could be later exposed to

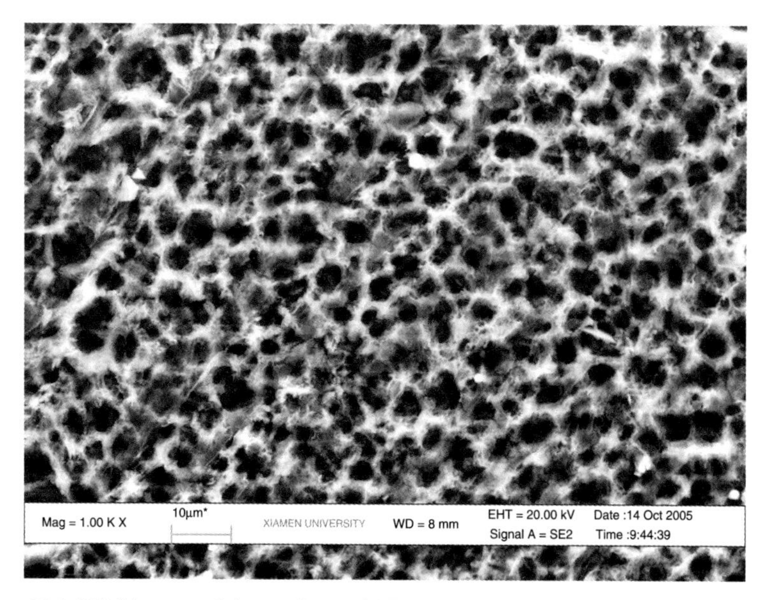

13.2 SEM image of the surface of AZ91D substrate after acid pickling.

the Ni-B plating solution and react with Ni ions, which implies that the pickled surface has some chemical activity for subsequent Ni deposition.

After acid pickling, the AZ91D substrate was placed in the optimized plating bath listed in Table 13.2 for electroless Ni-B plating. The topographic surface morphology of the Ni-B coating is shown in Fig. 13.3. It can be seen that the Ni-B coating obtained in the optimized plating bath was compact with no evident surface cracks. The Ni atoms nucleate around active sites on the Mg alloy surface and form a typical spherical structure. Compact particles with particle size about 2~5 μm are uniformly distributed on the magnesium alloy surface.

13.3.2 Coating thickness and morphology

The EL bath composition can influence the coating deposition rate and coating porosity. The compositions of the plating baths #1~#4 listed in Table 13.3 are slightly different. The thickness and B contents of the formed Ni-B coatings on AZ91D are presented in Table 13.4. The coating thickness was determined by a highly sensitive micrometer (sensitivity 1×10^{-3} mm). The boron content of coating was analyzed by inductively coupled plasma atomic emission spectrometry (ICP-AES, Baird PS-4, USA). The effect of nickel acetate concentration on plating rate can be found by comparing sample #1 with #3. Increasing nickel acetate concentration from 35 to 40 g/L

Table 13.3 Compositions of the four recipes (#1–#4)

Recipe	Composition
Ni-B coating(#1)	Ni% = 35 g/L $NaBH_4$% = 0.6 g/L
Ni-B coating(#2)	Ni% = 38 g/L $NaBH_4$% = 0.6 g/L
Ni-B coating(#3)	Ni% = 40 g/L $NaBH_4$% = 0.6 g/L
Ni-B coating(#4)	Ni% = 40 g/L $NaBH_4$% = 0.8 g/L

Note: Ni% corresponds to the concentration of nickel acetate in the plating solution; other unlisted compositions of the plating solution are the same as in Table 13.2.

Table 13.4 Coating thickness, boron content and bonding strength of the coatings after 120 min electroless plating in baths #1–#4

Sample	Coating thickness (μm)	Boron content (%)	Bonding strength (MPa)
Ni-B coating(#1)	10.6	5.86	20.82
Ni-B coating(#2)	11.2	5.21	21.36
Ni-B coating(#3)	11.5	4.81	21.59
Ni-B coating(#4)	12.1	5.54	22.32

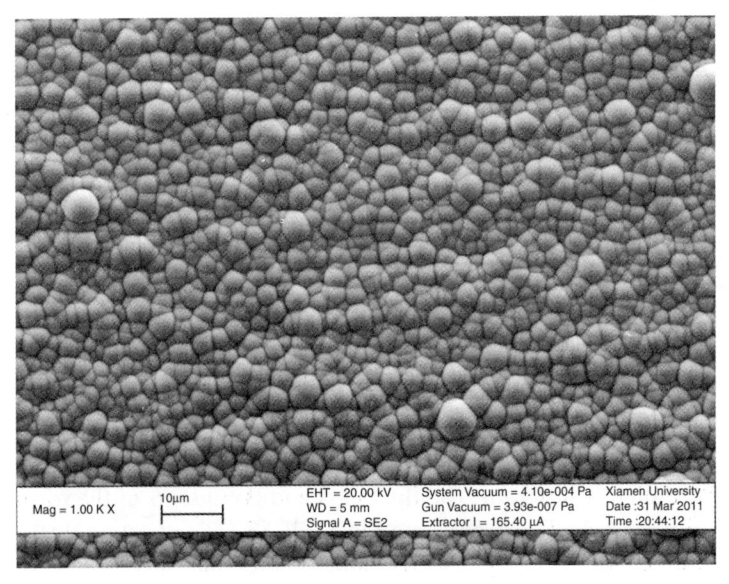

13.3 SEM image of the surface of Ni-B coating in optimized deposition conditions for 120 min electroless plating.

leads to a higher plating rate and thus a thicker coating. A leveling effect can be observed at higher nickel acetate concentration. Obviously, the addition of $Ni(CH_3COO)_2{\cdot}4H_2O$ increases the concentration of free Ni^{2+} ions in the bath, which is beneficial for higher plating rate. Compared with sample #3, the coating thickness in sample #4 is thicker, which should be caused by the difference in sodium borohydride concentration. A sodium borohydride molecule can provide eight electrons to ionic nickel in a reduction reaction. Therefore, increasing its concentration accelerates Ni-B deposition.

Surface morphologies of the Ni-B coatings prepared using different concentrations of nickel acetate are shown in Fig. 13.4. Very different surface morphologies are obtained. The increase in nickel acetate concentration causes the Ni-B nodules to coalesce together, which enhances the compactness of the deposits, leading to lower porosity. It is noted that the nodule sizes are similar in four cases, implying that nickel acetate concentration has no evident effect on the particle size. Surface morphologies of the coatings prepared in different concentrations of sodium borohydride are shown in Fig. 13.5. At 0.2 g/L sodium borohydride concentration, the coating has many voids and evident surface cracks. The incomplete coverage of coating is due to the low deposition rate due to the low sodium borohydride concentration. With the increase in sodium borohydride concentration, the coating becomes more compact.

13.3.3 Adhesion

Coating adhesion is an important issue concerning the reliability of the coating in practical applications. The adhesion can be tested according to ASTM-F1044 (see Fig. 13.6). The specimens were clamped and glued with FM 1000 adhesive film (Cytec Fiberite Inc., USA) and oven cured at 165°C for 2 h. The test was carried out with a universal testing machine (LR5K Plus, Lloyd Instrument, UK) using a 5 kN load cell and a crosshead speed of 1.0 mm/min. The value of the adhesive strength can be calculated from the fracture force over the stressed area. The results of the bonding strength of the coatings after 120 min EL in baths #1~#4 were shown in Table 13.4. They are higher than 20 MPa. This confirmed the good adhesion of the Ni-B coating to AZ91D substrate.

13.3.4 Post heat-treatment

In practical applications, Mg alloy parts may experience various heating processes. For example, in the auto industry, Mg parts have to be welded to other parts and they have to be cured after coating/painting. The heating may lead to some changes in coating microstructure. Phase transformation

13.4 SEM images of the surface of Ni-B deposit prepared by different nickel acetate concentration: (a) 30 g/L; (b) 35 g/L; (c) 40 g/L; (d) 45 g/L.

by heating was investigated using a differential scanning calorimeter (DSC-204, Netzsch, Germany) at a scanning rate of 20°C/min. Phase constituents of the coating at different temperatures were analyzed with X-ray diffraction technique (XRD, Philips PANalytical X'pert, Holland) in an argon atmosphere. Crystallization of the coating was detected. In fact, similar crystallization during heat-treatment has been widely reported.[21,38–40] Ni_3B,

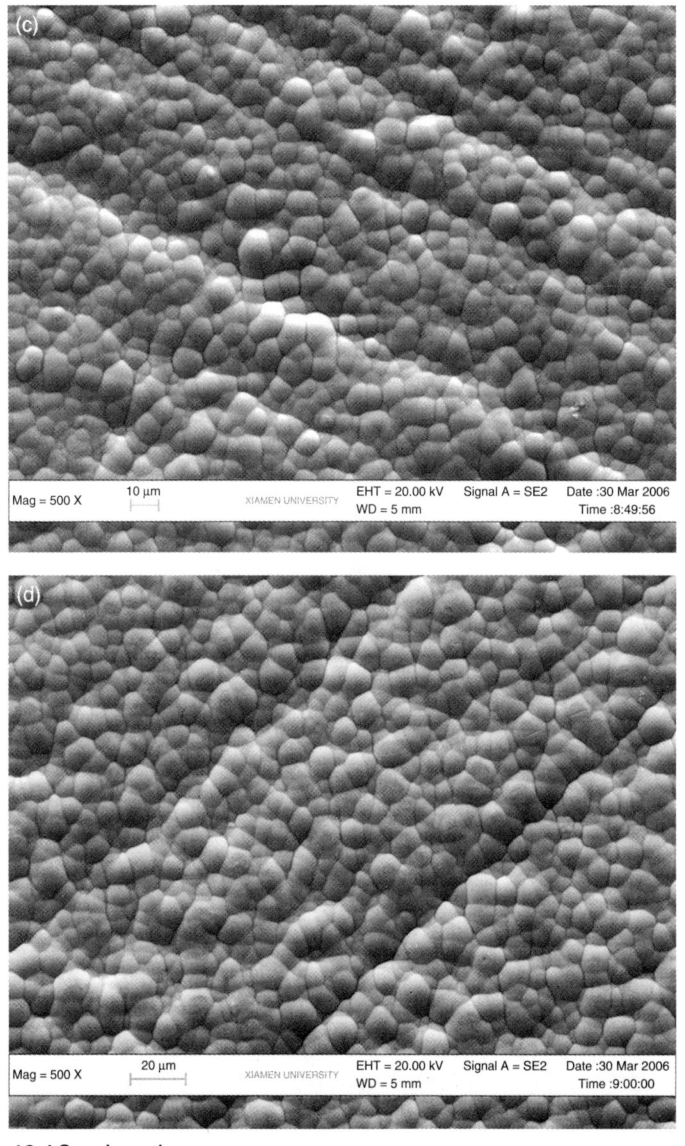

13.4 Continued

as the main crystallization product, has been described in previous studies.[13,21] Some other products can also be formed during heat-treatment, such as Ni_2B and Ni_7B_3.[27,41] Figure 13.7 shows a DSC spectrum obtained at a heating rate of 20°C/min. There are two exothermic peaks at 303°C and 428°C, respectively. The lower temperature peak should result from formation of N_3B phase and the higher one might be from a combination of several stable

13.5 SEM images of the surface of Ni-B deposit prepared by different sodium borohydride concentration: (a) 0.2 g/L; (b) 0.4 g/L; (c) 0.8 g/L; (d) 1 g/L.

metal phases. The positions of these exothermic peaks are slightly different from those reported recently in literature,[42] which could be due to a difference in boron content or instrument sensitivity.

An *insitu* XRD experiment confirmed the phase transition reactions and crystallization products (see Fig. 13.8). At room temperature (see Fig. 13.8

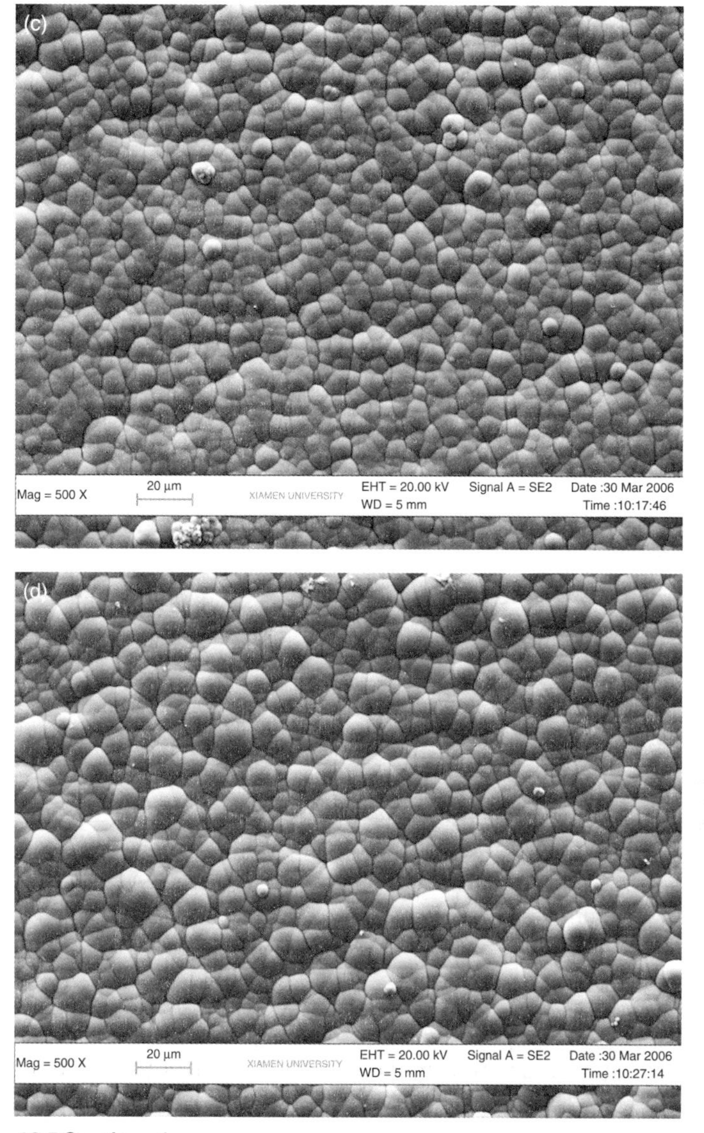

13.5 Continued

curve (a)), the plating layer formed on AZ91D is overwhelmingly an amorphous structure. When the temperature is raised to 300°C (Fig. 13.8 curve (b)), crystal Ni and Ni_3B are detected. These support the DSC result that a stable N_3B phase formed at 300°C (Fig. 13.7). At 400°C (see curve Fig. 13.8(c)), the peaks corresponding to Ni(111), Ni_3B and Ni_2B signals become much stronger. In the 480°C XRD pattern (Fig. 13.8 curve (d)) new diffraction peaks

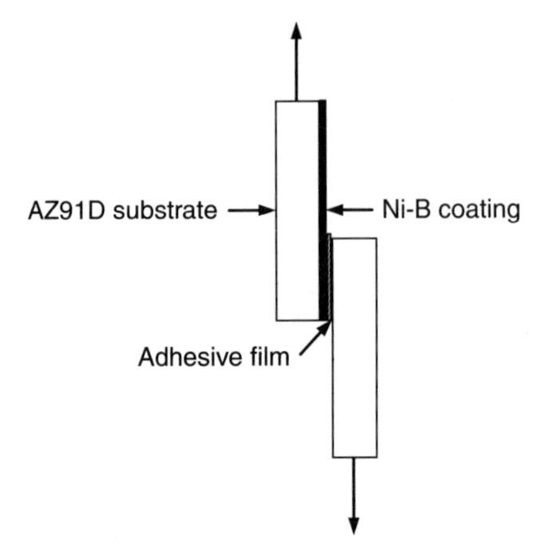

13.6 Schematic diagram of the adhesion test.

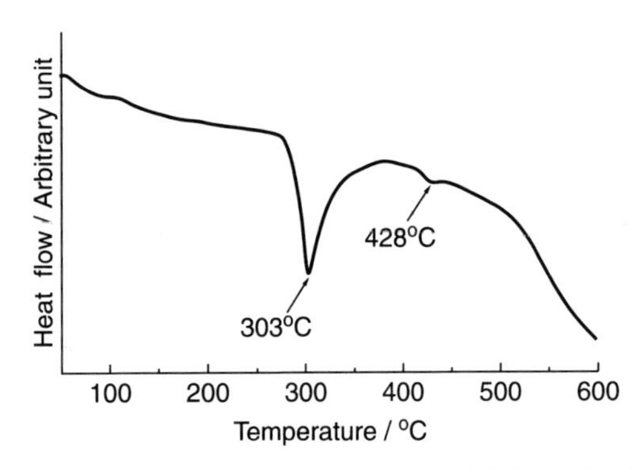

13.7 DSC spectrum of electroless Ni-B plating heated at 20°C/min up to 600°C.

appear and they can be attributed to formation of Ni_4B_3. The above results suggest that the Ni-B plating on AZ91D is not stable at high temperatures. The precipitated compounds in the plating may affect the corrosion performance of the Ni-B plating.

13.3.5 Corrosion resistance

Corrosion protection is one of the most important purposes of forming an electroless Ni-B plating layer on Mg alloy. The corrosion performance of Ni-B plating can be assessed by potentiodynamic polarization curve and electrochemical impedance spectroscopy (EIS) measurements in 3.5 wt.% sodium chloride solution. Figure 13.9 shows the potentiodynamic polarization curves of bare AZ91D and the samples after 120 min EL in the 4

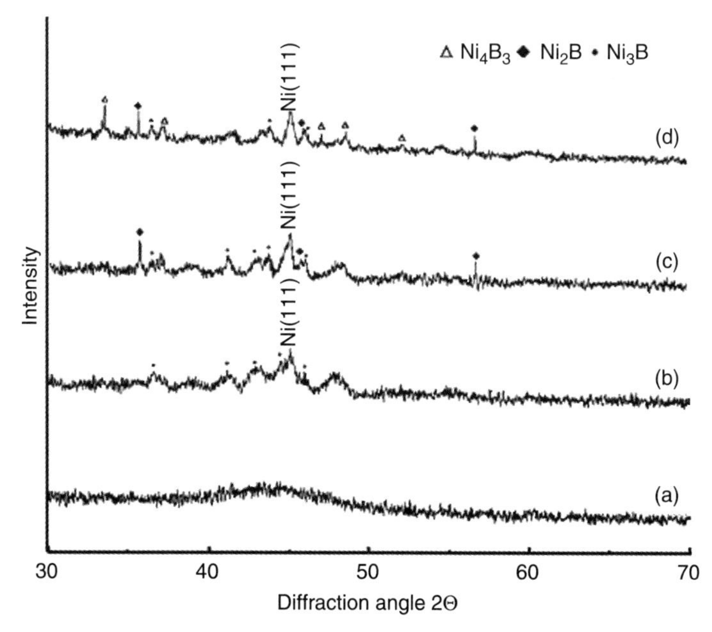

13.8 In situ XRD patterns of electroless Ni-B plating at various heat-treatment temperatures: (a) 25°C; (b) 300°C; (c) 400°C; (d) 480°C.

different baths (#1–#4 as listed in Table 13.4). The significantly positively shifted corrosion potential E_{corr} and the strikingly decreased corrosion current density i_{corr} for the coated samples relative to the bar AZ91D substrate imply that the EL greatly improves the corrosion resistance of the Mg alloy in the NaCl aqueous solution. The electroless Ni-B plating layers on AZ91D have good corrosion protection performance.

It should be noted that the polarization curves of the coated samples are very different from that of the bare AZ91D. The anodic polarization current density increases rapidly with increasing anodic potential for bare AZ91D (Fig. 13.9a), while a very narrow plateau presents in the potential range slightly more positive than the corrosion potential for those Ni-B coating samples (Fig. 13.9b). The big difference in polarization curves suggests the deposited plating layers completely cover the AZ91D surface. Hence, on bare and coated sample surfaces, there are different electrochemical or corrosion reactions. The primary reactions during cathodic and anodic polarization of the Ni-B coatings are reduction of oxygen and dissolution of nickel,[13,42,43] while the anodic and cathodic reactions on bare AZ91D include dissolution of Mg, 'anodic' hydrogen evolution and cathodic hydrogen evolution.[44]

The four coated samples #1–#4 have a similar anodic polarization curve shape. Below the breakdown potential (E_b), there is a small passive plateau.

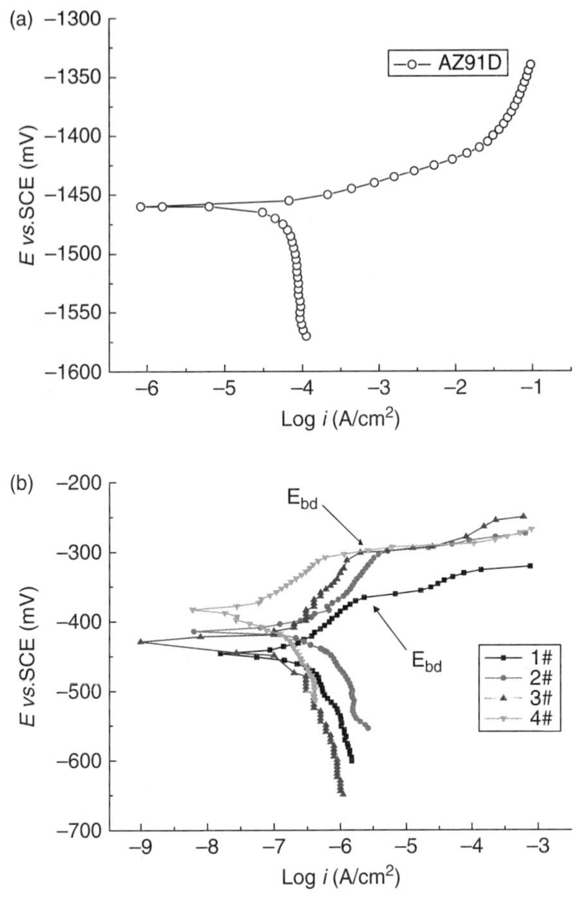

13.9 Potentiodynamic polarization curves of (a) bare AZ91D substrate and (b) Ni-B coatings prepared from different recipes (1# to 4#) for 120 min immersed in 3.5 wt.% NaCl solution.

This signifies that the samples are in a kind of passive state around their corrosion potentials. There is a passive film formed on the Ni-B plating surfaces. E_b is a parameter characterizing the resistance of a passive film against pitting corrosion. Figure 13.9 shows that the breakdown potential slightly varies from sample to sample. It increases in the following order: #4 > #3 > #2 > #1. It means that the stability of the four Ni-B coatings in the chloride solution decreases in that order.

Based on the measured polarization curves and through Tafel extrapolation, corrosion current densities of these samples can be estimated and listed in Table 13.5. All of the four coated samples have corrosion current densities about 2 orders of magnitude lower than the bare AZ91D, confirming the great corrosion protection performance or corrosion resistance of

Table 13.5 Estimated corrosion potential, corrosion current density and corrosion rates from polarization curves

Sample	Corrosion potential E_{corr} (mV vs. SCE)	Corrosion current density i_{corr} (μA/cm²)	Corrosion rate (mm/year)
AZ91D substrate	−1460	28.548	0.3091
Ni-B coating (#1)	−446	0.631	0.0068
Ni-B coating (#2)	−414	0.256	0.0028
Ni-B coating (#3)	−429	0.457	0.0049
Ni-B coating (#4)	−385	0.158	0.0017

the Ni-B coatings on AZ91D. The corrosion resistance of the coatings can be ranked in the following order based on the estimated corrosion rates: #4 > #2 > #3 > #1.

To further confirm the estimated corrosion behavior of the Ni-B coatings, EIS measurement was also conducted. Figure 13.10 shows Nyquist plots of the bare AZ91D substrate and the four Ni-B coatings in 3.5 wt.% NaCl solution. Their EIS results are basically one-semicircle spectra though the diameters of the semicircles are different. These means that corrosion has a simple and similar process on these sample surfaces, but corrosion rates differ between samples.[2,45] The one-semicircle capacitive spectrum is typical of an EIS result for a good quality coating,[46,47] implying that the EIS behavior can be simulated by using a simple equivalent circuit as shown in Fig. 13.11, where R_s is a solution resistance, CPE represents a constant-phase element corresponding to the capacitance of the electrode surface and R_c is electrode polarization resistance. The resistance values listed in Table 13.6 are obtained through curve-fitting of Nyquist spectra in Fig. 13.10 using the equivalent circuit (Fig. 13.11). All the four coatings have much higher polarization resistance than the bare AZ91D substrate. The corrosion resistance of the coated samples can be ranked in order: #4 > #2 > # > #3 > #1. This is in good agreement with the polarization curve results.

The four coatings were formed under the same EL condition in slightly different bath solutions. They are assumed to have similar porosity. Hence, coating thickness and coating composition should be the main influencing factors. An increase in boron content can lead to a decrease in crystallinity of the Ni-B plating layer, resulting in less grain boundaries. Thus, a coating with a high boron content usually has low electrical conductivity and a smaller number of possible paths for corrosive species or defect sites for corrosion attack.[48] Consequently, the coating will have enhanced corrosion resistance. This explains the corrosion resistance of coatings #2, #3 and #4. According to Table 13.4, the B contents in coatings #2, #3 and #4 are in the order #4 > #2 > #3, which is consistent with their corrosion resistance ranking (Table 13.6). However, this explanation contradicts the corrosion

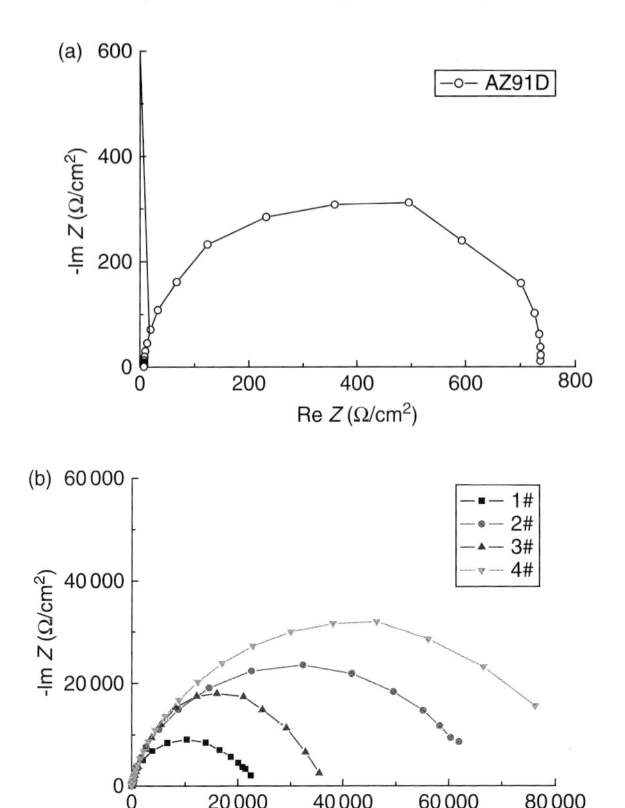

13.10 Nyquist plots of (a) bare AZ91D substrate and (b) Ni-B coatings prepared from different recipes (1# to 4#) for 120 min immersed in 3.5 wt.% NaCl solution.

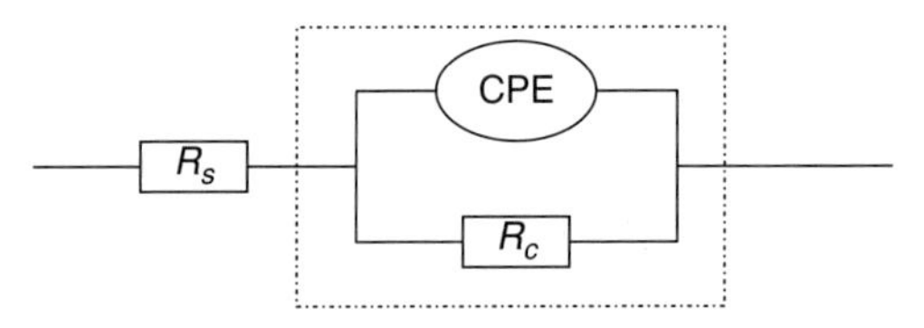

13.11 The equivalent circuit for curve-fitting of the EIS results presented in Fig. 13.10.

performance of coatings #1 and #2. The former has a higher B content than the latter, but its corrosion rate is higher than for the latter. If thickness of coatings #1 and #2 is taken into consideration, the 'abnormal' corrosion resistance of coatings #1 and #2 may become understandable. Coating #1 is the thinnest of the four coatings. If a coating is not thick enough, the

Table 13.6 Resistance obtained from AC impendence measurements

Samples	R_c (Ω/cm^2)
AZ91D substrate	7.296×10^2
Ni-B coating (#1)	2.378×10^4
Ni-B coating (#2)	6.176×10^4
Ni-B coating (#3)	3.904×10^4
Ni-B coating (#4)	7.776×10^4

corrosion species could penetrate the coating relatively easily. This could be the case for coating #1. Therefore, it appears that the corrosion resistance of Ni-B coatings is determined by both coating thickness and boron content.

13.4 Applications and future trends

This study presents a direct electroless Ni-B plating onto magnesium alloy substrate. The traditional acid etching and fluoride surface activation procedures are replaced with a safe and environmentally friendly acetic acid pickling step. This new electroless Ni-B coating has a uniform surface, good corrosion resistance and good adhesion. In view of these advantages, as well as the small capital investment required, this innovative plating technology will promote the corrosion and wear resistance of magnesium alloy and may find applications in automotive and aerospace industries.

Despite the outstanding merits of the electroless Ni-B plating, some further research is required to develop this technology. In general, in the plating industry, the first and also the most serious concern with this EL is the lifetime of plating baths and the waste generation from the plating. This should still be a primary consideration in this new plating technique development. Secondly, the plating rate of the electroless Ni-B coating is slower than that of an electroless Ni-P coating. Therefore, improving plating rate of the electroless Ni-B coatings is another focus of future research. Thirdly, the pretreatment and EL processes need to be further optimized. Also, lowering the coating cost cannot be overstressed. Finally, the coating corrosion resistance is expected to be further improved for applications in more aggressive environments.

13.5 References

1. Song G-L, '"Electroless" deposition of a pre-film of electrophoresis coating and its corrosion resistance on a Mg alloy', *Electrochim. Acta* (2010), **55**, 2258.
2. He M F, Liu L, Wu Y T, Tang Z X and Hu W B, 'Corrosion properties of surface-modified AZ91D magnesium alloy', *Corros. Sci.* (2008), **50**, 3267.

3. Mordike B L and Ebert T, 'Magnesium: Properties-applications-potential', *Mater. Sci. Eng. A: Struct.* (2001), **302**, 7.
4. Baghni I M, Wu Y S, Li J Q, Du C W and Zhang W, 'Corrosion behavior of magnesium and magnesium alloys', *Trans. Nonferr. Met. Soc. China* (2003), **13**, 253.
5. Wang Z C, Jia F, Yu L, Qi Z B, Tang Y and Song G-L, 'Direct electroless nickel-boron plating on AZ91D magnesium alloy', *Surf. Coat. Technol* (2012), **206**, 3676.
6. Aghion E, Bronfin B and Eliezer D, 'The role of the magnesium industry in protecting the environment', *J. Mater. Process. Tech.* (2001), **117**, 381.
7. Friedrich H and Schumann S, 'Research for a "new age of magnesium" in the automotive industry', *J. Mater. Process. Tech.* (2001), **117**, 276.
8. Gray J E and Luan B, 'Protective coatings on magnesium and its alloys-a critical review', *J. Alloys Compd.* (2002), **336**, 88.
9. Sorkhabi H A and Rafizadeh S H, 'Effect of coating time and heat treatment on structures and corrosion characteristics of electroless Ni–P alloy deposits', *Surf. Coat. Technol.* (2004), **176**, 318.
10. Mathieu S, Rapin C, Hazan J and Steinmetz P, 'Corrosion behaviour of high pressure die-cast and semi-solid cast AZ91D alloys', *Corros. Sci.* (2002), **44**, 2737.
11. Chong K Z and Shih T S, 'Conversion-coating treatment for magnesium alloys by a permanganate-phosphate solution', *Mater. Chem. Phys.* (2003), **80**, 191–200.
12. Song G-L, 'An irreversible dipping sealing technique for anodized ZE41 Mg alloy', *Surf. Coat. Technol.* (2009), **203**, 3618–3625.
13. Huang C A, Wang T H, Weirich T and Neubert V, 'Electrodeposition of a protective copper/nickel deposit on the magnesium alloy (AZ31)', *Corros. Sci.* (2008), **50**, 1385–1390.
14. Fairweather W A, 'Electroless nickel plating of magnesium', *Trans. IMF* (1997), **75**, 113–117.
15. Sharma A K, Suresh M R, Bhojraj H, Narayanamurthy H and Sahu R P, 'Electroless nickel plating on magnesium alloy', *Met. Finish.* (1998), **96**, 10–18.
16. Shahin G E (2002), *Corrosion and wear resistance of electroless nickel on magnesium alloys*, Minerals, Metals and Materials Society, Seattle, WA, United States.
17. Ambat R and Zhou W, 'Electroless nickel-plating on AZ91D magnesium alloy: effect of substrate microstructure and plating parameters', *Surf. Coat. Technol.* (2004), **179**, 124–134.
18. Huo H W, Li Y and Wang F H, 'Corrosion of AZ91D magnesium alloy with a chemical conversion coating and electroless nickel layer', *Corros. Sci.* (2004), **46**, 1467–1477.
19. Zhao H, Huang Z H and Cui J Z, 'A new method for electroless Ni–P plating on AZ31 magnesium alloy', *Surf. Coat. Technol.* (2007), **202**, 133–139.
20. Rao Q L, Bi G, Hua Q H, Wang H W and Fan X L, 'Microstructure evolution of electroless Ni-B film during its depositing process', *Appl. Surf. Sci.* (2005), **240**, 28.
21. Krishnaveni K, Narayanan T S N S and Seshadri S K, 'Electroless Ni–B coatings: preparation and evaluation of hardness and wear resistance', *Surf. Coat. Technol.* (2005), **190**, 115.
22. Amendola S C, Onnerud P, Kelly M T, Petillo P J, Sharp-Goldman S L and Binder M J, 'A novel high power density borohydride-air cell', *Power Sources* (1999), **84**, 130.

23. Liu B H, Yang J Q and Li Z P, 'Concentration ratio of [OH⁻]/[BH4⁻]: A controlling factor for the fuel efficiency of borohydride electro-oxidation', *Int. J. Hydrogen Energ.* (2009), **34**, 9436.
24. Baril G, Blanc C, Keddam M and Pebere N, 'Local electrochemical impedance spectroscopy applied to the corrosion behavior of an AZ91 magnesium alloy', *J. Electrochem. Soc.* (2003), **150**, B488.
25. Song G L and Atrens A, 'Corrosion mechanisms of magnesium alloys', *Adv. Eng. Mater.* (1999), **1**, 11.
26. Song G L, Bowles A L and StJohn D H, 'Corrosion resistance of aged die cast magnesium alloy AZ91D', *Mater. Sci. Eng.* (2004), **366**, 74.
27. Delaunois F and Lienard P, 'Heat treatments for electroless nickel–boron plating on aluminium alloys', *Surf. Coat. Technol.* (2002), **160**, 239.
28. Gorbunova K M, Ivanov M V and Moissev V P, 'Electroless deposition of nickel-boron alloys mechanism of process, structure and some properties of deposits', *J. Electrochem. Soc.* (1973), **120**, 613.
29. Duncan R N and Arney T L, 'Operation and use of sodium-borohydride-reduced electroless nickel', *Plat. Surf. Finish.* (1984), **71**, 49.
30. Liu Z M and Gao W, 'Electroless nickel plating on AZ91 Mg alloy substrate', *Surf. Coat. Technol.* (2006), **200**, 3553.
31. Tsuru Y, Nomura M and Foulkes F R, 'Effects of boric acid on hydrogen evolution and internal stress in films deposited from a nickel sulfamate bath', *J. Appl. Electrochem.* (2002), **32**, 629.
32. Lang K, 'Ein neues Verfahren zur stromlosen Vernickelung', *Metalloberfläche* (1965), **19**, 257.
33. Rao Q L, Wang H W, Fan X L and Zhou Y H, 'Morphology and Formation Mechanism of Electroless Ni-B Alloy Coating', *Journal of Shanghai Jiaotong University* (2003), **12**, 1965.
34. Hu W B, Xiang Y H, Liu X K, Liu L and Ding W J, 'A study on surface state during the pretreatment of electroless nickel plating on magnesium alloys', *Journal of Chinese Society for Corrosion and Protection* (2001), **21**, 22.
35. Lo Y L and Hwang B J, 'Decomposition of sodium borohydride in an electroless nickel bath', *Ind. Eng. Chem. Res.* (1994), **33**, 56.
36. Lafront A M, Zhang W, Jin S, Tremblay R, Duke D and Ghali E, 'Pitting corrosion of AZ91D and AJ62x magnesium alloys in alkaline chloride medium using electrochemical techniques', *Electrochim. Acta* (2005), **51**, 489.
37. Mathieu S, Rapin, Hazan J and Steinmetz P, 'Corrosion behaviour of high pressure die-cast and semi-solid cast AZ91D alloys', *Corros. Sci.* (2002), **44**, 2737.
38. Cheong W J, Luan B L and Shoesmith D W, 'Protective coating on Mg AZ91D alloy–The effect of electroless nickel (EN) bath stabilizers on corrosion behaviour of Ni–P deposit', *Corros. Sci.* (2007), **49**, 1777.
39. Anik M, Körpe E and Sen E, 'Effect of coating bath composition on the properties of electroless nickel-boron films', *Surf. Coat. Technol.* (2008), **202**, 1718.
40. Hamid Z A, Hassan H B and Attyia A M, 'Influence of deposition temperature and heat treatment on the performance of electroless Ni–B films', *Surf. Coat. Technol.* (2010), **205**, 2348.
41. Mallory G O (1990), Electroless Plating: Fundamentals and Applications, AESF, New York, United States.
42. Dong H, Sun Y and Bell T, 'Enhanced corrosion resistance of duplex coatings', *Surf. Coat. Technol.* (1997), **90**, 91.

43. Luo J L and Cui N, 'Effects of microencapsulation on the electrode behavior of Mg2Ni-based hydrogen storage alloy in alkaline solution', *J. Alloys Compd.* (1998), **264**, 305.
44. Song G-L, Chapter 1 'Electrochemistry of Mg and its Alloys' in Guang-Ling Song (ed.), *Corrosion of Magnesium Alloys*, Woodhead Publishing Limited, UK (2011).
45. Galicia G, Pébére B and Vivier V, 'Local and global electrochemical impedances applied to the corrosion behaviour of an AZ91 magnesium alloy', *Corros. Sci.* (2009), **51**, 1789.
46. Anik M and Gelikten G, 'Analysis of the electrochemical reaction behavior of alloy AZ91 by EIS technique in H_3PO_4/KOH buffered K_2SO_4 solutions', *Corros. Sci.* (2007), **49**, 1878.
47. Ardelean H, Frateur I and Marcus P, 'Corrosion protection of magnesium alloys by cerium, zirconium and niobium-based conversion coatings', *Corros. Sci.* (2008), **50**, 1907.
48. Narayanan T S N S, Krishnaveni K and Seshadri S K, 'Electroless Ni–P/Ni–B duplex coatings: preparation and evaluation of microhardness, wear and corrosion resistance', *Mater. Chem. Phys.* (2003), **82**, 771.

14

Electrodeposition of aluminum on magnesium (Mg) alloys in ionic liquids to improve corrosion resistance

W.-T. TSAI and I.-W. SUN,
National Cheng Kung University, Taiwan

DOI: 10.1533/9780857098962.3.393

Abstract: To improve the corrosion resistance of magnesium (Mg) alloys, surface modification is applied in an attempt to produce a corrosion resistant barrier. In this chapter, a new method to deposit aluminum (Al) film on a Mg alloy surface is demonstrated. Electrodeposition of Al on AZ91D Mg alloy, using an acidic aluminum chloride–1-ethyl-3-methyl-imidazolium chloride ionic liquid ($AlCl_3$–EMIC), has been shown to be feasible. The existence of Al coating can cause a substantial increase in corrosion resistance, reducing the susceptibility of Mg alloy to aqueous corrosion. The results of electrochemical impedance spectroscopy show that Al coating leads to an increase in the polarization resistance of a bare Mg alloy by one order of magnitude in 3.5 wt% NaCl solution. Furthermore, the potentiodynamic polarization results show that Al-coated Mg alloy can be passivated, and a wider passive region with a lower passive current density can be obtained if the Al is electrodeposited at a lower applied current or a low cathodic overpotential. The passivity of the co-deposited Al/Zn film is slightly inferior to that of the pure Al coating.

Key words: magnesium alloy, ionic liquid, electrodeposition, aluminum, corrosion.

Note: This chapter has been adapted from Chapter 14 'Electrodeposition of aluminium (Al) on magnesium (Mg) alloys in ionic liquids' by W.-T. Tsai and I.-W. Sun, originally published in *Corrosion of magnesium alloys*, ed. Guang-Ling Song, Woodhead Publishing Limited, 2011, ISBN: 978-0-85709-41-3.

14.1 Introduction

Magnesium (Mg) and its alloys are increasingly implemented into a number of components where weight reduction is of great concern. However, the poor corrosion resistance of magnesium and its alloys has limited their applications in corrosive environments.[1–5] Extensive reviews on the forms of magnesium corrosion and the influences of composition and microstructure have been summarized by Song.[5] To mitigate the susceptibility to corrosion, surface treatment to form a protective barrier layer is always required. In general, the methods for surface treatment can be classified into two categories, dry and wet processes. A number of possible protective coating techniques for magnesium and its alloys have been reviewed by Gray and Luan.[6] Certainly, each method has its own advantages and disadvantages, regarding the material properties and the manufacturing processes, etc. Since the wet methods generally have a high yield, they have merit and potential for large-scale production.

Among the various wet processing techniques, anodization[7–10] and conversion coating[11–17] are most commonly applied. The coatings resulting from these treatments are mainly oxides or inorganic compounds. When electromagnetic shielding is required, metallic coating is recommended. Since electro- and electroless deposition are well established and have been applied to various substrates for surface modification, they are considered for metallic coating on Mg alloys. Electroless plating of Ni or its alloys such as Ni–P and Ni–B, employing aqueous electrolytes, has been attempted for Mg alloy metallization.[18–24] Electrodeposition of pure metal and/or alloys such as Cu, Ni, Zn, Cu–Ni, Zn–Sn and Zn–Ni on Mg alloys, using aqueous electrolytes, has also been investigated.[20,25–28] In order to improve the cohesion of the deposited layer, electroless Ni coating and/or Zn immersion process is always applied before electrodeposition. Nevertheless, the active nature of magnesium may always cause the formation of loose MgO or Mg(OH) on surface before electrodeposition when aqueous electrolyte is used. The liberation of hydrogen molecules during electrodeposition of metals from aqueous electrolytes may also produce pinholes in the deposits. To overcome these problems, non-aqueous solutions with organic solvents may be used. In a review article,[29] Simka and her co-workers indicate that several metals and alloys can be electrodeposited from conventional organic solvents. However, to the authors' knowledge, metal coating on magnesium substrate from organic baths has yet to be established.

For combined corrosion resistance and electromagnetic shielding, Al coating is of interest because of the following reasons. First, being a light metal, an Al coating does not significantly increase the overall density of the material. Second, both anodization and electrolysis coloration techniques for Al are already well established, they can be directly applicable to the Al-coated Mg alloys. Third, the low electric resistivity and electromagnetic shielding property of the Al-coated Mg alloy can be maintained even after the anodization of the surface Al film. Finally, as one of the major alloying elements, Al coating is more chemically compatible with the substrate.

Various methods are available for Al coating such as sputter deposition, vapor deposition, thermal spraying, hot dipping and electrodeposition. Among these methods, electrodeposition offers some advantages including easy control of the coating thickness through the experimental parameters, lower operating temperature and relatively low cost. However, Al cannot be electrodeposited from aqueous solution because the reduction potential of Al metal ion is well below that of the hydrogen evolution. Therefore, aprotic electrolytes such as organic solvents or molten salts must be employed for Al deposition. Electrodeposition of Al has been successfully performed using three classes of organic solvents: ethers, aromatic hydrocarbons, and di-methylsulfone.[30–35] Diethyl ether containing $AlCl_3$ and $LiAlH_4$ was first employed in the NBS (National Bureau of Standards) bath.[36] Although this bath has been adopted in industrial application, it possessed several drawbacks including low Al anode current efficiency, limited lifetime, and hydrogen evolution.[30] Tetrahydrofuran (THF) was used to replace the diether to improve the anode current efficiency and longer bath lifetim.[30] On the other hand, owing to its high solubility compared with AlF_3 and $AlCl_3$, $AlBr_3$ has been used as the Al(III) source in plating baths using aromatic hydrocarbons such as benzene, toluene, xylene and their mixtures and derivatives.[31–33] The electroactive species responsible for the deposition of Al is $Al_2Br_7^-$. Alkali, quaternary ammonium or pyridinium halide salts were added to improve the conductivities and current efficiencies of the baths. Process using aromatic solvents containing alkylaluminum has also been reported to be successful.[37] Nevertheless, alkylaluminum compounds possessed the drawbacks of self-ignition in air. The utility of the aromatic hydrocarbons can be, however, limited by their toxicity. Another organic solvent that has been studied for electrodeposition of Al is dimethylsulfone (DMSO).[34] DMSO bath shows good conductivity, thermal stability, and ability to dissolve metallic salts. The complex compound responsible for Al deposition from $AlCl_3/LiCl/DMSO_2$ bath is $Al[(CH_3)_2SO_2]_3^{3+}$. A recent study indicates that good Al coatings can be obtained in a temperature range of 110–140°C.[35]

Although Al can be successfully electrodeposited from organic solvents, the volatility, toxicities and flammability of the organic electrolytes create safety concerns. Such safety concerns can be circumvented by using inorganic molten salts such as mixtures of NaCl and $AlCl_3$ which are non-volatile and non-flammable. The inorganic molten salts, however, normally require a high operating temperature (> 130°C) due to their high melting points. On the other hand, ambient temperature molten salts that are liquid below 100°C can be obtained by replacing the inorganic salts with organic salts. To distinguish them from the high-temperature inorganic molten salts, the ambient temperature molten salts are classified as ionic liquids. The ionic liquids have proven to be versatile electrolytes for electrodeposition due to their advantageous properties such as wide electrochemical window, low melting temperature, high thermal stability, non-volatility, good ionic conductivity and non-flammability. In this chapter, the background of low-temperature electroplating of Al in aluminum chloride–1-ethyl-3-methylimidazolium chloride ionic liquids ($AlCl_3$–EMIC) is introduced. The deposition of Al and co-deposition of Al and Zn on an Mg alloy, namely AZ91D from the $AlCl_3$–EMIC baths are demonstrated, and the associated advantages in enhancing the corrosion resistance are described.

14.2 Basics for ionic liquid plating

Although numerous kinds of ionic liquids have been developed, as far as Al deposition is concerned, the most studied ionic liquids are the chloroaluminates which are obtained by reaction of $AlCl_3$ with a quaternary ammonium chloride salt (RCl) such as 1-ethyl-3-methylimidazolium chloride (EMIC),[38,39] 1-butyl-3-methylimidazolium chloride (BMIC),[40–42] trimethylphenylammonium chloride (TMPAC)[43,44] and benzyltrimethyl ammonium chloride (BTMAC).[45] The fundamental studies reported by Osteryoung an co-workers,[46,47] Lai and Skyllas-Kazacos[48] and Hussey and co-workers,[49] have classified the chloroaluminates into Lewis basic, neutral and acidic depending on the relative mole fraction of the organic salt (X_R) and $AlCl_3$ (X_{Al}). In the Lewis basic ($X_{Al} < 0.5$) and the neutral ($X_{Al} = 0.5$) melts the major aluminum species is $AlCl_4^-$ in which Al(III) is fully coordinated by Cl^- ions, rendering the reduction potential of Al(III)/Al couple more negative than the cathodic electrochemical window of the melts and thus, metallic Al cannot be electrodeposited from the basic melts. On the other hand, in the Lewis acidic melts ($0.5 < X_{Al} < 0.67$), the major aluminum species is $Al_2Cl_7^-$ of which the number of Cl^- ions coordinated with Al(III) is less than that of $AlCl_4^-$, making the reduction of Al(III)/Al shift toward much positive potential so that the electrodeposition of Al metal is possible according to the equation:

$$4Al_2Cl_7^- + 3e^- \leftrightarrow Al + 7AlCl_4^- \quad [14.1]$$

The deposition process normally involves with either instantaneous or progressive three-dimensional nucleation with hemispherical diffusion-controlled growth depending on the substrate. The morphology of the Al deposit is affected, in addition to the deposition potential or current, by the type of cation forming the ionic liquid. For example, compact, shining and adherent Al films have been successfully electrodeposited from EMI–$AlCl_3$[38] but the Al obtained from TMPAC–$AlCl_3$ is fairly coarse.[43] Reviews on the electrodeposition of Al and alloys from chloroaluminates are available.[50,51]

Recently nanocrystalline Al has been electrodeposited from air- and water-stable ionic liquids based on the bis(trifluoromethylsulfonyl) amide (Tf_2N) anion with cations such as *N*-butyl-*N*-methyl pyrrolidinum

Table 14.1 Examples of ionic liquids containing $Al_2Cl_7^-$ anion for electrodeposition of aluminum

Ionic liquid	References
$EMI^+ Al_2Cl_7^-$	38, 39, 46–49, 56–59
$BMI^+ Al_2Cl_7^-$	40–42
$TMPA^+ Al_2Cl_7^-$	43, 44
$BMP^+Tf_2N^-/AlCl_3$ (saturated)	38, 52–54
$P_{14,6,6,6,6}{}^+Tf_2N^-/AlCl_3$ (saturated)	52

($BMP–Tf_2N$)[38,52–54] and trihexyl-tetradecyl phosphonium ($P_{14,6,6,6,6}–Tf_2N$)[52] saturated with $AlCl_3$. The cation of the ionic liquid influences profoundly the chemical and electrochemical behavior; biphasic behavior was observed for the $EMI–Tf_2N$ and $BMP–Tf_2N$ ionic liquids upon addition of $AlCl_3$ within certain concentration ranges at temperature < 80°C, and Al films could be deposited from the upper phase. Shining, dense, and adherent nanocrystalline Al were deposited from $AlCl_3$-saturated $BMP–Tf_2N$, but coarse cubic-shaped Al particles were obtained from the $AlCl_3$-saturated $EMI–Tf_2N$ ionic liquid. The $P_{14,6,6,6,6}–Tf_2N$ ionic liquid does not show biphasic behavior upon addition of $AlCl_3$. However, Al could only be electrodeposited from this ionic liquid–$AlCl_3$ mixture containing more than 4M of $AlCl_3$ after being heated to 150°C. More recently, a preliminary study shows that Al can also be electrodeposited from an *N*-butyl-*N*-methyl pyrrolidinum dicyanamide (BMP–DCA) ionic liquid containing $AlCl_3$.[55] It is noteworthy that in all these processes, the highly water and air-sensitive $AlCl_3$ is used exclusively as the Al(III) source, and thus, the experiments need to be performed in a glove box filled with inert gas such as Ar and N_2. Some typical ionic liquids employed for electrodeposition of Al are summarized in Table 14.1.

Although the deposition of Al from ionic liquids has been intensively studied, corrosion behavior of the as-deposited Al films has not yet been fully explored. Recently, we have initiated a study on the deposition of Al and Al–Zn on Mg alloy, AZ91D, and demonstrated that the corrosion behavior of Al-coated AZ91D is greatly improved.[56–59] The results from these studies are described in the following section.

14.3 Electrochemical characteristics of $AlCl_3$–EMIC ionic liquids

Figure 14.1 shows the cyclic voltammogram of a tungsten electrode performed in room temperature $AlCl_3$–EMIC with a molar ratio of 60:40 ionic liquid. For brevity, the $AlCl_3$–EMIC mixture, consisting of a 60% molar fraction of $AlCl_3$, is denoted as 60 m/o ionic liquid. Similarly, the ionic liquids with $AlCl_3$–EMIC molar ratios of 50:50, 53:47 and 57:43 are denoted in this

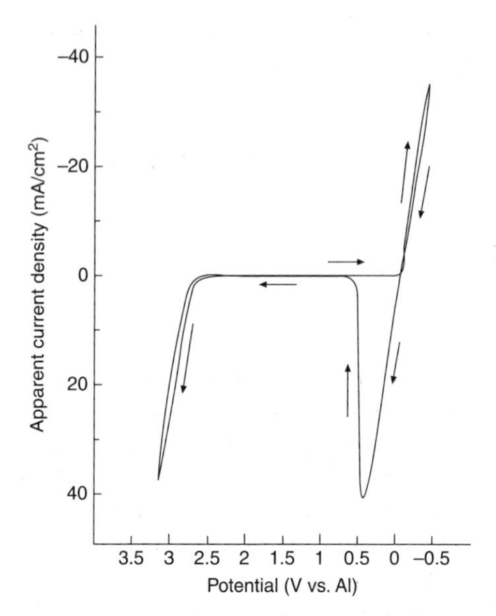

14.1 Cyclic voltammogram of a tungsten electrode in the 60 m/o $AlCl_3$–EMIC ionic liquid, room temperature, potential scan rate: 50 mV/s.

chapter as 50 m/o, 53 m/o and 57 m/o ionic liquid, respectively. As the potential is scanned from the open-circuit potential (approximately 0.9 V) towards the negative direction, the only cathodic reaction takes place around –0.2 V, corresponding to the reduction of $Al_2Cl_7^-$ precursor in the ionic liquid to form Al, according to reaction [14.1] specified above.[49,60–63] On reverse scanning, the sharp anodic peak corresponds to Al dissolution. Further scanning to the more positive potential, the oxidation of $AlCl_4^-$ according to the following reaction occurs at 2.6 V:

$$4AlCl_4^- \rightarrow 2Al_2Cl_7^- + Cl_2 + 2e^- \qquad [14.2]$$

By decreasing the concentration of $AlCl_3$ to 53 m/o, the cathodic peak can still be observed but with a decreasing intensity. At 50 m/o $AlCl_3$, the absence of the cathodic peak indicates that Al cannot be deposited at such composition. It is also noted that the Al deposition potential shifts to a more positive value with increasing $AlCl_3$ molar fraction, indicating a thermodynamic favor of Al deposition.

Co-deposition of Al and Zn is also feasible. Figure 14.2 shows the cyclic voltammograms of a glassy carbon electrode obtained at room temperature in the 60 m/o ionic liquid with 1 wt% $ZnCl_2$ addition. Curve **a** in Fig. 14.2 displays the cyclic voltammogram obtained in 60 m/o $AlCl_3$–EMIC without $ZnCl_2$ addition, in the potential range of –0.5 to 2.5V, consistent with that shown in Fig. 14.1.

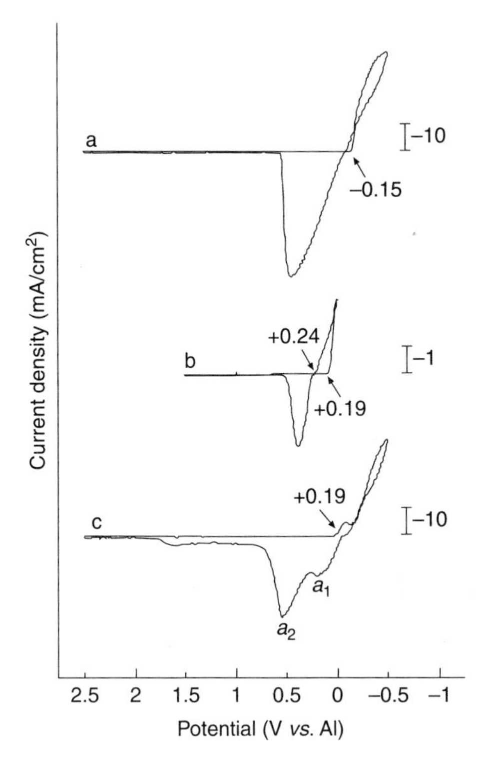

14.2 Cyclic voltammograms on a glassy carbon electrode at room temperature and at a scan rate of 50 mV/s: curve a: 60 m/o $AlCl_3$–EMIC (scan range: –0.5 ~ +2.5 V); curve b: 1 wt% $ZnCl_2$ + 60 m/o $AlCl_3$–EMIC (scan range: 0 ~ +1.5V), and curve c: 1 wt% $ZnCl_2$ + 60 m/o $AlCl_3$–EMIC (scan range: –0.5 ~ +2.5V).

Curve b in Fig. 14.2 depicts the cyclic voltammogram obtained at potentials ranging from 0 to +1.5V. The cathodic scan shows that an abrupt increase in current density occurs at about +0.19V, indicating the reduction of Zn. The intercept of the reverse scan at zero current density is found at +0.24 V, above which Zn cannot be reduced. The peak in the anodic scan is associated with the dissolution of Zn. The cyclic voltammogram for an enlarged potential scan range of –0.5 to +2.5 V is shown as curve c in Fig. 14.2. Two cathodic peaks are observed showing the sequential reduction of Zn and Al. The initial potential for Zn reduction, as indicated in curve c is +0.19 V, identical to that revealed in curve b. The second peak at the more negative potential, where the current density begins to rise again, corresponds to Al reduction. The results shown in Fig. 14.2 suggest that co-deposition of Zn and Al can occur at potentials below –0.08 V. The two anodic peaks at a_1 and a_2 represent the sequential anodic reactions for Al and Zn respectively.

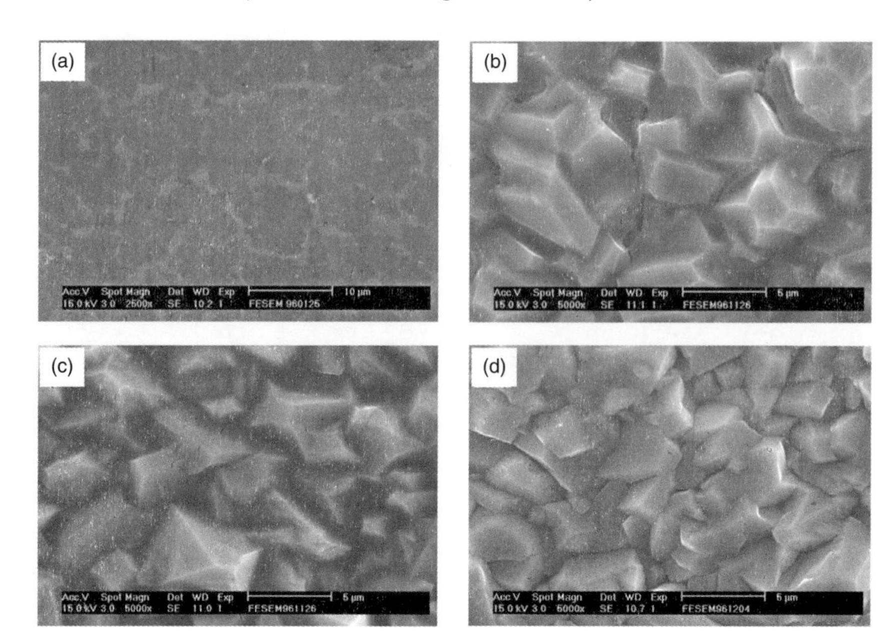

14.3 SEM top-view micrographs of (a) the bare AZ91D Mg alloy; (b), (c) and (d) the Al-coated Mg samples deposited at –0.2 V (*vs.* Al) in the 53, 57 and 60 m/o ionic liquids, respectively.

14.4 Material characteristics

Electrodeposition of Al, either under constant potential or constant current density condition, greatly depends on the acidity or chemical composition of the ionic liquid employed. In the $AlCl_3$–EMIC ionic liquid with 50 m/o of $AlCl_3$, electrodeposition of Al is hindered. As the concentration of $AlCl_3$ increases above 53 m/o, Al coating can take place. The surface morphologies of the uncoated and Al-coated AZ91D Mg alloy, observed under a scanning electron microscope (SEM), are demonstrated in Fig. 14.3. Figure 14.3a shows a micrograph of the bare AZ91D Mg alloy, demonstrating its dual-phase microstructure. Figure 14.3b–14.3d display the Al-coated sample deposited at –0.2 V (*vs.* Al) in the 53, 57 and 60 m/o ionic liquids; respectively. It is found that the entire surface of the Mg alloy, regardless of α or β phase, is covered by a layer of Al with granular appearance. The properties of deposited film are greatly affected by the deposition condition. Figure 14.4 shows the effect of deposition potential on the surface morphology and cross-section micrograph of the Al-coated AZ91D Mg alloy. Figure 14.4a gives the SEM micrograph revealing a fine and uniform granular surface morphology of Al film formed at –0.2 V. At a more negative potential, namely –0.4 V, the Al film exhibits a coarse and non-uniform surface feature with crevices existing in the film (Fig. 14.4b).

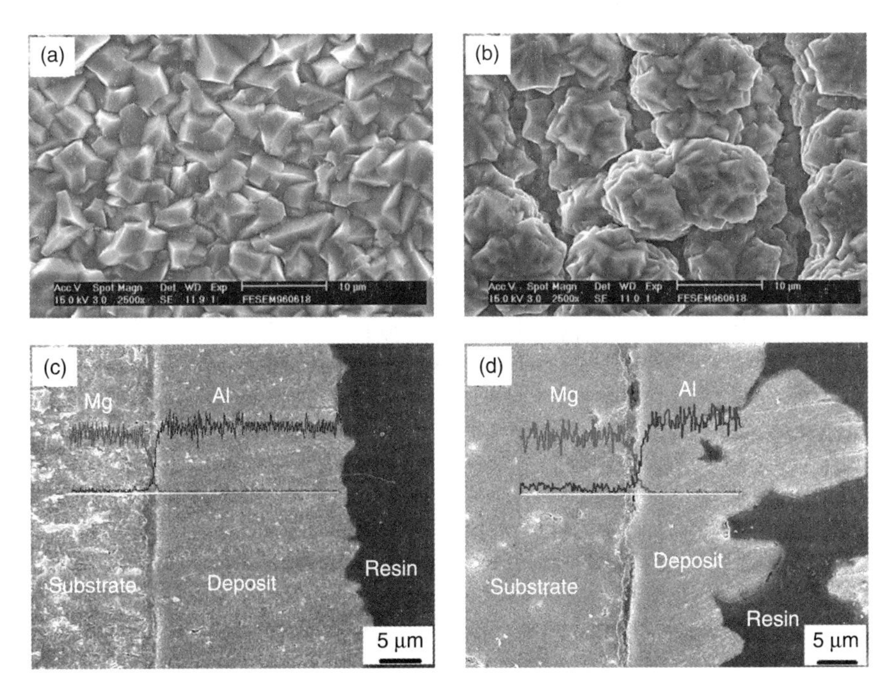

14.4 SEM surface micrographs of Al-coated AZ91D Mg alloy, electrodeposited at (a) –0.2 V (*vs.* Al) and (b) –0.4 V; and the cross-section images and energy dispersive spectroscopy (EDS) line scans of the Al film formed at (c) –0.2 V and (d) –0.4 V.

The corresponding cross-section SEM micrographs of the Al-coated AZ91D Mg alloy are displayed in Fig. 14.4c and 14.4d. The energy dispersive spectroscopy (EDS) line scans showing in each figure confirm the formation of Al film on the Mg alloy surface. The thickness of Al film formed at –0.2 V is rather uniform as compared with that formed at –0.4 V. More defects, such as voids and crevices, etc., are found in the film electrodeposited at a more negative potential. The quality of the Al film also depends on the deposition current density. As reported elsewhere,[57] a porous and less compact Al film is observed if it is electrodeposited from an ionic liquid at a higher current density.

The deposition of Al from various $AlCl_3$–EMIC ionic liquids on the Mg alloy surface is confirmed by EDS and X-ray diffraction analysis (XRD). The EDS result for each of the above Al-coated AZ91D Mg alloy only reveals the spectrum for Al element. The absence of the alloying elements of the substrates indicates that the coated layer can be quite thick. Figure 14.5 compares the XRD patterns of the Mg substrate (curve a) with those electroplated in the ionic liquid containing 53 m/o (curve b), 57 m/o (curve c) and 60 m/o (curve d) $AlCl_3$. The characteristic peaks for

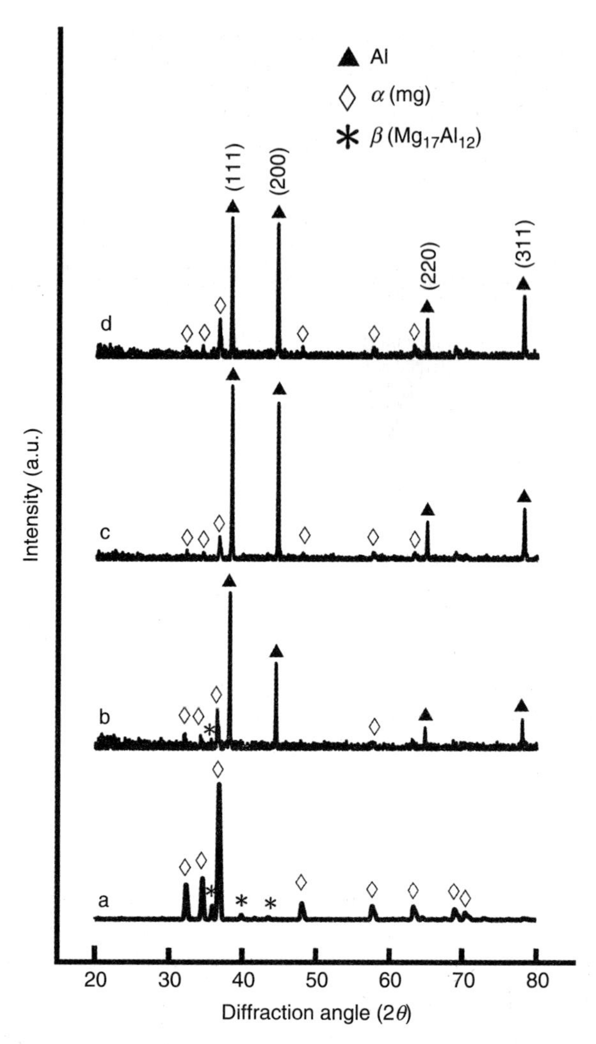

14.5 X-ray diffraction patterns of the various samples. Curve a presents the bare AZ91D Mg alloy. Curves b, c and d present the Al-coated Mg samples deposited in the ionic liquids containing 53, 57 and 60 m/o $AlCl_3$, respectively.

metallic Al are clearly displayed in curves c and d in Fig. 14.5. Similar results have been reported elsewhere.[56]

Co-deposition of Al and Zn can be obtained by electrodeposition from $AlCl_3$–EMIC ionic liquids containing $ZnCl_2$. Figure 14.6a demonstrates the SEM micrograph showing the surface morphology of the AZ91D alloy after electrodeposition in 1 wt% $ZnCl_2$ + 60 m/o $AlCl_3$–EMIC ionic liquid at −0.2 V. The corresponding cross-section SEM micrograph shown in Fig. 14.6b

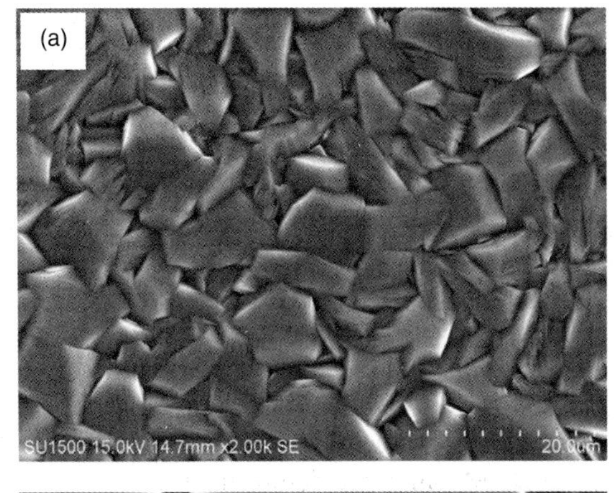

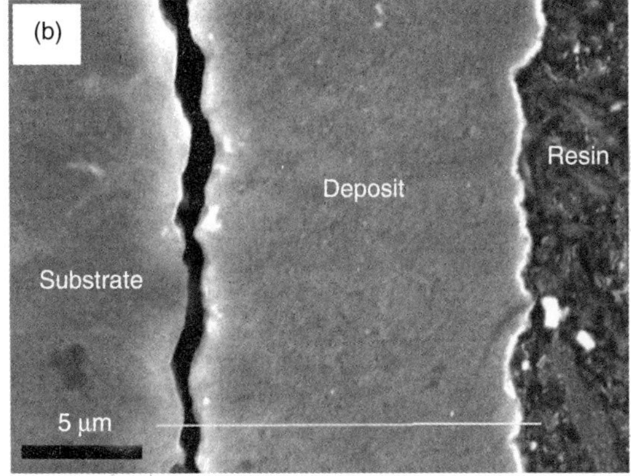

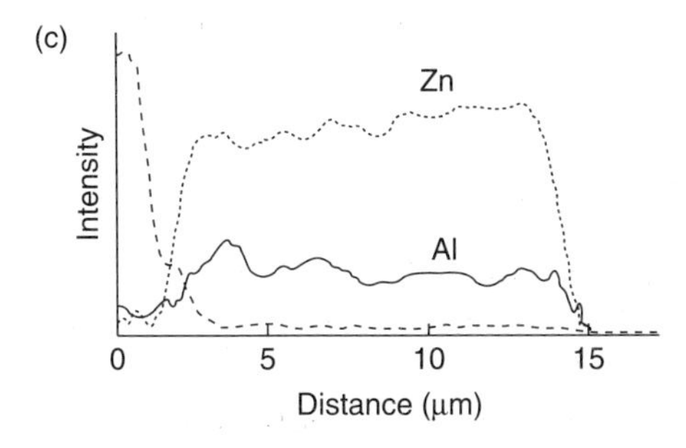

14.6 (a) SEM micrograph, (b) cross-section image and (c) EDS line scans for the deposit formed in 1 wt% $ZnCl_2$ + 60 m/o $AlCl_3$–EMIC ionic liquid at −0.2 V.

indicates that the deposit is uniform in thickness. EDS results show that the distributions of Al, Zn and Mg across the whole thickness are displayed in Fig. 14.6c. As shown in this figure, co-deposition of Al and Zn with uniform chemical composition can be obtained by electrodeposition in the above ionic liquid at an appropriate controlled potential. The surface roughness and the uniformity of chemical composition of the electrodeposited film are affected by the deposition potential as reported elsewhere.[59]

14.5 Electrochemical and corrosion resistance of aluminum (Al) and aluminum/zinc (Al/Zn)-coated magnesium (Mg) alloys

The corrosion resistance of various Al- and Al/Zn-coated AZ91D Mg alloys has been evaluated by salt spray and electrochemical methods. In the following, the results from salt spray test, polarization curve and electrochemical impedance measurements manifesting the beneficial effects of Al and/or Al–Zn coating on AZ91D Mg alloy are demonstrated.

14.5.1 Salt spray test

The salt spray test was performed in accordance with the specification of the American Society of Testing and Materials B117. For the as-polished AZ91D Mg alloy, the digital micrograph showing the shiny surface appearance is depicted in Fig. 14.7a. After exposing in the salt spray chamber for 1 h, fast corrosion occurs on the bare Mg alloy surface causing the loss of metallic luster, as can be seen in Fig. 14.7b. The digital micrograph of the Al as-coated AZ91D Mg alloy is demonstrated in Fig. 14.7c. After 8 h salt spray test, the Al-coated Mg sample maintains its integrity and silver-gray surface appearance as revealed in Fig. 14.7d. The improved corrosion resistance by Al coating is clearly demonstrated. More detailed results can be found in a previous investigation.[57]

14.5.2 Polarization behavior

The polarization behavior of the bare and Al-coated AZ91D Mg alloy has been compared in 3.5 wt% NaCl solution, as can be seen in Fig. 14.8. As illustrated in curve a of Fig. 14.8, the anodic curve of the bare Mg alloy exhibits an active dissolution behavior. The passive region can hardly be seen. The plateau region at very high current density corresponds to the limiting current density. For the Mg alloy with Al electrodeposited at –0.2 V either from the 53 m/o or the 60 m/o ionic liquid, a wide passive region with sufficient low passive current density is observed as demonstrated in curves b and c in Fig. 14.8. The passivity obtained for the Al-coated Mg alloy is attributed to the formation of aluminum oxide or hydroxide on the surface of Al film, which can consequently provide adequate corrosion protection of the Mg alloy substrate. As shown in Fig. 14.8, curve b has a high passive

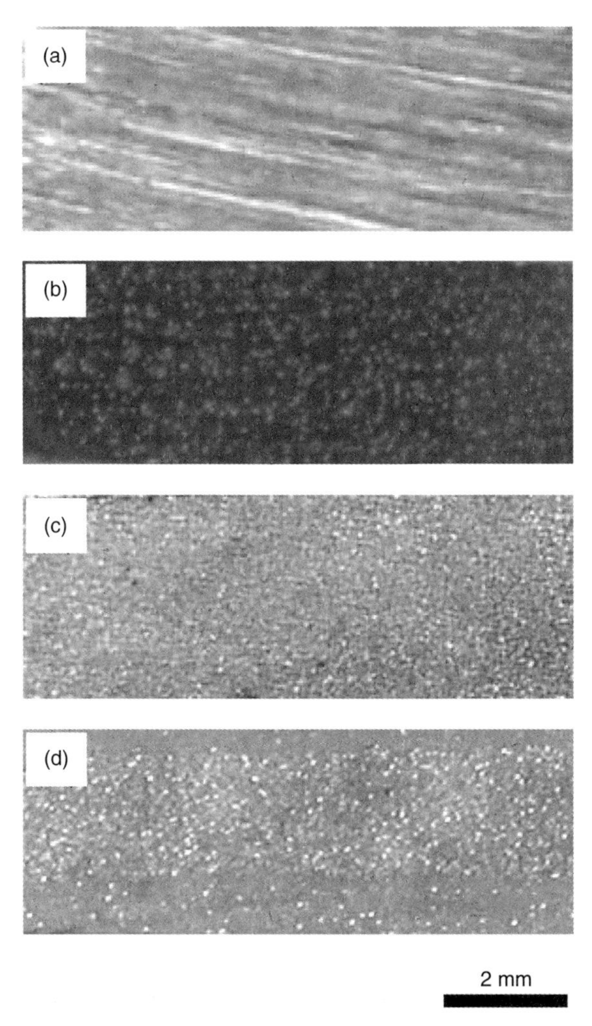

14.7 Digital micrographs of AZ91D Mg alloy (a) as-polished, (b) after salt spray for 1 h; and Al-coated AZ91D Mg alloy (c) as-coated, (d) after salt spray for 8 h.

current density than that of curve c. The difference is attributed to the more protective film formed in 60 m/o ionic liquid as compared with that formed in 53 m/o ionic liquid. The passivity of Al-coated Mg alloy can vary if Al film formed in ionic liquid is electrodeposited at different conditions. It has been found that a wider passive region with a lower passive current density can be obtained if the Al is electrodeposited at a lower applied current[57] or a low cathodic overpotential.[56] Under such conditions, the deposition rate is slow, leading to the formation of a more protective film with low defect concentration as manifested in Fig. 14.4.

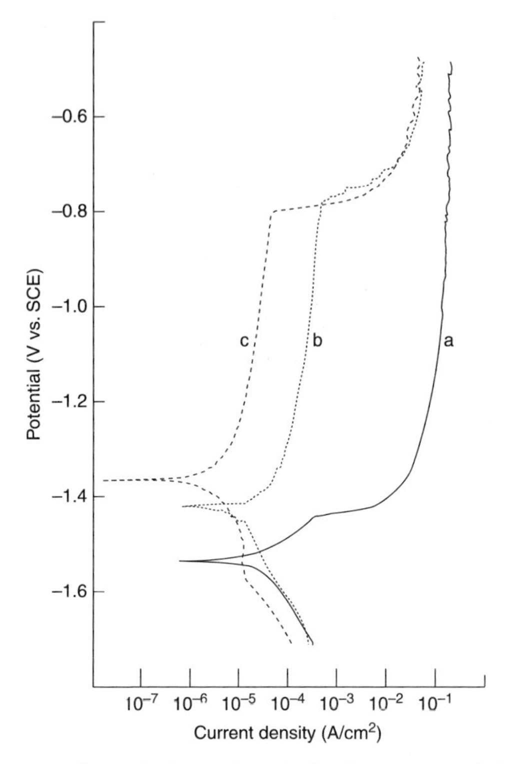

14.8 Potentiodynamic polarization curves of the bare Mg alloy (curve a) and the Al-coated Mg samples deposited in the 53 m/o (curve b) and 60 m/o (curve c) ionic liquids, respectively.

The potentiodynamic polarization curve of an Al/Zn-coated AZ91D Mg alloy in 3.5 wt% NaCl solution is demonstrated in Fig. 14.9, in comparison with those of the bare and the Al-coated Mg alloys. The Al/Zn coating was formed under the same condition as shown in Fig. 14.6. The results show that Al/Zn film can also be passivated in 3.5 wt% NaCl solution. However, the passive region becomes narrower and the passive current density is higher. The presence of Zn in the coating causes a slightly deterioration of the passivity as compared with that of a pure Al film.

14.5.3 Electrochemical impedance spectroscopy (EIS)

The Electrochemical impedance spectroscopy (EIS) results for the Mg alloy without and with surface Al coated from 53 to 60 m/o ionic liquid, respectively, are depicted in Fig. 14.10. For bare Mg alloy, the polarization resistance was about 470 Ωcm^2. A substantial increase in the polarization resistance, as evidenced by an enlarged diameter of the semicircle of the

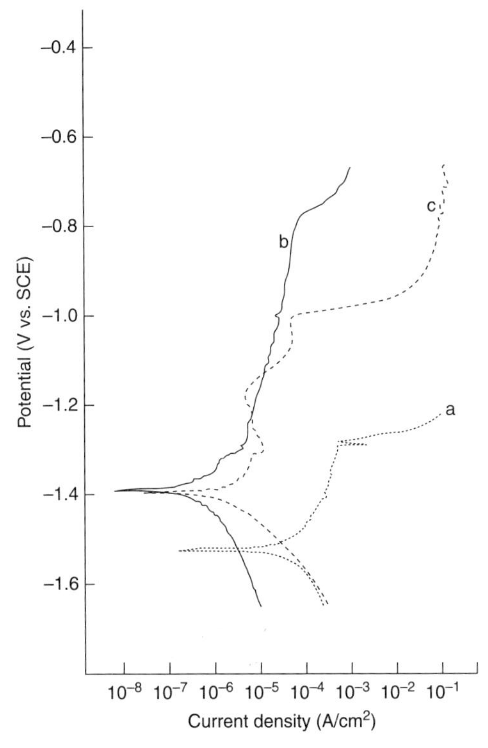

14.9 Comparison of the potentiodynamic polarization curves measured in 3.5 wt% NaCl solution: curve a, AZ91D alloy; curve b, Al deposit formed in $AlCl_3$–EMIC at −0.2 V (vs. Al); curve c, Al–Zn co-deposited at −0.2 V (*vs.* Al) in $AlCl_3$–EMIC + 1 wt% $ZnCl_2$.

Nyquist plot, can be obtained for Mg alloy if it is electroplated with Al. For those with surface Al electrodeposited at −0.2 V from 53 to 60 m/o ionic liquid, the polarization resistance in 3.5 wt% NaCl solution are 3000 and 5200 Ωcm^2, respectively. The results were consistent with those revealed in the polarization curves demonstrated in Fig. 14.8. The improved polarization resistance of AZ91D Mg alloy with Al coating from ionic liquid is clearly demonstrated. However, the passivity or the polarization resistance of the Al-coated Mg alloy depends on the deposition conditions. The Al film formed in more acidic $AlCl_3$–EMIC and at a lower deposition rate renders a better passivation behavior.

14.6 Conclusions

The $AlCl_3$–EMIC ionic liquid with proper chemical composition can be used as a successful electrolyte for Al electrodeposition on Mg alloy. Above the critical chemical composition, the deposition efficiency increases with

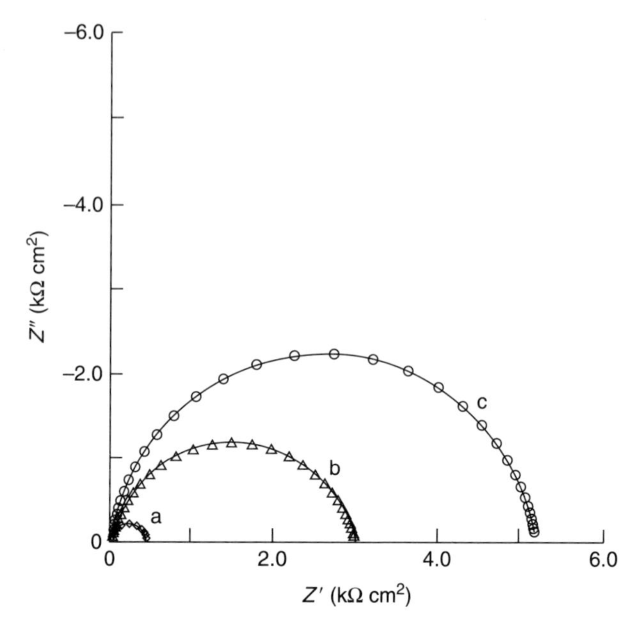

14.10 Nyquist plots of the bare Mg alloy (curve a) and the Al-coated Mg samples deposited in the 53 m/o (curve b) and 60 m/o (curve c) ionic liquids, respectively.

increasing acidity of the ionic liquid. With the presence of Zn(II) ion in the $AlCl_3$–EMIC ionic liquid, co-deposition of Al and Zn on the Mg alloy can be achieved. The formation of a barrier Al and/or Al/Zn surface film can provide adequate corrosion resistance of the Mg alloy substrate. In 3.5 wt% NaCl solution, Al coating can cause an increase in the polarization resistance of AZ91D Mg alloy from 470 to 5200 Ω cm^2, about one order of magnitude higher. The potentiodynamic polarization curves show that the passivity of the Al-coated Mg alloy depends on the deposition condition. A wider passive region with a lower passive current density can be obtained if the Al is electrodeposited at a lower applied current or a low cathodic overpotential. The passivity of the Al-coated Mg alloy is slightly degraded with the co-existence with Zn.

In conclusion, the electrodeposition of Al on AZ91D from the EMI–$AlCl_3$ ionic liquid is feasible. Since the deposition is performed at ambient temperature, the thermal stress in the substrate can be avoided. Without the presence of water, the oxidation of Mg and the formation of hydrogen pinhole during electrodeposition can be prevented. With these advantages, the formation of a more compact and adherent Al and/or Al–Zn coating on Mg alloy surface can be achieved. Recently, electroless plating of Al on glass substrate in EMI–$AlCl_3$ ionic liquid has been reported.[64,65] Furthermore, it has been demonstrated that corrosion protective Mg–Al intermetallic surface layer can be formed on AZ91D by immersing the AZ91D specimens in a NaCl–$AlCl_3$ molten salt at 300°C for 7 h.[66] Taken together, the electrodeposition combining with electroless plating and conversion coating may provide directions for developing corrosion protective Al and alloy coatings on Mg alloys using ionic liquids.

14.7 Acknowledgement

The authors would like to thank Dr Jeng-Kuei Chang and Dr Szu-Jung Pan for assisting the preparation of this manuscript.

14.8 References

1. G.L. Makar and J. Kruger, 'Corrosion studies of rapidly solidified magnesium alloys', *Journal of the Electrochemical Society*, **137** (1990), 414–421.
2. G. Song, A. Atrens, D. St John, X. Wu, and J. Nairn, 'The anodic dissolution of magnesium in chloride and sulphate solutions', *Corrosion Science*, **39** (1997), 1981–2004.
3. R. Ambat, N.N. Aung and W. Zhou, 'Evaluation of microstructural effects on corrosion behaviour of AZ91D magnesium alloy', *Corrosion Science*, **42** (2000), 1433–1455.
4. G. Song, A. Atrens and M. Dargusch, 'Influence of microstructure on the corrosion of diecast AZ91D', *Corrosion Science*, **41** (1999), 249–273.
5. G. Song, 'Recent progress in corrosion and protection of magnesium alloys', *Advanced Engineering Materials*, **7** (2005), 563–586.
6. J.E. Gray and B. Luan, 'Protective coatings on magnesium and its alloys – a critical review', *Journal of Alloys and Compounds*, **336** (2002), 88–113.
7. Y. Zhang, C. Yan, F. Wang, H. Lou and C. Cao, 'Study on the environmentally friendly anodizing of AZ91D magnesium alloy', *Surface and Coatings Technology*, **161** (2002), 36–43.
8. H.Y. Hsiao and W.T. Tsai, 'Characterization of anodic films formed on AZ91D magnesium alloy', *Surface Coatings and Technology*, **190/2–3** (2005), 299–308.
9. H.Y. Hsiao, H.C. Tseng and W.T. Tsai, 'Anodization of AZ91D magnesium alloy in silicate-containing electrolytes', *Surface and Coatings Technology*, **199/2–3** (2005), 127–134.
10. H.Y. Hsiao and W.T. Tsai, 'Effect of heat treatment on anodization and electrochemical behavior of AZ91D magnesium alloy', *Journal of Materials Research*, **20** (2005), 2763–2771.
11. M.A. Gonzalez-Nunez, C.A. Nunez-Lopez, P. Skeldon, G.E. Thompson, H. Karimzadeh, P. Lyon and T.E. Wilks, 'A non-chromate conversion coating for magnesium alloys and magnesium-based metal matrix composites', *Corrosion Science*, **37** (1995), 1763–1772.
12. H. Huo, Y. Li and F. Wang, 'Corrosion of AZ91D magnesium alloy with a chemical conversion coating and electroless nickel layer', *Corrosion Science*, **46** (2004), 1467–1477.

13. L.Y. Niu, Z.H. Jiang, G.Y. Li, C.D. Gu and J.S. Lian, 'A study and application of zinc phosphate coating on AZ91D magnesium alloy', *Surface and Coatings Technology*, **200** (2006), 3021–3026.
14. X. Cui, Y. Li, Q. Li, G. Jin, M. Ding and F. Wang, 'Influence of phytic acid concentration on performance of phytic acid conversion coatings on the AZ91D magnesium alloy', *Materials Chemistry and Physics*, **111** (2008), 503–507.
15. X. Cui, Q. Li, Y. Li, F. Wang, G. Jin and M. Ding, 'Microstructure and corrosion resistance of phytic acid conversion coatings for magnesium alloy', *Applied Surface Science*, **255** (2008), 2098–2103.
16. Z. Yong, J. Zhu, C. Qiu and Y. Liu, 'Molybdate/phosphate composite conversion coating on magnesium alloy surface for corrosion protection', *Applied Surface Science*, **255** (2008), 1672–1680.
17. X. Chen, G. Li, J. Lian and Q. Jiang, 'An organic chromium-free conversion coating on AZ91D magnesium alloy', *Applied Surface Science*, **255** (2008), 2322–2328.
18. C. Gu, J. Lian, G. Li, L. Niu and Z. Jiang, 'Electroless Ni–P plating on AZ91D magnesium alloy from a sulfate solution', *Journal of Alloys and Compounds*, **391** (2005), 104–109.
19. J. Chen, G. Yu, B. Hu, Z. Liu, L. Ye and Z. Wang, 'A zinc transition layer in electroless nickel plating', *Surface and Coatings Technology*, **201** (2006), 686–690.
20. C. Gu, J. Lian, J. He, Z. Jiang and Q. Jiang, 'High corrosion-resistance nanocrystalline Ni coating on AZ91D magnesium alloy', *Surface and Coatings Technology*, **200** (2006), 5413–5418.
21. H. Zhao, Z. Huang and J. Cui, 'A new method for electroless Ni–P plating on AZ31 magnesium alloy', *Surface and Coatings Technology*, **202** (2007), 133–139.
22. N. El Mahallawy, A. Bakkar, M. Shoeib, H. Palkowski and V. Neubert, 'Electroless Ni–P coating of different magnesium alloys', *Surface and Coatings Technology*, **202** (2008), 5151–5157.
23. W.X. Zhang, Z.H. Jiang, G.Y. Li, Q. Jiang and J.S. Lian, 'Electroless Ni-P/Ni-B duplex coatings for improving the hardness and the corrosion resistance of AZ91D magnesium alloy', *Applied Surface Science*, **254** (2008), 4949–4955.
24. H.H. Elsentriecy and K. Azumi, 'Electroless Ni–P deposition on AZ91D magnesium alloy prepared by molybdate chemical conversion coatings', *Journal of the Electrochemical Society*, **156** (2009), D70–D77.
25. Y.F. Jian, L.F. Liu, C.Q. Zhai, Y.P. Zhu and W.J. Ding, 'Corrosion behavior of pulse-plated Zn-Ni alloy coatings on AZ91D magnesium alloy in alkaline solutions', *Thin Solid Films*, **484** (2005), 232–237.
26. Y.F. Jian, C.Q. Zhai, L.F. Liu, Y.P. Zhu and W.J. Ding, 'Zn–Ni alloy coatings pulse-plated on magnesium alloy', *Surface and Coatings Technology*, **191** (2005), 393–399.
27. L. Zhu, W. Li and D. Shan, 'Effects of low temperature thermal treatment on zinc and/or tin plated coatings of AZ91D magnesium alloy', *Surface and Coatings Technology*, **201** (2006), 2768–2775.
28. C.A. Huang, T.H. Wang, T. Weirich and V. Neubert, 'Electrodeposition of a protective copper/nickel deposit on the magnesium alloy (AZ31)', *Corrosion Science*, **50** (2008), 1385–1390.
29. W. Simka, D. Puszczyk and G. Nawrat, 'Electrodeposition of metals from non-aqueous solutions', *Electrochimica Acta*, **54** (2009), 5307–5319.

30. Y. Zhao and T.J. VanderNoot, 'Electrodeposition of aluminium from non-aqueous organic electrolytic systems and room temperature molten salts', *Electrochimica Acta*, **42** (1997), 3–13.
31. E. Peled and E. Gileadi, 'The electrodeposition of aluminum from aromatic hydrocarbon', *Journal of the Electrochemical Society*, **123** (1976), 15–19.
32. W. Kautek and S. Birkle, 'Aluminum-electrocrystallization from metal–organic electrolytes', *Electrochimica Acta*, **34** (1989), 1213–1218.
33. S.P. Shavkunov and T.L. Strugova, 'Electrode processes upon electrodeposition of aluminum in aromatic solvents' *Elektrokhimiya*, **39** (2003), 714–721.
34. L. Legrand, A. Tranchant and R. Messina, 'Aluminium behaviour and stability in $AlCl_3/DMSO_2$ electrolyte', *Electrochimica Acta*, **41** (1996), 2715–2720.
35. T. Jiang, M.J. Chollier Brym, G. Dubé, A. Lasia and G.M. Brisard, 'Studies on the $AlCl_3$/dimethylsulfone ($DMSO_2$) electrolytes for the aluminum deposition processes', *Surface and Coatings Technology*, **201** (2007), 6309–6317.
36. D.E. Couch and A. Brenner, 'A hydride bath for the electrodeposition of aluminum', *Journal of the Electrochemical Society*, **99** (1952), 234–244.
37. K. Ziegler and H. Lehmkuhl, 'Elektrolytische abscheidung von aluminium', *Angewandte Chemie*, **67** (1955), 424–425.
38. Q.X. Liu, S. Zein El Abedin and F. Endres, 'Electroplating of mild steel by aluminium in a first generation ionic liquid: a green alternative to commercial Al-plating in organic solvents', *Surface and Coatings Technology*, **201** (2006), 1352–1356.
39. T. Jiang, M.J. Chollier Brym, G. Dubé, A. Lasia and G.M. Brisard, 'Electrodeposition of aluminium from ionic liquids: Part I – electrodeposition and surface morphology of aluminium from aluminium chloride ($AlCl_3$) – 1-ethyl-3-methylimidazolium chloride ([EMIm]Cl) ionic liquids', *Surface and Coatings Technology*, **201** (2005), 1–9.
40. C.L. Aravinda, B. Burger and W. Freyland, 'Nanoscale electrodeposition of Al on n-Si(111): H from an ionic liquid', *Chemical Physics Letters*, **434** (2007), 271–275.
41. S. Caporali, A. Fossati, A. Lavacchi, I. Perissi, A. Tolstogouzov and U. Bardi, 'Aluminium electroplated from ionic liquids as protective coating against steel corrosion', *Corrosion Science*, **50** (2008), 534–539.
42. G. Yue, X. Lu, Y. Zhu, X. Zhang and S. Zhang, 'Surface morphology, crystal structure and orientation of aluminium coatings electrodeposited on mild steel in ionic liquid', *Chemical Engineering Journal*, **147** (2009), 79–86.
43. T. Jiang, M.J. Chollier Brym, G. Dubé, A. Lasia and G.M. Brisard, 'Electrodeposition of aluminium from ionic liquids: Part II – studies on the electrodeposition of aluminum from aluminum chloride ($AlCl_3$) – trimethylphenylammonium chloride (TMPAC) ionic liquids', *Surface and Coatings Technology*, **201** (2006), 10–18.
44. Y. Zhao and T.J. VanderNoot, 'Electrodeposition of aluminium from room temperature $AlCl_3$–TMPAC molten salts', *Electrochimica Acta*, **42** (1997), 1639–1643.
45. A.P. Abbott, C.A. Eardley, N.R.S. Farley, G.A. Griffith and A. Pratt, 'Electrodeposition of aluminium and aluminium/platinum alloys from $AlCl_3$/benzyltrimethylammonium chloride room temperature ionic liquids', *Journal of Applied Electrochemistry*, **31** (2001), 1345–1350.

46. J. Robinson and R.A. Osteryoung, 'The electrochemical behavior of aluminum in the low temperature molten salt system *n* butyl pyridinium chloride: aluminum chloride and mixtures of this molten salt with benzene', *Journal of the Electrochemical Society*, **127** (1980) 122–128.
47. B.J. Welch and R.A. Osteryoung, 'Electrochemical studies in low temperature molten salt systems containing aluminum chloride', *Journal of Electroanalytical Chemistry*, **118** (1981), 455–466.
48. P.K. Lai and M. Skyllas-Kazacos, 'Aluminium deposition and dissolution in aluminium chloride – n-butylpyridinium chloride melts', *Electrochimica Acta*, **32** (1987), 1443–1449.
49. Q. Liao, W.R. Pitner, G. Stewart, C.L. Hussey and G.R. Stafford, 'Electrodeposition of aluminum from the aluminum chloride – 1-methyl-3-ethylimidazolium chloride room temperature molten salt + benzene', *Journal of the Electrochemical Society*, **144** (1997), 936–943.
50. G.R. Stafford and C.L. Hussey, 'Electrodeposition of transition metal-aluminum alloys from chloroaluminatr molten salts', in *Advances in Electrochemical Science and Engineering,* Vol. 7, R.C. Alkire and D.M. Kolb (Ed.), Wiley-VCH Verlag GmbH, (2002).
51. F. Endres, A.P. Abbott and D.R. MacFarlane (Ed.), *Electrodeposition from Ionic Liquids,* Wiley-VCH Verlag GmbH (2008).
52. S. Zein El Abedin, E.M. Moustafa, R. Hempelmann, H. Natter and F. Endres, 'Electrodeposition of nano- and microcrystalline aluminium in three different air and water stable ionic liquids', *ChemPhysChem*, **7** (2006), 1535–1543.
53. Q.X. Liu, S. Zein El Abedin and F. Endres, 'Electrodeposition of nanocrystalline aluminum: Breakdown of imidazolium cations modifies the crystal size', *Journal of the Electrochemical Society*, **155** (2008), D357–D362.
54. P. Eiden, Q. Liu, S. Zein El Abedin, F. Endres and I. Krossing, 'An experimental and theoretical study of the aluminium species present in mixtures of $AlCl_3$ with the ionic liquids [BMP]Tf_2N and [EMIm]Tf_2N', *Chemistry – A European Journal*, **15** (2009), 3426–3434.
55. M.J. Deng, P.Y. Chen, T.I. Leong, I.W. Sun, J.K. Chang and W.T. Tsai, 'Dicyanamide anion based ionic liquids for electrodeposition of metals', *Electrochemistry Communications*, **10** (2008), 213–216.
56. J.K. Chang, S.Y. Chen, W.T. Tsai, M.J. Deng and I.W. Sun, 'Electrodeposition of aluminum on magnesium alloy in aluminum chloride ($AlCl_3$) – 1-ethyl-3-methylimidazolium chloride (EMIC) ionic liquid and its corrosion behavior', *Electrochemistry Communications*, **9** (2007), 1602–1606.
57. J.K. Chang, S.Y. Chen, W.T. Tsai, M.J. Deng and I.W. Sun, 'Improved corrosion resistance of magnesium alloy with a surface aluminum coating electrodeposited in ionic liquid', *Journal of the Electrochemical Society*, **155** (2008), C112–C116.
58. J.K. Chang, I.W. Sun, S.J. Pan, M.H. Chuang, M.J. Deng and W.T. Tsai, 'Electrodeposition of Al coating on Mg alloy from Al chloride – 1-ethyl-3-methylimidazolium chloride ionic liquids with different Lewis acidity', *Transactions of the Institute of Metal Finishing*, **86** (2008), 227–233.
59. S.J. Pan, W.T. Tsai, J.K. Chang and I.W. Sun, 'Co-deposition of Al–Zn on AZ91D magnesium alloy in $AlCl_3$ – 1-ethyl-3-methylimidazolium chloride ionic liquid', *Electrochimica Acta*, **55** (2010), 2158–2162.

60. P.K. Lai and M. Skyllas-Kazacos, 'Electrodeposition of aluminium in aluminium chloride – 1-methyl-3-ethylimidazolium chloride', *Journal of Electroanalytical Chemistry*, **248** (1988), 431–440.
61. R.T. Carlin and R.A. Osteryoung, 'Aluminum anodization in a basic ambient temperature molten salt', *Journal of the Electrochemical Society*, **136** (1989), 1409–1415.
62. T.J. Melton, J. Joyce, J.T. Maloy, J.A. Boon and J.S. Wikes, 'Electrochemical studies of sodium chloride as a Lewis buffer for room temperature chloroaluminate molten salts', *Journal of the Electrochemical Society*, **137** (1990), 3865–3869.
63. R.L. Perry, K.M. Jones, W.D. Scott, Q. Liao and C.L. Hussey, 'Densities, viscosities, and conductivities of mixtures of selected organic cosolvents with the Lewis basic aluminum chloride + 1-methyl-3-ethylimidazolium chloride molten salt', *Journal of Chemical and Engineering Data*, **40** (1995), 615–619.
64. I. Shitanda, A. Sato, M. Itagaki, K. Watanbe and N. Koura, 'Electroless plating of aluminum using diisobutyl aluminum hydride as liquid reducing agent in room-temperature ionic liquid', *Electrochimica Acta*, **54** (2009), 5889–5893.
65. N. Koura, H. Nagase, A. Sato, S. Kumakura, K. Takeuchi, K. Ui, T. Tsuda and C. K. Loong, 'Electroless plating of aluminum from a room-temperature ionic liquid electrolyte', *Journal of the Electrochemical Society*, **155** (2008), D155–D157.
66. M. He, L. Liu, Y. Wu, Z. Tang and W. Hu, 'Improvement of the properties of AZ91D magnesium alloy by treatment with a molten $AlCl_3$–NaCl salt to form Mg–Al intermetallic surface layer', *Journal of Coating Technology Research*, **6** (2009), 407–411.

15

Cold spray coatings to improve the corrosion resistance of magnesium (Mg) alloys

V. K. CHAMPAGNE and B. GABRIEL,
US Army Research Laboratory, USA and
J. VILLAFUERTE,
CenterLine (Windsor) Ltd, Canada

DOI: 10.1533/9780857098962.3.414

Abstract: Magnesium alloys are widely used in aircraft and automotive components because of their inherent light weight compared to other engineering metals. However, premature corrosion is one of the challenges associated with magnesium. In the aerospace industry, many expensive castings cannot be reclaimed adequately because current methods are often inappropriate due to excessive porosity, oxidation and thermal damage. Cold spray is a solid-state coating process that uses a supersonic gas jet to accelerate small particles against a substrate to produce metal bonding by rapid plastic deformation of the impacting particles. Significantly, magnesium components can be repaired, restored and protected by cold spray deposition of pure aluminum which has thus already been specified by some aerospace users as a standard technique for magnesium repair. In this chapter, the state of the art of commercial cold spray technologies for magnesium repair will be discussed, as well as their advantages and limitations compared to traditional thermal processes.

Key words: cold spray, high velocity deposition, supersonic spray, magnesium alloys, powder spray, corrosion protection, commercial equipment.

15.1 Introduction

Magnesium is the lightest of all structural metals, being approximately 35% lighter than aluminum and 78% lighter than steel with exceptional stiffness and damping capacity (Yamauchi *et al.*, 1991). Magnesium's low density and good castability are ideal for mass-production and weight sensitive applications, such as automobile and aircraft components.

15.1.1 Magnesium applications in the automotive and aerospace industries

In the automotive industry, the percentage of magnesium used in vehicles has been traditionally low, with an average of about 5 kg in a typical domestic model in unexposed interior components. However, over the last few years, the demand for lighter, more fuel-efficient vehicles, has spurred interest and extended the use of magnesium to components, such as engine cradles and transmission housings, which are typically exposed to the corrosive conditions of roads. The widespread use of magnesium in aircraft and rotorcraft began during the 1960s to reduce weight and increase performance. Ever since, magnesium alloys have been used for the fabrication of various aircraft components including transmission and gearbox housings, flap drive gearboxes and pilot yokes.

Magnesium is electrochemically very active being anodic to all other structural metals. It will corrode preferentially when coupled with virtually any other metal in a corrosive medium (Gonzales *et al.*, 1995). Therefore, magnesium must be protected against galvanic corrosion in mixed-metal systems. As an example, many of the corrosion problems associated with magnesium helicopter components occur at the contact points between inserts or mating parts, where ferrous metals are located, creating galvanic couples (Gorman and Woolsey, 2003). In addition, magnesium alloys are also susceptible to surface damage due to impact, which occurs frequently during manufacture and/or overhaul and repair. Scratches from improper handling or tool marks can result in preferential corrosion sites. The United States Department of Defense (DoD) and the aerospace industry have expended much effort over the last two decades to develop specific surface treatments to prevent corrosion, increase surface hardness and combat impact damage for magnesium alloys as an effort to prolong equipment service life (Griffin and Zuniga, 2005). Yet, many modern magnesium corrosion protection strategies still face environmental, reliability and/or economical challenges. Additionally, the means to provide dimensional restoration to large areas on components, where deep corrosion occurs, had remained a challenge until recent advancements in cold spray technology.

15.1.2 Cost of magnesium corrosion

The USA DoD alone reportedly spends millions of dollars to mitigate magnesium corrosion in their various aircraft fleets (Champagne, 2007). Many expensive magnesium components are decommissioned because of the lack of an appropriate repair technique to provide dimensional restoration after corrosion has been ground down (Champagne, 2007). The Army and Navy helicopter fleet is comprised of more than 4500 aircraft, each having numerous

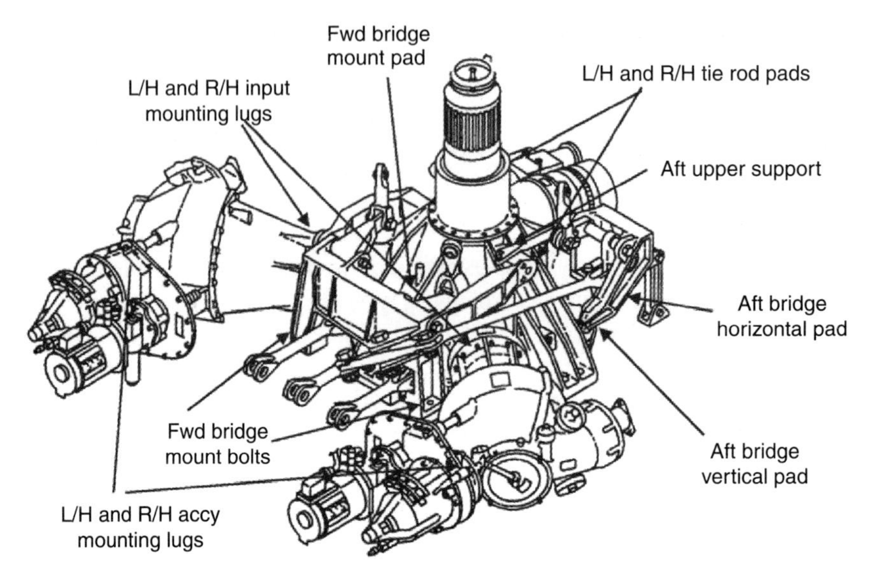

15.1 Schematic of the main transmission housing for the H-60 helicopter showing the areas most susceptible to corrosion.

components manufactured from magnesium alloys, generally consisting of transmission and gearbox housings. In addition to helicopters, magnesium alloy gearboxes are used extensively in fixed-wing aircraft for the military and are currently designed into the Joint Strike Fighter (JSF). Add to this the thousands of commercial aircraft flying around the world and this would bring up to the ten of thousands the number of magnesium components currently in service. The cost of each component ranges from several thousand to over $800000 dollars. In 2001, the Fleet Readiness Center conducted an extensive review of the cost of corrosion on the main transmission of one type of helicopter; they reported that, from 1991–2000, the total estimated cost for both unscheduled maintenance and module replacement was about $41 million. Figure 15.1 shows a schematic of the main transmission housing for the H-60 helicopter showing the areas most susceptible to corrosion. Figure 15.2 shows the extent of corrosion damage on one of the mounting pads after one overhaul cycle. Because of the localized nature of the corrosion, surface treatments intended to mitigate the problem would only have to be applied in these specific areas. In summary, the susceptibility of magnesium components to corrosion and damage, even with the most current surface protection schemes, is still very significant as shown by the very high repair/overhaul and maintenance costs.

15.2 Cold spray technology

The technology associated with very-high-velocity, low-temperature spray deposition of coatings falls under the terminology of cold gas-dynamic

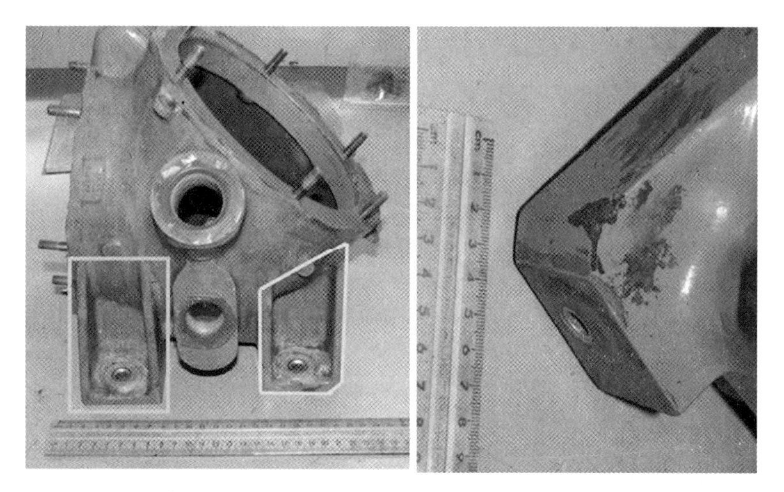

15.2 Extent of corrosion damage on one of the mounting pads after one overhaul cycle.

spraying or 'Cold Spray' as it is referred to in the published literature and throughout the industry. This technique was first demonstrated in the mid-1980s at the Institute of Theoretical and Applied Mechanics in Novosibirsk, Russia, where coatings were produced by injecting solid metal particles into a gas stream accelerated to supersonic velocities. Research in the U.S. began in 1994 under a project sponsored by the US National Center for Manufacturing Sciences (NCMS). The principal organizations conducting R&D work on the technology in the initial years were the Applied Research Laboratory at Penn State University and Sandia National Laboratories, followed by the US Army Research Laboratory (ARL). Meanwhile, other R&D consortiums were established in Germany in partnership with the Helmut Schmidt University of Hamburg.

15.2.1 Description

Cold spray is a materials deposition process whereby combinations of metallic and non-metallic particles are consolidated to form coatings or free-standing structures by means of ballistic impingement upon a suitable substrate (Steenkiste, 1999; Papyrin, 2001; Stoltenhoff *et al.*, 2002). The particles utilized are in the form of commercially available powders, typically ranging in size from 5 to 100 μm, which can be accelerated from 300 to over 1200 m/s by injection into a high-velocity stream of gas. The high velocity gas stream is generated via the expansion of a pressurized, preheated gas through a converging-diverging DeLaval nozzle. The pressurized gas is expanded to supersonic velocity, with an accompanying decrease

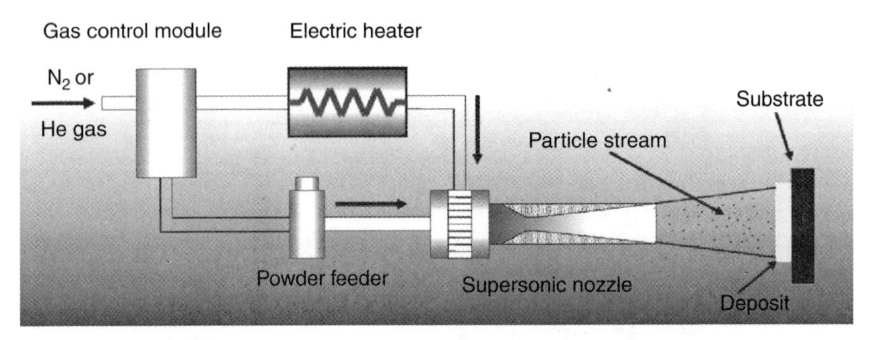

15.3 Schematic of the cold spray process (upstream injection).

in pressure and temperature (Dykhuisen and Smith, 1998; Kosarev *et al.*, 2003; Grujicic *et al.*, 2004). The particles, initially carried by a separate gas stream, are injected into the nozzle either prior to the throat of the nozzle (upstream injection) or downstream of the throat (downstream injection). The particles are subsequently accelerated by the main nozzle gas flow and impacted onto a substrate after exiting the nozzle (Fig. 15.3). Upon impact, the solid particles deform and create a bond with the substrate (Dykhuisen *et al.*, 1999; Grujicic *et al.*, 2003). As the process continues, particles continue to impact the substrate and form bonds with the consolidated material resulting in a uniform deposit with little porosity and high bond strength. The term 'cold spray' has been used to describe this process due to the expected low temperatures (–250°C to 100°C) of the expanded gas stream as it exits the nozzle. The latter is illustrated in Fig. 15.4, which uses a cold spray model developed by Helfritch *et al.* (Champagne, 2007) to calculate the temperature of the gas as it expands from the nozzle as a function of inlet temperature and pressure.

15.2.2 Advantages and limitations

The temperature of the gas stream is always below the melting point of the particulate material being sprayed. The resultant consolidated material is formed in the solid state and because of this, the mechanical characteristics of the cold sprayed material are unique in many regards. The low temperatures associated with the cold spray process are particularly desirable for use on thermally sensitive materials like magnesium as the risks associated with grain growth, phase transformations, oxidation and tensile thermal stresses are minimal or non-existent. Cold spray may produce coatings with a cohesive (particle–particle bond strength) and adhesive (coating–substrate bond strength) property superior to that of the magnesium substrate. Further advancement of techniques to manufacture 'cold spray

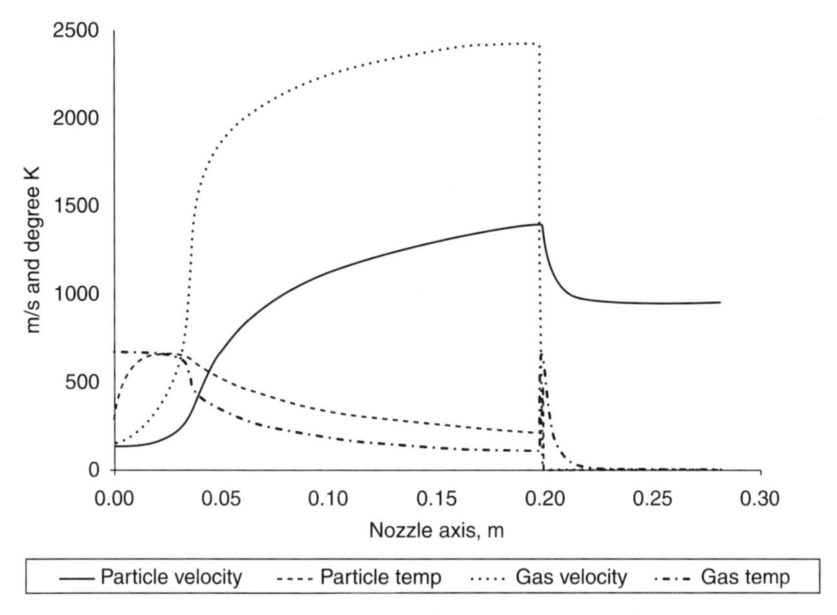

15.4 Model showing the temperature of the gas as it expands from the nozzle as a function of inlet temperature and pressure (Champagne, 2007).

grade' powders have increased the ability to produce a variety of coatings with greater mechanical strength when compared to conventional thermal spray, powder metallurgy and wrought material.

The density of cold spray material also compares favorably to theoretical values of material produced by conventional ingot metallurgy techniques. Figures 15.5 and 15.6 show typical examples of 6061 aluminum and tantalum, respectively, that have been deposited by the cold spray process and have achieved 100% theoretical density.

The particle impact associated with the cold spray process is conducive to the formation of beneficial compressive stresses as surface impact acts like shot peening (Fig. 15.7). This is ideal for the use of cold spray coatings for repair as compressive stresses impede crack growth. Additionally, compressive stresses beneath the surface are preferred since the corresponding tensile stresses are not located near the coating interface. Other advantages of the cold spray process include:

- High deposition rates, which can be equivalent or superior to other thermal spray processes.
- High deposition efficiencies, which can be equivalent or superior to other thermal spray processes.
- Ability to produce functionally graded coatings.

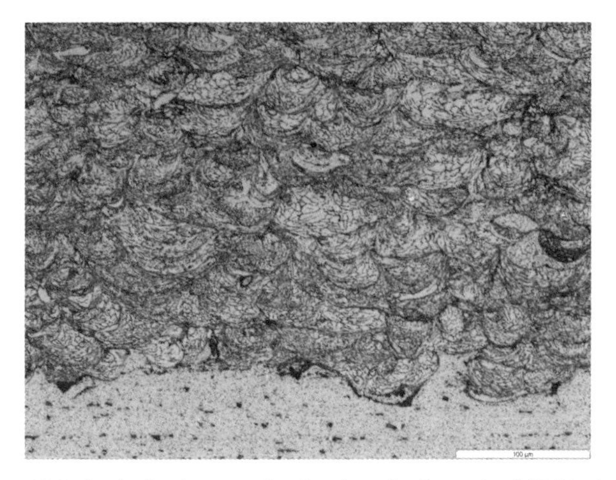

15.5 Optical micrograph of a chemically etched 6061 aluminum alloy cold spray coating deposited on a 6061-T6 aluminum substrate.

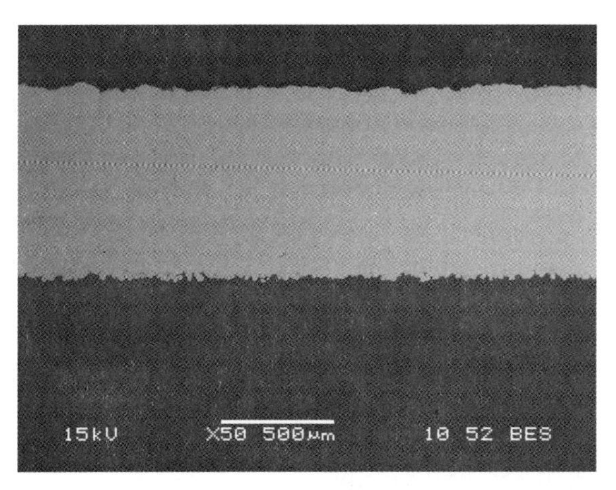

15.6 Scanning electron microscope micrograph of a tantalum cold spray coating deposited on aluminum alloy substrate.

- Low heat input that allows coating of a broad range of materials as well as over ultra thin substrates.
- No toxic gases, radiation or chemical reactions.
- Localized, well-defined deposition with minimum over-spray which eliminates most requirements for masking.
- Ability to produce extremely thick coatings and free-standing forms.

Some of the limitations of the technology are:

- Line-of-sight can be limiting for application on deep internal diameters or complex geometries.

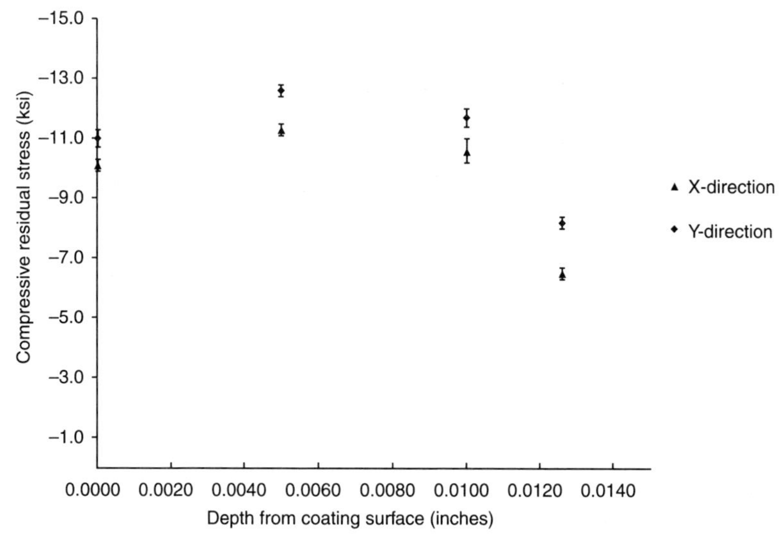

15.7 X-ray diffraction (XRD) residual stress measurements of 6061 and CP-aluminum cold spray coatings, 300 μm thick, showing state of compressive stresses at and under the surface of the coating.

- Coating properties may be degraded when the particle angle-of-incidence to the substrate is other than 90° ± 15°.
- Coatings materials must contain particles with some minimum ductility or mix with particulates that can undergo plastic deformation.

15.2.3 Comparison of cold spray to thermal spray

Cold spray technology is the logical choice for applications where the high temperatures associated with conventional thermal spray technology can result in undesirable thermal effects, as explained earlier. These characteristics make cold spray suitable for corrosion repair of magnesium and its alloys.

Tensile residual stresses arising from thermal contraction during solidification and cooling are some of the disadvantages of conventional thermal spray processes which promote delamination and limit the maximum thickness that can be achieved. This problem is compounded when the substrate material differs from the coating material in terms of hardness, density and coefficient of thermal expansion. In contrast, with cold spray there is very little thermally-induced dimensional change since consolidation of the particles takes place in the solid-state. Also, the significantly high impact velocities of the solid particulates are very effective at peening the underlying material and producing deposits which are typically in a state of compressive stress (Champagne, 2007).

In comparison with conventional thermal spray techniques, the materials produced by cold spray are typically less porous, with higher hardness and conductivity and lower oxide concentration. In some cases, the Young's moduli of cold sprayed deposits have been reported to be greater than 80% of bulk values (Amateau and Eden, 2000). This is ascribed to the fact that high velocity impact creates dense, well consolidated material. The superior qualities of cold sprayed materials are required by certain applications. For example, the high heat transfer coefficient and electrical conductivity of cold spray materials favor their use in electronic applications (McCune *et al.*, 2000).

15.2.4 Carrier gases

Helium, nitrogen or compressed air are the most commonly used gases with cold spray, with helium providing maximum particle velocities and subsequently the possibility of obtaining better coating quality. In fact, the consolidation of higher melting point and/or less ductile materials is often achieved more easily with the use of helium. However, nitrogen or compressed air are generally preferred because of their favorable economics. By optimizing temperature and pressure, adequate adhesion, cohesion, density and corrosion resistance may be also achieved with nitrogen, in particular when spraying commercially pure Aluminum for the repair of magnesium components. Helium may still be best for spraying powders made of 5XXX, 6XXX and 7XXX series aluminum alloys, in order to meet stringent aerospace quality standards.

For high value assets, using expensive powders and in a production environment, the extra cost associated with the use of helium may be justifiable. Today, there are commercial helium recovery systems capable of recovering up to 90% of the helium used (Johnson, 2004). Recent studies have also suggested the use of mixtures of nitrogen with recycled helium for general industrial applications as these blends may significantly improve quality and productivity at a more reasonable cost.

15.2.5 Nozzle design

The ultimate objective of a cold spray nozzle is to create favorable gas flow conditions at the nozzle exit that maximize the ability of particles to produce quality coatings upon impact with the substrate. Often, a convergent diverging type (DeLaval) nozzle is used to generate favorable supersonic gas flow conditions at the nozzle exit. The characteristics of the supersonic

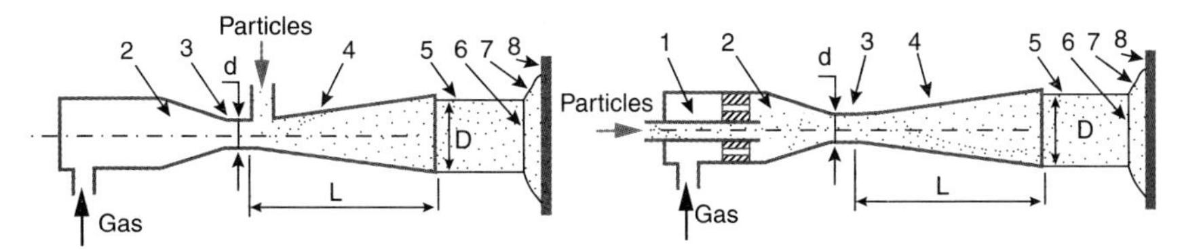

15.8 Downstream and upstream injection nozzle parameters. 1 – Tube for injection of particles, 2 – pre-chamber, 3 – nozzle throat (d = diameter), 4 – divergent supersonic section, 5 – free jet, 6 – bow-shock wave, 7 – compressed layer, 8 – substrate (Papyrin *et al.*, 2007).

flow, including its kinetic and thermal energy content, are a function of the nozzle design and gas parameters (type, pressure and temperature). The amount of energy (thermal and kinetic) actually transferred to the particles strongly depends on the physical characteristics of the powder (e.g. density, shape, size distribution) as well as where and how the particles are injected into the gas stream.

Some important parameters of the nozzle geometry (Fig. 15.8) are the divergence ratio (D/d), divergent shape and length of the diverging section. These geometrical features determine the characteristics of the gas flow inside and at the nozzle exit. As the gas jet leaves the nozzle other conditions prevail including bow-shock wave and compressed layers which may affect particle impact conditions on the substrate. In practice, round nozzle exit diameters range from 2 to 12 mm while throat diameters from 1.0 to 3.0 mm. Large areas are often achieved by a raster approach of subsequent passes with a step-over, which is typically 25% of the total width of the pass. Because of the sharp threshold for stick versus re-bounce conditions, the width of any 'over-spray' is relatively small. Subsequently, a spray pass generally displays well-defined sharp edges with a thickness that is close to the nozzle exit diameter.

Repeatable use of nozzles may eventually cause nozzle clogging and/or internal erosion damage. Some commercial nozzles in use today are designed for quick change, anti-clogging and durable performance. Nozzle clogging is predominant with high-purity and low-melting point metals such as tin, aluminum and indium. In order to minimize clogging, some manufacturers fabricate nozzles using high-temperature polymers that display anti-clogging characteristics. The use of polymers naturally limits the maximum operating gas temperature. Alternatively, nozzle clogging may be avoided by continuous water-cooling of metallic nozzles, though this brings about additional complexities to the design, maintenance and cost of the equipment.

Production spraying of hard materials may eventually result in erosion damage. In upstream injection systems, erosion is predominant at the throat section where the passage of hard particle-gas mixture is constricted.

Therefore, the nozzle throat in upstream injection must be made from erosion-resistant materials. Downstream systems experience erosion somewhere in the divergent side of the nozzle after the injection point. Subsequently, commercial nozzles intended for spraying hard materials are generally made of erosion-resistant materials.

15.2.6 Process optimization

The choice of carrier gas, pressure and gas temperature determine the energy level or enthalpy available in the gas to accelerate particles. The ultimate particle velocity is a function of the nozzle design and characteristics of the powder. It has been well established that impacting particles must exceed a 'critical velocity' in order to deposit instead of bouncing off. The magnitude of the critical velocity can be estimated through the use of empirical relationships, which generally depend on particle material characteristics, such as density, ultimate strength and melting point, as well as the particle temperature immediately before impact (Schmidt *et al.*, 2006). Typically, velocities and temperatures of particles prior to impact are calculated as functions of particle diameter. Particles with velocities higher than critical vclocity will deposit.

Particle velocity can be calculated or measured experimentally. Both techniques are non-trivial and require complex calculation and equipment. Compressible, iso-kinetic flow equations can be used to predict the gas flow for a given nozzle geometry; modified drag and heat transfer coefficients can then be used to iteratively calculate the resulting particle velocities and temperatures. Champagne *et al.* (2005) utilized one-dimensional, frictionless, gas-dynamic calculations in order to predict gas flow velocities for various cold spray operating conditions. Li and Li (2005) made use of commercial computational fluid dynamics (CFD) to optimize nozzle geometry for maximum particle velocity. Pardhasaradhi *et al.* (2008) compared laser illuminated, time of flight velocity measurements with the empirical model given by Alkhimov *et al.* (2001). Jodoin *et al.* (2006) utilized the Reynolds average Navier Stokes equations within a computational platform to model nozzle flow with similar boundary conditions. Samareh *et al.* (2009) used computational fluid mechanics to describe the effect of particle concentration on gas velocity for two nozzle geometries. An example of a computational calculation for the aluminum/helium nozzle acceleration is shown in Fig. 15.9. The particle size for this calculation is 20 μm, and the initial gas pressure and temperature are 2.75 MPa and 20°C. Calculated particle velocities are verified experimentally by means of a dual-slit, laser velocimeter.

A number of researchers have studied the bonding mechanisms associated with cold spray, concluding that cold spray bonding phenomena are

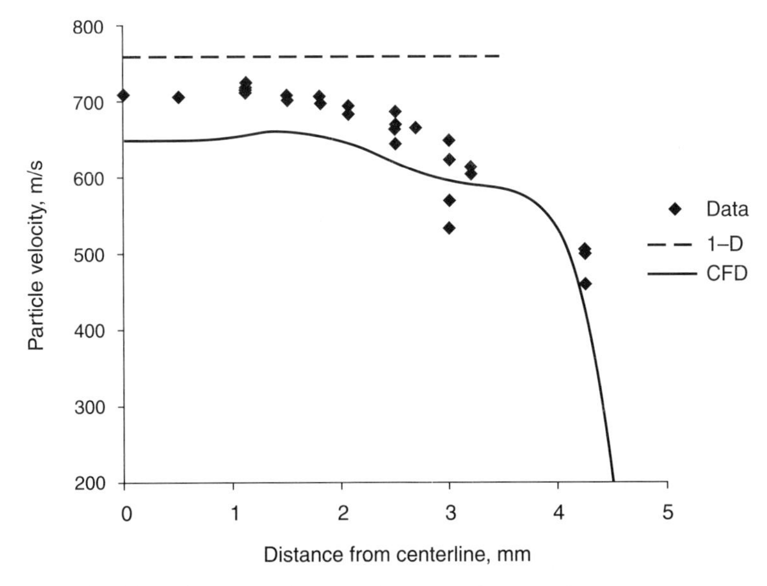

15.9 Particle velocity *vs* distance from nozzle centerline.

similar to those identified in explosive welding, where bond formation relies on deformation under high pressures (Zhang *et al.*, 2005).

15.2.7 Cold spray grade powders

Perhaps the most crucial variable in the cold spray process is the powder itself. Particle morphology, size distribution, porosity, and even grain size have been shown to influence final cold spray consolidation properties (Fuknuma *et al.*, 2006; Helfritch and Champagne, 2006; Hall *et al.* 2008; Gabriel *et al.*, 2011; Kim *et al.* 2011). As an example, Table 15.1 illustrates the hardness values of two types of cold sprayed 5083 aluminum alloy. The composition of these two powders is practically identical, but the grain size of n-5083 is finer due to secondary grinding with a planetary grinding mill. This powder processing difference resulted in significant hardness of the coatings (Helfritch *et al.* 2006; Hall *et al.* 2008; Gabriel *et al.*, 2011; Kim *et al.* 2011).

In the late 1990s, cold spray-specific powders were virtually non-existent on the market; a handful of thermal spray powders developed for plasma spray and high velocity oxy-fuel (HVOF) were used. Early R&D took advantage of these commonly available powders but their limitations for cold spray were soon discovered. Today, there is a wide variety of materials that can be sprayed with the cold spray process. Table 15.2 lists a few of the most commonly deposited materials.

Table 15.1 Vicker's Microhardness of 5083 and n-5083

Description	Powder processing methodology	Coating Vicker's hardness
5083	Atomized	126
n-5083	Atomized then custom planetary mill grinding	261

Table 15.2 Common cold spray materials

Copper	Steels including stainless
Nickel	Bronze and brass
Aluminum	Silver
Tantalum	Nickel and cobalt based super alloys
Zinc	Metal matrix composite
Tin	Amorphous alloys
Titanium	Nano-structured aluminum alloys
Niobium	Quasi-crystalline alloys

Table 15.3 Cold spray powder manufacturers

ACuPowder International, LLC
F.J. Brodmann & Co., LLC
Valimet, Inc
CenterLine (Windsor) Ltd
Praxair, Inc
Sulzer Metco, Inc
H.C. Starck North America Trading, LLC
Accushape, Inc

The development of cold spray-specific powders is an on-going collaborative effort between various universities, institutions, and industries. These efforts have led to the introduction of new cold spray grade powders to the market. In turn, this has helped cold spray transition from R&D to production-ready technology. Table 15.3 provides a partial list of vendors that can provide cold spray compatible powders.

A crucial property for a cold spray compatible powder is good plastic deformation properties. Metals such as copper, aluminum and nickel are naturally among the most commonly deposited materials. Metal matrix composites (MMCs) have been identified as compatible with the cold spray process. These hybrid systems can offer specific benefits over traditional alloy systems including higher hardness, machinability and improved wear resistance. Another interesting aspect is the development of cold spray powders that are sprayable at even lower operating gas pressures and temperatures; in particular, sprayability can be enhanced by the addition of a ceramic component (such as aluminum oxide) to the powder blend as, during deposition, the ceramic particles superimpose a peening action that further densifies the deposited metal. A fraction of the ceramic particles become entrapped in the deposit, further strengthening the matrix due to the presence of well dispersed ceramic particles. This type of composite microstructure also displays enhanced machinability (Leyman and Champagne, 2009; Gabriel *et al.*, 2011).

15.3 Cold spray commercial equipment

Based on the two patented methods (downstream and upstream) of injecting the powder, two families of commercial cold spray equipment, namely downstream injection (also known as low-pressure) and upstream injection (also known as high-pressure) were commercially implemented in the 1990s. One method injects the powder downstream into the low-pressure side, at the diverging side of the DeLaval nozzle (Kashirin *et al.*, 2002). The other method injects the powder upstream into the high-pressure side of the DeLaval nozzle (Alkhimov *et al.*, 1994). The capacity, portability, applicability and cost for each type of commercial equipment result from the specific design requirements dictated by the injection point and required operating pressures and temperatures.

15.3.1 Commercial downstream injection equipment

In downstream injection, the spray powder is introduced radially into the gas jet downstream of the nozzle throat. These systems are capable of preheating air, nitrogen or helium to up to 550°C and from low to medium pressures (4–34 bar/60–500 psi).

Driven by opportunities for field repair, commercial downstream injection systems were initially designed to operate at gas pressures as low as 4 bar (60 psi) and relatively low gas temperatures. Downstream injection systems operating at low pressures work without the need for a pressurized powder feeder because, at low pressures, there is negative pressure created at the divergent side of the nozzle. When designed for low operating pressures and temperatures, downstream injection systems can be compact, portable and economical. However, at low gas pressures the maximum attainable particle velocities may limit the range of sprayable materials to low-melting point, ductile metals such as aluminum, zinc and tin. In recent years, manufacturers of downstream injection equipment have raised the operating gas pressures which has widened the range of sprayable materials to include harder metals such as stainless steels, titanium and nickel-based alloys.

The first commercial downstream injection portable system was produced in Russia by the Obninsk Center for Powder Spraying Ltd. under the brand

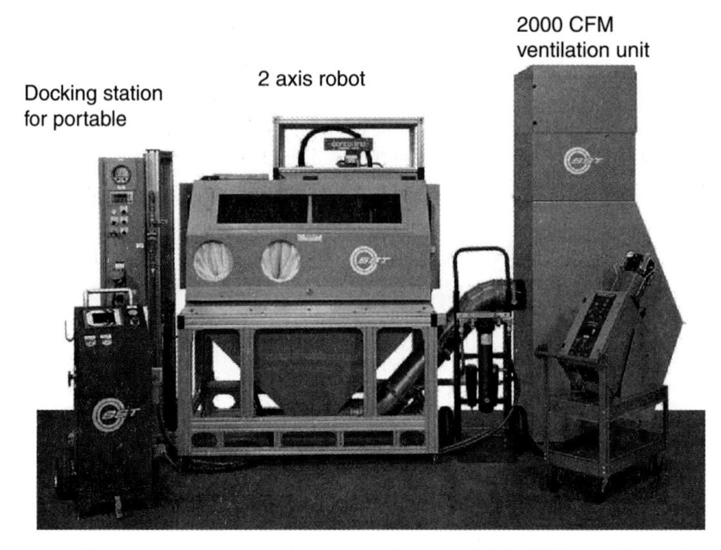

15.10 Downstream injection 17 bar (250 psi) SST™ SERIES P integrated with cabinet, ventilation, and auxiliary systems for hand-held and/or mechanized operation. Photo courtesy of CenterLine (Windsor) Ltd.

name DYMET (http://www.dymet.info) in the late 1990s for localized corrosion repair and dimensional restoration in the field. In the early 2000s DYMET technology was introduced to North America by R. Maev (Maev and Leshchynsky, 2008) for applications within the automotive industry. In 2005, CenterLine (Windsor) Ltd, acquired exclusive rights to commercialize this technology for the North American market (www.supersonicspray.com).

The CenterLine range of SST™ industrial units are production-ready, downstream injection machines based on Kashirin's patent (Kashirin *et al.*, 2002) which can operate at low to medium pressures (4–34 bar/60–500 psi) with air, nitrogen or helium. These systems are engineered to comply with all North American industrial safety standards and are commercially used for corrosion repair, dimensional restoration, metallization and other applications either in the field or in a controlled workshop environment (Figs 15.10 and 15.11) including full integration for mechanized or robotic mass production. Programmable motion is desirable for enhanced consistency and repeatability, as described later in Section 15.3.3.

15.3.2 Commercial upstream injection (high-pressure) equipment

Upstream injection equipment is typically less mobile than downstream injection hardware, simply because of the complexity of the equipment

15.11 SST™ modular straight and 90° nozzle assemblies for internal diameters. Each holder can take standard nozzles tubes made of abrasion resistant material or anti-clogging material. Photo courtesy of CenterLine (Windsor) Ltd.

required to operate at higher pressures and temperatures. Consequently, upstream injection has traditionally been used as a stationary tool to spray specialty materials requiring high impact velocities for bonding. Manufacturers of upstream injection equipment have also introduced portability by downgrading design pressures and temperatures at the expense of limiting the diversity of coating materials that can be sprayed at these operating parameters.

In upstream injection cold spray, helium or nitrogen at medium to high pressures (20–50 bar/300–700 psi) can be preheated to 400–1000°C. The powder feedstock is axially introduced into the gas jet upstream of the nozzle throat, using a specially designed pressurized powder feeder. The powdered material is pre-mixed with the high pressure/high temperature carrier gas. Because of the early interaction with the carrier gas, the conditions for energy transfer to particles are maximized. Additionally, because of the relatively high gas enthalpy at higher pressures and temperatures, high particle velocities can be attained, depending on the available energy as dictated by the gas type, pressure and temperature.

Sulzer Metco AG (Switzerland) (www.coldspraying.info; www.sulzermetco.com) offers commercial upstream injection cold spray equipment including KINETIKS® 2000, KINETIKS® 4000 and KINETIKS® 8000 (Fig. 15.12).

Plasma Giken (http://plasma.co.jp/en/products/index.html) offers commercial upstream injection cold spray equipment including ColdSpray PCS-800 and ColdSpray PCS-1000. These systems are intended for production cold spraying in mechanized or stationary configurations (Fig. 15.13).

15.3.3 Cold spray automation

For accurate control of coating thickness (especially <500 μm) and/or high production rates, automation rather than hand-held operation is often

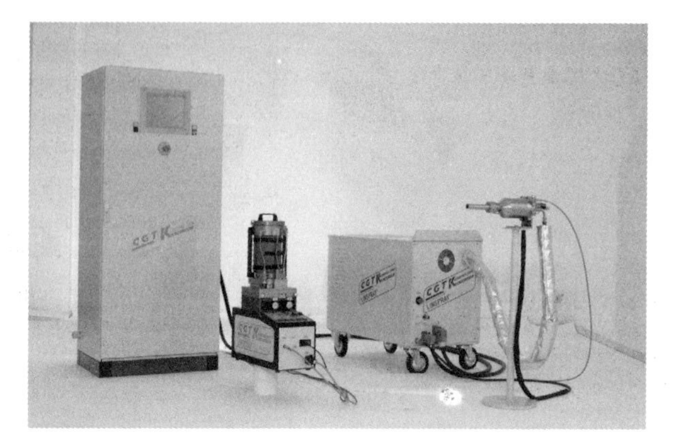

15.12 47KW upstream 40 bar (600 psi) cold spray equipment (KINETIKS® 4000) made by Sulzer Metco AG (Switzerland) showing control unit, powder feeder, 17 KW spray gun, and external 30 KW gas heater. Photo courtesy of Sulzer Metco AG (Switzerland).

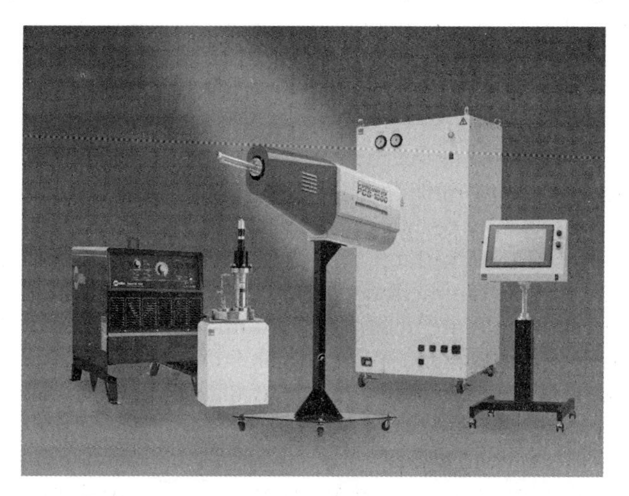

15.13 PCS-1000 upstream 50 bar (700 psi) cold spray system by Plasma Giken. Includes a 70 KW spray gun with a water-cooled nozzle. Maximum gas operating temperature is 1000°C. Photo courtesy of Plasma Giken Co.

preferred. The automation requirements of cold spray are generally the same as per any other thermal spray processes such as plasma spray. Cold spray does not require consideration of regulations pertaining to hazards from combustion, electric arcs and masking for protection against molten particles. Yet, cold spray does require special consideration pertaining handling, collection and disposal of metal dust. For example, in the US and Canada, the cold spraying area should comply with the National Fire

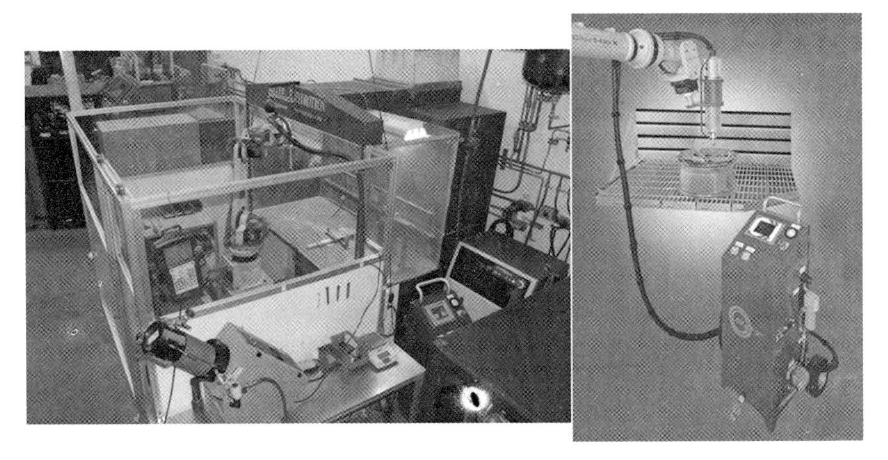

15.14 Seven-axis automation of cold spray using 10 and 100 kg payload industrial robots.

Protection Association (NFPA) codes (NFPA 654, NFPA484-2009, NFPA 69 and NFPA 70) per Class II, Div 1, Group E, depending on the requirements by the local authorities.

There are a number of different ways that cold spray equipment can be automated, including fixed-mount spray gun with moveable part and fixed-part with movable spray gun. In the fixed-mount spray gun configuration, the part can be mounted on single or multiple axis holders or turntables. Cylindrical and shaft type part geometry can be easily processed by rotating the part using a rotating holder. The fixed-mount gun can be (a) stationary or (b) movable along the centerline of the cylindrical part. The latter can be practical for applications that require coating of the inner walls of cylinders or tubes.

In the fixed-part case, the spray gun (5–100 kg) can be deployed by simple XY automation or industrial multiple axis robot (Fig. 15.14). Some cold spray systems are able to adapt to small robot sizes – around 10 kg or less payload – while others require medium to large payloads (20–100 kg). Some cold spray control units have the ability to communicate directly with motion control equipment supplied with commercial robots.

15.3.4 Manual cold spray operation

Portable hand-held cold spray equipment facilitates operation in the field, which is highly desirable for localized cold spray repair of large components that cannot be easily transported to a repair depot (Fig. 15.15). Manual portable repair can be used for large parts at multiple repair sites. Field application is best for non-structural cosmetic repair where quality relies

15.15 Hand-held cold spray repair of localized damage in a large component.

on operator skill. Where a given quality standard must be met, the operator must be certified or qualified to produce the expected quality. Because of the required compactness and ergonomics of any hand-held equipment, maximum gas temperatures and pressures are limited (compared to automated or mechanized configurations). Therefore, the range of materials that can be sprayed manually in the field is generally limited to ductile metals such as aluminum, zinc, copper, tin and their various blends with ceramics and/or other materials.

15.4 Corrosion protection by cold spray

Being the lightest of all structural metals and abundant in the planet, magnesium is ideal for the production of weight-sensitive components. However, one of magnesium's most critical challenges is its service corrosion performance. Although there exist a number of conventional corrosion methods for magnesium, they are not without challenges and limitations. Cold spray technology represents a novel method for localized corrosion protection and dimensional restoration of magnesium components.

15.4.1 Magnesium corrosion

Magnesium is the most reactive (anodic) metal in the electrochemical series with a standard potential of Mg^{2+}/Mg of –2.356 V at 25°C. Therefore, magnesium is prone to severe galvanic corrosion when in contact with other engineering metals. The increasing demand for magnesium in aerospace, military and automotive applications has traditionally driven the development of corrosion-resistant magnesium alloys as well as novel protective coating strategies for magnesium. Over the years, there have been significant advances in magnesium alloy development leading to new and improved commercial alloys. In particular, the addition of aluminum, zinc and manganese, for better corrosion resistance as well as additions of zirconium, rare earths, thorium and silver for better elevated temperature mechanical properties, in combination with the reduction of harmful impurities (e.g. iron, nickel, copper) during the alloy making process, have paved the way to the production of corrosion-resistant, high performance magnesium alloys.

15.4.2 Conventional protection methods for magnesium

Besides the development of more corrosion-resistant magnesium alloys, current methods for general corrosion protection of magnesium include conversion and organic coatings. Although any of these protection techniques may be acceptable to prevent general corrosion of magnesium, they lack the ability to locally protect magnesium in the area of galvanic attack (Zheng *et al.*, 2005). Galvanic corrosion typically occurs within 5 mm of fasteners or dissimilar interfaces. Therefore, one of the methods to combat galvanic attack is to use isolation materials to prevent direct electrical contact between bare magnesium and the dissimilar metal, increasing the electrolytic resistance of the corrosion cell. Where a high torque load is required, such isolation materials must be made of special metals or inorganic substances that can sustain loading without failure or creep. In fact, the use of aluminum washers or shims in dissimilar joints has been a standard practice to stop galvanic corrosion of magnesium.

In conversion coating, the surface of a magnesium component is forced to chemically react in a special chemical bath to produce a uniform and continuous film that protects the material underneath from further corroding. Conversion coatings can be achieved by electrochemical reactions, chemical immersion or by heat treatments. One of these methods is anodizing, where the formation of complex magnesium oxide films is induced under controlled high-voltage anodic polarization conditions. There are a number of proprietary commercial anodizing techniques including Tagnite, Keronite, Magoxide and Anomag.

Chromate coatings are excellent for general corrosion protection, however, with the recent restrictions on the use of hexavalent-chromium during processing, most of these solutions have been banned, and alternative chromate-free chemistries have evolved (US Automotive Materials Partnership, 2006). Low-cost conversion coatings include Alodine and Magpass, which produce protective chemistries that are similar to those with phosphates and chromates; these are obtained by immersion in specially formulated chemical solutions. Any of these techniques can also be combined with a finishing sealer and polymeric top-coat. For less demanding service, organic coatings are often utilized including epoxy, poly-amide, poly-ester, acrylic,

Latex, polyurethane and paraffin based products, which can be applied as powders or as water-based paintable solutions. Unfortunately, organic coatings are prone to localized failure due to poor workmanship or chipping damage, which may result in severe, localized corrosion.

15.4.3 Surface treatment of magnesium and environmental concerns

Both conversion and organic coatings often require stringent surface preparation (water rinsing, alkaline treatment, acid pickling) and post-treatments (neutralization, water rinsing, drying) which pose environmental and health concerns (Avedesian and Baker, 1998).

Much of the industry still relies on hard anodizing using a process designated Dow 17 or a chromate conversion process designated MIL-M-3171, which involve the use of hexavalent-chromium. The electrolyte used to apply the coating comprises sodium dichromate, ammonium acid fluoride and phosphoric acid which are hazardous to the environment and difficult to recycle. These operations also produce harmful vapors and hazardous waste in the form of contaminated wastewaters and solid waste, which can be costly to dispose of. There are also health and safety issues to contend with, such as the formation of hazardous vapors that contain metal salts and carcinogenic substances. These rise up from heated bath solutions and require additional worker protection measures. These surface treatment processes are followed with the application of a phenolic resin sealer, then a chromated primer and top-coat for most surfaces. Robinson (1985) has reported that this surface treatment regimen, which incorporates the use of sealers, can increase the resistance of magnesium to salt spray corrosion. However, regardless of the progress made to improve the corrosion resistance of magnesium alloys, millions of dollars are expended each year on repair and replacement of magnesium alloy components that have corroded or have been damaged. In addition, emerging legislation will be reducing the hexavalent-chromium permissible exposure limit from 52 to 1 $\mu g/m^3$ making the replacement of processes using this compound mandatory. The qualification of the cold spray repair process for magnesium alloy components would represent a significant milestone in the elimination of hexavalent-chromium-containing compounds currently used to provide surface protection.

Plasma and HVOF thermal spray have been investigated for deposition of aluminum, but the results have generally been unsatisfactory due to inconsistent coating integrity. Poor adhesion and delamination of the coatings are typically the cause for high rejection rates. Both of these processes involve the use of high thermal energy to melt or partially melt the coating material before it is accelerated onto the surface of the substrate. Molten or partially melted particles rapidly solidify upon impact with the substrate then contract forming tensile residual stresses on coatings. Besides, the thermal spray pattern is very wide so that it would be difficult to apply the coatings to localized areas requiring repair or rebuild.

15.4.4 Cold sprayed aluminum coatings for corrosion protection

Several major US aerospace manufacturers along with the US DoD determined that the deposition of aluminum and certain aluminum alloys followed by phenolic resin and sealant had the highest probability of success in reducing corrosion and impact damage of magnesium. The development and qualification of the cold spray process to deposit aluminum and aluminum alloys was originally proposed by the Center for Cold Spray Technology at the Army Research Lab (ARL) in 2003, for providing dimensional restoration and corrosion protection to magnesium components, primarily for the aerospace and automotive industries.

The technical challenge was to identify a method to deposit the aluminum alloys onto the magnesium while meeting all service requirements for bond strength and corrosion resistance without sacrificing the structural integrity of the substrate. Additionally, the processing of production parts as well as field repair capability were important in the selection of a viable process.

Large collaborative efforts were established between all US DoD services (Navy, Air Force and Army) as well as aerospace original equipment manufacturers (OEMs) such as Sikorsky Aircraft Corporation and executed by ARL, including the Environmental Security Technology Certification Program (ESTCP) that extended from 2005. The ESCTP program culminated with the qualification of the cold spray process for use on the UH-60 Blackhawk, in collaboration with Sikorsky Aircraft Company and the establishment of the first dedicated cold spray repair facility at the Navy Fleet Readiness Center (FRC-East), Cherry Point, North Carolina (NC). This program served as an international benchmark for the adaptation of cold spray for the aerospace industry. The cold spray process is now viewed as the best possible method for depositing the aluminum alloys to provide dimensional restoration to magnesium components and significantly improving performance and reducing life-cycle costs.

The protective capability of pure aluminum cold sprayed onto magnesium has also been demonstrated by others (McCune and Ricketts, 2004; Balani *et al.*, 2005; Gärtner *et al.*, 2006; Zheng *et al.*, 2006) (Fig. 15.16). In all cases, the corrosion potentials of cold sprayed magnesium coupons approached those of commercially pure aluminum. Such polarization behaviors were promising since there is no galvanic protection strategy that is reasonable

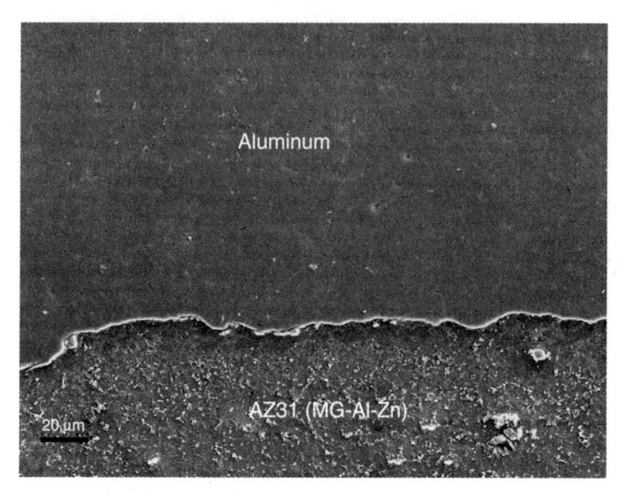

15.16 Scanning electron micrograph illustrating a high-density aluminum cold spray deposit on magnesium alloy AZ31.

15.17 (a) AE44 with the central area cold sprayed with pure aluminum, after 100 hours of ASTM B117 exposure (Courtesy of CANMET- MTL Natural Resources Canada). (b) AM60 with the area surrounding the fastener cold sprayed with high-purity aluminum, after 1000 B117 hours. (Courtesy of NRC IMI Canada).

for magnesium given its strong thermodynamic potential for oxidation. In galvanic corrosion, only small areas surrounding the dissimilar interface require protection, for which cold spray represents an innovative alternative to the use of washers and insulating bushings (Fig. 15.17).

Unscribed B117 salt spray test coupons exceeded 7000 h total exposure with no pinhole defects. An example of a cold spray coating at 500 h exposure

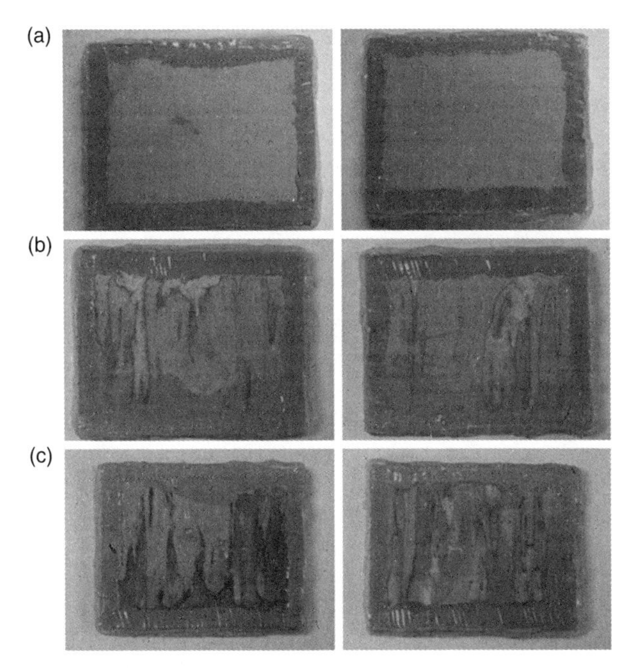

15.18 ASTM B117 salt spray corrosion test specimens of CP-Al cold spray on ZE41A-T5 cast magnesium at (a) no exposure time, (b) 134.5 h, and (c) approximately 500 h. No failure detected at 500 h.

15.19 Before and after images of an aluminum cold spray coating deposition on a Sikorsky SH-60 Intermediate Gearbox Housing.

is shown as Fig. 15.18. Figure 15.19 shows a before and after photograph of a typical cold spray repair on a helicopter magnesium housing. Initial trials by the cold spray facility established at the Navy Depot, FRC-East, NC using the process parameters established by ARL proved successful.

Several aluminum alloy powders have been investigated with cold spray over the years including commercially pure Al (CP-Al), Al-12Si, 5056, 5083,

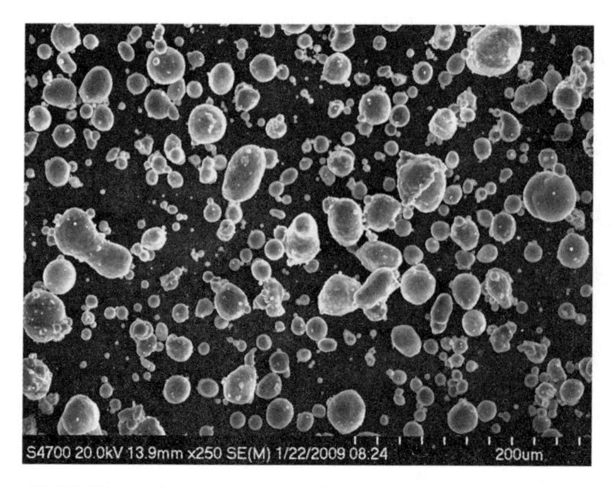

15.20 Scanning electron microscope micrograph 6061 aluminum alloy powder.

5356 aluminum alloys, 7005, 7075, 6061 aluminum alloys and high-purity Al (HP-Al) (Champagne *et al.*, 2008; DeForce *et al.*, 2011). ARL has performed extensive research with CP-Al and 6061. These two compositions have good compatibility with the process and they are readily available on the commercial market.

Characteristics such as particle size and distribution are important for aluminum spraying. Typically aluminum powder size distribution is between 5 and 44 μm. 6061 powder provides an advantage over pure aluminum in terms of hardness and strength, however, at an extra cost compared to CP-Al. A scanning electron microscope micrograph of 6061 inert gas atomized powder is provided as Fig. 15.20.

One of the benefits of the cold spray process is that high strength aluminum coatings can be produced due to a combination of work hardening and grain refinement. Tensile test results for a 6061 cold spray coating are shown in Fig. 15.21, according to ASTM E8 (ASTM E8, 2009). The coating displays a very high strength in the as-sprayed condition; the yield strength (YS) and ultimate tensile strength (UTS) of the cold spray deposit exceed the properties of wrought 6061-T6 (ASM Handbook, Vol 02) with a percent elongation at failure of only 3%. The percent elongation increased to approximately 17% after an in-process annealing heat treatment. The YS and UTS dropped to values typical of wrought 6061 with T4 tempering (ASM Handbook Vol 02).

15.4.5 Cold spray performance and optimization for magnesium components

In 2008, the Defense Standardization Program (DSP) Office granted approval of the first military specification, MIL-STD-3021 for cold spray.

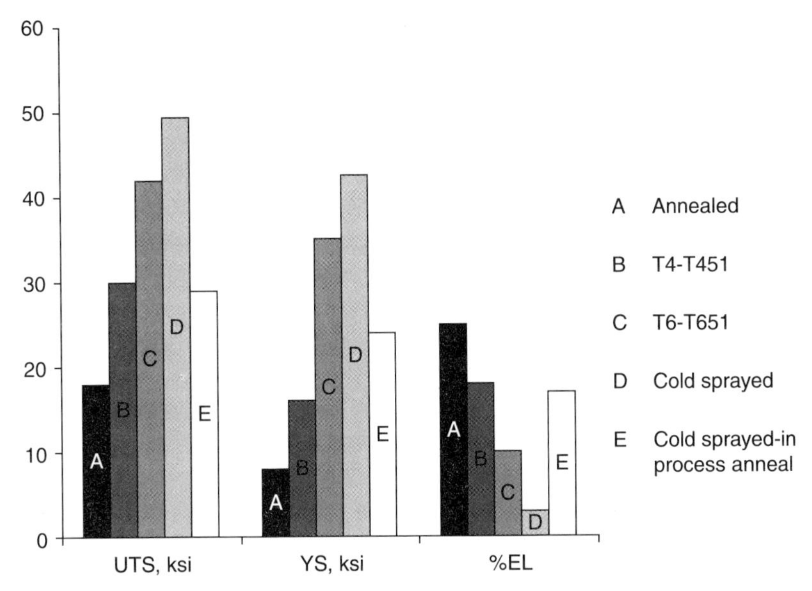

15.21 UTS, YS, and percent elongation at failure for as deposited 6061 He cold spray ('cold sprayed'), in-process annealled 6061 He cold spray ('cold sprayed-in process anneal') versus wrought 6061 ('annealed').

The procedures covered by this standard are intended to ensure that cold spray coating operations, either manual or automated, meet prescribed requirements. The cold spray process is significantly different from other established magnesium gearbox coating systems such as Rockhard and Dow 17. Both Rockhard and Dow 17 are thin corrosion protection coatings which show negligible mechanical strength or thickness compared to the properties of the magnesium gearbox. On the other hand, cold spray can be directly applied to damaged areas at thicknesses from a few fractions of a millimeter to 12 or more millimeters thick.

Quantitative and qualitative performance objectives required to qualify cold spray for repair of magnesium aerospace components are provided in Table 15.4. These performance objectives were set by NAVAIR and ARL with input from Sikorsky Aircraft Company based on the Joint Test Protocol (JTP), which was designed to define the upper and lower technical ca pabilities of a repair process. The JTP protocol evaluates mechanical performance, corrosion resistance, level of porosity and visual appearance of the deposits. The establishment of such protocol followed the same path as the development of hard chrome plating replacement projects managed by the Hard Chrome Action Team (HCAT) to qualify the HVOF process for use on aircraft landing gear. Concurrent Technologies Corporation developed a preliminary JTP titled 'JTP for Validation of Corrosion Protection

Table 15.4 JTP performance objectives

Performance objective	Data requirements	Success criteria	Results
Quantitative performance objectives			
Deposition rate	Coating thickness measurement to an accuracy of ±0.0005 inch	Ability to deposit coatings at a rate of at least 0.005 inch per hour with coating quality such that they pass the acceptance criteria specified in the JTP	Passed
Coating thickness uniformity	Coating thickness measurement to an accuracy of ±0.0005 inch	Cold spray coating thickness shall be uniform within ±20% for deposition onto various surfaces that simulate Mg alloy components	Passed
Microstructure	Examined with optical microscopy	A uniform microstructure, especially for alloy coatings, must be achieved	Passed
Microhardness	American Society for Testing and Materials (ASTM) E384-10e2	Vicker's microhardness of as-deposited coatings shall be no less than 50 VHN	Passed
Fatigue	R.R. Moore high speed rotating beam rotating	No debit as compared to baseline non-coated specimens as specified in the JTP	Passed
Stress/strain testing; ductility	Microtensile testing	Monotonic stress/ strain testing shall be conducted in a standard tensile tester. This will evaluate strain tolerance	Passed
Residual stress	XRD	Applied coating must be in either a compressive or neutral stress state	Passed 13 ksi Compression
Adhesion in tension	ASTM C633	Coating must meet or exceed 8.0 ksi	Passed >10 000 psi
Shear adhesion	MIL-SPEC MIL-J-24445A	Coating must meet or exceed 8.0 ksi	Passed up to 20 000 psi

Table 15.4 Continued

Performance objective	Data requirements	Success criteria	Results
Fretting fatigue	Conducted by United Technologies Research Center	No debit as compared to baseline non-coated specimens as specified in the JTP	Passed- coating system dependent
Salt spray corrosion	ASTM B117	Minimum of 336 h exposure without penetration of salt spray through coating to the substrate as described in the JTP	Passed >7000 h
Cyclic corrosion	General Motors (GM) 9540 Specification	Minimum of 500 h exposure without penetration of salt spray through coating to the substrate as described in the JTP	Passed
Appearance	Visual inspection	Coatings are continuous, smooth, adherent, uniform in appearance, free from blisters, pits, nodules and other apparent defects	Passed
Porosity	Examined with optical microscopy	Porosity of cold spray coatings should be less than 1%	Passed
Beach corrosion	Conducted by Navy at Cape Canaveral, FL	No observable penetration or pitting through the coating and into the magnesium	Passed
Galvanic corrosion	ASTM G71-81	No defined criteria. Used for comparison to HVOF Al-12 Si baseline specimen	Passed
Crevice corrosion	ASTM G78	No observable corrosion product	Passed

for Magnesium Alloys.' The development of the JTP involved the input from a variety of individuals from Army AMCOM, PEO Aviation, Army RDEC Materials, US Army Corrosion Office and Sikorsky. The JTP was not directed towards qualification of aluminum alloy coatings but was meant to be generic for qualification of any type of corrosion protection scheme. This JTP can be used as a basis to qualify other similar applications.

15.5 Conclusions

The cold spray process has matured from an emerging technology to a viable advanced alternative for the restoration and corrosion protection of magnesium alloys. It has become a promising cost-effective and environmentally acceptable technology to impart corrosion protection and restore dimensional tolerances to magnesium components.

Early in the development of cold spray technology, two methods of injecting the spray materials into the nozzle were patented leading to what is known as upstream and downstream injection cold spray making it possible to deposit a wide range of materials from very ductile, to cermets, to Nickel-based alloys over metal, ceramic and/or polymeric substrates.

Corrosion of costly magnesium castings for military and commercial aircraft is one major issue. The use of cold spray aluminum coatings has proven promising to reduce subsequent corrosion damage in these components as well provide effective dimensional restoration.

As cold spray technology moves forward, the methods of propelling materials are being advanced so that high-demand materials, such as stainless steels, Inconels and titanium-based, can be deposited with minimum thermal penalty to substrate materials, and at high rates of deposition and efficiencies.

15.6 References

Alkhimov A.P., Kosarev V.F. and Klinkov S.V. (2001) 'The Features of Cold Spray Nozzle Design'. *Journal of Thermal Spray Technology*, **10**(2), 375–381.

Alkhimov A.P., Papyrin A.N., Kosarev V.F., Nesterovich N.I. and Shushpanov M.M. (1994) 'Gas-dynamic spraying method for applying a coating', US Patent 5,302,414, April 12.

Amateau M.F. and Eden T.J. (2000) 'High Velocity Particle Technology'. *iMast Quarterly*, **2**, 3.

ASM Handbook (1990), Vol 02 'Properties and Selection: Nonferrous Alloys and Special-Purpose Materials', ASM International, Materials Park, OH, 10th Ed., p. 49.

ASTM E8 / E8M – 09. (2009) *Standard Test Methods for Tension Testing of Metallic Materials*, ASTM International, West Conshohocken, PA, DOI: 10.1520/E0008_E0008M-09, www.astm.org.

Avedesian M. and Baker H. (Eds) (1998), *ASM Specialty Handbook: Magnesium and Magnesium Alloys*. ASM International, Materials Park, OH.

Balani K., Laha T., Agarwal A., Karthikeyan J. and Munroe N. (2005) 'Effect of Carrier Gases on Microstructural and Electrochemical Behavior of Cold

Sprayed 1100 Aluminum Coating'. *Surface and Coatings Technology*, **195**(2–3), 272–279.

Champagne V.K. (Ed.) (2007) *'The Cold Spray Materials Deposition Process': Fundamentals and Applications*. Woodhead Publishing Limited, Abington Hall, Abington, Cambridge CB21 6AH, England, p. 57.

Champagne, V.K., Helfritch, D.J., Leyman, P.F., Lempicki, R. and Grendahl, S. (2005) 'The Effects of Gas and Metal Characteristics on Sprayed Metal Coatings'. *Modelling and Simulation in Materials Science and Engineering*, **13**, 1–10.

Champagne V.K., Leyman P.F. and Helfritch D.J. (2008) Magnesium Repair by cold spray, ARL Technical Report ARL-TR-4438, May, p. 34.

DeForce, B. S., Eden T.J. and Potter J.K. (2011) 'Cold Spray Al-5% Mg Coatings for the Corrosion Protection of Magnesium Alloys'. *Journal of Thermal Spray Technology*, **20**(6), 1352–1358

Dykhuisen R. and Smith M. (1998) 'Gas Dynamic Principles of Cold Spray'. *Journal of Thermal Spray Technology*, **7**(2), 205.

Fuknuma H., Ohno N., Son B. and Huang R. (2006) 'In-flight Particle Velocity Measurements with DPV-2000 in Cold Spray', *Surface and Coatings Technology*, **201**(5), 1935–1941.

Gabriel B.M., Champagne V.K. and Helfritch D.J. (2011) 'CS Repair of Magnesium – Technical Program Review', Cold Spray Action Team Kickoff Meeting, May 17–18, Worcester Polytechnic Institute, Worcester MA, 2011., http://coldspray-team.org/, Accessed December 5, 2011.

Gärtner F., Stoltenhoff T., Schmidt T. and Kreye H. (2006) 'The Cold Spray Process and its Potential for Industrial Applications'. *Journal of Thermal Spray Technology*, **15**(2), 223–232.

Gorman W. and Woolsey E., (2003) 'Selective Anodization Process for Repair of Magnesium Helicopter Components', Tri Service Corrosion Conference.

Griffin R. and Zuniga D. (2005) 'Evaluation of Coatings on Mg Alloy ZE41A Used in Helicopter Rotor Gearboxes', Tri Service Corrosion Conference.

Grujicic M., Saylor J.R., Beasley D.E., Derosset W.S. and Helfritch D. (2003) 'Computational Analysis of the Interfacial Bonding between Feed-Powder Particles and the Substrate in the Cold-Gas Dynamic-Spray Process', *Applied Surface Science*, **219**, 211.

Grujicic M., Zhao C.L., Tong C., De Rosset W.S. and Helfritch D. (2004) 'Analysis of the Impact Velocity of Powder Particles in the Cold-Gas Dynamic-Spray Process', *Materials Science and Engineering*, **A368**, 222.

Hall A.C., Brewer L.N. and Roemer T.J. (2008) 'Preparation of Aluminum Coatings Containing Homogenous Nanocrystalline Microstructures Using the Cold Spray Process', *Journal of Thermal Spray Technology*, **17**(3), 352–359.

Helfritch D.J. and Champagne V.K. (2006) 'Optimal Particle Size for the Cold Spray Process', Presented at 2006 International Thermal Spray Conference and Exposition, May 15–18, 2006, Seattle, WA., Presentation Online at http://www.arl.army.mil/www/default.cfm?page=374.

Jodoin B., Raletz F. and Vardelle M. (2006) 'Cold Spray Modeling and Validation Using an Optical Diagnostic Method', *Surface and Coatings Technology*, **200**, 4424–4432.

Johnson A. (2004) Helium Recycle – A Viable Industrial Option for Cold Spray. Cold Spray Conference, Akron, OH, ASM International.

Kashirin A.I., Klyuev O.F. and Buzdygar T.V. (2002) 'Apparatus for gas-dynamic coating', US Patent 6,402,050, June 11

Kim G.E., Champagne V.K., Trexler M., Sohn Y. and Whang S.H. (Eds) (2011) *Nanostructured Metals and Alloys: Processing, Microstructure, Mechanical Properties and Applications*, Woodhead Publishing Limited, Cambridge.

Kosarev V.F., Klinkov S.V., Alkhimov A.P. and Papyrin A.N. (2003) 'On Some Aspects of Gas Dynamic Principles of Cold Spray Process', *Journal of Thermal Spray Technology*, **12**(2), 265.

Leyman P.F. and Champagne V.K. (2009) 'Cold Spray Process Development for the Reclamation of the Apache Helicopter Mast Support', ARL Technical Report ARL-TR-4922.

Li W. and Li C. (2005) 'Optimal Design of a Novel Cold Spray Gun Nozzle at a Limited Space', *Journal of Thermal Spray Technology*, **14**(3), 391–396.

Maev R. and Leshchynsky V. (2008) *Introduction to Low Pressure Gas Dynamic Spray: Physics & Technology*. WILEY-VCH Veriag Gmbh & Co KGaA, Weinheim.

McCune R.C., Donlon W.T., Popoola O.O. and Cartwright E.L. (2000) 'Characterization of Copper Layers Produced by Cold Gas-Dynamic Spraying', *Journal of Thermal Spray Technology*, **9**(1), 73.

McCune R. and Ricketts M. (2004) 'Selective Galvanizing by Cold Spray Processing' in 'Cold Spray 2004', Akron, OH, ASM International – TSS.

NFPA 69 Standard on Explosion Prevention Systems http://www.nfpa.org/categoryList.asp?categoryID=124&URL=Codes%20&%20Standards.

NFPA 70 The National Electric Code®.

NFPA 484 Standard for Combustible Metals-2009.

NFPA 654 Standard for the Prevention of Fire and Dust Explosions from the Manufacturing, Processing, and Handling of Combustible Particulate Solids 43http://www.nfpa.org/categoryList.asp?categoryID=124&URL=Codes%20&%20Standards.

Papyrin A. (2001) 'Cold Spray Technology', *Advanced Materials & Processes*, **9**, 49.

Papyrin A., Kosarev V., Klinkov S., Alkimov A. and Fomin V. (2007) *Cold Spray Technology*. Elsevier Science, Oxford.

Pardhasaradhi S.P., Venkatachalapathy V., Joshi S.V. and Govindanet S. (2008) 'Optical Diagnostics Study of Gas Particle Transport Phenomena in Cold Gas Dynamic Spraying and Comparison with Model Predictions', *Journal of Thermal Spray Technology*, **17**(4), 551–563.

Robinson A. (1985) 'Evaluation of Various Magnesium Finishing Systems', Proceedings of International Magnesium Association, 42nd World Conference, N.Y.

Samareh B., Stier O., Luthen V. and Dolatabadi A. (2009) 'Assessment of CFD Modeling via Flow Visualization in Cold Spray Process', *Journal of Thermal Spray Technology*, **18**(5–6), pp 934–942.

Schmidt T., Gartner F., Assadi H. and Kreye H. (2006) 'Development of a Generalized Parameter Window for Cold Spray Deposition', *Acta Materialia*, **54**, 729–742.

Steenkiste Van T.H. (1999) 'Kinetic Spray Coatings', *Surface and Coatings Technology*, **111**, 62.

Stoltenhoff T., Kreve H. and Richter H. (2002) 'An Analysis of the Cold Spray Process and Its Coatings', *Journal of Thermal Spray Technology*, **11**(4), 542.

U.S. Automotive Materials Partnership (USAMP), DOE/USAMP Cooperative Research and Development Agreement (2006), Structural Cast Magnesium Development, Contract No.: FC26-02OR22910.

Yamauchi M., Seki J., Sakita E., Miyata Y. and Arita K. (1991) 'Corrosion Resistant Composite Layer on Magnesium Alloys', *Advanced Composite Materials*, **1**(1), 3–10.

Zhang D., Shipway P. and McCartney D. (2005) 'Cold Gas Dynamic spraying of Aluminum: The Role of Substrate Characteristics in Deposit Formation', *Journal of Thermal Spray Technology*, **14**(1), 109–116.

Zheng W., Derushie C., Lo J. and Essadigi E. (2006) 'Corrosion Protection of Joining Areas in Magnesium Die Cast and Sheet Products', *Materials Science Forum*, **546–549**, 523–528.

Zheng W., Osborne R., Derushie C. and Lo J. (2005) Corrosion protection of structural Magnesium Alloys: Recent Developments, SAE paper 2005-01-0732, SAE.

16
Electroless electrophoresis coatings to improve the corrosion resistance of magnesium (Mg) alloys

G.-L. SONG, General Motors Corporation, USA

DOI: 10.1533/9780857098962.3.446

Abstract: A thin E-coating can be rapidly deposited on the surface of a Mg alloy in a cathodic E-coating bath without applying any current or potential. The selectively self-deposited coating results from the high surface alkalinity of Mg. The 'electroless' E-coating pre-film can offer adequate corrosion protection for Mg alloys. The stability of the film can be further improved significantly after curing. The coating can be either a temporary or permanent layer. It may also be used to pre-treat a Mg alloy surface for a normal E-coating or powder coating. Such a simple dipping step can even seal porous phosphating or anodizing coatings on Mg alloys.

Key words: Mg, corrosion, coating, electroless deposition, electrophoretic coating, surface treatment.

16.1 Introduction

Mg alloys have many attractive properties, such as high mechanical strength, non-toxicity, great damping capability, excellent casting fluidity, low heat capacity and negative electrochemical potentials.[1–11] They are potentially great structural and functional materials and should have had much wider uses in industry than they have today. Unfortunately, many of their ambitious applications are currently limited due to their poor corrosion performance. To enable the use of Mg alloys in industries, some suitable corrosion protection techniques are usually required to mitigate the corrosion damage of Mg alloy parts in their service environments. Undoubtedly, coating is one of the most cost-effective corrosion protection methods.[12,13] Industry is currently looking for robust coatings for magnesium alloys.

Many surface treatments and coating techniques have been proposed or developed for Mg alloys,[3,4,13] such as surface conversion (e.g. chromating and phosphating), anodizing (Dow 17, HAE, Anomag, Keronite, Tagnite, Magoxid); galvanizing/plating (Zn, Cu, Ni, Cr), chemical vapor deposition (CVD), physical vapor deposition (PVD), flame or plasma spraying, laser/

electron/ion beam treatment, hot-diffusion alloying, sol-gel coating, powder coatings and E-coatings. Among these, organic powder coating and E-coating are the two most popular industrial processes. Although these organic coatings are very successful on conventional metals, their applications on Mg alloys have rarely been reported.

Generally speaking, painting or electroless plating is the simplest surface treatment, but it does not offer sufficient corrosion protection for Mg alloys. Other surface treatments and coatings, although relatively corrosion resistant, are complicated and expensive. It would be of great significance from a practical point of view to have a corrosion-resistant coating formed on Mg alloy parts simply through a dipping process like a painting or electroless plating process.

E-coating is known as an electrocoating or electrophoresis process, because a current has to be applied to enable the coating deposition. Cathodic E-coating is currently a popular process in industry due to its excellent corrosion resistance and great covering ability on complex metallic components. Traditionally, E-coating is a great oxygen and moisture barrier,[14] used as one of the most important layers in coating systems. However, E-coating requires precise electrical control and expensive bath solution maintenance. An undesired current or voltage can result in a 'throwing power' or 'cratering' problem.

To reduce the cost, complicated operational steps and current related coating defects, a coating that can self-deposit on magnesium alloys without application of current or voltage is highly desired. If a coating similar to an E-coating can be formed on a metallic part without current or voltage, a significant cost reduction will result and all the current/voltage related issues in the traditional E-coating process will be avoided. Recently, an electroless E-coating technique has been proposed and developed for Mg alloys. It has been theoretically and experimentally verified to have good corrosion resistance.[15,16] This chapter will briefly review and summarize this innovative electroless E-coating formation and performance on Mg alloys.

16.2 Electroless electrophoresis coating on Mg

The electroless E-coating refers to a thin E-coating layer formed on a Mg alloy surface through a dipping process without application of a current or voltage. It is a unique E-coating technique, or a special chemical dipping method or surface treatment for Mg and its alloys.

16.2.1 Coating formation process

A cleaned and dried Mg alloy coupon is dipped in a cathodic E-coat bath solution. For example, the bath may contain aminoepoxy which becomes

16.1 Electroless E-coating formation on a Mg alloy (AZ91D) dipping in a cathodic E-coating bath solution at different times (seconds counted from the beginning of dipping): 0 s – initial shiny surface of AZ91D after polishing, right before dipping; 1.5 s – dipping into the bath; 3 s – pulling out from the bath; 4 s –completely out of the bath; 6 s – excess bath solution flowing to the bottom of the specimen, the top part drying and film being formed there; 17 s – the excess bath solution continuously flowing back to the beaker, the dried area spreading down and the film being formed there; 29 s – whole surface completely dried and covered by a formed film.

insoluble and deposits at a high pH value (a typical bath has a main composition: water 71–82 wt.%, epoxy resin 16–26 wt.%, titanium dioxide 1.3 wt.%). The dipping time (typically, less than 5 s) can be adjusted according to the coating thickness requirement. The Mg alloy specimen is then slowly pulled out from the bath solution to let the E-coating bath solution flow from the specimen surfaces back to the bath, leaving a uniformly coated surface without bath solution stain spots remaining there. Certainly, carefully controlled flowing air can also remove excess both solution drips on the surface during or after the specimen being pulled out of the solution. A film immediately forms on the Mg alloy surface while the specimen is being pulled out of the bath. Figure 16.1 shows the surface changes of a Mg alloy specimen during electroless E-coating.

After the film is formed on the Mg alloy surface, it is hydrophobic. Water cannot stay on the film covered surface. The film cannot be washed away by water. The film can only form on a Mg alloy surface. On conventional

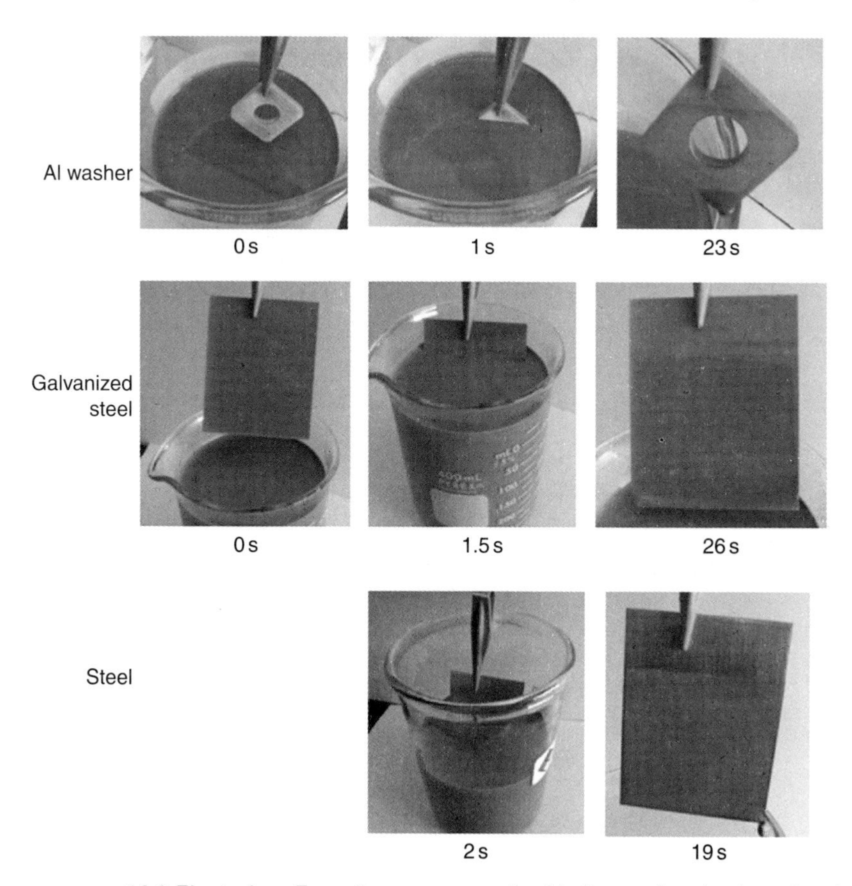

16.2 Electroless E-coating processes for Al alloy, galvanized steel and steel coupons in a cathodic E-coating bath solution (seconds counted from the beginning of dipping).

metals, such as Al, steel, Zn, etc., no film can be formed under the same dipping condition. Figure 16.2 shows the same dipping process for an Al washer, steel coupon, and galvanized steel specimen in the same E-coating bath. The coupon surfaces do not become dry as quickly as the AZ91D Mg alloy sample after being pulled out from the E-coating bath. They are wet for a long period. After washing in distilled water, their surfaces return to their original states; no coating or film remains on their surfaces.

If a Mg alloy is dipped into the bath together with a non-Mg metal, then the film is selectively formed on the Mg alloy surface. Figure 16.3 is a comparison between galvanized steel and Mg alloy surfaces during such a dipping process. The dipped AZ91D surface area dries quickly and a film is formed, while the steel surface is still completely wet and no film deposited there.

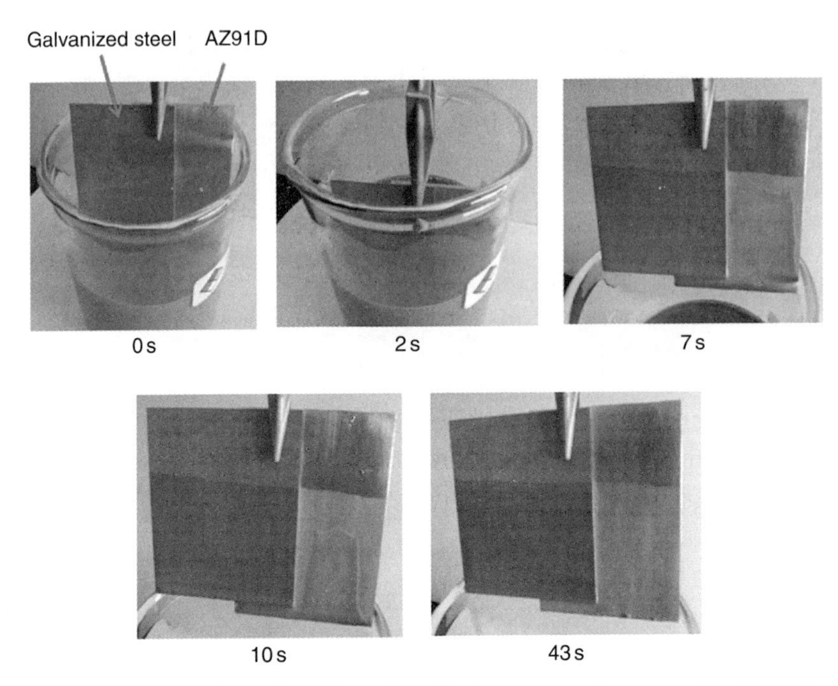

16.3 An electroless E-coating dipping process for a combination of galvanized steel and AZ91 Mg alloy coupons in a cathodic E-coating bath solution (seconds counted from the beginning of dipping).

The rapidly formed film after pulling out from the bath is a pre-film of E-coating. It can be cured at a high temperature. For example, curing can be performed in an oven at 171°C for 25 min. After curing, the electroless E-coating pre-film is fully polymerized and has the same properties as a traditional E-coating.

16.2.2 Coating deposition mechanism

It is well known that the traditional cathodic E-coating is deposited on a metal surface through the following reactions:[17–20]

$$2H_2O + 2e \rightarrow 2OH^- + H_2 \uparrow \qquad [16.1]$$

$$R\text{–}NH_3^+ + OH^- \rightarrow R\text{–}NH_2 \downarrow + H_2O \qquad [16.2]$$

in which $R\text{–}NH_3^+$ is an amine epoxy resin dissolved in the E-coating bath; $R\text{–}NH_2$ is insoluble amine. Cathodic reaction [16.1] is an essential step in the traditional coating process, which generates hydroxyls, resulting in an increased pH value at the cathode surface. The most critical step is reaction

[16.2], which is responsible for deposition or coagulation of the E-coating pre-film. In this reaction, the ammonia bearing organic $R{-}NH_3^+$ reacts with OH^- from reaction [16.1] and becomes an insoluble organic deposit ($R{-}NH_2$) on the cathode surface. The E-coating pre-film can polymerize and turn into a stable insoluble hydrophobic film at a high temperature.

It is very clear that OH^- is a very important species determining the coating formation. Only if a sufficient amount of OH^- is generated or provided, can reaction [16.2] occur and continue, and an E-coating pre-film be formed. To generate enough OH^-, a cathodic current should be applied. In other words, on conventional metals the traditional E-coating is a coating formation process (Equation [16.2]) resulting from an electric current driven alkalization effect over the cathode surface (Equation [16.1]).

Mg is very different from conventional metals in electrochemistry;[2] its surface always has a high alkalinity in an aqueous environment because of its reaction with water:

$$Mg + 2H_2O \rightarrow Mg^{2+} + 2OH^- + H_2 \uparrow \qquad [16.3]$$

This alkalization reaction can rapidly generate a sufficiently high concentration of OH^- on Mg surface, which can trigger reaction [16.2], leading to deposition of an E-coating pre-film there. In this case, no reaction [16.1] is required, and thus the coating process is electroless. It should be noted that electroless only refers to applied external current or voltage. In fact, reaction [16.3] can be the result of anodic and cathodic electrochemical reactions,[2,6] and there are electrons transferring or currents flowing between anodic and cathodic areas on the Mg surface. Such currents are distributed within the electrode system and cannot be directly measured externally.

Figure 16.4 schematically illustrates the electroless E-coating process on Mg. Reactions [16.2] and [16.3] can occur on bare Mg surface and the electroless E-coating pre-film is rapidly deposited there.[15] After the Mg surface of has been full covered by the pre-film, reaction [16.3] will occur at the interface of the substrate Mg and pre-film. The generated OH^- can transport from there to the outer surface of the pre-film, where reaction [16.2] takes place and more $R{-}NH_2$ deposition occurs, resulting in growth of the pre-film.[16]

The above coating formation mechanism is proposed and discussed for pure Mg. It can actually be applied to all Mg alloys, as Mg is the main composition of a Mg alloy and all the Mg alloys have a surface alkalization effect. According to this electroless E-coating mechanism, the coating will not be deposited on steel, galvanic steel or Al, as they cannot generate sufficient OH^- if no current is applied. This has been verified experimentally (Figs 16.2 and 16.3). The model also predicts that the pre-film growth should follow a power function of dipping time, which has also been supported by experimental measurements and will be demonstrated later.

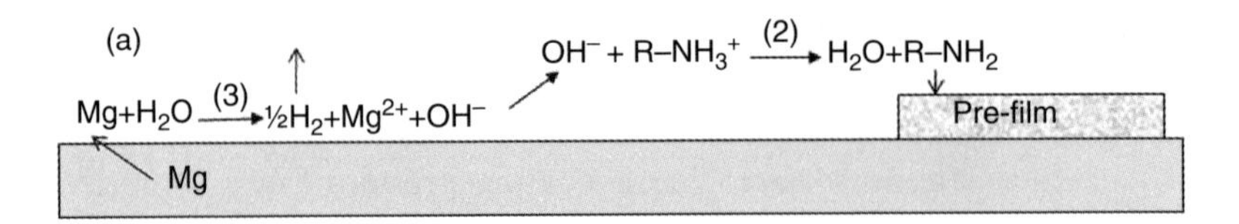

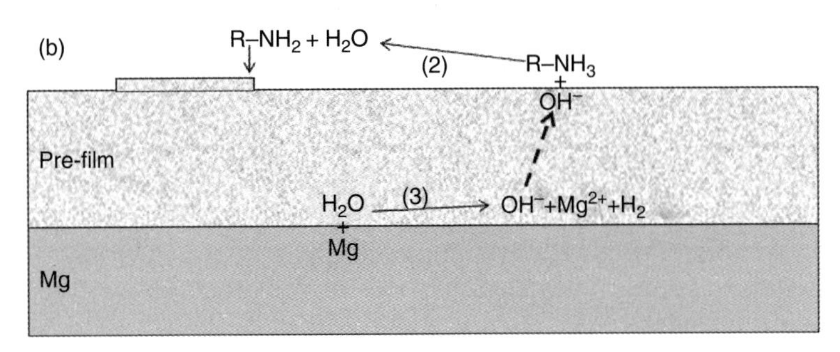

16.4 Schematic illustration of the electroless E-coating process on Mg: (a) pre-film deposition on a bare Mg surface; (b) pre-film growth over a pre-film covered surface.[15,16]

16.2.3 Coating appearance, thickness and microstructure

The electroless E-coating film covered Mg alloy surface is uniformly duller than its original surface (see Fig. 16.1). The formed pre-film is very thin and nearly transparent, which cannot even conceal the original scratches on the substrate. Curing does not change these features. Figure 16.5 is a comparison between a pre-film before curing and a final E-coating after curing.[15,21] Although the naked eye cannot tell any difference between them, the former feels sticky to the touch while the latter has a completely dry surface.

The thickness of the deposited pre-films and their cured E-coatings on a Mg alloy after dipping in an E-coating bath solution for various periods of time is plotted in Fig. 16.6.[21] The pre-film grows, and its thickness (L) is a power function of dipping time (t). Based on the proposed electroless E-coating deposition mechanism, it has been theoretically deduced that the film thickness before or after curing can be expressed as:[15,16]

$$L = Kt^{1/4} \qquad [16.4]$$

where K is a constant depending on the diffusion coefficient of OH^- in the pre-film, saturated OH^- concentration at the interface of the pre-film and the substrate, film porosity and pore sinuosity. For a cured film, K is also curing condition dependent. This equation describes very well the experimentally measured film growth behavior; the theoretical power of the function is ¼, very close to the experimental values 0.24 and 0.27, respectively for the pre-film and cured coating (see Fig. 16.6).

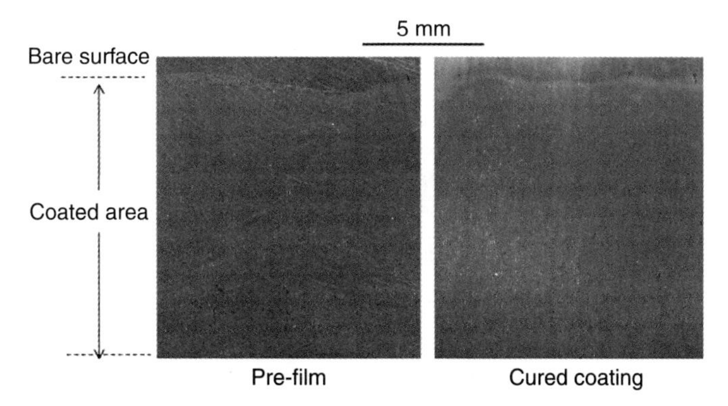

16.5 Comparison of uncured pre-film and cured film of the electroless E-coating on a Mg alloy surface (AZ91D; dipping time 5 s; curing at 271°C for 25 min).[15,16]

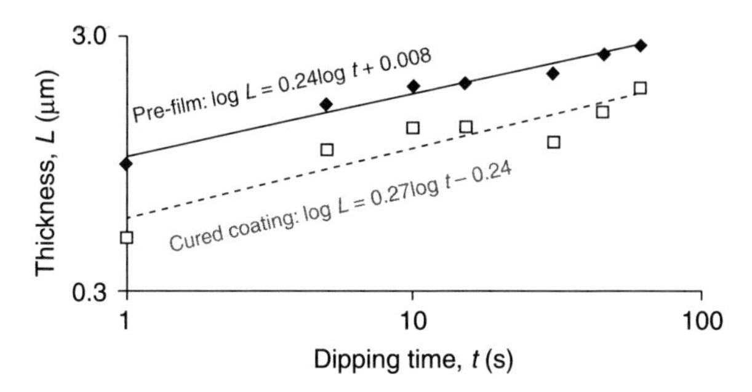

16.6 Dependence of film thickness on dipping time.[15,16,21]

It should be noted that the polymerization of the pre-film during the curing process can result in evaporation of solvent and release of H_2, which can lead to contraction of the film, but cannot change the growing kinetics of the pre-film. Therefore, the cured film follows the same power function growth mechanism as the pre-film, but it is thinner (see Fig. 16.6).

The microstructures of the uncured and cured electroless E-coating are shown in Fig. 16.7. The pre-film before curing does not appear to be dense in the cross-section. However, the pores rarely extend through the film directly from the outer surface to the substrate. Its topographic surface is relatively uniform with some visible dirt or defects. It seems that polymerization after curing does not cause a significant change to these characteristics. The cured film is also porous. These micropores and defects in the uncured pre-film and cured final coating are likely to be the sites for corrosion attack.[1]

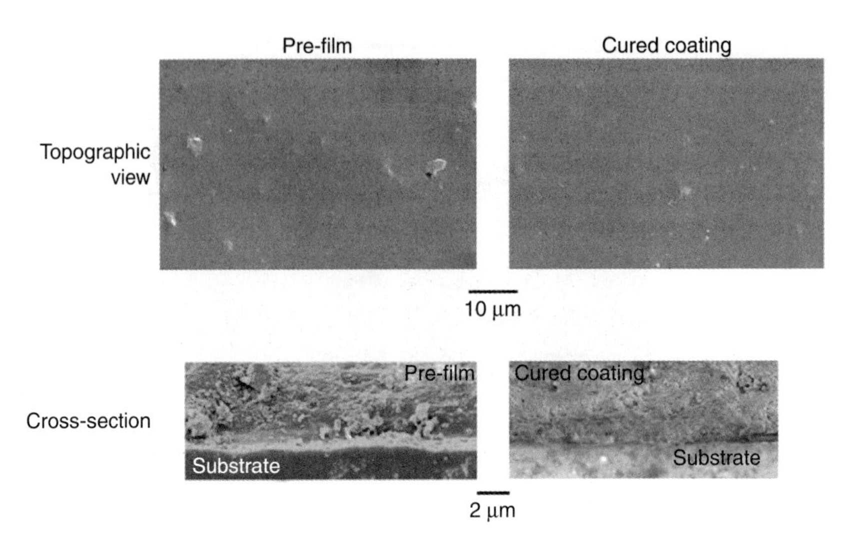

16.7 Topographic and cross-section SEM images of the uncured pre-film and cured coating on a Mg alloy (AZ31).

16.3 Coating protection performance

The presence of an electroless E-coating on a Mg alloy surfacc will significantly change the corrosion behavior of the Mg alloy. At the very least, it acts as a barrier to separate the substrate from corrosive environments.

16.3.1 Corrosion behavior

Even though the deposited pre-film is porous and may be slowly dissolved in an aqueous solution, it can to a great extent retard the ingress of aggressive species. After curing the polymerized coating becomes very stable in most natural solutions and has dramatically enhanced the barrier effect further. More importantly, both the uncured pre-film and the cured coating are insulating layers, not an electronic conductor. They will not act as a cathode to accelerate the corrosion of the substrate Mg alloy through a galvanic effect when the coating is damaged. It has been found[21] that a scratched pre-film covered AZ91D specimen after 24 h of immersion in a 5 wt.% NaCl solution does not show severe localized corrosion damage; curing significantly improves the corrosion resistance of such coated AZ91D, which only exhibits very tiny and insignificant corrosion along the scratches after immersion.

Generally speaking, after immersion in 5 wt.% NaCl, a pre-film coated surface can gradually show slight discoloration. After a certain period, gas bubbles can be observed from some tiny, black colored areas where the

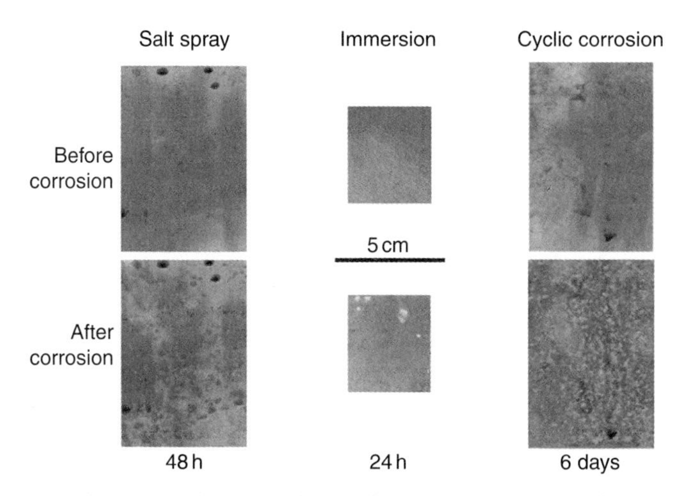

16.8 Corrosion damage of a pre-film covered Mg alloy (AZ31B) after ASTM B117 salt spray, 5 wt.% NaCl immersion and GMW14872 cyclic corrosion.

coating breaks down and corrosion of the substrate occurs. If the immersed sample is exposed to air for a while, the coating breakdown areas will change to white. The white powder is the corrosion product $Mg(OH)_2$ coming from the substrate Mg alloy. Similar corrosion damage can also be observed under salt spray and cyclic corrosion conditions.

Figure 16.8 presents the corrosion damaged surfaces of pre-film coated AZ31Mg alloy samples after exposure in 5 wt.% NaCl solution, salt spray and a wet-dry cyclic corrosive environment. In addition to obvious white color corrosion zones and black spots, many coating areas change color.

If the pre-film is cured and becomes a final stable coating, then its corrosion resistance will be significantly further improved. Figure 16.9 shows that after 24 h of immersion in 5 wt.% NaCl, the cured coating on a Mg alloy (AZ91D) is still intact while the uncured coating has a considerably large area of corrosion damage. It should be noted that coated Mg alloys either by a pre-film or a cured coating are much more corrosion resistant than their uncoated bare counterparts.

16.3.2 Corrosion protection mechanism

The corrosion protection performance of the electroless E-coating can be explained based on the coating micropore structure as illustrated in Fig. 16.10.[16] After an electroless E-coating pre-film is formed on a Mg alloy surface and dried in air, a certain amount of water evaporates from the film pores. If the film is cured, all the water in the pre-film will be

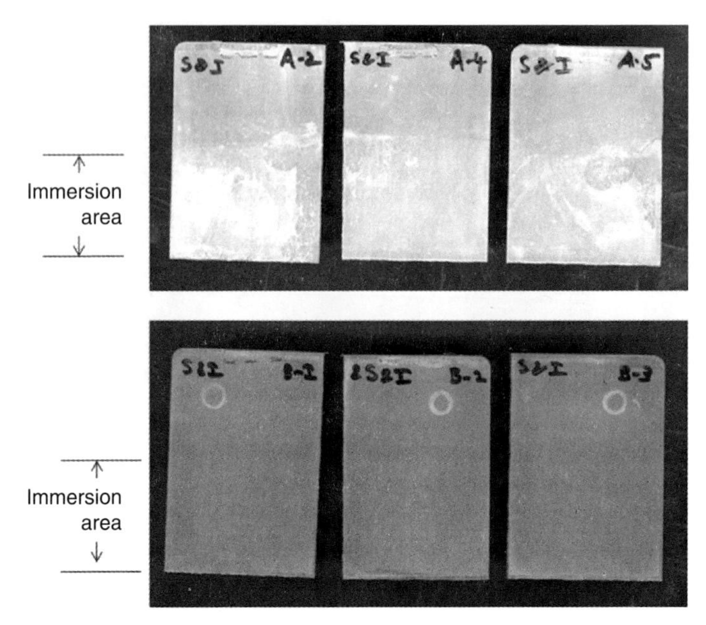

16.9 Comparison of corrosion damage of uncured pre-film (top) and cured coating (bottom) on a Mg alloy (AZ91D) after immersion in 5 wt.% NaCl for 24 h.

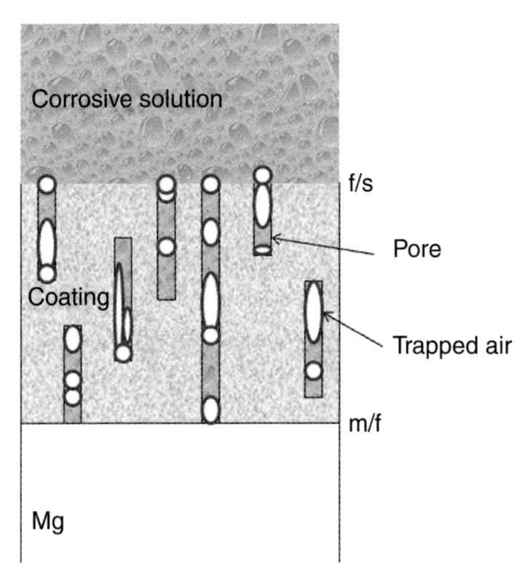

16.10 Schematic illustration of the micropores in a dried pre-film or cured coating in a corrosive environment.

removed. This will leave a large amount of air in the micropores of the film. When the specimen is immersed in a corrosive solution, such as 5 wt.% NaCl, the continuity of the corrosive solution from the environment to the substrate through the micropores is interrupted due to the presence of the trapped air in the pores,. Even if some of the original pore solution from the coating bath remains in the pores due to the trapped air, it is non-corrosive to the substrate Mg alloy. As the corrosive ions from the environment cannot be transferred through the trapped air in the micropores, the ingress of aggressive species from the environmental solution to the substrate is stopped. Therefore, an electroless E-coating, even though it is porous, can offer a certain level of corrosion protection for the substrate Mg alloy.

The pre-film and the cured coating in particular are hydrophobic. Air can easily be trapped in the pores and the non-porous areas are impermeable. It will take a relatively long time to dissolve or remove the trapped air and to wet the micropores. Hence, the film can resist corrosion attack for a certain period of time.

Nevertheless, with time the corrosive solution will gradually become continuous in the micropores and finally reach the substrate at the pore bottoms to initiate corrosion there. It is possible that hydrogen bubbles generated from the corroding substrate will be trapped in the pores, which can quickly block the liquid paths again. Thus, the corrosion is unlikely to continue in the same pores and the corrosion damage does not tend to continue locally. New corrosion has to be initiated in other pores where the corrosive solution has reached the substrate. Certainly, air trapping and pore wetting in the pores can be affected by the corrosion products in neighboring pores. Consequently, the micropores next to a corroded area are more likely to be degassed and wetted. As a result, corrosion will spread out from the first corrosion site. Moreover, as the pre-film and cured coating are insulating, there is no galvanic effect between the coatings and substrate. The undamaged coat will not accelerate the corrosion of the substrate in the broken coating areas. Therefore, the corrosion damage will not be concentrated locally in the corroded area. Figure 16.11 shows an example for such corrosion damage spreading out. Under the optical microscope, there are many black dots in the original pre-film, which are not necessarily micropores (film pores are much smaller and cannot be detected by an optical microscope).

In summary, the initiation and development of corrosion in an uncured pre-film or a cured coating is similar to that of a traditional E-coating on a Mg alloy:[22]

1. corrosive electrolyte penetrates the coating through the micropores in the coating and reaches the substrate;
2. the substrate starts corrosion; meanwhile
3. corrosion products are generated, Mg hydroxides and hydrogen bubbles underneath the coating push the coating apart from the substrate; this results in

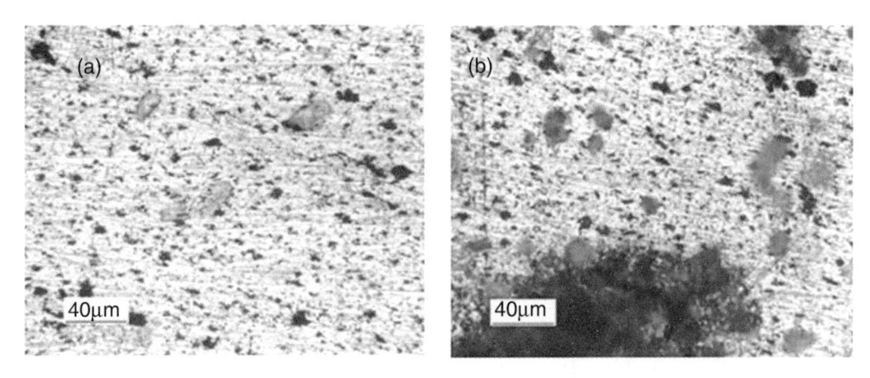

16.11 Optical Microscopic images of a pre-film covered Mg alloy (AZ91D) (a) before and (b) after 32 h immersion in 5 wt.% NaCl solution.

4. the coating delaminating, blistering or cracking in the corroding area which further leads to;
5. continued, more severe corrosion underneath the coating, giving rise to undermining or undercutting damage in a larger area; and finally
6. large amounts of corrosion products building up and hydrogen evolution becomes visible there.

As the electroless E-coating is much thinner than a traditional E-coating, step (1) is relatively quicker and steps (4) and (5) are not very evident.

16.4 Factors influencing the formation of electroless electrophoresis coatings

Many factors can affect the electroless E-coating formation and thus its corrosion performance.

16.4.1 Coating operational conditions

Coating/dipping time is the first operational parameter that can influence coating quality. The coating thickness has been demonstrated to be a function of dipping time (see Fig. 16.6). The variation in coating thickness can to some extent influence the coating protection performance. In the lab, measurements have shown that an increase in coating (dipping) time (in the E-coating bath) leads to a more corrosion resistant-film formed on a Mg alloy. This is understandable, because corrosive solution has to take a longer time to penetrate a thicker coating to reach the substrate.

Bath temperature is the second factor to consider. According to the coating formation mechanism, a low temperature is not beneficial to the

OH^- generation reaction [16.3], while too high a temperature will slow down the deposition of generated $R–NH_2$ (Reaction [16.2]). Thus, neither too low nor too high temperature is recommended in practice. Moreover, bath solution concentration, pH value and stirring condition, as well as dissolved Mg^{2+} concentration in the bath can all influence the coating process. For example, if the bath concentration is diluted, the corrosion resistance of the deposited film decreases. Therefore, all these parameters should be controlled carefully.

16.4.2 Substrate alloy

Under the same coating operational conditions, the coating formation and performance on different Mg alloys may be slightly different. Different Mg alloys have different chemical activities and thus the rates of reactions [16.2] and [16.3] may vary. For example, the coating thickness has been measured to be slightly different on eight Mg alloys (AZ91D, MRI153M, MRI2300, AZ61, AXJ530, AM50, AE44, AZ31).[23] The pre-film covered Mg alloys show significantly different corrosion performance in 5 wt.% NaCl. The most severe corrosion damage occurs on pre-film coated AXJ530; coated AZ31, MRI230D and AM50 have less evident corrosion damage; the other metals AZ91D, MRI153M, AZ61 and AE44 have a corrosion-resistant coating with insignificant corrosion damage. The corrosion resistance of the substrate alloys appears to influence the corrosion performance of the coated samples to a great extent. The ranking order of the corrosion resistance of the coated samples coincides very well with that of the corrosion performance of the substrate alloys.[23] Figure 16.12 shows that the corrosion damage areas of the films on the eight Mg alloys have a good relationship with the corrosion weight loss of the bare substrates.

In theory, all the substrate chemical compositions may have an influence on the coating formation and thus coating performance. This is because the alloying elements of the substrates may be dissolved and included in the deposited film and thereby may affect the corrosion protection behavior of the film. However, the above corrosion results of the eight Mg alloys indicate that the corrosion resistance of the substrate is the most significant factor, rather than the composition. In fact, according to the corrosion mechanism proposed earlier, after corrosive solution penetrates the coating and reaches the substrate Mg alloy, the corrosion resistance of the substrate will to a great degree determine the corrosion performance of the coated sample. Therefore, the dependence of the corrosion protection performance of the electroless E-coating on the corrosion resistance of the substrate Mg alloy is understandable in this sense. A similar correlation in corrosion resistance of Mg alloys before and after anodizing has actually been observed;

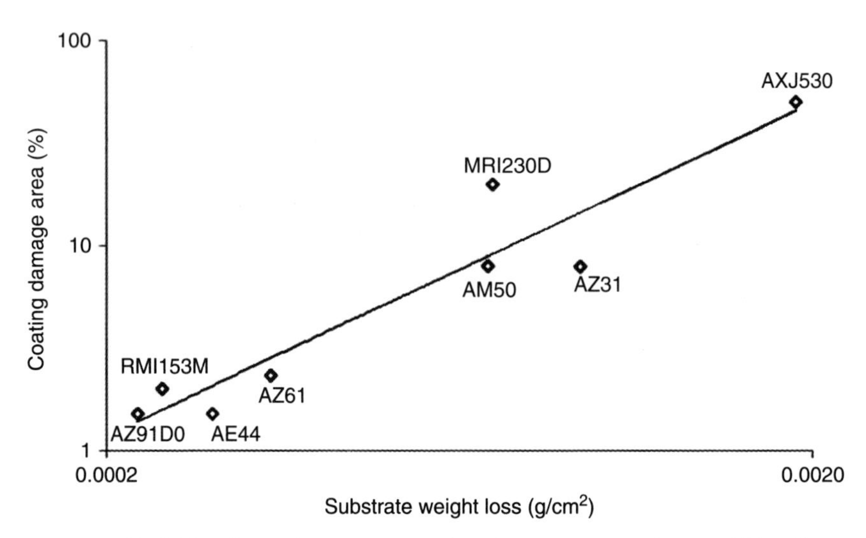

16.12 The dependence of corrosion damage areas of coated Mg alloys after 56 h immersion in 5 wt.% NaCl solution on the weight loss of corresponding substrate bare Mg alloys after 32 h immersion in the same NaCl solution.[23]

it is for the same reason that the corrosion resistance of the substrate at the bottoms of the pores in an anodized film determines to a great degree the corrosion performance of the anodized Mg alloy.[24]

16.4.3 Substrate pre-treatment

For a given Mg alloy, its surface condition has a significant contribution to the electroless E-coating quality. For example, dry-abrading can produce a clean and active surface for rapid formation of an electroless E-coating. Sandblasting of the substrate is detrimental to the coating quality and protection performance due to surface contamination by the sand and micro-stress storage by blasting impact. Polishing in non-aqueous media does not have an evident beneficial effect on the coating formation. Acid etching appears to be favorable to the coating quality. Figure 16.13 shows that wet-abraded and acid etched Mg alloy (AZ31) coupons with and without a treatment at 350°C for 40 min and then electroless E-coated (pre-film covered) have different corrosion resistance in 5 wt.% NaCl solution.

The results suggest that the heat-treatment of the substrate has a beneficial effect on coating protection performance. Since heat-treatment deteriorates the corrosion resistance of the substrate,[8] the beneficial heating effect on the coating cannot be associated with the corrosion resistance variation of the substrate. It must be caused by a change in the surface oxide film

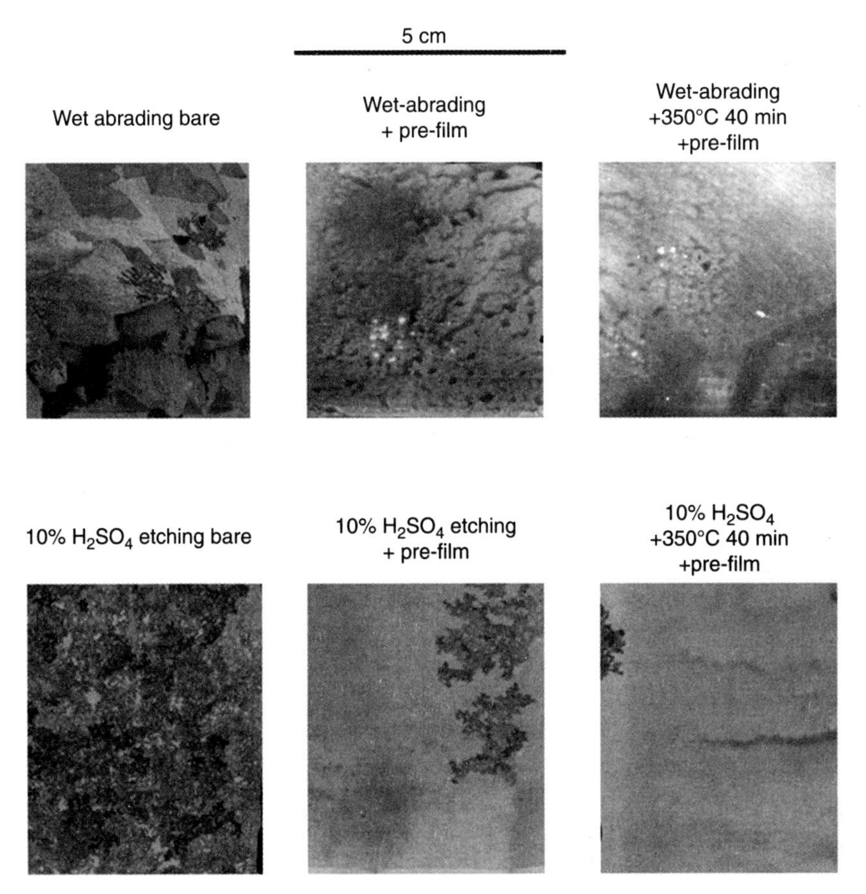

16.13 Corrosion morphologic photographs of Mg alloy (AZ31) surfaces that have been pre-treated differently, then electroless E-coated, and finally immersed in 5 wt.% NaCl solution for 24 h.

during heating. An acid etching + heating practice involves less labor duty and lower operational costs than an abrading process. Since the former can also produce a good substrate surface for electroless E-coating, it may become a practical surface treatment for coating production in industry.

16.5 Potential applications of electroless electrophoresis coatings

The new electroless E-coating with traditional E-coating properties but much lower cost and greater ease in operation can find many practical applications in industry. For example, the pre-film may be used as a temporary protective layer in Mg alloy component production or act as a base for further corrosion resistant top-coating on Mg alloy components. It can also be further cured to become a permanent coating for a long-term protection purpose. In addition, the electroless dipping process may even serve as a sealing step for some porous surface layers.

16.5.1 Temporary and permanent protection

The above results have shown that the electroless E-coating pre-film after curing is in nature a thin E-coating. It can offer adequate corrosion protection for a Mg alloy in a corrosive environment. Therefore, this coating can obviously be used as a permanent coating like an E-coating on some Mg alloy parts in mild service environments, such as the inner of a car body/closure, indoor articles, etc.

A wider application of this new coating should be as a temporary layer. If a Mg alloy part is involved in a phosphating process together with other metals, the dissolved Mg^{2+} ions from the Mg alloy part in the phosphating acid will contaminate the bath, which will ultimately deteriorate the phosphating quality of the other metals. In this case, temporary protection for the Mg alloy in the acidic bath is required. For example, in auto production lines, a vehicle body (body-in-white) which has steel, galvanized steel and Al components, has to go through phosphating and then E-coating baths. If a Mg part is included in the body, its dissolution in the phosphating acid bath will be a major concern. The electroless E-coating technique may solve this problem. If an electroless E-coating pre-film can be selectively formed on the Mg part in a vehicle body before phosphating, the dissolution of the Mg part during phosphating will be avoided. It has been observed that a phosphating solution remains clear after a pre-film covered Mg alloy is dipped in it for 4 h, while the solution becomes cloudy for a bare Mg alloy specimen under the same condition.[21] The amount of dissolved magnesium ions in the phosphating solution from the coated Mg alloy was significantly lower than that from the bare specimen.

In practice, the electroless E-coating may be used on Mg alloys in two possible ways: (1) Mg parts are separately electroless E-coated. After assembling in a car body together with steel and Al alloy components, they go through normal phosphating and E-coating processes. (2) All the Mg alloy, steel and Al alloy components after assembling are dipped together in an electroless E-coating deposition bath, in which a pre-film is formed on the magnesium components only, not on the steel and Al alloy part. After washing, they go through normal phosphating and E-coating production lines. In either way, the electroless E-coating pre-film offers temporary protection for the Mg alloy component in the phosphating bath while steel and Al alloy components are being phosphated. Subsequently, in the normal

E-coating process, while an E-coating pre-film layer is being deposited on the steel and Al alloy components, the existing electroless E-coating pre-film on the Mg alloy component is repaired and a new E-coating layer can also be deposited on the electroless E-coating pre-film, resulting in a thicker coating. After curing, a stable, permanent E-coating can be obtained on all the Mg alloy, steel and Al alloy components.

16.5.2 Base for top-coating

Electroless E-coating before or after curing can be a surface treatment for Mg alloys prior to further coating. In theory, the electroless E-coating is porous and thin. In an E-coating bath it can grow further if a current is applied. There is no question that an electroless E-coating can act as the first layer or a base for a normal (traditional) E-coating to grow on its top. Powder coating is different from E-coating in deposition mechanism, composition, microstructure and performance. Nevertheless, since an electroless E-coating is so thin, the static electricity mechanism, which is responsible for the deposition of a powder coating, can still operate on an electroless E-coating coated Mg alloy surface. Thus, a powder coating can also be easily deposited on it. In experiments, the deposition of a powder coating on an electroless E-coating coated Mg alloy does occur. Figure 16.14 illustrates the cross-section of such coated Mg alloy sample.

It can be seen that the powder coating and the electroless E-coating have good adhesion without any delamination at their interface and the interface along the substrate. Compared with a commercial surface conversion treatment, the electroless E-coating treatment appears to be superior for the powder coating in terms of corrosion performance. Figure 16.15 shows a comparison between powder coated Mg alloy samples that have different surface treatments. The commercially conversion treated and powder coated sample is badly corroded, especially along the scratch, while no corrosion damage can be detected on the electroless E-coating treated and then powder coated sample, even along the scratch.

16.5.3 Sealing

Some porous Mg conversion or anodized coatings are usually highly alkaline in their coating pores. Since the electroless E-coating tends to deposit at highly alkaline areas, it can be used to seal the pores. For example, if an anodized ZE41 is dipped in an E-coating bath solution for a few seconds, the specimen has considerably improved corrosion resistance after baking.[25] The sealing mechanism is similar to the electroless E-coating deposition,[16,25] but mainly takes place at the bottoms of the film micropores. The appearance

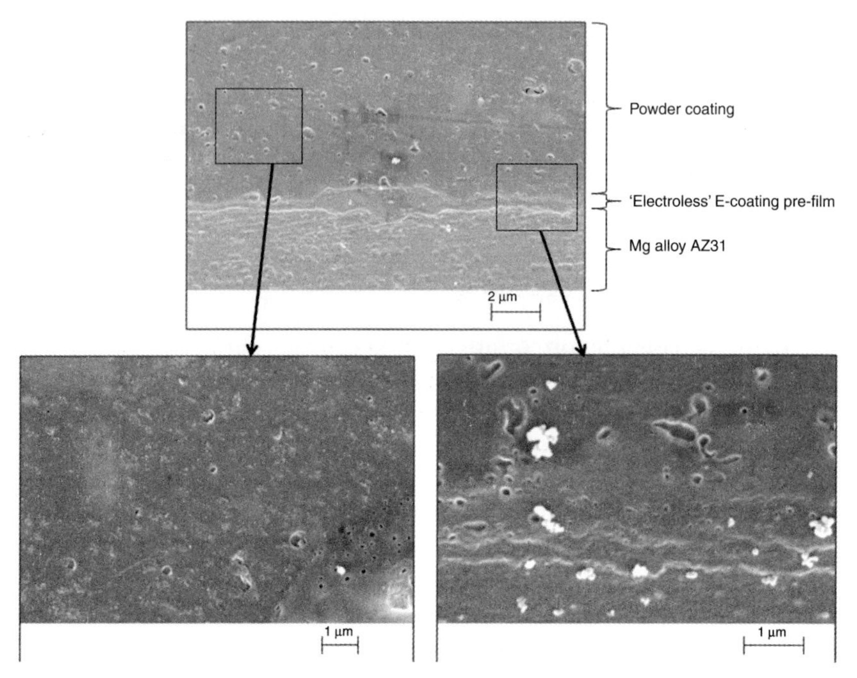

16.14 SEM cross-section images of a powder coating on an electroless E-coating pre-film covered Mg alloy (AZ31).

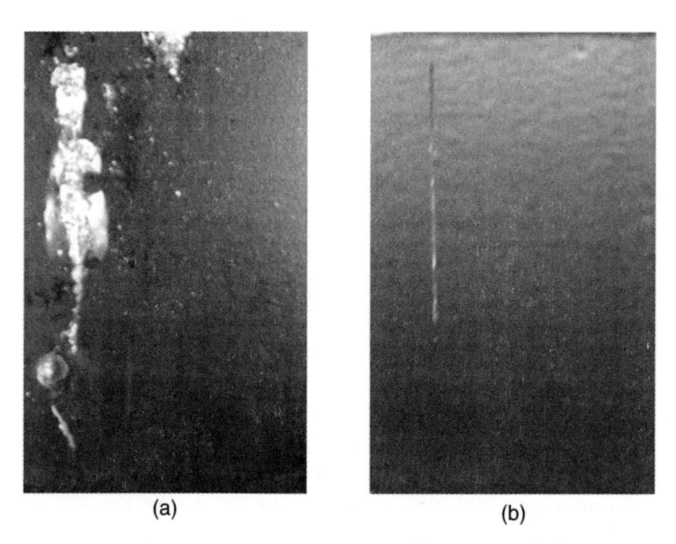

16.15 Corrosion damage of (a) conversion treated + powder coated and (b) electroless E-coating treated + powder coated Mg alloy (AZ31) coupons after 300 h ASTM B117 standard salt spray test.

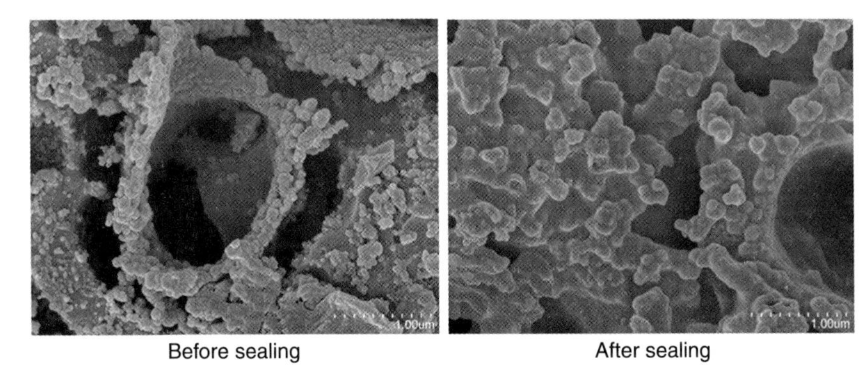

16.16 Microstructures of the anodized ZE41 before and after electroless E-coating sealing.

to the naked eye is that the sealed surface is slightly darker. Under SEM, no difference can be detected on the sealed and unsealed samples at a low magnification. If the magnification is increased above 10^4, some differences in morphology between the unsealed and sealed coatings are revealed (see Fig. 16.16). The anodized coating after sealing seems to be smoother at that magnification; those tiny particles in the unsealed original coating appear to have 'melted' or 'merged' together after sealing, which may to a great degree close up many gaps between the particles and tiny pores at the bottoms of those visible pores. This will effectively block the penetration of corrosive solution through the porous layer and thus improve the corrosion resistance of the anodized Mg alloy. A similar beneficial sealing effect has also been observed on a phosphated Mg alloy. The phosphated sample has remarkably enhanced corrosion resistance after electroless E-coating sealing. After 2 days of immersion in 5 wt.% NaCl solution, the phosphated and sealed area is not corroded, while the whole phosphated area without sealing has been very badly corroded.[26]

16.6 Conclusions and future trends

The innovative electroless E-coating has the following characteristics in terms of its formation and protection performance:

- A thin E-coating layer can be rapidly self-deposited on magnesium alloys in an E-coating bath solution without applying a current or voltage. The Mg surface alkalization effect is responsible for the coating deposition.
- The electroless E-coating can offer adequate corrosion protection for Mg alloys in corrosive solutions. The stability and protectiveness of the

coating can be significantly improved after curing. The coating is neither anodic nor cathodic to the substrate Mg alloys and thus coating damage will not lead to severe localized corrosion.

- Many operational parameters can influence the coating formation and performance. The corrosion protection performance of a coated Mg alloy depends on the corrosion resistance of the substrate Mg alloy. The surface cleaning or pre-treatment of the substrate can also critically affect the coating quality and corrosion resistance. Acid etching and then heating may be a practical pre-treatment process for electroless E-coating on Mg alloys.
- An uncured electroless E-coating can be either used as a temporary protective layer or further cured to become a permanent film for long-term corrosion protection.
- The electroless E-coating can also serve as a surface treatment for Mg alloys. The electroless E-coating treated Mg alloys can be further E-coated or powder coated. A Mg alloy which has been treated and then powder coated has higher corrosion resistance than a commercially conversion-treated sample with the same powder coating.
- The electroless E-coating dipping process can also seal phosphated or anodized Mg alloy surface to significantly improve its corrosion resistance.

The low cost, corrosion resistant and easy operated electroless E-coating technique has a wide application in industry. To enable its practical uses, further optimization of this process should be considered:

- The electroless E-coating deposition is a pH dependent process. So far, the coating is obtained from existing E-coating baths. These baths were not designed for an electroless process. If the bath composition and pH value can be modified to further optimize the pH dependent deposition on Mg alloys, the electroless E-coating should have better quality and performance.
- Mg dissolution is inevitable in this electroless coating process. The dissolved Mg^{2+} ions will sooner or later contaminate the coating bath and deteriorate the coating. Some additives in the bath may be considered to combine with the dissolved Mg^{2+} ions, such that Mg^{2+} containing compounds will be easily separated from the bath and removed, or be included in the deposited pre-film to further strengthen the film and improve its corrosion resistance.
- Better surface treatment techniques should be developed to further enhance the electroless E-coating formation rate and improve its adhesion and corrosion resistance.

16.7 Acknowledgments

The author would like to thank Dr Yar-Ming Wang, Dr Harry Kuo, Mr Sundaresan Avudaiappan, Dr Zhenqing Xu and Dr Minghong Liu for their support.

16.8 References

1. G.-L. Song, 'Preface – Corrosion of Magnesium Alloys' in Guang-Ling Song (ed.), *Corrosion of Magnesium Alloys*, Woodhead Publishing Limited, UK, (2011).
2. G.-L. Song, Chapter 1 'Electrochemistry of Mg and its Alloys' in Guang-Ling Song (ed.), *Corrosion of Magnesium Alloys*, Woodhead Publishing Limited, UK, (2011).
3. J. Gray and B. Luan, 'Protective coatings on magnesium and its alloys – a critical review', *Journal of Alloys and Compounds*, **336** (2002) 88–113.
4. C. Blawert, W. Dietzel, E. Ghali and G-L. Song, 'Anodizing treatments for magnesium alloys and their effect on corrosion resistance in various environments,' *Advanced Engineering Materials*, **8** (2006) 511–533.
5. G. Song, 'Control of biodegradation of biocompatible magnesium alloys', *Corrosion Science*, **49** (2007)1696–1701.
6. G. Song, 'Recent progress in corrosion and protection of magnesium alloys', *Advanced Engineering Materials*, **7**(7) (2005) 563–586.
7. G. Song and D. StJohn, 'Corrosion of magnesium alloys in commercial engine coolant', *Materials and Corrosion*, **56**(1) (2005)15–23.
8. G.-L. Song and Z. Xu, 'Effect of microstructure evolution on corrosion of different crystal surfaces of AZ31 Mg alloy in a chloride containing solution', *Corrosion Science*, **54** (2012) 97–105.
9. G.-L. Song, R. Mishra and Z.Q. Xu, 'Crystallographic orientation and electrochemical activity of AZ31 Mg alloy', *Electrochemistry Communications*, **12** (2010) 1009–1012.
10. G.-L. Song and Z.Q. Xu, 'The surface, microstructure and corrosion of magnesium alloy AZ31 sheet', *Electrochimica Acta*, **55** (2010) 4148–4161.
11. G.-L. Song, 'Effect of tin modification on corrosion of AM70 magnesium alloy', *Corrosion Science*, **51** (2009) 2063–2070.
12. R.-G. Hu, S. Zhang, J.-F. Bu, C.-J. Lin and G.-L. Song, 'Recent progress in corrosion protection of magnesium alloys by organic coatings', *Progress in Organic Coatings*, (accepted for publication on 15 Oct, 2011).
13. B. Luan, D. Yang, X.Y. Liu and G.-L. Song, Chapter 15 'Conversion and electrophoretic coatings for Mg alloys' in G.-L. Song (ed.), *Corrosion of Magnesium Alloys*, Woodhead Publishing Limited, UK (2011).
14. W. Funke, Chapter 4 'Organic coatings in corrosion protection', in A.D. Wilson, J.W. Nicholson and H.J. Prosser (eds), *Surface Coatings 2*. Elsevier Applied science, London (1988), p.107
15. G.-L. Song, 'Electroless E-coating: An innovative surface treatment for magnesium alloys', *Electrochemical and Solid-State Letters*, **12**(10) (2009) D77–D79.
16. G.-L. Song, 'Electroless deposition of a pre-film of electrophoresis coating and its corrosion resistance on a Mg alloy,' *Electrochimica Acta*, **55** (2010) 2258–2268.

17. F. Beck, 'Fundamental aspects of electrodeposition of paint', *Progress in Organic Coatings*, **4** (1976) 1–10.
18. N. Vatistas, 'Initial mechanisms of the electrocoating process', *Industrial and Engineering Chemistry Research*, **37** (1998) 944–951.
19. E. Okada, H. Hosono, T. Nito and Y. Kojima, 'Development of High Throw power E-coat', 2005 SAE World Congress, Detroit, SAE Technical Paper #2005-01-0619 (2005).
20. N. Sato, 'Interfacial control between phosphate films and electrodeposition films on coated steels', SAE technical paper #900838, International Congress and Exposition, Detroit, MI (1990).
21. G.-L. Song, 'A dipping E-coating for Mg alloys', *Progress in Organic coatings*, **70** (2011) 252–258
22. S. Song, G.-L. Song, W. Shen and M. Liu, 'Corrosion and electrochemical evaluation of E-coated and powder coated magnesium alloys', *Corrosion* (to be published in 2012).
23. G.-L. Song, 'The effect of Mg alloy substrate on "Electroless" E-coating performance', *Corrosion Science*, **53** (2011) 3500–3508.
24. Z. Shi, G. Song and A. Atrens, 'Corrosion performance of anodized magnesium alloys', *Corrosion Science*, **48** (2006) 3531–3546.
25. G.-L. Song, 'An irreversible dipping sealing technique for anodized ZE41 Mg alloy', *Surface and Coatings Technology*, **203** (2009) 3618.
26. G.-L. Song, 'An electroless E-coating bath sealing technique for a phosphated magnesium alloy', *Materials Science Forum*, **618–619** (2009) 269.

17

Sol-gel coatings to improve the corrosion resistance of magnesium (Mg) alloys

Q. LI, Southwest University, China

DOI: 10.1533/9780857098962.3.469

Abstract: Sol-gel coatings have been widely used to improve corrosion resistance of Mg alloys. This chapter summarizes the history of development of sol-gel coatings, the mechanism and research results of sol-gel coatings improving the corrosion resistance of Mg alloys and some relevant applications. In addition, the prospects of the future development of sol-gel coatings on Mg alloy are presented.

Key words: sol-gel, Mg alloy, corrosion resistance, technique.

17.1 Introduction

The preparation of inorganic refractory materials often requires high-temperature solid state reactions.[1] For thousands of years, this method has been used to obtain ceramic and glassy materials from natural raw materials. The temperature of synthesis, sintering or melting is the main factor in each process, although many specific technological conditions can also significantly influence the reactions. As a low temperature alternative to the preparation of glasses and ceramics, sol-gel technology has developed rapidly in the past 40 years.

In 1845, Ebelman first reported the formation of a transparent material which was a result of slow hydrolysis of an ester of silicic acid. Roy and Roy proposed a method for preparation of more homogeneous melts and glasses using a sol-gel process.[2] The number of publications at the beginning of the 1980s showed a sustainable exponential growth in research. By the end of the twentieth century, increasing attention was paid to the organic-inorganic hybrids and their applications in the sol-gel processes.[3,4] The development of sol-gel science and technology is impressive. The process can be seen as a nanotechnology because gel products may contain nanoparticles or nanocomposites. The sol-gel process can prepare powders, fibers, coatings, bulk monolithic products, etc., from the same initial composition. Its applications are shown schematically in Fig. 17.1.[5]

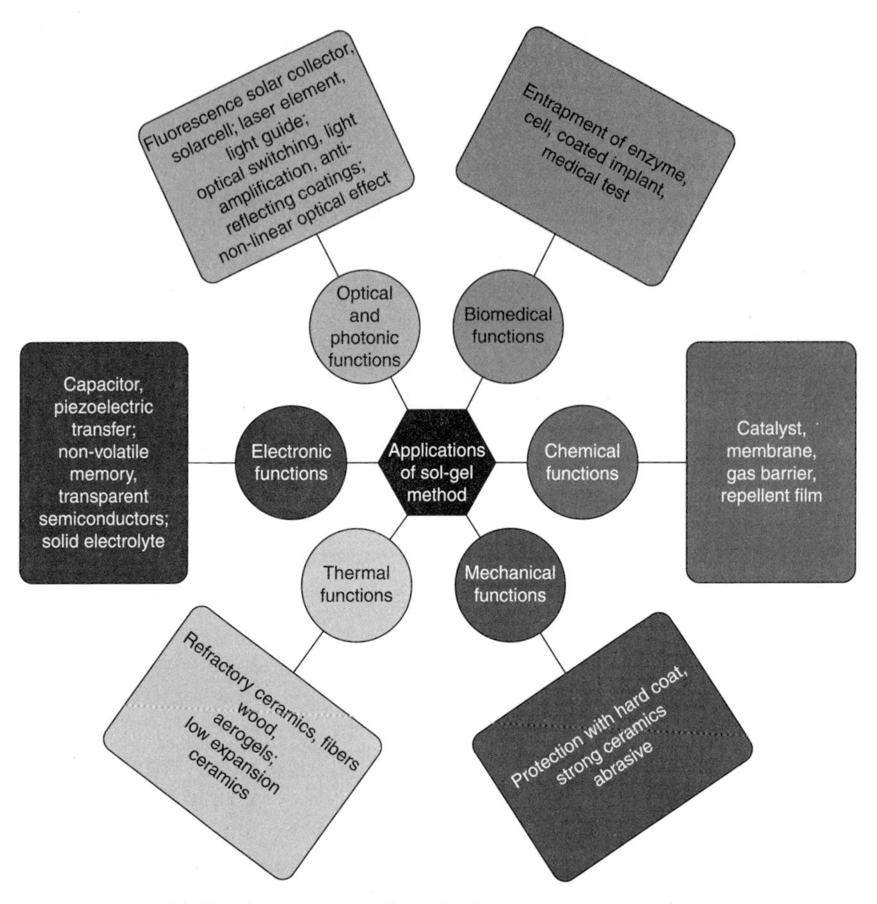

17.1 Applications of sol-gel method.

Since corrosion is one of the main destructive processes leading to huge economic losses, improved chemical resistance of Mg alloys is an increasing requirement in a variety of Mg alloy applications.[6,7] Today's high-performance coating systems always pose a compromise between being environmentally friendly and having good chemical resistance. There is a definite need for new types of coating raw materials. Sol-gel based coatings exhibit great potential as a substitute for environmentally unfriendly surface pretreatment.[8] As an effective surface pretreatment to provide a barrier against corrosion species, sol-gel based coating systems also offer good adhesion to metallic surfaces via chemical bonding and also good physical bonding to organic top coats subsequently applied to the cured sol-gel. Furthermore sol-gel technology offers other important advantages such as cost-effectiveness, low life-cycle environmental impact and simple application procedures which may be easily adapted by industry.[9]

The sol-gel process can generate an oxide network by progressive condensation reactions of molecular precursors in a liquid medium.[10] Depending on the initial precursors, the synthesis process can be divided into three types: (1) aqueous solutions of metal salts; (2) metal alkoxide precursors; (3) mixed organic and inorganic precursors. Using aqueous solutions of metal salts as the precursors, the process is quite complex. The precursors have high reactivity in water which plays a double role as ligand and solvent. Moreover, there is a large number of reaction parameters that have to be strictly controlled, such as hydrolysis and condensation rate of the precursors, pH, temperature, method of mixing, oxidation rate, the nature and concentration of anions, etc.

Nowadays, the most widely used method is the non-aqueous sol-gel approach, instead of the aqueous method, using a colloidal suspension and gelation of the sol to form a network in continuous aqueous phase. The non-aqueous sol-gel process starts with a solution of metalloid alkoxide precursors $M(OR)_x$ or with organic precursors in an alcohol or other low-molecular mass organic solvent. M represents a network-forming element, such as Si, Ti, Zr, Al, Fe, B, etc.; and R is typically an alkyl group. This method is not only able to overcome some of the major limitations of aqueous systems but can also supply oxygen for the oxide formation.

The sol-gel process involves three stages: (1) hydrolysis; (2) condensation and polymerization of monomers to form chains and particles; (3) drying and aging. In fact, both the hydrolysis and condensation reactions occur simultaneously once the hydrolysis reaction has been initiated. They can be presented as follows:

$$M(OR)_x + H_2O \rightarrow M(OR)_{x-1}OH + ROH \qquad [17.1]$$

$$2M(OR)_{x-1}OH \rightarrow (OR)_{x-1}M-O-M(OR)_{x-1} + H_2O \qquad [17.2]$$

$$M(OR)_{x-1}OH + M(OR)_x \rightarrow (OR)_{x-1}M-O-M(OR)_{x-1} + ROH \qquad [17.3]$$

Both hydrolysis and condensation steps could generate low-molecular mass by-products such as alcohol and water. These small molecules will be driven off by drying. Further condensation may occur, leading to shrinkage of the network. These processes are influenced by the reaction conditions such as pH, temperature, molar ratios of reactants and solvent composition, etc.[6,11–13] Tetraethyl orthosilicate (TEOS) can serve as an example: with controllable hydrolysis of TEOS, an M–O–Si chemical linkage is established between surface metallic atoms and TEOS, followed later by polymerization, and finally formation of a three-dimensional network via siloxane bond formation (Si–O–Si) with an increasing TEOS concentration and degree of hydrolysis.[14] However, only very thin and pure inorganic sol-gel coatings can be obtained using TEOS.

To expand the applications of Mg alloys, surface treatments based on organic functional solutions have become an attractive method to enhance adhesion and reduce the corrosion rate of metallic substrates. Co-hydrolysis and co-condensation with organic alkoxysilanes can produce organically modified hybrid sol-gel coatings with reduced brittleness and significantly increased coating thicknesses. Various sol-gel procedures for magnesium have been reported in the literature.[15–17] Organic functional silane[18] and self-assembled monolayer[19] have been tested as pretreatments for magnesium alloys. These treatments provide good corrosion protection besides surface functionality, which improves the compatibility of the metallic substrate with the painting systems. This offers a linkage between a metal-surface and a polymer primer through covalent bonding of a hydrolysable silicate group.

17.2 Sol-gel coatings and how they improve the corrosion resistance of Mg alloys

Sol-gel coatings exhibit great potential as substitutes for the environmentally unfriendly chromate-based pretreatment methods. They offer good adhesion between metals and organic paint. Figure 17.2 depicts possible ways to obtain an enhanced chemical compatibility of a paint system to a Si sol-gel pretreatment by functionalizing the organic component with some groups of paint. On one hand, they can offer strong chemical bonding between metal and sol-gel coating. On the other, compared with chromate pretreatments (where adhesion is based on mechanical interlocking, dispersion forces and hydrogen bonds),[20] chemical bonding is possible between the sol-gel film and the top coat conferring enhanced adhesion.

To date, sol-gel derived thin films have been extensively investigated as promising coatings for various metallic substrates. The typical metals and alloys which have been studied with application of sol-gel film are shown in Table 17.1. Metal oxide sol-gel coatings (SiO_2, ZrO_2. Al_2O_3, TiO_2 and CeO_2) all have very good chemical stability and can provide effective protection. With further development, hybrid films are very promising because they combine properties of the metal oxide material and properties of the ceramic. Incorporation of inorganic nanoparticles can also be a way to include corrosion inhibitors, which create an 'inhibitor reservoir' for 'self-repairing' coatings that slowly release the inhibitor. The presence of nanoparticles also reduces the negative effect of inhibitors on the stability of the sol-gel matrix.

Magnesium alloys have a great development potential for 'green' engineering in the twenty-first century and are being paid increasing attention by researchers. The numbers of related patents has been increasing year on year. The highly reactive nature of magnesium and intermetallic particles

Table 17.1 The typical metal or alloys which have studied applied with sol–gel film.

Substrate	Coating or precursors	Coating method	Year and reference
Al	TEOS-MTMOS-PTMOS	Electrodeposition	2003 [21]
Al 2024	Biotic sol-gel coating	Dip-coating	2008 [22]
Al 2024	Hybrid sol, CeO_2 filled	Dip-coating	2009 [23]
Al 2024	Ormosil coating	Spraying	2001 [24]
Al 3003 H14	Silica film with chromium acetylacetonate	Dip-coating	2009 [25]
Al 6061	Polysilozne coating	Spin coating	2007 [26]
Mg AZ31	Hybrid coating with 8HQ	Dip-coating	2010 [27]
Mg AZ31B	GPTMS-TPOZ-tTMSph	Dip-coating	2008 [20]
Mg ZK30	Anodized oxide coating-Hybrid sol	Dip-coating	2009 [28]
AM60B	Anodic oxidation Sol-gel coating	Spraying	2010 [29]
Stainless steel	Polysiloxane hybrid coating	Dip-coating	2010 [30]
316L	TiO_2-Al_2O_3	Dip-coating	2011 [31]
316L SS	Glass-ceramic	Spin coating	2007 [32]
AISI 304	SiO_2/Na_2O	Electrophoretic deposition	2002 [33]
Co-Cr-Mo	Hybrid coating	Dip-coating	2005 [34]
CrCoMo	TEOS-MTES CaO-SiO_2-P_2O_5	Dip-coating	2004 [35]
Ti-6Al-4V	Hydroxyapatite	Dip-coating	2011 [36]
NiTi alloy	SrO-SiO_2-TiO_2	Dip-coating	2011 [37]
Copper	MTES-MeOH	Dip-coating	2011 [38]
Brass	GPTMS-MTMS	Brushing	2003 [39]

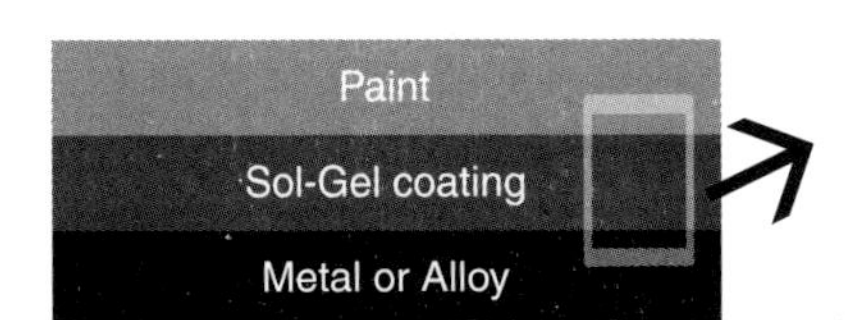

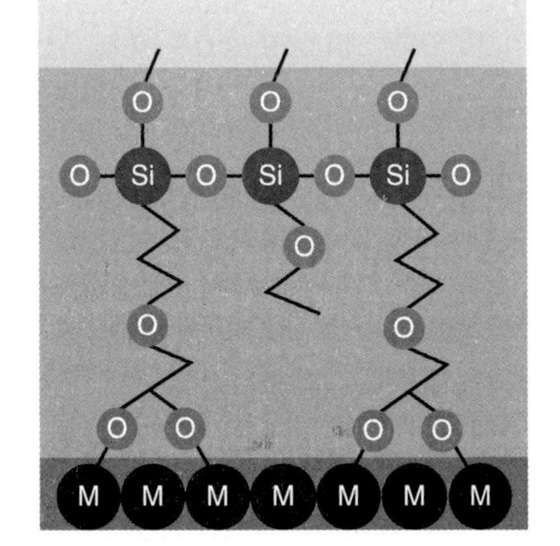

17.2 Schematic representation of enhanced compatibility of paint system with Si sol-gel coating.

during fabrication make magnesium alloys highly susceptible to corrosion. In addition, the volume ratio of the magnesium oxide and the metal unit cells falls under the limited value of 1. As a result, some sols possessing low pH as well as containing corrosive ions such as Cl^- could not be directly applied on magnesium alloy: they will react with substrate, leading to poor adhesion of the layer and subsequent corrosion. Various pre-coatings such as anodized, fluoride and molybdate conversion coatings have been used as an interlayer. Lamaka *et al.*[20] developed a new complex anti-corrosion protection system based on an anodic oxide layer applied prior to sol-gel coatings. Zhong *et al.*[40] describe a novel approach to improve the corrosion resistance of magnesium alloys with a combination of fluoride coating and CeO_2 thin film.

17.2.1 Synthesis of sol-gel coatings for Mg alloys

Metal oxide coatings are often obtained by hydrolysis and condensation of metalloid alkoxide precursors. The cerium oxide (CeO_2) sol is prepared by using cerous nitrate ($Ce(NO_3)_3 \cdot 6H_2O$) as the precursor. It is dissolved in the ethanol and celloidin is added drop-wise as the disperser. The volume ratio between ethanol and celloidin is 4:1 and the concentration of the precursor is 0.3 M. In addition, $Zr(NO_3)_4 \cdot 5H_2O$ dissolved in methanol, with AcAc (acetylacetonate) as the modifier, is used to obtain ZrO_2 sol. The mixture is stirred vigorously and aged 48 h before deposition. The molar ratio of $Zr(NO_3)_4 \cdot 5H_2O:CH_3COCH_2COCH_3:CH_3OH$ is equal to 1:4:8 and the concentration of Zr is 0.2 M.

The silicon sol is produced by a two step process which includes hydrolysis via acid catalysis and condensation via base catalysis. Sol is prepared by hydrolyzing TEOS and 3-glycidoxypropyltrimethoxysilane (GPTMS) (1.16 mol/l Si in ethanol and TEOS/GPTMS in a 1:3 molar ratio) by addition of acetic acid acidified water ($pH < 6$) with the water being added drop-wise into the mixture. Initially the mixture of two kinds of silane alkyl and ethanol is stirred at 60°C for 30 min to achieve a hydrolysis reaction between TEOS and GPTMS. Then ammonia is added to accelerate the condensation reaction (at pH 7) of the precursors at 60°C for 60 min. The sol is aged for 1 day at room temperature before deposition on the magnesium alloy.

17.2.2 Techniques for applying sol-gel coatings

A dip-coating technique is most often used in applying sol-gel coatings. The substrate to be coated is immersed in a liquid and then withdrawn at a well-defined withdrawal speed. The schematics of the dip-coating process are shown in Fig. 17.3a. The resulting film has to be densified by thermal treatment; the densification temperature depends on the composition of sol.

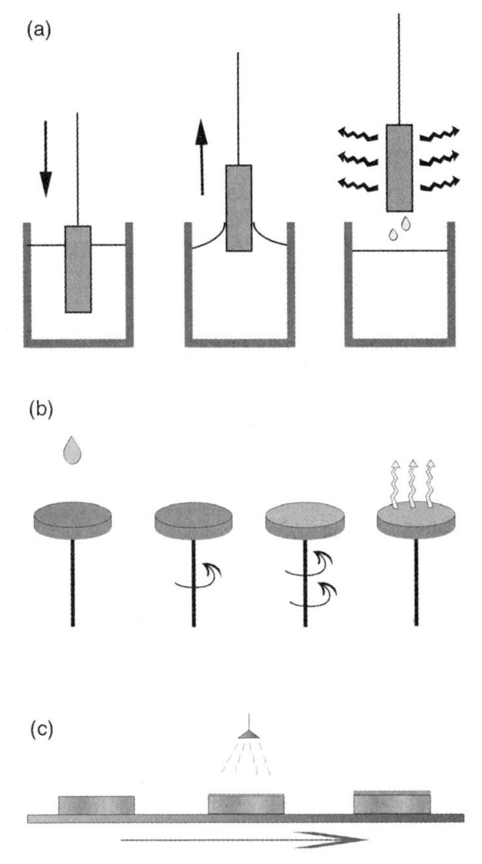

17.3 Schematic diagram of the stages of the (a) dip-coating process; (b) spin coating process; (c) spray coating process.

The coating thickness is defined mainly by the withdrawal speed, the solid content and the viscosity of the sol. There are six forces acting on the coating during withdrawal:[41] (1) viscous drag upward on the liquid by moving substrates; (2) gravity; (3) resultant force of surface tension in the concavely shaped meniscus; (4) inertial force of the boundary layer liquid arriving at the deposition region; (5) surface tension gradient; and (6) the disjoining or conjoining pressure (important for films less than 1 μm thick).

In the spin coating process, the substrate spins around an axis which should be perpendicular to the coating area. The stages of the spin coating process are shown schematically in Fig. 17.3b. The spin coating process is carried out in four stages: deposition of the sol, spin up, spin off and gelation by solvent evaporation. Spin coating is different from dip-coating, in that the deposited film becomes thin through centrifugal draining and tends to be uniform due to the balance between the centrifugal force which drives flow radially outwards and viscous force (friction) that acts radially inwards. The quality of the coating depends on parameters such as temperature and solvent concentration of the coating liquid.

Spray coating techniques are widely used in industry for organic lacquers. Figure 17.3c shows a schematic diagram of a normal spray technique.

Sheffer *et al.* prepared sol-gel films on aluminum by an electrochemical method.[21] The deposition is based on altering the pH on the Al surface, which is driven by a cathodic reaction of water reduction. A constant negative potential (–1.8 V, Hg/Hg_2SO_4) was applied on the substrate to produce an environment full of electrons. It is clear that for corrosion inhibition, electrochemically depositing the film is superior as compared with the conventional dip or spin coating methods as it allows coating of structures with complex geometries. SEM images showed that the edges of the dip-coated plate are not covered with sol-gel film whereas the electrodeposited sample plate exhibited uniform covering with the protective film.

17.3 Applications of sol-gel coatings

The preparation methods of CeO_2 sol, ZrO_2 sol and silicon sol will be introduced in this section.

17.3.1 Metal oxide coatings

All the metal oxide coatings based on ZrO_2, Al_2O_3, TiO_2 and CeO_2, etc., have very good chemical stability and can provide effective protection to a Mg alloy substrate.

ZrO_2 has a high expansion coefficient very close to many bulk metals, which can reduce the formation of cracks during high-temperature curing.[42,43] ZrO_2 also shows good chemical stability and high hardness,[44] making it a good protective material. Li *et al.*[45] developed a nanocrystalline ZrO_2 ceramic film on AZ91D magnesium alloy using the sol-gel dip-coating technique. The ZrO_2 ceramic film was mainly composed of tetragonal phase ZrO_2 and the corrosion potential (E_{corr}) increased by 310 mV compared with the bare magnesium alloy.

Al_2O_3 is a well-known insulator. It has very low conductivity for transmitting electrons, an ideal property for protective coatings. Zhong *et al.*[46] synthesized Al_2O_3 coatings on the magnesium alloy substrate by mixing aluminum isopropoxide and water in defined ratios. The results showed that the coating began to transform from amorphous to crystalline phase when it was sintered at 280°C. The sol-gel coated samples sintered at 380°C showed the best corrosion resistance properties in comparison to the sol-gel coated samples sintered at 120°C and 280°C. Meanwhile, a phytic acid conversion coating was provided as an interlayer, to reduce the potential corrosion of magnesium alloy during the coating preparation stage.

TiO_2 has excellent chemical stability, heat resistance and low electron conductivity, making it an excellent anti-corrosion material. However, it is difficult to apply TiO_2 thin films to the surface of magnesium alloys directly by sol-gel process because the TiO_2 sol has a relatively low pH and magnesium alloys would be eroded by the sol, leading to poor coating adhesion and subsequent corrosion of the substrates. Therefore, there is little reported in the literature about the corrosion resistance of TiO_2 thin films on magnesium alloys. Fan *et al.*[47] prepared anti-corrosive composite CeO_2/TiO_2 thin films by applying cerium oxide thin films as the inner layer with a sol-gel process. CeO_2 is widely used in optics, catalyst chemistry, pigments, superconductors and sensors. It is popular in hybrid sol-gel coatings used as corrosion inhibitors.[48] Chen *et al.*[49] and Zhang *et al.*[50] reported sol-gel-based CeO_2 films with hydrofluoric acid pretreatment and phytic acid pretreatment, respectively. Zhong *et al.*[51] described a novel approach to improve the corrosion resistance of magnesium alloys with a combination of fluoride coating and CeO_2 thin film. It was found that small amounts of MgO and MgF_2 were encapsulated in the CeO_2 thin film. The corrosion resistance of this film was improved by approximately two orders of magnitude compared with that of the bare substrate due to the presence of CeO_2 component. The EIS results are shown in Fig. 17.4. Zhang *et al.*[52] developed two sol-gel films to improve the anti-corrosion performance of AZ91D magnesium alloy. The schematic for the development of films is shown in Fig. 17.5. A most surprising result is found in potentiodynamic polarization tests. The polarization resistance R_p of the CM composite films is about 30 times higher than that of the MC composite films. According to the SEM images of CM film, the whole interface between Ce and Mg film is cross-linking of Mg

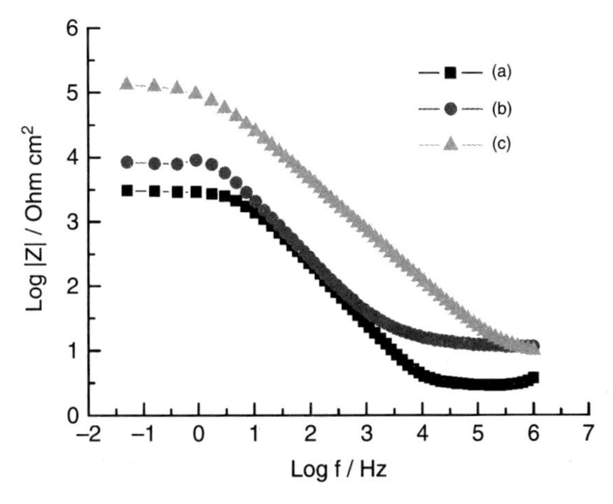

17.4 EIS plots of samples after an initial immersion of 1 h in 3.5 wt.% NaCl solution, (a) the AZ91D Mg alloy, (b) the fluorinated sample, (c) the fluorinated sample with CeO_2 thin film.

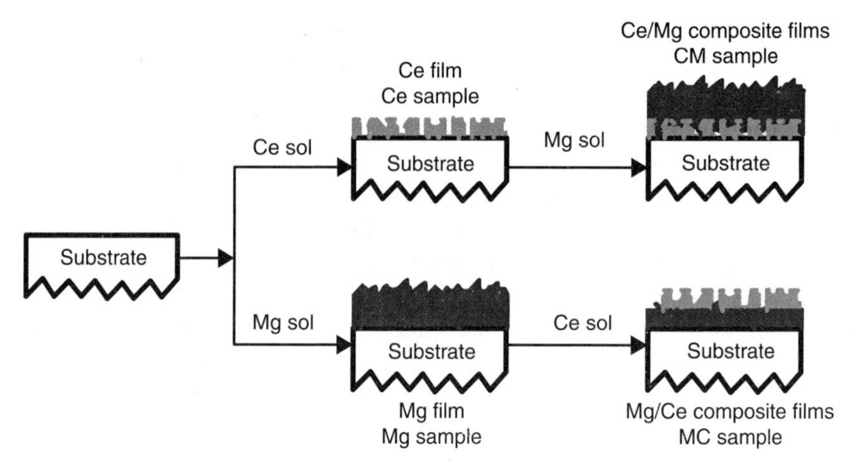

17.5 Schematic flow chart for the preparations of samples.

compound and CeO_2 particles. This combination of cross-linking is significant for the anti-corrosion properties of the CM composite films. With the MC composite films, when the CeO_2 sol is deposited onto the Mg film, Mg compounds such as $Mg_3(OH)_4(NO_3)_2$ are formed during the sintering process.

17.3.2 Organic–inorganic hybrid sol-gel coatings

To overcome the limitations of pure inorganic sol-gel coatings (too thick, with cracks, requiring high temperatures, etc.) many organic species have been successfully incorporated within inorganic networks by various methods. Hu *et al.*[53] prepared SiO_2 sol which can be directly applied to the surface of magnesium alloy using TEOS and triethoxyvinylsilane (VTEO) as the precursors. Figure 17.6 shows the flow chart of the main reaction between TEOS and VTEO.

17.3.3 Inhibitor doped sol-gel coatings

Sol-gel based coating systems can offer good adhesion to metallic surfaces via chemical bonding, good physical bonding to organic top coats, cost-effectiveness, low life-cycle environmental impact and simple application procedures which are easily adaptable for industry. Whilst it exhibits great potential as a candidate for the substitution of the environmentally unfriendly chromate surface pretreatments, even small defects appearing in the sol-gel coating system can allow corrosive agents to penetrate to the metal surface so the coating system cannot stop the corrosion process entirely. Since the coating contains micropores, cracks and areas of low

$$\begin{array}{c} OR \\ | \\ C_2H_5O\text{-}Si\text{-}OC_2H_5 + xH_2O \\ | \\ OR \end{array} \xrightarrow{\text{acetic acid}} \begin{array}{c} OR \\ | \\ RO\text{-}Si\text{-}OH + xC_2H_5OH \\ | \\ OR \end{array} \qquad (1)$$

$$\begin{array}{c} OC_2H_5 \\ | \\ H_2C\text{–}CH\text{-}Si\text{-}OC_2H_5 + xH_2O \\ | \\ OC_2H_5 \end{array} \xrightarrow{\text{acetic acid}} \begin{array}{c} OR \\ | \\ H_2C{=}CH\text{-}Si\text{-}OH + xC_2H_5OH \\ | \\ OR \end{array} \qquad (2)$$

$$\begin{array}{c} OR \qquad\quad OR \\ | \qquad\qquad | \\ H_2C{=}CH\text{-}Si\text{-}OH + HO\text{-}Si\text{-}OR \\ | \qquad\qquad | \\ OR \qquad\quad OR \end{array} \xrightarrow{\text{ammonia}} \begin{array}{c} OR \quad OR \\ | \qquad | \\ H_2C{=}CH\text{-}Si\text{-}O\text{-}Si\text{-}OR + HO\text{-}R \\ | \qquad | \\ OR \quad OR \end{array} \qquad (3)$$

$$R{=}H \text{ or } C_2H_5$$

17.6 The flow chart of the main reaction between tetraethyl orthosilicate (TEOS) and triethoxyvinylslane (VTEO).

cross-link density, pathways are provided for diffusion of corrosive species. The development of an environmentally friendly approach which can stop the development of corrosion processes and heal the corroded areas will be an important issue in industries where adequate corrosion protection is needed. Nowadays, the most popular and fast developing strategy is to develop new sol-gel protective coatings which combine good barrier properties with active corrosion protection originating from 'self-healing' of corroded areas.

Two general trends for design and preparation of this kind of coating are observed. On the one hand, corrosion inhibitors such as phosphate and Ce^{3+} can be directly added into the sol-gel matrix to guarantee the self-healing function. Zhong *et al.*[54] investigated the effect of cerium concentration on anti-corrosion performance of cerium-silica hybrid coatings on magnesium alloy. Cerium nitrate hexahydrate was added as dopant into the silica coatings in five different concentrations. The increase of cerium concentration resulted in the decomposition of Si–O–Si chains and the increase of Si–OH as well as water adsorption. These demonstrate that with increasing cerium concentration, the degree of polycondensation of coatings decreases and the wettability increases. An optimum cerium concentration of 0.01 mol/L was demonstrated. Further increasing the cerium concentration decreases the anti-corrosion effect of the cerium-silica hybrid coating. This is attributed to an excess of cerium causing decomposition of the silane chains.

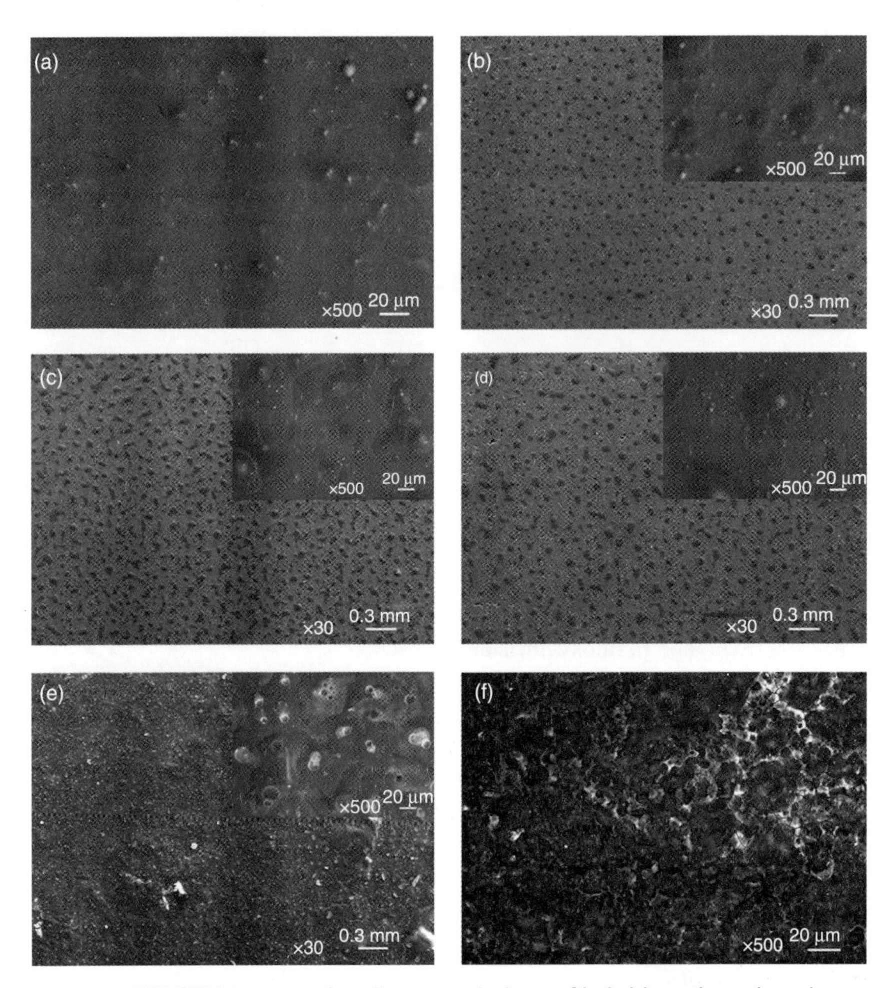

17.7 SEM images of surface morphology of hybrid coatings doped with (a) 0 mol/L, (b) 0.005 mol/L, (c) 0.01 mol/L, (d) 0.02 mol/L, (e) 0.1 mol/L and (f) 0.2 mol/L of cerium nitrate.

The agglomerations present in the coatings disappear with the increase of cerium concentration. Meanwhile, the particles trapped in the coating became larger and more numerous as cerium concentration increased from 0 to 0.01 mol/L. Above this concentration, some pores and cracks appeared in the coatings, indicating poor corrosion resistance. Figure 17.7 shows the SEM images of surface morphology of the hybrid coating doped with different concentrations of cerium nitrate. Clearly, a low concentration of cerium nitrate can cause a depression of agglomerations and the appearance of many particles trapped in the coatings without any micro-scale pores and cracks. High concentrations can result in the presence of pores and cracks in the coatings due to the decomposition of silane chains.

Alternatively, the corrosion inhibitors can be incorporated into the coatings in different ways using nanocontainers, nanoporous layers and oxide nanoparticle reservoirs for controlled inhibitor release. Thus, sol-gel coatings can obtain a corrosion protection system with good barrier properties and an effective self-healing function. This method can decrease the negative effect of inhibitor on the stability of the sol-gel coating. These works originated from both direct and indirect incorporation of corrosion inhibitor into the sol-gel matrix, accelerating the applications of sol-gel methods for long-term corrosion protection of mild steel and aluminum alloys.

As development of a magnesium alloy-based engine block with a cooling system becomes increasingly desirable in the automotive, aerospace and electronic industries, the approach described above of adding inhibitors into corrosion electrolyte makes such designs more realistic from the point of view of practical application.[55,56] As a result, the method is also a kind of inhibitor doped for sol-gel coatings. Zhong *et al.*[57] and Luo *et al.*[48] developed this approach by healing Si sol-gel films coated on Mg alloy using zinc nitrate and cerium nitrate. These can not only stop the development of the corrosion process but also repair the partially destroyed sol-gel coating, probably through the formation of precipitation covering the micron scale cracks or defects without degeneration of the sol-gel coating.

17.4 Future trends

Since the use of chitosan and its derivatives as protective coatings[58] were first reported, Zheludkevich *et al.*[59] have reported a 'green' self-healing coating incorporating chitosan. A chitosan pre-layer has been successfully used as a reservoir for the storage of the corrosion inhibitor. As Mg alloys gain increasing attention for their vast potential application in medical and biotechnological fields, the development of novel sol-gel coatings based on ecological approaches is increasingly necessary.

In addition, whilst the self-healing sol-gel coatings for Mg alloys have been studied quite widely, the investigation of nanosensors for self-repair is an interesting area for further research. These sensors can detect a corrosion signal before aggressive species reach the substrate surface and simultaneously release an inhibitor.

17.5 References

1. Y. Dimitriev, Y. Ivanova and R. Iordanova, 'History of sol-gel science and technology', *J. U. C. T. M.* (2008), **43**, 181–92.
2. D.M. Roy, R. Roy, 'An Experimental Study of the Formation and Properties of Synthetic Serpentines and related Layer Silicate Minerals', *Am. Mineral.* (1954), **39**, 957–75.

3. J. Wen and G.L. Wilkes, 'Organic/Inorganic Hybrid Network Materials by Sol-Gel Approach', *Chem. Mater.* (1996), **8**, 1667–81.
4. H. Schmidt, G. Jonssschker, S. Goedicke and M. Menning, 'The sol-gel process as a basic technology for nanoparticle-dispersed inorganic-organic composites', *J. Sol-Gel Sci. Technol.* (2000), **19**, 39–51.
5. S. Sakka (ed.), *Handbook of Sol-Gel Science and Technology, Processing Characterization and Application*, Kluwer Academic Publishers, Boston/Dordrecht/London (2005).
6. S.H. Cho, S.R. White and P.V. Braun, 'Self-healing polymer coatings', *Adv. Mater.* (2009), **21**, 645–9.
7. D.G. Shchukin, M. Zheludkevich, K. Yaskau, S. Lamaka, M.G.S. Ferrira and H.Möhwald, 'Layer-by-layer assembled nanocontainers for self-healing corrosion protection', *Adv. Mater.* (2006), **18**, 1672–8.
8. M.L. Zheludkevich, M.I. Salvado and M.G.S. Ferreira, 'Sol-gel coatings for corrosion protection of metals', *J. Mater. Chem.* (2005), **15**, 5099–111.
9. H. Wang and R. Akid, 'Encapsulated cerium nitrate inhibitors to provide high-performance anti-corrosion sol-gel coatings on mild steel', *Corros. Sci.* (2008), **50**, 1142–8.
10. C.J. Brinker and G.W. Scherer, *Sol–Gel Science: The Physics and Chemistry of Sol–Gel Processing*, Harcourt Brace Jovanovich (Academic Press, Inc.), Boston (1990).
11. J.D. Wright and N.A.J. Sommerdijk, *Sol–Gel Materials Chemistry and Applications*, CRC Press, OPA (Overseas Publishers Association), Florida (2001).
12. M. Guglielmi, 'Sol-gel coatings on metals', *J. Sol–Gel Sci. Technol.* (1997), **8**, 443–9.
13. L.L. Hench, *Sol–Gel Silica, Properties, Processing, Technology Transfer*, Noyes Publications, New Jersey (1998).
14. Q. Liu, Z. Xu, J.A. Finch and R. Egerton, 'A novel two-step silica-coating process for engineering magnetic nanocomposites', *Chem. Mater.* (1998), **10**, 3936–40.
15. Q. Wang, N. Liu, X. Wang, J. Li and X. Zhao, 'Conductive hybrids from water-borne conductive polyaniline and (3-glycidoxypropyl)trimethosy silane', *Macromolecules* (2003), **36**, 5760–4.
16. A.R. Phain, F.J. Gammel and T. Hack, 'Structural, mechanical and corrosion resistance properties of Al_2O_3-CeO_2 nanocomposites in silica matrix on Mg alloys by a sol-gel dip coating technique'. *Surf. Coat. Technol.* (2006), **201**, 3299–306.
17. A.R. Phain, F.J. Gammel, T. Hack and H. Haefke, 'Enhanced corrosion resistance by sol-gel-based ZrO_2-CeO_2 coatings on magnesium alloys', *Mater. Corros.* (2005), **56**, 77–82.
18. F. Zucchi, V. Grassi, A. Frignani, C. Monticelli and G. TrabanellI, 'Influence of a silane treatment on the corrosion resistance of a WE43 magnesium alloy', *Surf. Coat. Technol.* (2006), **200**, 4136–43.
19. C.F. Cheng, H.H. Cheng, P.C. Cheng and Y.L. Lee, 'Effect of reactive channel functional groups and nano porosity of nanoscale mesoporous silica on properties of polyimide composite', *Macromolecules* (2006), **39**, 7583–90.
20. S.V. Lamaka, M.F. Montemor, A.F. Galio, M.L. Zheludkevich, C. Trindade, L.F. Dick and M.G.S. Ferreira, 'Novel hybrid sol-gel coatings for corrosion protection of AZ31B magnesium alloy', *Electrochim. Acta* (2008), **53**, 4773–83.

21. M. Sheffer, A. Groysman and D. Mandler, 'Electrodeposition of sol-gel films on Al for corrosion protection', *Corros. Sci.* (2003), **45**, 2894–904.
22. R. Akid, H. Wang, T.J. Smith, D. Greenfield and J.C. Earthman, 'Biological functionalization of a sol-gel coating for the mitigation of microbial-induced corrosion', *Adv. Funct. Mater.* (2008), **18**, 203–11.
23. M. Schem, T. Schmidt, J. Gerwann, M. Wittmar, M. Veith, G.E. Thompson, I.S. Molchan, T. Hashimoto, P. Skeldon, A.R. Phani, S. Santucci and M.L. Zheludkevich, 'CeO_2-filled sol-gel coatings for corrosion protection of AA2024-T3 aluminium alloy', *Corros. Sci.* (2009), **51**, 2304–15.
24. R.L. Parkhill, E.T. Knobbe and M.S. Donley, 'Application and evaluation of environmentally compliant spray-coated ormosil films as corrosion resistant treatments for aluminum 2024-T3', *Prog. Org. Coat.* (2001), **41**, 261–5.
25. L. Davydenko, Y. Plyuto and E.M. Moser, 'Characterisation of sol-gel silica films doped with chromium (III) acetylacetonate', *Thin Solid Films* (2009), **517**, 3625–8.
26. K.H. Wu, C.M. Chao, T.F. Yeh and T.C. Chang, 'Thermal stability and corrosion resistance of polysiloxane coatings on 2024-T3 and 6061-T6 aluminum alloy', *Surf. Coat. Technol.* (2007), **201**, 5782–8.
27. A.F. Galio, S.V. Lamaka, M.L. Zheludkevich, L.F.P. Dick, I.L.Müller and M.G.S. Ferreira, 'Inhibitor-doped sol-gel coatings for corrosion protection of magnesium alloy AZ31', *Surf. Coat. Technol.* (2010), **204**, 1479–86.
28. S.V. Lamaka, G. Knörnschild, D.V. Snihirova, M.G. Taryba, M.L. Zheludkevich, M.G.S. Ferreira, 'Complex anticorrosion coating for ZK30 magnesium alloy', *Electrochim. Acta* (2009), **55**, 131–41.
29. M. Bestetti, A.D. Gorno, P.L. Cavallotti, P. Gronchi and F. Barlassina, 'Anodic oxidation and sol-gel coatings for corrosion and wear protection of AM60B allloy', *Trans. Inst. Met. Finish.* (2010), **88**, 57–62.
30. V.H.V. Sarmento, M.G. Schiavetto, P. Hammer, A.V. Benedetti, C.S. Fugivara, P.H. Suegama, S.H. Pulcinelli and C.V. Santilli, 'Corrosion protection of stainless steel by polysioxane hybrid coatings prepared using the sol-gel process', *Surf. Coat. Technol.* (2010), **204**, 2689–701.
31. H. Vaghari, Z. Sadeghian and M. Shahmiri, 'Investigation on synthesis, characterisation and electrochemical properties of TiO_2-Al_2O_3 nanocomposite thin film coated on 316L stainless steel', *Surf. Coat. Technol.* (2011), **205**, 5414–21.
32. U. Vijayalakshmi and S. Rajeswari, 'Synthesis and characterization of sol-gel derived glass-ceramic and its corrosion protection on 316L SS', *J. Sol-Gel Sci. Technol.* (2007), **43**, 251–8.
33. Y. Castro, A. Durán, R. Moreno and B. Ferrari, 'Thick sol-gel coatings produced by electrophoretic deposition', *Adv. Mater.* (2002), **14**, 505–8.
34. L.E. Amato, D.A.López, P.G. Galliano and S.M. Ceré, 'Electrochemical characterization of sol-gel hybrid coatings in cobalt-based alloys for orthopaedic implants', *Mater. Lett.* (2005), **59**, 2026–31.
35. A. Durán, A. Conde, A.G. Coedo, T. Dorado, C. García and S. Ceré, 'Sol-gel coatings for protection and bioactivation of metals used in orthopaedic devices', *J. Mater. Chem.* (2004), **14**, 2282–90.
36. S. Jafari, M.M. Taheri and J. Idris, 'Thick hydroxyapatite coating on Ti-6Al-4V through sol-gel method', *Adv. Mater. Res.* (2012), **341–342**, 48–52.
37. C.Y. Zheng, F.L. Nie, Y.F. Zheng, Y. Cheng, S.C. Wei, L. Ruan and R.Z. Valiev, 'Enhanced corrosion resistance and cellular behavior of ultrafine-grained

biomedical NiTi alloy with a novel SrO-SiO2-TiO2 sol-gel coating', *Appl. Surf. Sci.* (2011), **257**, 5913–8.

38. A.V. Rao, S.S. Latthe, S.A. Mahadik and C. Kappenstein, 'Mechanically stable and corrosion resistant superhydrophobic sol-gel coatings on copper substrate', *Appl. Surf. Sci.* (2011), **257**, 5772–6.
39. E. Besher and J.D. Mackenzie, 'Sol-gel coatings for the protection of brass and bronze', *J. Sol–Gel Sci. Technol.* (2003), **26**, 1223–6.
40. X.K. Zhong, Q. Li, J.Y. Hu and Y.H. Lu, 'Characterzation and corrosion studies of ceria thin film based on fluorinated AZ91D magnesium alloy', *Corros. Sci.* (2008). **50**, 2304–09.
41. C.J. Brinker, A.J. Hurd, P.R. Schunk, G.C. Frye and C.S. Ashley, 'Review of sol-gel thin film formation', *J. Non-Cryst. Solids* (1992), **147–148**, 424–36.
42. M. Atik, P. Neto, L.A. Avaca and M.A. Aegerter, 'Sol-gel thin films for corrosion protection', *Ceram. Int.* (1995), **21**, 403–6.
43. H. Li, K. Liang, L. Mei, S. Gu and S. Wang, 'Corrosion protection of mild steel by zirconia sol-gel coatings', *J. Mater. Sci. Lett.* (2001), **20**, 1081–3.
44. L. Fedrizzi, F.J. Rodriguez, S. Rossi, F. Deflorian and R.D. Maggio, 'The use of electrochemical techniques to study the corrosion behavior of organic coatings on steel pretreated with sol-gel zirconia films', *Electrochim. Acta* (2001), **46**, 3715–24.
45. Q. Li, B. Chen, S.Q. Xu, H. Gao, L. Zhang and C. Liu, 'Structural and electrochemical behavior of sol-gel ZrO_2 ceramic film on chemically pre-treated AZ91D magnesium alloy', *J. Alloys Compd.* (2009), **478**, 544–9.
46. X.K. Zhong, Q. Li, B. Chen, J.P. Wang, J.Y. Hu and W. Hu, 'Effect of sintering temperature on corrosion properties of sol-gel based Al_2O_3 coatings on pre-treated AZ91D magnesium alloy', *Corros. Sci.* (2009), **51**, 2950–58.
47. J.M. Fan, Q. Li, W. Kang, S.Y. Zhang and B. Chen, 'Composite cerium oxide/titanium oxide thin films for corrosion protection of AZ91D magnesium alloy via sol-gel process', *Mater. Corros.* (2009), **60**, 438–43.
48. F. Luo, Q. Li, X.K. Zhong, H. Gao, Y. Dai and F.N. Chen, 'Corrosion electrochemical behaviors of silane coating coated magnesium alloy in NaCl solution containing cerium nitrate', *Mater. Corros.* (2012), **63**, 148–54.
49. B. Chen, Q. Li, H. Gao, J.M. Fan and X. Tan, 'Microstructural characteristics and corrosion property of non-chromate surface treatments on AZ91D magnesium alloy', *Mater. Corros.* (2009), **60**, 521–26.
50. S.Y. Zhang, Q. Li, B. Chen and X.K. Yang, 'Preparation and corrosion resistance studies of nanometric sol-gel-based CeO2 film with a chromium-free pretreatment on AZ91D magnesium alloy', *Electrochim. Acta* (2010), **55**, 870–7.
51. X.K. Zhong, Q. Li, J.Y. Hu and Y.H. Lu, 'Characterization and corrosion studies of ceria thin film based on fluorinated AZ91D magnesium alloy', *Corros. Sci.* (2008), **50**, 2304–09.
52. S.Y. Zhang, Q. Li, J.M. Fan, W. Kang, W. Hu and X.K. Yang, 'Novel composite films prepared by sol-gel technology for the corrosion protection of AZ91D magnesium alloy', *Prog. Org. Coat.* (2009), **66**, 328–35.
53. J.Y. Hu, Q. Li, X.K. Zhong and W. Kang, 'Novel anti-corrosion silicon dioxide coating prepared by sol-gel method for AZ91D magnesium alloy', *Prog. Org. Coat.* (2008), **63**, 13–17.
54. X.K. Zhong, Q. Li, J.Y. Hu, X.K. Yang, F. Luo and Y. Dai, 'Effect of cerium concentration on microstructure, morphology and corrosion resistance of

cerium-silica hybrid coatings on magnesium alloy AZ91D', *Prog. Org. Coat.* (2010), **69**, 52–56.

55. J.E. Gray and B. Luan, 'Protective coatings on magnesium and its alloys-a critical review', *J. Alloys Compd.* (2002), **336**, 88–113.
56. G. Song and D. StJohn, 'Corrosion behavior of magnesium in ethylene glycol', *Corros. Sci.* (2004), **46**, 1381–99.
57. X.K. Zhong, Q. Li, J.Y. Hu, S.Y. Zhang, B. Chen, S.Q. Xu and F. Luo, 'A novel approach to heal the sol-gel coating system on magnesium alloy for corrosion protection', *Electrochim. Acta* (2010), **55**, 2424–29.
58. B. Ghosh and M.W. Urban, 'Self-Repairing Oxetane-Substituted Chitosan Polyurethane Networks', *Science* (2009), **323**, 1458–60.
59. M.L. Zheludkevich, J. Tedim, C.S.R. Freire, S.C.M. Fernandes, S. Kallip, A. Lisenkov, A. Gandini and M.G.S. Ferreira, 'Self-healing protective coatings with "green" chitosan based pre-layer reservoir of corrosion inhibitor', *J. Mater. Chem.* (2011), **21**, 4805–12.

Part IV

Case studies

18

Magnesium (Mg) corrosion protection techniques in the automotive industry

G. S. COLE, LightWeightStrategies LLC, USA

DOI: 10.1533/9780857098962.4.489

Abstract: This chapter discusses magnesium corrosion with particular reference to how magnesium automotive components, exposed to the automotive corrosion environment, can be protected and provide durable function. The chapter introduces the four types of corrosion that occur with automotive magnesium alloy components. The major focus is on the galvanic events associated with manufacturing and fastening between magnesium components and the steel/aluminum structures used in automotive construction. Examples of new applications, taken from the literature, are discussed with details of the protective modalities that were developed to ensure long-term vehicle durability.

Key words: automotive, mass reduction, magnesium, corrosion, corrosion protection.

18.1 Introduction

There is an increasing global realization of the need for fuel efficient, low emission vehicles. It is universally accepted that reducing vehicle mass (by 10%) can improve fuel economy (by from 3% to 6%, depending on vehicle and powertrain size and type). The North American (NA) automotive industry expects that to meet the 2025 U.S. emission reduction standards of 163 g/mile of greenhouse gas emissions, and the equivalent corporate average fuel economy (CAFE) automotive fuel efficiency standards of 4.5 L/100 km (49.6 miles per gallon (mpg)), 25% lighter compared to their current mass. An effective way that this can be achieved is through substituting magnesium (Mg), the lightest structural metal, for the current heavier construction materials, aluminum, steel and cast iron.

Low mass magnesium components have many potential advantages for the automotive industry:

- Reduced fuel consumption and CO_2 emissions. Each liter of gasoline-reduction reduces CO_2.output by 2.2 kg, and 10% mass reduction can achieve ~3–6% fuel efficiency improvement.

- Improved performance and safety. Mass reduction enhances acceleration and deceleration. Reducing vehicle mass in the front allows the center of gravity to be moved rearward thus improving steering and cornering response. Reducing mass in the roof and doors lowers the center of gravity, and improves vehicle stability. Additionally, cast Mg components – all current components are high pressure die cast (HPDC) – can be designed with varied wall thickness, to control vehicle stiffness where improved crash response is required, unlike sheet steel fabrications.
- Improved design, allowing unique customer features. Magnesium castings allow package improvements via part consolidation. Compared to a stamped and welded steel instrument panel (IP) cross-car beam, a cast Mg beam allows components to be cast-in-place (air conditioning ducts, air bag housing and instrument clusters). Lighter structures can be removed and/or reconfigured (e.g., a third-row Mg seat is 18 kg lighter than a comparable stamped steel fabrication). With Mg components, additional comfort/convenience, entertainment and safety options can be made available to customers while maintaining the vehicle axle loading at its allowable Gross Vehicle Weight.
- Reduced costs, improved value. Magnesium castings can have reduced manufacturing cost compared with steel at low production volumes (near 100 000 units pa) since tooling investment is lower. For example, a 30-part steel IP cross-car beam requires 30 expensive tools/gauges; while the cast Mg version has only six. This improves fit and finish, by reducing dimensional error compared with stamped/welded and assembled steel parts. Magnesium parts have improved machinability compared with Al and steel, and its lower specific heat and latent heat can reduce melting costs.
- Improved mechanical and physical properties in comparison with aluminum and plastics. Magnesium has higher strength, stiffness, thermal stability and thermal conductivity compared to plastics and higher specific strength, ductility and impact resistance compared to HPDC aluminum.

Almost 160 kg of different Mg parts have been 'approved' in chassis, interior, body/exterior and powertrain applications, through laboratory and proving ground mechanical and corrosion tests, according to appropriate acceptance standards; USCAR (2006). There is still only 5–6 kg of Mg components on the average NA vehicle, (primarily on the larger sports utility vehicles). One of the major concerns that has limited the growth of automotive Mg applications is corrosion, especially galvanic corrosion associated with fastenings.

18.2 Corrosion of magnesium

Corrosion is one of the most pressing technical and economic issues that has limited Mg component use in vehicles. As shown elsewhere in this volume, magnesium corrodes through ions transferred within an electrolytic cell, from the Mg anode to the cathode; this is termed galvanic corrosion. In the absence of an electrolyte, and/or when there is no contact with the cathode, there is no corrosion. There are four types of galvanic corrosion relevant to automotive construction: endogenous or atmospheric corrosion, crevice corrosion, exogenous surface corrosion and localized galvanic corrosion.

18.2.1 Endogenous or atmospheric corrosion

Industrial Mg alloys contain 5–9% Al. Since the 1980s they have been produced to have a low content of the cathodic-forming impurity elements, Ni, Fe and Cu (in the range of 10–20, 40–50 and 80–300 ppm, respectively) which promote dissolution of the anodic magnesium. Such high purity Mg alloys have better environmental corrosion resistance than either carbon HPDC recycled aluminum alloys, which usually contain around 2% Fe and 2–4% Cu impurities.

Endogenous atmospheric corrosion is a result of a micro-scale galvanic cell being created when there is a localized, high concentration build-up of Ni, Fe and Cu in the cast Mg component to form a precipitated atom cluster. In the presence of atmospheric moisture, a localized electrolytic cell is created between the more pure Mg and the less pure material, causing the higher impurity region to preferentially dissolve and form a corrosion pit. The problem will be exacerbated by the shape-forming process. Almost all Mg alloys are produced by HPDC, which is a high solidification-rate process. Consequently there are very fine grains and dendritic solidification structures which do not grow large enough to build-up a critical concentration of the corroding elements, at grain and sub-grain boundaries. The size and distribution of impurity-rich Ni, Fe and Cu phases (i.e. the phases that cause Mg to corrode) and the more corrosion-resistant $Mg_{17}Al_{12}$ β phase precipitates have a large influence on corrosion. HPDC components freeze rapidly, have finer structures, smaller inclusions and a higher tolerance to impurities (especially at the rapidly solidified surface in contact with a die) than do sand or permanent mold castings which freeze several orders of magnitude more slowly. In the higher Al, Mg-9% Al alloys, the $Mg_{17}Al_{12}$ β phase presents a continuous Al layer which forms a protective oxide layer resisting atmospheric corrosion. However, under slow solidification conditions, as well as during heat treatment, the size and shape of precipitates induced by the presence of Ni, Fe and Cu impurity can increase, increasing corrosion-causing pitting. At the same time, the β phase is discontinuous and dispersed, so that it does not provide complete surface coverage and corrosion prevention. Even in the same alloy, cold worked surfaces (such as shot blasted) can be anodic to unworked areas, leading to corrosion (Busk, 1987).

For automotive interior parts, corrosion protection is not an issue since there is limited exposure to foreign materials such as water, mud, salt and road debris. There is also no exposure to wind-borne sand, rocks and stone-pecking which can remove protective treatment/coatings and expose the bare metal surface to potential corroding conditions. Of course, corrosion issues are less severe in external areas of the vehicle exposed to heat from the engine or brakes since the electrolyte will rapidly evaporate.

18.2.2 Crevice corrosion

Mud, dirt and other road debris can form a poultice on a Mg component, which can retain moisture. Poultices may be a source of chlorides, sulfates and other corrosive ions (including Fe^{2+}, Fe^{3+}, Cu^{+}, Cu^{2+}, etc.), allowing a corrosion cell to form that can promote local corrosion and cause pitting. This is termed crevice corrosion. Crevice corrosion can be also associated with a stagnant electrolyte which tends to occur in shielded crevices such as those formed under gaskets, washers, insulation materials, fastener heads, surface deposits, disbonded coatings, threads, lap joints, clamps and so forth. Crevice corrosion can also be initiated by changes in local chemistry within the crevice, since the electrolyte is not able to flow out of position. Interior castings are usually dry, and except for unique locations (where Mg stanchions meet the steel floor), will rarely display such corrosion concerns.

Another manifestation of crevice corrosion occurs when two pieces of a Mg alloy are in contact with electrolytes of different composition: as the more anodic corrodes and dissolves, the more cathodic will be protected. Thus, if an electrolyte seeps into a narrow gap, its composition changes as dissolved elements are released. Since the electrolyte cannot be replenished, a concentration difference may be set up, causing a voltage difference and a flow of ions which leads to corrosion. Such corrosion occurs quite slowly and is important to eliminate it, in order to ensure long-term vehicle durability.

18.2.3 Exogenous surface corrosion

For most applications, the corrosion resistance of HPDC Mg alloys is adequate without any protective treatments. However, under certain service conditions (such as marine environments) surface protection is required. Kurze (2006) has written an excellent review of surface protection using conversion coatings. Conversion coatings provide temporary corrosion protection; however, their main function is to provide a primer for the additional coating of paint which is required to provide improved protection and durability.

Kurze (2006) also shows that the oldest conversion process, chromating, is the best. But with the advent of ISO 14001 environmental standards, Cr(VI)-based coatings, which are carcinogenic, are no longer acceptable. There are many conversion coating surface treatment processes being developed globally, which the Kurze article details.

Another concern is the presence of metallic iron particles, or other physical cathode-forming, pit-forming particles that are mechanically present on the surface of magnesium parts. These particles can occur when the HPDC casting is ejected from the tool, or from forging, extrusion, rolling and stamping, when the Mg part is dragged across the steel die surface. It can also take place from shot blasting with iron-containing shot, from die lubricants that contain molybdenum disulphide and carbon, and from pickling solutions if they contain heavy-metal salts. Minute amounts of particulate-inducing corrosion can be deposited this way and if not removed by an aggressive chemical acid etching, can form a local galvanic cell when an electrolyte is present.

18.2.4 Localized galvanic corrosion

Magnesium components are subject to galvanic corrosion when portions of the surface are electrically connected to another metal (such as steel, and certain aluminum alloys), either through physical contact, or when immersed in the same conducting liquid electrolyte, such as acidic-water, chloride-containing water or moisture condensate (see Fig. 18.1). Corrosion occurs when there is an electrochemical potential difference between the two metals. Current flows when the dissolving Mg part (Mg being the anode) has a greater negative voltage compared with the other metal, the Fe or bare Al (acting as the cathode). Mg is anodic to many Al alloys, so that galvanic corrosion should be expected when Al and Mg components are in electrolytic contact. However, most Al alloys form a tenacious and dense oxide layer which electrically insulates and protects the Al alloy surface from developing an electrically conductive corrosion couple with the Mg part. Mg alloys also form oxide layers, though the oxide is non-protecting, since it is soft and not well-bound to the Mg surface; this allows the oxide to be easily scratched, exposing the active Mg surface to corrosion. Unlike Al oxide which expands on the Al surface, Mg oxide shrinks and is porous; thus the unprotected Mg surface is always present, allowing corrosion. The most common source of

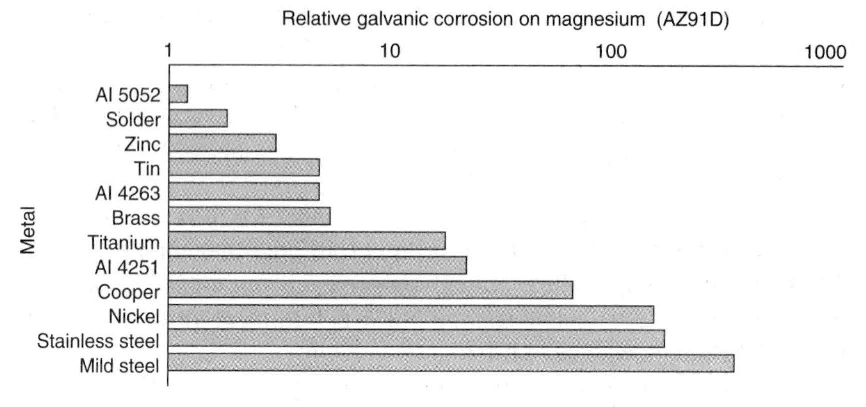

18.1 Relative galvanic corrosion of different materials on magnesium (Hydro Magnesium, 1997).

galvanic corrosion occurs from fastening. Hydro Magnesium (1997) produced an extensive brochure on the subject which defines how to prevent most joint designs from corroding. However, there are situations where unique designs must be invented, as discussed later.

18.3 Preventing corrosion

There are several ways to eliminate corrosion:

- Ensure that the base Mg alloy has low levels of the key corroding elements, Ni, Fe and Cu.
- Separate the metallic connection between the Mg part and the ferrous (or alloyed Al part) using washers and shims of the appropriate Al alloy with defined spacing and/or overlap between washers, nuts, bolts and the component.
- Design to ensure no electrolyte (liquid) can seep into the joint.
- Design for drainage to eliminate sites/pockets where road debris and moisture can collect.
- Use aluminum alloy fasteners.
- Protect the Mg and/or the ferrous and Al alloy parts with a coating.

That is why potentially effective Al alloy washers and fasteners require consideration of whether there might be impurities from the manufacturing process that could influence corrosion. The many ways to accomplish this will be discussed below.

18.3.1 Corrosion and Mg alloy contamination

Galvanic corrosion is hard to predict since some metals polarize in the electrolyte (form a protective oxide), so where a galvanic couple is expected, none occurs. But if there are Ni, Fe and/or Cu impurities in the cathodic material, they can dominate the polarization so that galvanic corrosion may still occur, even at very low levels of impurity due to solidification segregation, as discussed previously. AXJ530 Mg alloy (5% Al, 3% Ca) was developed by researchers at General Motors (GM) several years ago (Powell and Rezhets, 2001). Its excellent high temperature creep resistance, castability and low cost made it an attractive material for powertrain applications. The automotive industry requires that the inventor of a new alloy cross-licenses at least three suppliers who can produce the alloy; in case one supplier runs into difficulties (financial and/or production issues such as a strike or fire), there will always be an approved supplier as back-up. When the GM engineers surveyed potential Mg alloy suppliers, they observed inconsistent corrosion data. In particular, specimens cast by one vendor showed 20 times the environmental corrosion of specimens cast elsewhere, with no correlation among the usual corrosion causes, associated with Fe, Ni or Cu contamination. Without a clear understanding of the corrosion variability, GM's confidence in using AXJ530 was undermined, and this prevented its initial application in automotive components.

It is well-known that the solubility of Fe in Mg is related to Mn content (Reichek *et al.*, 1985; Hillis and Reichek, 1986; Hillis and Shook, 1989; Mercer and Hillis, 1992). Recently Liu *et al.* (2009), using phase diagram calculations, determined that the presence of submicron Fe-rich phases, precipitated during solidification, could serve as active catalysts for electrochemical attack. GM researchers reassessed the initial problem and recalculated the phase diagram of alloy AXJ530 with different Mn content. They determined the Fe-rich phase-precipitation region in the presence of different Mn levels, and measured the corrosion performance of HPDC samples cast at different temperatures. A small Mn chemical composition-difference and/or casting temperature variation changed the precipitation behavior of the Fe-rich phases and significantly changed corrosion behavior of the alloy. Increasing the Mn content and/or lowering the casting temperature (which allowed Fe-rich phases to precipitate and settle out of solution to the bottom of the melting pot), improved the corrosion resistance of the AXJ530 magnesium alloy. A new series of cast AXJ530 alloy samples based on these suggestions showed consistently low corrosion rates, and will be the basis for a re-evaluation of the alloy regarding its readiness for production.

18.3.2 Washers and shims

Steel bolts and nuts cannot be used to connect magnesium castings directly to steel or aluminum vehicle components. Plastic washers (with sufficient

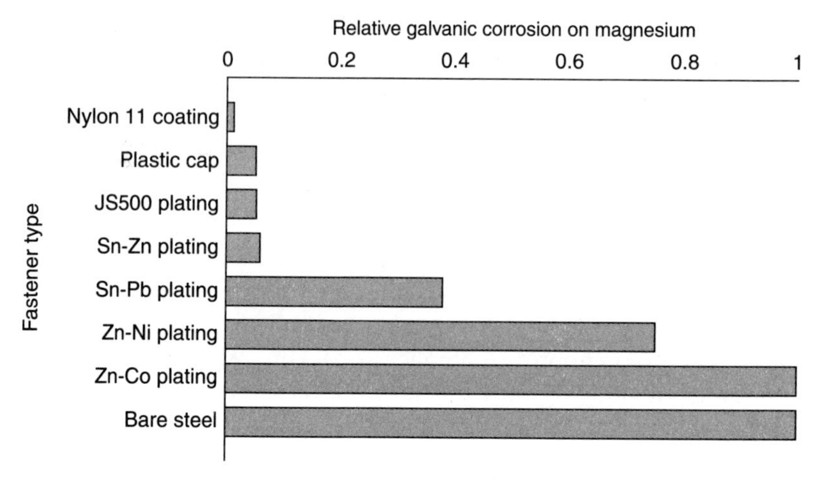

18.2 Relative galvanic corrosion of plastic separators compared with metals (Hawke and Ruden, 1995).

compressive strength so that they do not crack under the high fastening loads used in the NA automotive industry), and polymer capped nut driver heads, can break the electrical connection between a bolt and/or nut and the Mg to eliminate galvanic corrosion.

But there are really no perfect solutions. Hawke and Ruden (1995) examined caps on socket heads and hex-head bolts commonly used to fasten Mg to steel. They looked at different coatings such as Zn, Zn/Ni, Zn/Co and Sn/Zn. They also studied the used of polymer coatings on the shaft and plastic (Nylon 11). Zn and Sn/Zn coatings reduced (but did not eliminate) corrosion. The best solution is to encapsulate the fastener heads with substantial polymer coatings or molding caps onto the fastener heads, as shown in Fig. 18.2. This of course, adds additional cost to the magnesium assembly.

A typical assembly of a magnesium component fastened to an cathodic metal (steel or alloyed Al containing impurities) using a washer to separate the two metals is shown in Fig. 18.3 (Hydro Magnesium, 1997). The size of the washer should be such that the total distance between the bolt head and the magnesium is at least 10 mm (excess diameter plus washer thickness). To avoid water entrapment, proper clearance between the bolt head and the die casting walls is required. The clearance should be at least 5 mm, but there are no consistent standards, since it depends on the road splash/debris and the drainage configurations.

Champrenault *et al.* (2007) examined the engineering and corrosion issues in magnesium brackets that were mounted in direct contact to a cast Al powertrain structure: engine block and transmission housing. Since HPDC

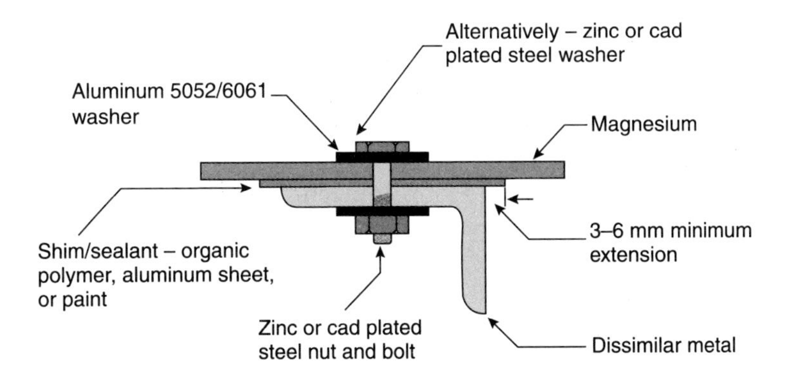

18.3 A typical assembly showing the use of washers and coated bolts to eliminate galvanic corrosion (Hydro Magnesium, 1997).

aluminum powertrain components contain Fe and Cu as impurities, they will be cathodic to the magnesium leading to galvanic corrosion if not properly isolated. When electrolyte was present at or around the joints, the parts had to be separated, since redesign was not a possible option. The authors used conventional steel fasteners with a Zn-rich coating during the initial testing, with 5000/6000 series Al washers as shown in Fig. 18.3. The authors preferred Al fasteners; however they did not have time to prove these out during the writing of their paper. Fortunately, the fastener shank touched the inside of the magnesium mounting holes, and on tightening, effectively sealed off both ends of the hole eliminating all electrolyte so that no galvanic cell was able to develop. The authors were obviously successful with their anticorrosion strategy since these parts are now installed on the 2013 Cadillac ATS sedan.

Wang *et al.* (2001) also examined galvanic corrosion associated with fasteners, spacers and washers. Steel fasteners coated with JS500 yellow chromate zinc plating + silicate sealer was shown to be twice as effective as ordinary zinc plated steel fasteners. However, while Al AA6082 and Al 2024 alloy washers showed some corrosion, Al AA6013 fasteners were a better choice with no galvanic corrosion occurring. Hawke (1995) also studied the galvanic compatibility of coated steel fasteners and washers. The critical issues were the washer material, its thickness and its diameter compared with the bolt head diameter; 2–3 mm thick and 20 mm diameter Al AA6082 washers (with M8 hex-head bolts) worked well. The authors also showed that plastic washers were a better solution than the Al ones. But in applications where the system was exposed to phosphate E-coat (and its associated ~180°C curing temperature for 30 min), all plastics embrittled and thus were unsuitable in this application.

18.3.3 Design

Some components require design to eliminate the presence of moisture, either a design that ensures no electrolyte can seep into the joint and/or that provides drainage paths to eliminate sites/pockets where road debris and moisture can collect and form an electrolytic cell.

Understanding where the component is situated with respect to the 'splash-line' of the vehicle will help the engineer design for drainage. As well, it is important to understand HPDC engineering principles so that the designed drainage shapes can be cast-in, without additional cost and quality implications.

18.3.4 Aluminum fasteners

Reducing the electrochemical potential between the Mg component and the fastener using coatings and washers was discussed previously; however a better way is to utilize high strength Al alloy fasteners. Steel fasteners are standard assembly practice in NA where the fastening loads are near 90 Nm; Al alloy fasteners are not approved for such loads. In Europe, fasteners loads are lower (below 70 Nm) so that Al alloy fasteners can be used. High strength Textron aluminum alloy fasteners can achieve 400 MPa tensile strength/350 MPa yield strength when fabricated from the specially grain-refined and heat treated. Al alloy AA6056. With steel fasteners, when the temperature increases, the part expands (but the ferrous fastener stays fixed), the compressive stresses increase beyond the component's yield strength so that upon cooling the clamp loads have relaxed. When heating a joint containing an Al fastener, both the Mg part and the fastener expand on heating, but the yield stress is never reached and the initial clamp loading remains to hold the part together. Friedrich (2004) details the corrosion protection benefits of securing Mg components with Al fasteners, compared to conventional steel fasteners. Skar *et al.* (2006) also showed that the least galvanic corrosion occurred using Al alloy bolts. Importantly, the authors also documented the differences between accelerated laboratory testing and field testing of various fastener systems. They showed that accelerated laboratory tests failed to replicate the galvanic corrosion observed in the field; the more aggressive the laboratory test, the poorer the correlation. Additionally, too long an exposure in a laboratory test may be detrimental to correlating laboratory testing with field data. The paper also demonstrated that when road debris is present, mud must also be included in the accelerated laboratory corrosion testing to effectively simulate the galvanic interactions. The presence of such debris must be balanced by the aggressiveness of the electrolyte.

18.4 The use of coatings in corrosion prevention

The following two paragraphs describe how coatings protect Mg from corrosion.

18.4.1 Conversion Coatings

Wang *et al.* (2001) wrote an excellent review of the general methodology for protecting large, external Mg automotive components using conversion coatings. The authors are associated with Meridian Technologies, the world leader in producing large automotive Mg components. Their paper presents the requirements for a corrosion-resisting conversion coating pretreatment, using the ASTM B117 240-hour salt spray test as the criterion. Chromate was the best coating, but it is no longer suitable because of environmental concerns. Of the many non-chromate conversion coatings that have been developed, Alodine 5200 (an organic-inorganic Henkel Surfaces Technology product) examined by Dolan and Reghi (1994), was found to be the best, followed by Zn phosphate, phosphate permanganate (Sklar and Albright, 2000), iron phosphate (Hawke and Gaw, 1992), fluorotitanate (Guerci, 2000) and fluorozincate (Gehmecker, 1994). It is important to note that these coatings are not sufficiently corrosion-protecting until at least a 40 μm coating of polymer or epoxy has been applied to the treated surface. Gao and Ricketts (2002) also showed that an epoxy powder coating on an Alodine 5200 conversion coating provided an effective galvanic corrosion-resistant barrier. Importantly, this coating procedure also prevented Mg dissolution during phosphate e-coating. The electrophoretic (e-coat) painting process includes a series of operations that concludes with processing the BIW in a zinc phosphate bath, (followed by a 10–40 μm e-coat application of paint which is then baked). Any magnesium part that is connected to the BIW will dissolve in the moderately acidic (pH 3.0–3.5) phosphate bath. This could contaminate the bath with Mg^{2+} ions as well as potentially accelerating corrosion, by virtue of the phosphate deposition.

18.4.2 Hard coatings

In addition to conversion coatings, anodizing is also used as a protective modality. Sealing is required (+ polymer coating or paint), since the oxide produced during the anodizing process grows as columnar Mg oxide crystals which lead to pores and open paths from the coating surface to the virgin magnesium surface. Moisture can diffuse through the porosity onto the Mg surface leading to corrosion if the oxide is not properly sealed. While sealed coatings provide acceptable corrosion resistance, they are more expensive. However, they are hard and wear-resistant, and less likely to be damaged during vehicle assembly compared to organic-based conversion coatings. There are many plasma electrolytic hard oxide coating processes currently available on the market: HAE, Keronite, Magoxid (Jurey, 1993), Tagnite (Hawkins, 1993) and Anomag (Ross and MacCulloch, 1998). The basic idea is to place the Mg part on a frame that passes through a series of wash baths and then into the appropriate electrolytic bath; connect a power source; and grow a 5–25 μm coating which is then washed, dried and sealed using a suitable organic coating.

18.5 Applications of corrosion mitigation for magnesium automotive parts

Corrosion mitigation strategies for internal and external Mg automotive components have been discussed.

18.5.1 Interior application

Transmission shifter mechanism

Typically, internal Mg components do not require corrosion protection since, while there may be Mg-steel contact through fastening to the vehicle, there is no electrolytc present. However, Mg transmission shifters are installed on the carpeted floor, where moisture and road salt may enter the vehicle, from driver and passenger foot traffic. Ruden and Lenda (2004) examined the corrosion concerns of this application and developed a suitable corrosion protection protocol. They employed a two part corrosion mitigation strategy. For the fasteners, zinc-rich inorganic-coated steel fasteners were used for weld studs and flange head nuts, following General Motors standard GM-7114M. This is a non-electrolytically applied, non-oily, corrosion-protective coating consisting of a zinc phosphate treatment, a zinc-rich inorganic resin coating and an aluminum-rich organic topcoat. The carpeting was also a concern; it was installed to a distance of approximately 10 mm from the assembly so as not to wick moisture into the Mg-steel fastener interface.

Low volume production instrument panel cross-car beam

Uncoated HPDC instrument panel (IP) beams have been installed in many different types of vehicles for over 30 years. Li *et al.* (2005) were involved in the production and installation of a Mg IP cross-car beam for a very low volume, aluminum space frame sports car, the 2005 Ford GT. Because of the low production volume, HPDC would have been too costly to tool and a low cost, sand casting process was employed instead. This introduced a potential corrosion issue. Under normal HPDC conditions, general corrosion for an interior Mg casting is not an issue. But as was discussed earlier, the slow solidification, grain and sub-grain boundary segregation from sand casting can lead to pitting upon exposure to even minor amounts of electrolyte. To prevent this and protect the surface finish of the IP casting, the authors used a protective coating: Henkel Alodine 5200 pretreatment followed by a ProTech epoxy urethane powder coating. It is of interest that the authors did not use structural fasteners to install the IP; instead they used Dow Betamate 1480 (a single-component epoxy adhesive), to join each of the two, one-half cross-car beams and cover plates. Mechanical fasteners were used only to stabilize the parts during the adhesive curing process.

18.5.2 Exterior application

Cast magnesium suspension crossmember for 2006 Corvette Z06

The Mg engine cradle technology for the GM Corvette was first developed under a program between the United States Department of Energy and the USCAR Automotive Materials Partnership (www.uscar.org). Aragones *et al.* (2005) summarized the metallurgy, engineering, manufacturing and R&D involved in bringing this 35% mass-reduced component to production. As a result of its location, the cradle is exposed to significant corrosion challenges: road debris, salt and water spray. Because of stone-pecking it would have been impossible to maintain an integral protective coating and it was installed uncoated onto the Al space-frame Corvette.

In order to prevent corrosion at the joints, three strategies were followed:

- All Mg casting surfaces were designed to be a minimum of 5.0 mm away from any interfacing steel fasteners; this was not possible for all the joints but interfacing components were modified to prevent accelerated corrosion.
- Al 60601-T6 isolators were placed between the crossmember and every steel component (brackets, fasteners, etc.).
- All through-holes were closed off to prevent corrosion at the threads.

Two durability vehicle tests were run at GM proving grounds equivalent to 10 years corrosion; an additional sample was run on a multi-axial test after 10 years of lab corrosion. No concerns were revealed in either test.

Fuel cell vehicle and its magnesium power distribution unit

Hybrid-electric vehicles are becoming increasingly common. Weight reduction has even more importance than with conventional internal combustion or diesel engines, since range, an important attribute, can be increased. A recent paper by Li *et al.* (2005) discussed using a 3.6 kg Mg casting in lieu of a 6 kg Al part to house the power distribution unit (PDU) of a fuel cell vehicle. The PDU provides the interface and distributes the electrical output from the fuel cell to the vehicle high voltage bus. A total vehicle accelerated corrosion test was performed at the Ford Arizona Proving Ground (APG) facility. This test consisted of 30 cycles over the APG circuit – dust, gravel, cobblestones, potholes, salt spray baths and controlled humidity, soaking and drying – and showed significant surface and galvanic corrosion sites are everywhere, from the steel brackets that mount the Mg PDU to the frame of the vehicle (in this case a Ford Focus), from the Al cover to the Mg PDU and from surface chemical sites. Much of this general corrosion occurred because of the corrosion-enhanced nature of the Mg sand castings used for this low volume production part. The design solutions described by the authors included the following:

- Nylon sleeve isolators and Al 6000 series washers were used over steel brackets and on studs/bolts at vehicle attachment points.
- Safety straps were nylon coated (except for a small clear patch for a grounding path).
- All Al parts in contact with the Mg PDU were made from high purity A356, rather than the Fe and Cu-containing A319 which promoted galvanic corrosion.
- All Mg surfaces were protected with a Tagnite hard coat anodizing coat.

Magnesium spare tire carrier

Duke and Logan (2007) were the two key engineers involved in developing the externally mounted magnesium spare tire carrier introduced on the rear door of the Chrysler 2007 Jeep Wrangler. This reduced weight by 3.4 kg (65%) compared with the previous 7.5 kg stamped steel tire mounting bracket and the thermoplastic composite, center high-mounted stop light (CHMSL). Galvanic corrosion between the steel fasteners, the steel wheel and the magnesium carrier was the primary concern; however there was also concern about the UV exposure at the top of the CHMSL, which could affect its appearance. The corrosion protection process developed by Chrysler included the following:

- The top wheel-mounting studs (previously welded to the steel door) were e-coated.
- The interface between the steel wheel and the Mg bracket was separated by a mylar sheet attached to the bracket.

- The steel bottom wheel mounting stud (double-ended stud with a shoulder threaded into a boss in the Mg bracket) was Zn-Ni plated and separated with an aluminum washer. While there was a pocket where electrolyte could be trapped, the mylar sheet provided sufficient isolation.
- The interface between the Mg carrier and the vehicle steel swing gate was e-coated, primed, painted and clear-coated prior to mounting the carrier.
- The bolts that attached the carrier to the swing gate employed plastic washers and were Zn-Al plated.
- The attachment of the plastic CHMSL bracket to the Mg carrier was a serious concern that required precoating the mounting screws with a Zn-Ni plating and a decorative black cathodic protective coating, followed by nylon washers. Nylon fasteners were considered but not required.

Vehicle closures – liftgate

The liftgate on the 2010 Lincoln MKT was constructed from a die cast Mg inner panel and a stamped Al outer panel. This allowed the assembly to be engineered for a 10 kg (40%) weight saving compared with standard steel and which allowed reduction in roof support and power lifting mechanism, while enhancing the corrosion protection of the inner panel. In addition, the use of magnesium and aluminum afforded more design flexibility than steel, enabling the MKT's swept-forward bustle-back styling without sacrificing interior rear-seat headroom. Every item of hardware and every contact point on the liftgate was designed and analyzed to ensure no corrosion could occur. Extensive full vehicle and component level testing proved out durability of the liftgate. The basic anticorrosion architecture included an Alodine 5200 submicron-thick chemical conversion coating, paint and a clearcoat. The painted Al outer panel was chemically bonded to the Mg with only a minimal number of fasteners to hold it in place during curing. The assembly passed all Ford APG corrosion durability tests as well as receiving a 5* crash safety rating.

Cast magnesium doors

Blanchard *et al.* (2005) analyzed the production and assembly of Mg doors for both a production Ford Contour vehicle as well as for an Aston Martin DB9 inner panel (with corresponding weight savings of 43%). The corrosion portion of the program focused on isolating the inner panel from the remainder of the subassembly using surface pretreatments and coatings based on a Ford test method, BI123-01 that simulated 100 cycles of a Ford proving ground test procedure. Test coupons were fabricated from cast Mg samples bonded to Al coupons to simulate the hem flange in the door assembly. Chrome-based conversion coatings were observed to be best; however, in 2004, when these tests were performed, Cr was allowed. They are not now. Results from a Cr-free conversion coating based on Ti/Zr were mixed, with some tests performing well and others offering considerably less protection. The large inconsistency in the results was attributed to poor control of the cleaning stages prior to applying the conversion coating. Choices of powder coating (such as epoxy and polyester) had only a secondary effect since differences in performance were diminished by the addition of the base paint followed by the clear coat. Anodizing performed exceptionally well when combined with a top coat of either a powder coat or an e-coat paint, and exceeded any of the conversion coatings. However, anodizing has a significant cost penalty.

Front end carrier

The Front End Carrier (FEC), or Magnesium Radiator Support, is one of the few Mg components considered part of the body-in-white (BIW) assembly. The excellent paper by Tippings and Lawson (2005) reviewed the different FEC architectures that have been globally installed on a range of vehicles, from the Ford F150 truck to the Chrysler Viper front of dash structure. Because the FEC is external and connected to the BIW steel rails and fenders it is exposed to a considerable range of particularly harsh environmental galvanic corrosion situations. With the new generation of Al BIW vehicles there would not be the same galvanic protection requirements. However, when the BIW structure is steel, the full suite of corrosion-prevention design and galvanic corrosion-protecting modalities is required for this component:

- Design for drainage holes so that debris and mud packing do not occur.
- Design to close off the void at joints (a void is formed at the bottom of a fastener and the magnesium part) to prevent electrolyte from entering the joint.
- Prepare the Mg surface to remove all traces of Fe using caustic etch, DI water wash, strong acid or acetic acid etch, DI water wash.
- Apply a conversion coating (Meridian Technologies use chrome-free patented fluorozincates and fluorotitanates) followed by organic polymers and powder coated paints.
- Wash and oven dry to degas the surface.

- Coat with ~40 μm polyester powder paint in non-galvanic regions (epoxy paints are harder to remove for repair) and ~100 μm in critical joint areas where galvanic corrosion can occur.
- For interfacing parts to a precoated exterior FEC section, plastic washers were found to be the best choice, followed by coated Al, bare Al and powder coated steel.
- Additional corrosion protection at galvanic joints requires Al 5000 or 6000 series washers or spacers to add ~10 mm to the distance between the anode (Mg) and cathode (steel).
- Additional protection to eliminate the potential for galvanic corrosion between the washer and Mg part may require an e-coat followed by an epoxy powder coat.

While anodizing is valuable it requires very careful adjustment of the electric current flow to minimize Faraday shadowing and non-uniform coating thickness and quality, especially near component edges which may be sharp. The major concern is cost.

18.6 Future trends

The future for magnesium has never been brighter. With the global emphasis on fuel efficiency and emissions reduction, mass reduction with this ultra-light metal can be a real enabler at meeting the 49.6 mpg MY2025 fuel standards (or the 54.5 mpg CAFE proposals) of the United States Government. The many issues involved in using more Mg automotive components were examined in a 2006 monograph prepared by the author (USCAR, 2006). Galvanic corrosion was a major concern that limited growth in the automotive use of Mg components and was discussed in detail. But as shown here, there are many choices for mitigating galvanic corrosion, keeping in mind the electrochemical kinetics of how Mg connects to more noble cathodic materials. Key points to note are:

- The rules for appropriate component design to enhance drainage and limit build-up of dirt and mud poultices are well documented. But as discussed by Skar (2006), it is exceedingly difficult to quantitatively compare test data to actual field experience, so that designing the cheapest, most functional system is not straightforward.
- The techniques for characterizing surface quality so that the process and materials properties for conversion coatings are optimized in a statistically elegant fashion need development.
- Anodizing is still the best process to apply a protective hard coat onto a Mg surface, but all coatings need to be sealed to ensure no 'electrochemical

leakage'. Every major global Mg R&D center is trying to develop a 'better' (i.e., faster, more dense, lower cost) anodizing process. Currently, anodizing adds considerable expense for chemicals, bath infrastructure, washing/drying, material handling and so forth. Al components do not have this problem and thus are more competitively priced.

- Paints and other organic coatings are well-characterized for preventing galvanic corrosion as long as they are thick enough and there is a good foundation for adhesion to the Mg surface. There is still concern about their durability; that is, will they crack and 'leak' after many years in service?
- Organic insulators and separators are the best choice at stopping the electric current that drives galvanic corrosion. But existing materials tend to flow and yield under compression loads, and crack at the temperature of paint curing ovens.
- Magnesium wrought products are just now entering the market. Twin roll casting of Mg sheet with a thin plastic bi-film overlayer could be a competitive way to protect the metal from corrosion for sheet and stamped applications.
- There is a considerable global R&D effort to use nanotechnology methods to reduce corrosion. The hope is that a simple surface dip treatment will promote a self-healing 'stainless-type' response on all magnesium alloys, and directly solve all corrosion issues ... but not yet.

18.7 References

Aragones J, Goundan K, Kolp S, Osborne R, Ouimet L and Pinch W (2005), 'Development of the 2006 Corvette Structural Cast Magnesium Crossmember', Transactions SAE, Paper # 2005-01-0340.

Blanchard P, Bretz G, Subramanian S and deVries J (2005), 'The Application of Magnesium Die Casting to Vehicle Closures', Transactions SAE, Paper # 2005-01-0338.

Busk R (1987), 'Corrosion Protection and Finishing', *Magnesium Products Design*, Marcel Dekker, New York, ISBN 0-8247-7576-7. 517–538.

Champrenault M, Maas C and Cunningham J (2007), 'Powertrain Mount Brackets: New Application of Materials Being Used in this Sub System for Mass Reduction', Transaction SAE Paper #2007-01-1031.

Christopher D and Logan S (2007), 'Lightweight Magnesium Spare Tire Carrier', Proceedings International Magnesium Association, Vancouver, BC, 75–80.

Dolan S and Reghi R (1994), 'Composition and Process for Treating Metals', US Patent 5281282 (Jan).

Duke C and Logan S (2007), 'Lightweight Magnesium Spare Tire Carrier', Proceedings International Magnesium Association, Vancouver BC, 75–80.

Emley E (1966), *Principles of Magnesium Technology*, Pergamon Press, Headington Hill Hall, London W1.

Erickson S (2004), presented at the 2004 International Magnesium Association Detroit, Automotive Magnesium seminar, with permission.

Friedrich H (2004), 'Reliable Light weight Fastening of Magnesium Components in Automotive Applications', Society of Automotive Engineers SAE Paper # 2004-01-0136.

Friedrich H and Mordike B (Eds.) (2005), *Magnesium Technology: Metallurgy, Design Data, Applications*, Springer, Berlin ISBN-10 3-540-20599-3.

Friedrich H and Schumann S (2002), 'Strategies for Overcoming Technological Barriers to the Increased Use of Magnesium in Cars', Transactions IMM, p C65, (May).

Gao G and Ricketts M (2002), 'Evaluation of Protective Coatings on Magnesium for Phosphate Process Compatibility and Galvanic Corrosion Prevention', SAE Transactions 2002-01-0081.

Gehmecker H (1994), 'Chrome Free Pretreatment of Magnesium Parts', Proceedings International Magnesium Association, Germany.

Guerci G, Mus C and Stewart K (2000), 'Surface Treatments for Large Automotive Magnesium Components', Magnesium 2000 (loc cit).

Hawke D and Gaw K (1992), 'Effects of Chemical Surface Treatments on the Performance of an Automotive Pan System on Die Cast Magnesium', Society of Automotive Engineers, SAE Paper 920074.

Hawke D and Ruden T (1995), 'Galvanic Compatibility of Coated Steel Fasteners with Magnesium', Society of Automotive Engineers, Paper # 950429.

Hawkins J (1993), 'Assessment of Protective Finishing Systems for Magnesium', Proceedings International Magnesium Association, Washington, DC.

Hillis J (2005), 'Corrosion', in *Magnesium Technology: Metallurgy, Design Data, Applications*, HE Friedrich and BL Mordicke editors, Springer-Verlag, p. 469–497.

Hillis JE and Reichek KN (1986) *High Purity Magnesium AM60 Alloy: The Critical Contaminant Limits and the Salt Water Corrosion Performance*, Society of Automotive Engineers, Paper 860288, Detroit, MI.

Hillis JE and Shook SO (1989) *Composition and Performance of an Improved Magnesium AS41 Alloy*, Society of Automotive Engineers, Inc., Warrendale, Pennsylvania, USA.

Hydro Magnesium Brochure (1997), *Corrosion and Finishing of Magnesium Alloys*, Porsgrun, Norway.

Jurey C (1993) *MAGOXID COAT- Hard Anodic Coating for Magnesium*, International Magnesium Association, Washington.

Kurze P (2006), 'Surface Treatments and Protection', in *Magnesium Technology: Metallurgy, Design Data, Applications,* HE Friedrich and BL Mordicke editors, Springer-Verlag, 430–468.

Li N, Chagas E and Zechel K (2005) 'Fuel Cell Vehicle and Its Magnesium Power Distribution Unit', Transaction SAE Paper #2005-01-0339.

Li N, Chen X, Hubbert T and Berkmortel R (2005), '2005 Ford GT Magnesium Instrument Panel Cross Car Beam', Transactions SAE (2005) Paper # 2005-01-0341.

Liu M, Uggowitzer PJ, Nagasekhar AV, Schmutz P, Easton M, Song GL, and Atrens A (2009), Calculated phase diagrams and the corrosion of die-cast Mg-Al alloys, *Corrosion Science*, **51**, 602–619.

Mercer WE and Hillis JE (1992), The critical contaminant limits and salt water corrosion performance of magnesium AE42 alloy, Society of Automotive Engineers, Inc., Warrendale, Pennsylvania, USA.

Powell B and Rezhets SV (2001), 'Creep-Resistant Magnesium Alloy Die Castings', US Patent 6264763 B1.

Reichek KN, Clark KJ and Hillis JE (1985), 'Controlling the Salt Water Corrosion Performance of Magnesium AZ91 Alloy', Paper 850417 Society of Automotive Engineers, Detroit, MI.

Ross P and MacCulloch J (1998), 'The Anodization of Magnesium at High Voltages', Society of Automotive Engineers, SAE Paper 980088.

Ruden T and Lenda D (2004), 'A Cast Study of a Die Cast Magnesium Structure Supporting Transmission Shifter Mechanisms and Interfaced with other Structural Systems', Society of Automotive Engineers, SAE 2004-01-0130.

Skar J, Isacsson M and . Tan Z (2006), 'Galvanic Corrosion of Die Cast Magnesium Exposed on Vehicles and in Accelerated Laboratory Tests', Transaction SAE Paper #2006-01-0255.

Sklar J and Albright D (2000), 'Phosphate-Permanganate: A Chrome Free Alternative for Magnesium Pretreatment', Magnesium 2000, Munich Germany, (25–26 September).

Tippings A and Lawson T (2005), 'Design and Development of a High Pressure DIE "FEC" (Front End Carrier) for Exterior Automotive Applications', International Magnesium Association 62nd Annual World Magnesium Conference, Berlin Germany.

USCAR (2006). Magnesium Vision 2020: A North American Automotive Strategic Vision for Magnesium, United States Materials Partnership, United States Council for Automotive Research, www.uscar.org.

Wang N, Stewart G, Berkmortel R, and Skar J, (2001) 'Corrosion Prevention for External Magnesium Automotive Components,' Society of Automotive Engineers, Paper # 2001-01-0421.

19
Control of biodegradation of magnesium (Mg) alloys for medical applications

K. YANG and L. TAN,
Chinese Academy of Sciences, China

DOI: 10.1533/9780857098962.4.509

Abstract: Magnesium-based metals, including pure magnesium and its alloys, are becoming increasingly popular in the medical industry due to the fact that they are biodegradable. This chapter briefly introduces the concept of biodegradable magnesium-based metals and their advantages as a new class of metallic biomaterials. Relevant research progress is summarized, especially the corrosion control of magnesium-based metals by appropriate surface modification, which is a key issue for clinical applications.

Key words: biomedical magnesium alloys, biodegradation, corrosion biodegradation techniques.

19.1 Introduction

In the late nineteenth century, magnesium-based metals began to be used as a form of implant material, after the discovery of elemental magnesium by Sir Humphrey Davy in 1808. Some clinical trials on pure magnesium and magnesium alloy implants – including ligatures, connectors for vessel anastomosis, wires for aneurysm treatment and osteosynthetic applications – were performed with some exciting results. For example, a 3-month-old child with proliferating hemangioma on its face, throat and shoulder was cured after 20 months of treatment by insertion of the magnesium arrows (see Fig. 19.1) (Witte, 2010). However as the problem of rapid corrosion of Mg *in vivo* was not sufficiently solved, surgeons preferred to use the more corrosion-resistant stainless steel, which was invented in the 1930s and, later, titanium alloys; these have better corrosion resistance, mechanical properties and biological performance. Since then, magnesium has no longer been intensively used as a biomaterial.

However, with widespread clinical application of stainless steels, titanium alloys and other bio-inert metals, certain problems have arisen. With

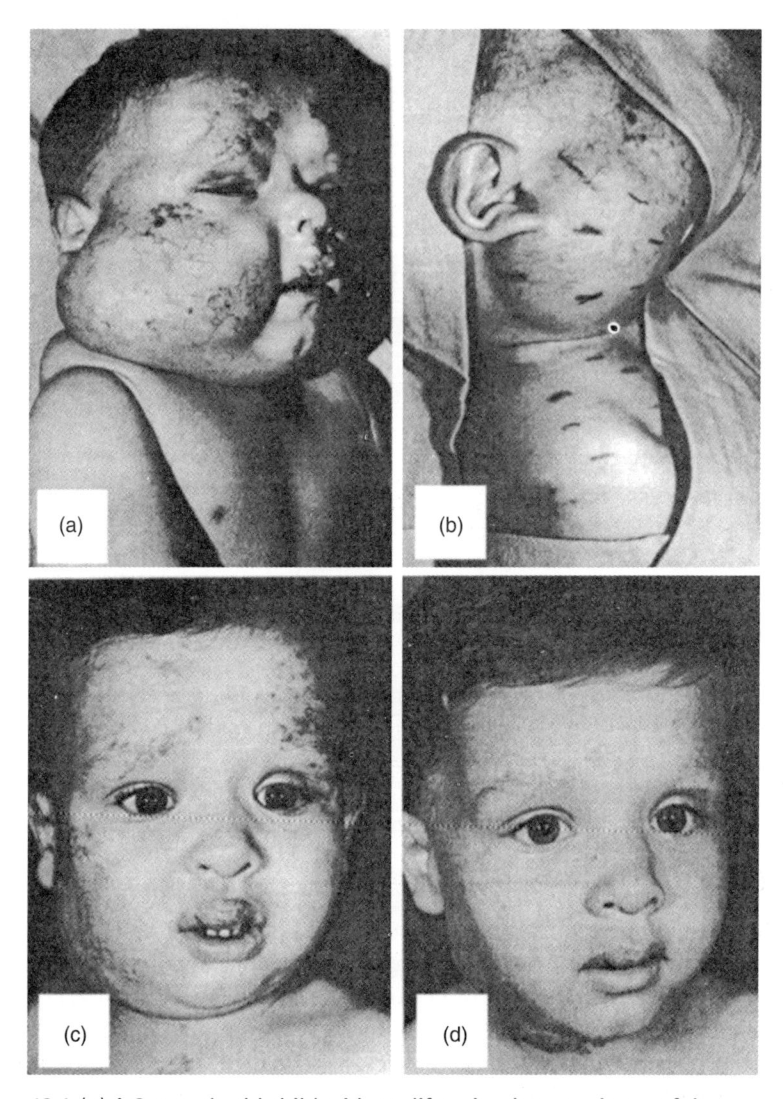

19.1 (a) A 3-month-old child with proliferating hemangioma of the face, throat and shoulder. (b) The same patient as in (a); the incisions for the insertion of the magnesium arrows are marked with ink on the skin. (c) After 3 weeks the treatment was repeated and 9 months after the beginning of the treatment a significant reduction of the hemangioma could be observed. (d) After an additional insertion of magnesium arrows at 3 months, the face of the 2 year old child became almost normal at 20 months after starting the magnesium treatment (Witte, 2010).

regard to medical implants, such as plates, screws and pins that are used to repair bone fractures, a subsequent surgical procedure to remove the implants from the human body after the tissues have healed is not always necessary. This is because repeated surgery generally increases health risks to the patient. Furthermore, the widely used bio-inert metals may induce a 'stress-shielding' effect owing to their much higher elastic modulus, compared to that of bones, leading to a risk of secondary fracture of bones. Stress shielding occurs when a bone is shielded by an implant from carrying a load. As a result, the bone tends to weaken over time, resulting in more damage. To minimize the effect of stress shielding on bones while still retaining enough strength, a soft lightweight metal is needed. Therefore an implant material that is degradable in the physiological environment is desirable. Biodegradable polymers, ceramics and bioactive glasses were consequently developed for this purpose; however their applications are limited due to their lower integrated mechanical properties as well as some biological problems. Magnesium-based metals with biodegradable behavior and superior strength-to-weight ratios are being reconsidered and are breaking the current paradigm for metallic biomaterials whose only use is to improve corrosion resistance.

Biodegradable magnesium-based metals have the potential to be used as new forms of biodegradable medical implant materials, since they possess many advantages over the current applied and developed biomaterials, such as:

- Good biological properties. Magnesium is an element essential to the human body. Mg^{2+} is the fourth most abundant cation in the human body and is largely stored in bone tissues. The recommended daily intake of Mg for a normal adult is around 300–400 mg. Mg is vital to the metabolism processes, a co-factor in many enzymes and a key component of the ribosomal machinery that translates the genetic information, encoded by mRNA, into polypeptide structures. The direct corrosion product of magnesium, Mg^{2+}, is easily absorbed or consumed by the human body and can be excreted in the urine (Staiger *et al.*, 2006).
- Good integrated mechanical properties. Magnesium and its alloys have obvious advantages over the currently used biodegradable materials, such as polymers, ceramics and bioactive glasses. These advantages are particularly prominent in load bearing applications that require higher tensile strength. Table 19.1 shows a comparison of the mechanical properties of different biomedical materials.
- Similar modulus to bones. Compared with Titanium alloys (about 110 GPa), stainless steels (about 200 GPa), and Cobalt-based alloys (about 230 GPa), the elastic modulus of magnesium alloys (about 45 GPa)

Table 19.1 Mechanical properties of different normal tissues and requirements in standard stent and orthopedic implants compared with available data for currently investigated magnesium alloys

Tissue/material	Compressive strength, MPa	Tensile strength, MPa	E-modulus, GPa	Tensile yield strength (YS), MPa	Elongation (A) at break, %
Cortical bone	164–240	35–283	5–23		1.07–2.10
Cancellous bone		1.5–38	10–1570 (MPa)		
Arterial wall		0.50–1.72	1 (MPa)		
Titanium (TiAl6V4, cast)		830–1025	114	760–880	12
Titanium (TiAl6V4, wrought)		896–1172	114	827–1103	10–15
Stainless steel 316L		480–620	193	170–310	30–40
Synthetic hydroxyapatite	100–900	40–200	70–120		
Bioactive glass		40–60	35–35		
DL-PLA		29–35	1.9–2.4		5–6
AZ91E-F sand cast	97	165	45	97	2.5
AZ91E-F HPDC	165	230	45	150	3
AZ91E-GAE		457	45	517	11.1
AZ31 extruded	83–97	241–260		165–200	12–16
AZ31 sheet	110–180	255–290		150–220	15–21
AZ31GAE		445		424	11.5
LAE442		247		148	18
WE43A-T6		250		162	2
WE43-B	345	220			2
WE43 extruded		277		198	17
WE43 tube		260		170	25
AZ91+2Ca-GAE		452		427	5.4
AZ91+2Ca		147			1.7
Mg0.8Ca		428			
Mg(0–4)Ca		210–240			
AM50A-F	113	210			10
AM60B-F	130	225			8

Source: Witte *et al.*, 2008b.

is closer to that of human bone. Hence, the stress-shielding effect induced by a serious difference in modulus between bones and metal implants is likely to be mitigated.

- Similar density to bones. The density of magnesium alloys (1.7–2.0 g/cm^3) is close to that of natural bones (1.8–2.1 g/cm^3), compared with Ti alloys (4.42 g/cm^3 for Ti_6Al_4V), stainless steels (about 7.8 g/cm^3), biodegradable polymers (about 1 g/cm^3 for poly-l-lactic acid (PLLA)) and hydroxyapatite (3.156 g/cm^3).

In light of these advantages, as well as the characteristic of biodegradation, magnesium-based metals have become one of the most intensely researched areas in metallic biomaterials over the past decade. Magnesium-based metals have been widely investigated both *in vitro* and *in vivo*, focusing largely on coronary stents, screws and plates for orthopedic fixations and porous scaffolds for bone filling materials.

19.2 Mg alloys studied for medical applications

Different magnesium alloys have been studied targeting for medical applications, including the currently commercialized alloys and newly developed alloys, since the early of this century when the new concept of biodegradation of magnesium alloys in physiological environment was proposed.

19.2.1 Commercial Mg alloys for medical applications

Currently, commercialized Mg alloys developed for engineering application purposes have relatively good mechanical and corrosion-resistant properties. As a result some commercial Mg alloys were originally specifically selected as the biodegradable magnesium alloys for investigations.

German company Biotronik adopted WE43 alloy, a rare earth (RE) element-strengthened magnesium alloy, as a biodegradable magnesium alloy for the development of bio-absorbable coronary stents. In 2005, the implantation of a 3 mm diameter stent was performed in a hybrid procedure on a baby weighing 1.7 kg (Zartner *et al.*, 2005). Despite their small size, the artery vessels of the baby tolerated the degradation process. The mechanical and degradation characteristics of the magnesium alloy stent proved adequate to secure the reperfusion of the previously occluded left pulmonary artery. AZ31B alloy, a Mg-Al-Zn alloy, has also been manufactured into bio-absorbable coronary stents with drug eluting coating (SEBMAS). The *in vivo* study showed that the SEBMAS degraded gradually after being implanted into a rabbit aorta, and the stent completely disappeared after about 120 days (Xu *et al.*, 2011). Witte *et al.* (2007) reported that the implantation of another Mg-Al-Zn alloy, AZ91, showed no significant harm to its neighboring tissues and also exhibited good biocompatibility. No skin sensitizing potential has been detected for commercial magnesium alloys currently under research, namely AZ31, AZ91, WE43 and LAE442 (Witte *et al.*, 2008a).

The potential to use pure magnesium as a biodegradable implant material has also been investigated, since it has single elemental composition, high purity and uniform degradation behavior (Kuwahara *et al.*, 2001; Ren *et al.*, 2005). The *in vitro* immersion, cytotoxicity assessment and *in vivo* bone formation study showed that pure magnesium may be more suitable to be used as a biodegradable implant material (Yun *et al.*, 2009).

19.2.2 Development of new biodegradable Mg alloys

Originally, commercial magnesium alloys were not designed for medical application, thus some potential problems may exist. Aluminum ions released from the AZ91 alloy could easily combine with inorganic phosphates, leading to a lack of phosphate in the human body, and an increased concentration of Al ions in the brain. This seems to be associated with Alzheimer's disease. Severe hepatotoxicity has been detected after the administration of RE elements such as cerium, praseodymium and yttrium (Hort *et al.*, 2010). Therefore in the past few years, some new magnesium alloys, better suited to medical applications, have been developed including the alloy systems of Mg-Ca, Mg-Zn, Mg-RE and Mg-Mn.

Mg-Ca alloy

Calcium (Ca) can accelerate the growth of, and is a major component in, human bones. Approximately 80–90% of bone mineral content is made up of calcium and phosphorus (Peng *et al.*, 2010a). Ca has a low density (1.55 g/cm^3), which gives the Mg-Ca alloy system the advantage of having a similar density to bone (Zheng *et al.*, 2008). Magnesium is necessary for the incorporation of Ca into the bone, which is usually helpful in the healing of bones, as it releases both Mg and Ca ions (Zheng *et al.*, 2008). For this reason, Mg-Ca based alloys have been studied by several research groups across the world as a new form of biodegradable magnesium alloys.

Binary Mg-xCa (x = 1–3 wt.%) alloys with various Ca contents under different processing conditions have also been studied (Zheng *et al.*, 2008). These are generally composed of two elements, alpha-Mg and Mg_2Ca. The yield strength (YS), ultimate tensile strength (UTS) and elongation (EL) of the alloy all decreased with the increase of Ca content. The UTS and EL of the as-cast Mg-1Ca alloy, from 71.38 ± 3.01 MPa and 1.87 ± 0.14%, were largely improved after either hot rolling – to 166.7 ± 3.01 MPa and 3 ± 0.78%, respectively – or hot extrusion, to 239.63 ± 7.21 MPa and 10.63 ± 0.64%, respectively. An increase of Mg_2Ca phase in the microstructure led to a higher corrosion rate for the alloy, whereas hot rolling and hot extrusion could reduce the corrosion rate.

Screws made of Mg-0.8Ca alloy were implanted into the tibiae of rabbits, and compared with 316L stainless steel screws (Erdmann *et al.*, 2010). Moderate inflammation was detected for both implant materials and was resolved to a minimum after the first week following implantation, indicating a comparable biocompatibility for Mg-0.8Ca alloy with 316L stainless steel. Uniaxial pull-out tests showed no significant difference between the pull-out forces for Mg-0.8Ca alloy and 316L two weeks after the surgery ($P = 0.121$), however significantly higher pull-out forces were detected for 316L from 4 weeks ($P < 0.001$) compared with those of Mg-0.8Ca alloy (Erdmann *et al.*, 2011). Since Mg-0.8Ca alloy showed good biocompatibility and mechanical properties – comparable with those of 316L in the first 2–3 weeks of implantation – its application as a biodegradable implant is certainly feasible. Guo and Salahshoor (2011) also developed an efficient and ecological machining method for a lab-made Mg-Ca alloy with 0.8 wt.% calcium, cutting at speeds of up to 47 m/s without coolant. Polycrystalline diamond inserts were applied and the possibilities of flank build-up formation, chip ignition and tool wear were sought during the cutting test, with the aid of a developed on-line and optical monitoring system.

Mg-1Ca alloy did not induce cytotoxicity to cells, and high activities of ostroblast and osteocytes were observed around the Mg-1Ca alloy pins implanted in rabbit femoral shafts (Zheng *et al.*, 2008), showing good biocompatibility and bioactivity. The performance of Mg-1Ca alloy was further improved by hot working (Harandi *et al.*, 2011). Increasing the forging temperature decreased the grain size of the alloy, which led to an enhanced hardness and improved plastic deformation property for the alloy, whereas no significant effect was observed by changing the forging rate. However, forging at higher temperatures led to an increase in the amount of Mg_2Ca phase at grain boundaries, which resulted in a higher corrosion rate for the alloy. So although the forging process improved the mechanical properties of the alloy, it still failed to fulfill the corrosion resistance criteria required for bone healing.

Melt-spinning techniques have also been adopted to prepare the rapidly solidified (RS) Mg-3Ca alloy ribbons, in order to obtain finer grain size and reduce both corrosion rate and uniform corrosion morphology (Zheng *et al.*, 2010a). The cytotoxicity evaluation revealed that the as-spun Mg-3Ca alloy ribbon extract did not induce toxicity to the L-929 cells, whereas the as-cast Mg-3Ca alloy ingot extract did.

Mg-Zn alloy

Zinc is an essential element in the human body and it also has a strengthening effect in magnesium alloys (Boehlert and Knittel, 2006). Zn could elevate both the corrosion potential and the Faraday charge transfer resistance of magnesium, and thus improve the corrosion resistance (Zhang *et al.*, 2009b). Up to this point, Mg-Zn based alloys have been developed with positive results.

A Mg alloy with 6 wt.% Zn fabricated using high-purity raw materials and a clean melting process proved to contain very low level of impurities

(Zhang *et al.*, 2009b, 2010d). After solid solution treatment and hot working, the grain size of this Mg-Zn alloy became finer and a uniform single-phase microstructure was created. The tensile strength and elongation of the alloy reached 279.5 MPa and 18.8%, respectively. Mg-6 wt% Zn alloy could be gradually absorbed *in vivo* at a degradation rate of about 2.32 mm/year, and was harmless to the L-929 cells and important organs in animals.

Rosalbino *et al.* (2010) examined the bio-corrosion behavior of Mg-2Zn-0.2X (X = Ca, Mn, Si) alloys in a Ringer's physiological solution that simulated body fluid (SBF), and compared it to that of an AZ91 alloy. The latter was proven to have better corrosion resistance than the former. The Mg-2Zn-0.2Mn alloy, however, transpired to have better corrosion resistance compared to the AZ91 alloy, due to the enhanced protective ability of the $Mg(OH)_2$ layer on its surface.

Furthermore, high pressure torsion (HPT) processing technique was attempted in order to improve the corrosion resistance of Mg-Zn-Ca alloy, (Guan *et al.*, 2011b). After this processing, the second phases in the alloy matrix became nano-sized particles and were distributed uniformly in grain interiors – instead of along grain boundaries – resulting in a homogeneous corrosion and decreased corrosion current density from 5.3×10^{-4} A/cm^2 down to 3.3×10^{-6} A/cm^2.

The Mg-based metal glass could offer an extended solubility for alloying elements, as well as a homogeneous single-phase microstructure, which might significantly alter the corrosion behavior. A Zn-rich Mg-Zn-Ca metal glass was also studied as an alternative biodegradable material. The animal study confirmed the alloy to have undergone a significant reduction in hydrogen evolution and to have similarly strong tissue compatibility to that of the crystallized Mg implants. Thus, the glassy $(60 + x)$Mg-$(35–x)$Zn-5Ca $(0 \leq x \leq 7)$ alloys can be said to show great potential as a new generation of biodegradable implants (Loffler *et al.*, 2009). 66Mg-30Zn-4Ca metal glass was also found to have more uniform corrosion morphology than the as-rolled, pure Mg and 70Mg-25Zn-5Ca metal glass (Zheng *et al.*, 2010b), with much smaller, uniformly distributed micro-scale pores beneath the corrosion product layer. The Mg-Zn-Ca metal glass extract was also found to have higher cell viability than that of the as-rolled pure Mg.

Mg-RE alloy

As Mg-RE alloys possess better mechanical performance and corrosion resistance, some Mg-RE alloys were studied as potential biodegradable magnesium alloys for medical application. An Mg–Y alloy was prepared by a zone solidification method, and showed improved corrosion resistance and mechanical properties (Peng *et al.*, 2010). The alloys Mg-Gd (Nakamura *et al.*,

1997), Mg-Ce, Mg-Nd, Mg-La, etc., were also studied for medical application, with the Mg-Nd alloy showing a much slower corrosion rate than the other alloys (Zheng and Gu, 2011). The Mg–Y–Zn alloy, moreover, showed an interesting combination of preferred microstructural, mechanical, electrochemical and biological properties, making it a very promising prospect for application as a biodegradable implant material (Uggowitzer *et al.*, 2009; Hänzi *et al.*, 2010). The corrosion resistance of the Mg-1.2%Nd-0.5%Y-0.5%Zr alloy was improved by addition of 0.4% Ca, though this did lead to problems with the stress corrosion behavior (Aghion and Levy, 2010).

Other alloys

Binary Mg-1X (wt%) (X = Al, Ag, In, Mn, Si, Sn, Y, Zn and Zr) alloys were studied to evaluate the biological properties compared to those of pure Mg (Gu *et al.*, 2009b). The cytotoxicity tests indicated that the Mg-1Al, Mg-1Sn and Mg-1Zn alloy extracts did have a significantly lower cell viability to that of fibroblasts (L-929 and NIH3T3) and osteoblasts (MC3T3-E1). The Mg-1A1 and Mg-1Zn alloy extracts, meanwhile appeared to have no negative effect on the respective viabilities of the blood vessel related cells, ECV304 and VSMC. The hemolysis and the amount of adhered platelets decreased for all the Mg-1X alloys, as compared to that seen in the pure magnesium control.

Mg-Si alloy was investigated for medical application due to the biological function of Si in the human body (Zhang *et al.*, 2010a). However, the Mg-Si alloy proved to have a low ductility, owing to the presence of a coarse Mg_2Si phase in matrix. The bio-corrosion resistance of Mg-Si alloy was improved by the addition of Ca due to the reduction and refinement of Mg_2Si phase. No improvement was observed in the strength and elongation, however. Tensile strength, elongation and bio-corrosion resistance were all improved significantly by the addition of 1.6% Zn, the most notable change being a 115.7% increase in elongation.

Finally, the biological properties of high-purity Mg-1.2Mn-1.0Zn were evaluated *in vivo* (Xu *et al.*, 2007). From 9 to 18 weeks after implantation, all the magnesium implants were tightly fixed and no inflammation was found. After 18 weeks, however, Zn and Mn were discovered to be homogeneously distributed in the residual Mg-Mn-Zn alloy implant, the degradation layer and the surrounding bone tissue, indicating that Zn and Mn were easily absorbed by the bio-environment.

19.2.3 Developments for medical application of biodegradable Mg alloys

Recent investigations into the application of coronary stents for commercial magnesium alloys including AE21 (Heublein *et al.*, 2003), WE43

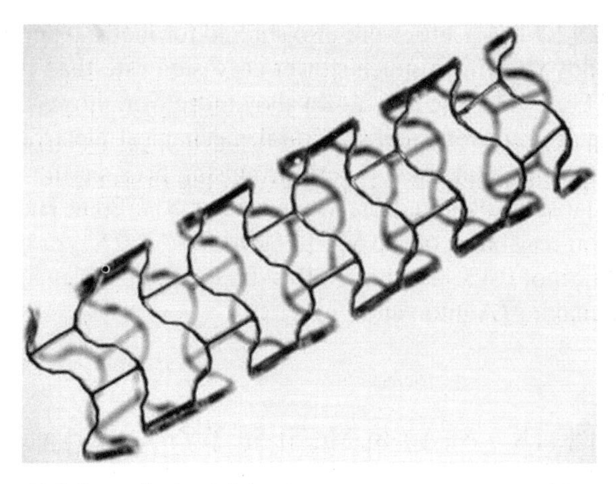

19.2 Strut design of the 3 mm magnesium stent (Zartner *et al.*, 2005).

(Zartner *et al.*, 2005; McMahon *et al.*, 2007) and AZ31B (Xu *et al.*, 2011), as shown in Fig. 19.2, have opened up new opportunities for biodegradable magnesium alloys. The stent is particularly useful as it is exposed to blood flow which encourages the diffusion of generated hydrogen gas (Yamamoto and Hiromoto, 2009). Clinical trials on the critical limb ischemia and coronary artery, moreover, are already underway in Europe, suggesting enormous potential in this area. As shown in Fig. 19.3, another potential use for biodegradable magnesium alloys is in the manufacture of orthopedic devices, the implants for either scaffolding on which new bone can grow or fixtures holding bones together long enough to allow natural healing to take place (Seal *et al.*, 2009). In both cardiovascular and orthopedic applications, implanted devices are expected to mechanically support the blood vessel wall or the fractured bones during their healing processes.

19.3 Corrosion/biodegradation of Mg alloys in the body environment

The corrosion behaviors of magnesium alloys as a class of engineering materials have been widely investigated in the previous studies. However the corrosion process of magnesium-based metals depends highly on the corrosive potential of the environment in which they are found. As a type of biomaterial that will be implanted *in vivo*, magnesium-based metals may therefore display a different corrosion behavior in the body than they would in an engineering environment, such as an industrial atmosphere and ocean air.

(a)

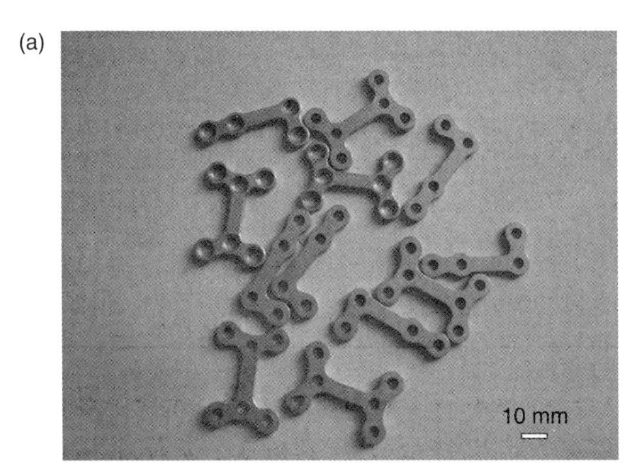

(b)

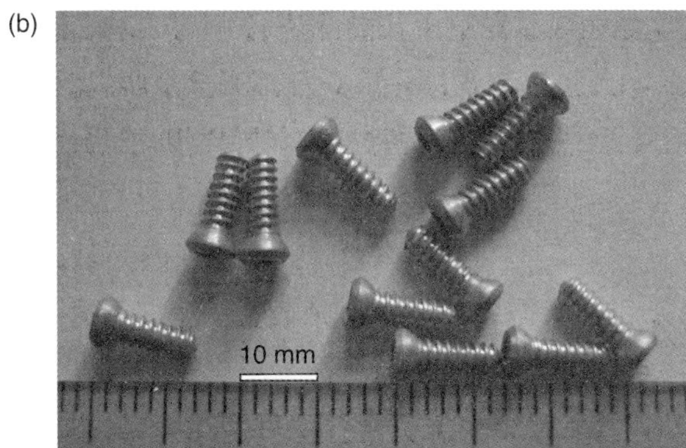

(c)

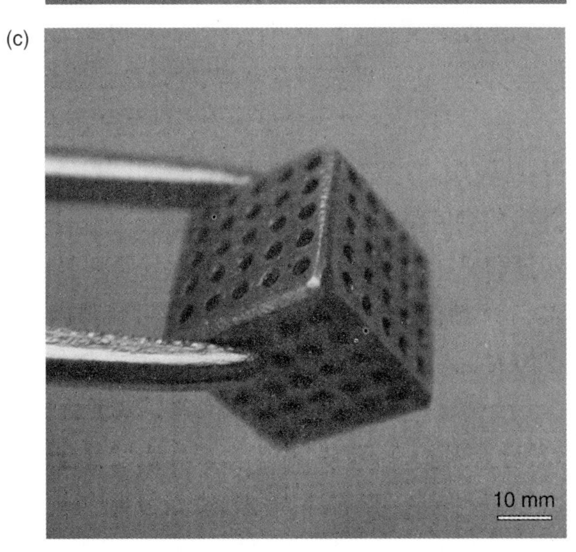

19.3 (a) Biodegradable bone plates, (b) screws for orthopedic fixation and (c) porous scaffold for bone filling prepared by AZ31B alloy. Scale bar = 10 mm.

19.3.1 Effect of the body environment composition on the corrosion behavior of Mg alloys

Magnesium and its alloys are generally known to be corroded in an aqueous environment via an electrochemical reaction, which produces magnesium hydroxide and hydrogen gas. Corrosions of magnesium-based metals are therefore relatively insensitive to the oxygen concentration in the aqueous solutions which surround the implants in different anatomical locations of human bodies. The overall corrosion reaction of magnesium in an aqueous environment is given below (Witte *et al.*, 2008b):

$$Mg_{(s)} + 2H_2O_{(aq)} = Mg(OH)_{2(s)} + H_{2(g)} \quad [19.1]$$

This overall reaction may include the following partial reactions:

$$Mg_{(s)} = Mg^{2+}_{(aq)} + 2e^- \text{(anodic reaction)} \quad [19.2]$$

$$2H_2O_{(aq)} + 2e^- = H_2(g) + 2OH^-_{(aq)} \text{(cathodic reaction)} \quad [19.3]$$

$$Mg^{2+}_{(aq)} + 2OH^-_{(aq)} = Mg(OH)_{2(s)} \text{(product formation)} \quad [19.4]$$

The environment in the human body is much more complicated. Aside from water – which accounts for more than 60% of the body weight – blood plasma and intercellular fluid, there are also neutral solutions containing inorganic ions such as Mg^{2+}, Ca^{2+}, Cl^-, HCO^{3-}, SO_4^{2-} and HPO_4^{2-}, as well as organic compounds such as amino acids and proteins.

Cl^- ions can induce porous pitting corrosions on Mg alloys. As shown above, magnesium hydroxides can, in water, be accumulated on the underlying magnesium matrix to form a corrosion protective layer. But when the chloride concentration in the corrosive environment rises above 30 mmol/L, magnesium hydroxide will react with magnesium to form a highly soluble magnesium chloride and thus the degradation rate is increased as follows (Staiger *et al.*, 2006):

$$Mg(s) + 2H_2O \rightarrow Mg(OH)_2(s) + H_2(g) \quad [19.5]$$

$$Mg(s) + 2Cl^-(aq) \rightarrow MgCl_2 \quad [19.6]$$

$$Mg(OH)_2(s) + 2Cl^- \rightarrow MgCl_2 \quad [19.7]$$

HPO_4^{2-} ions can decrease the corrosion rate of Mg alloys and the occurrence of pitting corrosion is significantly delayed due to the precipitation of magnesium phosphate. HCO^{3-} ions are known to stimulate the corrosion of Mg alloys during the early immersion stage, but they can also induce a rapid dissipation onto the surface, mainly resulting from the fast precipitation of magnesium carbonate in the corrosion product layer, that can subsequently completely inhibit the pitting corrosion. SO_4^{2-} ions have also been found to stimulate dissolution of magnesium (Xin *et al.*, 2008).

Proteins such as albumin have been demonstrated to form a corrosion blocking layer on the magnesium alloys in the *in vitro* experiment. This layer can be enriched with calcium phosphates that concomitantly participate in corrosion prevention (Witte *et al.*, 2008b). Organic compounds such as amino acids, however, catalyze the dissolution of magnesium (Yamamoto and Hiromoto, 2009).

The polarization curves of the magnesium alloys in SBF indicated a longer dissipation stage, as well as breakdown potential, indicating that a passive layer could be rapidly formed on the surface of magnesium alloy in the phosphate buffered SBF, which in turn could protect magnesium alloys from rapid corrosion (Xu *et al.*, 2008). An *in vivo* study showed that crystallized magnesium calcium phosphate was also found on the surface of the magnesium implant due to the reaction between the implant and blood or body fluid (Zhang *et al.*, 2009a).

19.3.2 Effect of *in vivo* stress on the corrosion behavior of Mg alloys

Any form of stress, when combined with corrosion, may result in a sudden, premature cracking, a phenomenon known as stress corrosion cracking (SCC). It is important to note here that magnesium alloys were reported to be susceptible to SCC in environments containing chloride (Kannan *et al.*, 2007; Winzer *et al.*, 2007). In 2011, Atrens summarized all previous work on SCC of Mg alloys in order to provide insight into possible SCC behaviors of Mg alloys for medical applications (Atrens *et al.*, 2011), indicating that it could occur at stress levels as low as 50% of the yield stress for a high Al-containing magnesium alloy under certain environmental conditions (see Table 19.2). Some commercial magnesium alloys such as AZ91, AZ31 and AM30 are susceptible to SCC even in mild environment like distilled water, suggesting that the SCC threshold could also be easily reached in an environment like the body. For magnesium-based implants in a loading-bearing application – such as coronary stents under the loading of blood vessel and blood flow, plates and screws for orthopedic fixation under the loading of body weight and movement – SSC should therefore be a major consideration. Wu *et al.* (2011) also reported that the stress concentration could reduce the recoil time of a magnesium coronary stent to nearly half

Table 19.2 SCC threshold stress for common Mg alloys

Alloy, environment	SCC (%YS)
HP Mg, 0.5% KHF_2	60
Mg2Mn, 0.5% KHF_2	50
MgMnCe, air, 0.001 N NaCl, 0.01 Na_2SO_4	85
ZK60A-T5, rural atmosphere	50
QE22, rural atmosphere	70–80
HK31, rural atmosphere	70–80
HM21, rural atmosphere	70–80
HP AM60, distilled water	40–50
HP AS41, distilled water	40–50
AZ31, rural atmosphere	40
AZ61, coastal atmosphere	50
AZ63-T6, rural atmosphere	60
HP AZ91, distilled water	40–50

Source: Winzer *et al.*, 2005.

the normal rate. The mechanical integrity of the AZ91 alloy in two different geometries, (i.e. circumferentially notched (CN), and circumferentially notched and the fatigue crack (CNFC)) tested by a constant extension rate tensile (CERT) method in air and in SBF also decreased substantially (by around 50%) in both the CN and CNFC samples exposed to SBF, and the secondary cracks suggested the SCC susceptibility of magnesium alloys in SBF (Kannan *et al.*, 2011).

Another major concern for Mg implants in service is their corrosion fatigue performance, for instance, for the orthopedic implants the ultimate implant failure is usually associated with corrosion fatigue, which is a synergetic effect of both electrochemical corrosion and cyclic mechanical loading. This issue is particularly pertinent in the medical field. Zheng *et al.* (2010c) proved that the corrosion rates of the die-cast AZ91D and extruded WE43 alloys increased under cyclic loading compared to the static immersion test.

19.3.3 The effect of other features on corrosion behavior of Mg alloys

Since the extent of corrosion in magnesium depends on the corrosive environment, the corrosion rates of magnesium alloys under standard, *in vitro* environmental conditions were compared to the corrosion rate in an animal model (Witte *et al.*, 2006). It was found that the *in vivo* corrosion was about four times less damaging than the *in vitro* corrosion of the tested alloys. The tendency of the corrosion rates obtained from *in vitro* corrosion tests, moreover, was sometimes in the opposite direction compared to those obtained from the *in vivo* study. Moreover, the degradation rate of the AZ31 alloy was found to be

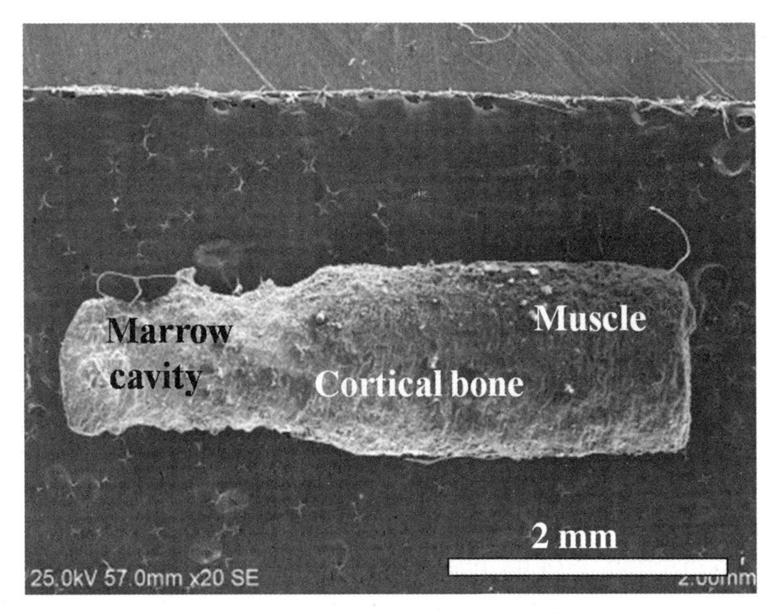

19.4 SEM photograph of AZ31 alloy after 20 weeks implantation in rabbit femoral shaft (Shu *et al.*, 2012).

different when in contact with different tissues (Yang *et al.*, 2009). The rate was at its highest in the marrow cavity, becoming exponentially slower in the muscle and the cortical bone, as shown in Fig. 19.4. These phenomena can be attributed to the difference in hydrogen diffusion coefficients between *in vitro* and *in vivo* environments, and even between the different tissues. Correlating the hydrogen diffusion coefficients from various biological media with fractional water contents (from about 68% to 100%) demonstrated that the diffusion coefficient of hydrogen could be increased exponentially with increasing the water fraction in the tissue (Vaupel, 1976).

Blood flow and temperature are different in various anatomical sites, and the flow of the corrosive media has a particularly significant effect on the corrosion rate of magnesium alloys (Mantovani *et al.*, 2008), while the effect of temperature in the human physiological range 35.8–37.2°C seems to be less important for corrosion of magnesium, which may influence the adsorption of proteins and thus the response of the biological environment (Witte *et al.*, 2008b; Zhang *et al.*, 2009a).

19.3.4 Importance of corrosion control for Mg alloys in medical applications

Many studies have shown that the degradation rates of magnesium alloys as biomedical implant materials were too high. This can lead to a number of different problems. Firstly, the implant must possess adequate strength for a sufficient period of time to allow healing to take place, and the corrosion rate must satisfy the need of healing process. Hard tissue repair typically requires an implantation of the fixture for at least 12 weeks (Staiger *et al.*, 2006). However the currently developed magnesium alloys tend to degrade faster than the bone heals (Tan *et al.*, 2010). Tan also found that the fluctuating pressing load of the AZ31B alloy when immersed in SBF at 37°C for 12 weeks, decreased to approximately 30% of the load as it stood before immersion, indicating that the mechanical integrity was lost before the tissue had sufficiently healed. In studies of *in vivo* corrosions of Mg-Ca alloy (Zheng *et al.*, 2008), Mg-Zn alloy (Zhang *et al.*, 2010d), LAE442 alloy (Witte *et al.*, 2005), etc., Mg alloys also exhibited rapid corrosion rate, inducing a detrimental relationship between the bone healing rate and the gas cavity between the implant and tissue. Such problems may lead the early loosening/disintegration of Mg alloy implants and increase the risk of further fractures.

As by-products of the degradation process, hydrogen gas evolution and an increase in alkalinity may induce biological problems when magnesium alloys degrade too fast (Witte *et al.*, 2005, 2007). Witte also observed subcutaneous gas bubbles which were believed to be hydrogen. Though no deleterious effects from these gas bubbles were recorded, they remain indicative of potential gas-gangrene, and should thus be eliminated if possible (Seal *et al.*, 2009). In addition, it was found that the hemolytic ratio of AZ31 alloy was as high as ~ 90%, induced by the high alkalinity of the magnesium alloy extract (Yang *et al.*, 2010). This figure is much higher than the value of 5%, which is recognized as the safety limit for implanted biomaterials according to ISO standard 10993–4, indicating that, without surface treatment, the AZ31 alloy could cause severe hemolysis if it came into contact with the blood. Gu *et al.* (2009b) also verified that some binary Mg alloys and pure Mg could not match the requirement of hemolysis. The high degradation rate also induced less adhesion of cells on the active surface of Mg alloy with H_2 evolution, high alkalinity and dynamic change of morphology (Yang *et al.*, 2009). There is, then, a high demand to reduce corrosion rates for the improvement of biocompatibility, as well as better synchronization between the degradation and new bone formation by various surface treatment techniques.

19.4 Corrosion control techniques for biodegradable Mg alloys

Since biodegradable magnesium alloys began to be developed towards the end of the twentieth century, surface treatments on magnesium alloys have been extensively studied, in order to help slow down the degradation rate and maintain mechanical integrity. The surface treatments for Ti alloy (used primarily to improve the surface bioactivity), as well as those for the engineering magnesium alloys (used to improve the corrosion resistance), have traditionally acted as a reference point for the surface treatment of biodegradable magnesium alloys. There have, however, been many new methods which have been developed specifically for biodegradable magnesium alloys, some of which are listed below.

19.4.1 Anodization

Pure Mg (Song, 2007; Yun *et al.*, 2011), AZ91D alloy (Kim *et al.*, 2008; Yun *et al.*, 2011) and AZ31B alloy (Kim *et al.*, 2008) were anodized and their corrosion resistances were studied. The results showed that anodization increased the corrosion resistance of both pure Mg and AZ91D alloy. No detectable hydrogen evolution from the anodized pure Mg was measured in the SBF solution over one month (see Fig. 19.5). Further observation showed the anodized layers on both pure Mg and AZ91D alloy consisted of Mg, O and Si in a mixture of MgO and Mg_2SiO_4 (Yun *et al.*, 2011). If the coating were to eventually break down, silicon oxides/hydroxides in the coating would be slowly dissolved into the body fluid. A trace amount of Si has been reported to be essential in mammals. Therefore, in the event

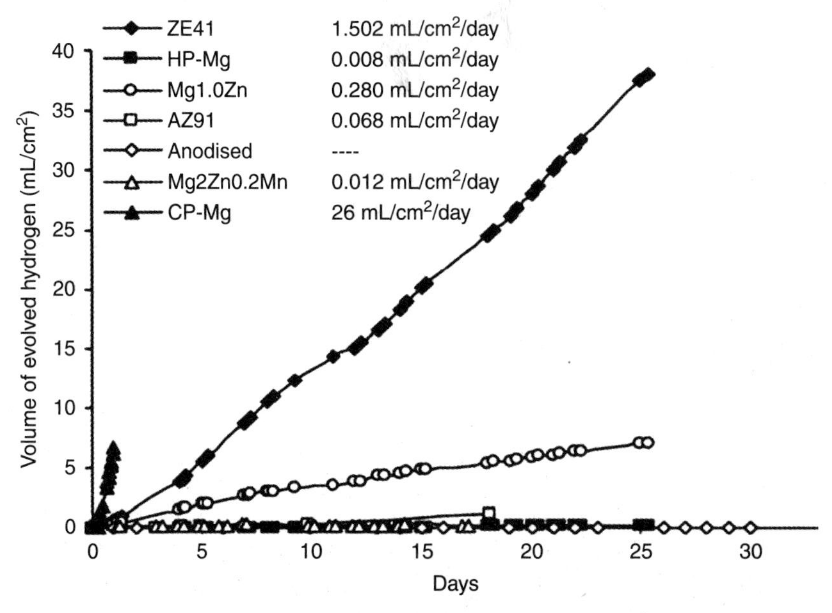

19.5 Hydrogen evolution and their average rates of specimens (Song, 2007).

of coating breakdown, the anodized coating is still non-toxic to the human body (Song, 2007).

19.4.2 Micro-arc oxidation

Ideally, a corrosion-resistant layer formed on a magnesium implant should also be wear resistant, to protect it from damage by scratching during the implantation. Micro-arc oxidation (MAO), a newly developed surface modification technique, is based on the *in situ* creation of ceramic coatings on magnesium alloys, and is aimed at increasing both corrosion- and wear resistances for engineering applications. With the development of biodegradable magnesium alloys, the MAO method was also used to make corrosion-resistant coatings on magnesium alloys with good wear resistance and adhesion strength for medical applications.

A MAO coating was fabricated on the surface of a Mg-Ca alloy and it was found that this MAO coating could better promote the adhesion, proliferation and differentiation of MG63 cells in comparison to the uncoated alloy, due to a significant reduction in Mg ion release and a pH value increase in the culture medium (Gu *et al.*, 2011). This discovery indicates that MAO coating is a prospective surface modification method for biodegradable magnesium alloys.

In order to promote bioactivity/osteointegration of the MAO coating on Mg alloys, a small amount of TiO_2 was introduced into the deposited coating, which mainly comprised MgO and Mg_2SiO_4 by addition of the titania sol into the alkaline silicate electrolytic bath (Wang *et al.*, 2009). A typical morphology of MAO coating is shown in Fig. 19.6. An increased corrosion resistance could be obtained when a proper amount of titania sol was contained in the electrolyte.

In addition, Lu *et al.* (2011) sealed the micro-cracks and micro-holes on the surface of an MAO coating using a PLLA coating containing a dosage of the drug PTX, which formed a MAO/PLLA composite coating on the AZ81 alloy substrate. The corrosion resistance of the alloy was increased and the degradation rate became controllable. The drug release rate of PTX from the coating exhibited a nearly linear sustained-release profile with no significant burst release that would occur from the uncontrollable corrosion of AZ81 alloy. It was also found that the alloy with this composite coating had better hemocompatibility than that of 316L stainless steel.

19.4.3 Electrodeposition

Electrochemical deposition, or electrodeposition, is a low cost and simple process that can be carried out at room temperature to form a uniform

(a)
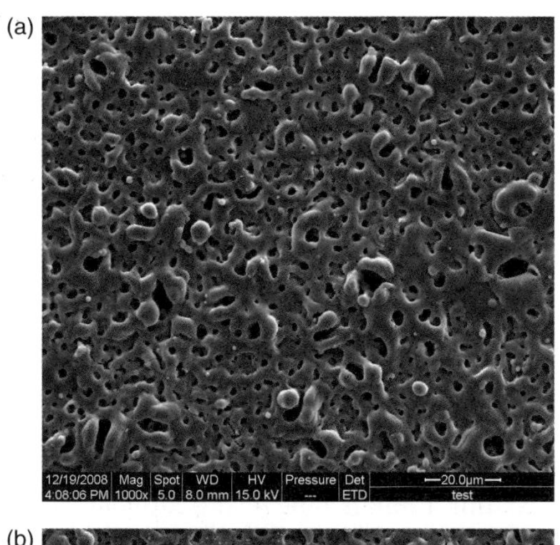

(b)
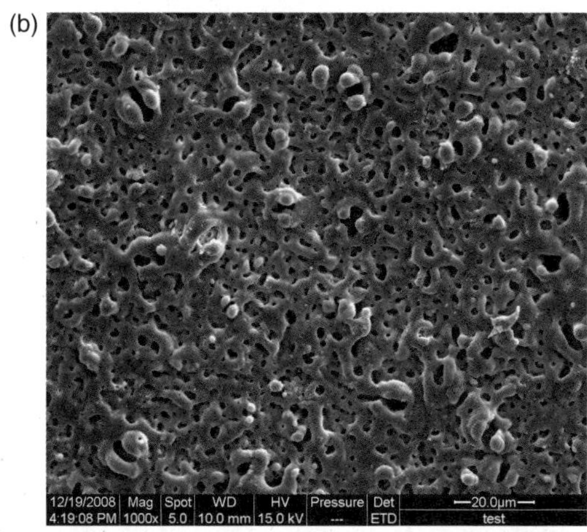

(c)
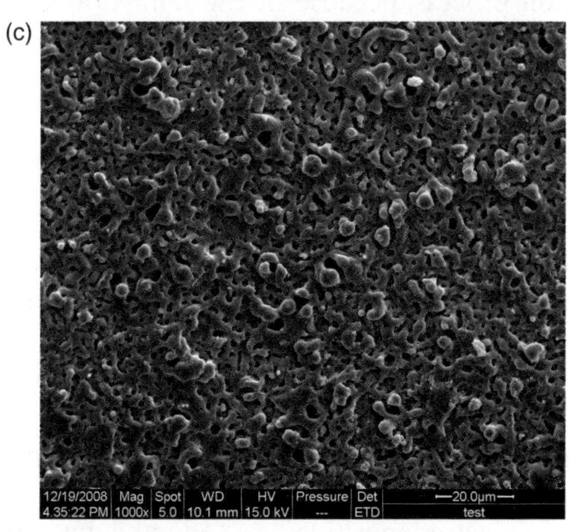

19.6 Effect of titania sol with different concentration on surface morphologies of micro-arc oxidation coating formed on AZ91D alloy. (a) MAO0; (b) MAO5; (c) MAO10 (Wang *et al.*, 2009).

coating. It is particularly useful in fabricating a hydroxyapatite (HA) coating that has been used as a biomaterial on surfaces of Mg alloys, exhibiting excellent biocompatibility and bioactivity due to its similar composition and structure to bone and tooth minerals. The thickness and composition of the HA coating can also be controlled by adjusting the electrodeposition conditions.

Guan *et al.* (2009) and Song *et al.* (2008) used electrodeposition to make HA coatings on the alloys AZ31 alloy and AZ91D respectively (as shown in Fig. 19.7) followed by a post-treatment in an alkaline solution to improve the corrosion resistance and bioactivity for implant application. The experimental results indicated that the deposited coating, consisting of dicalcium phosphate dehydrate (DCPD, $CaHPO_4{\cdot}2H_2O$) and beta-tricalcium phosphate (beta-TCP, $Ca_3(PO_4)_2$), could be transformed into hydroxyapatite (HA, $Ca_{10}(PO_4)_6(OH)_2$) coating after alkaline heat treatment. The HA coating could obviously slow down the biodegradation rate of the AZ91D alloy in the simulated body fluid.

It was noted, however, that the HA coating transformed from DCPD through alkaline heat treatment was fragile and less stable, and therefore its long-term corrosion resistance was not satisfactory (Zhang *et al.*, 2010c). So a more stable and better corrosion-resistant coating, fluoridated hydroxyapatite (FHA), was prepared (Zhang *et al.*, 2010c), again through electrodeposition including addition of $NaNO_3$ and NaF into the electrolyte. Both the HA and FHA coatings were able promote the nucleation of osteoconductive minerals (bone-like apatite or beta-TCP) for 1 month, while the FHA coating was able to improve the interfacial bioactivity of Mg-6 wt.% Zn alloy for the cellular proliferation and differentiation (Zhang *et al.*, 2010b). Moreover, Guan *et al.* (2011c) found that in the traditional electrodeposition process, because of the unfavorable effects of both the polarization of concentration difference and the H_2 evolution, the FHA coating became loose and porous, and could not ensure the long-term stability of the Mg alloy implants. So in order to further improve the corrosion resistance and bioactivity of the FHA coating, pulse electrodeposition and H_2O_2 were introduced into the electrodeposition process and a dense, uniform, nano-crystallined FHA coating could be obtained, as shown in Fig. 19.8. The potentiodynamic polarization measurement (Fig. 19.9) indicated that this new coating could effectively protect the Mg alloy substrate from corrosion. The immersion test in SBF, moreover, showed that the pulse electrodeposition coating could more effectively induce the precipitation of Mg^{2+}, Ca^{2+} and PO_4^{3-} in comparison with the traditional electrodeposition coating, because the nano-sized phase had a higher specific surface area.

In addition, the pulse electrodeposition was adopted by Guan *et al.* (2010) to fabricate a soluble Ca-deficient hydroxyapatite (Ca-def HA)

(a)

(b)

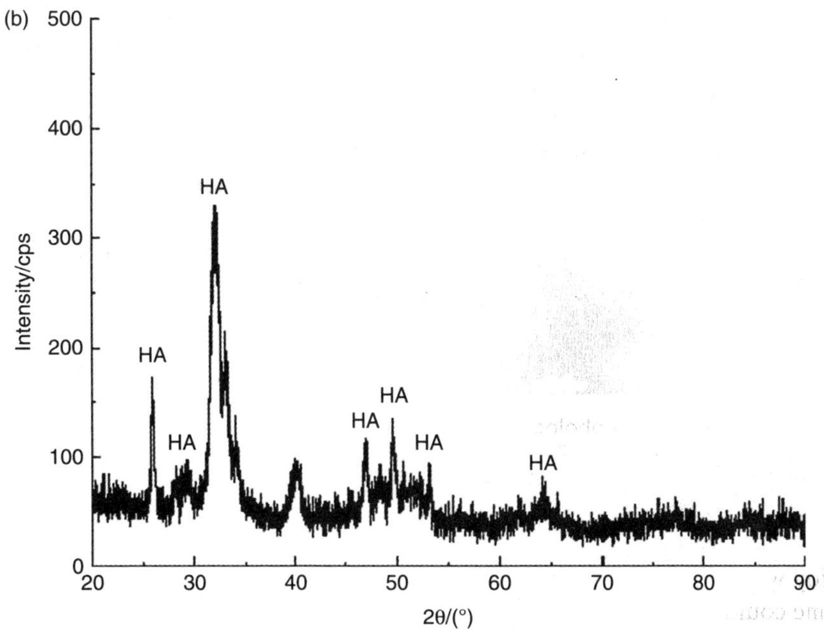

19.7 (a, b) Surface morphology and structure of HA coating prepared by electrodeposition (Song *et al.*, 2008).

coating on an Mg-Zn-Ca alloy, as shown in Fig. 19.10. The Ca-def HA coating showed better adhesion to the Mg-Zn-Ca alloy, whose lap shear strength was increased to 41.8 ± 2.7 MPa. The slow strain rate tensile (SSRT) testing results showed that the UTS and time of fracture for the coated Mg-Zn-Ca

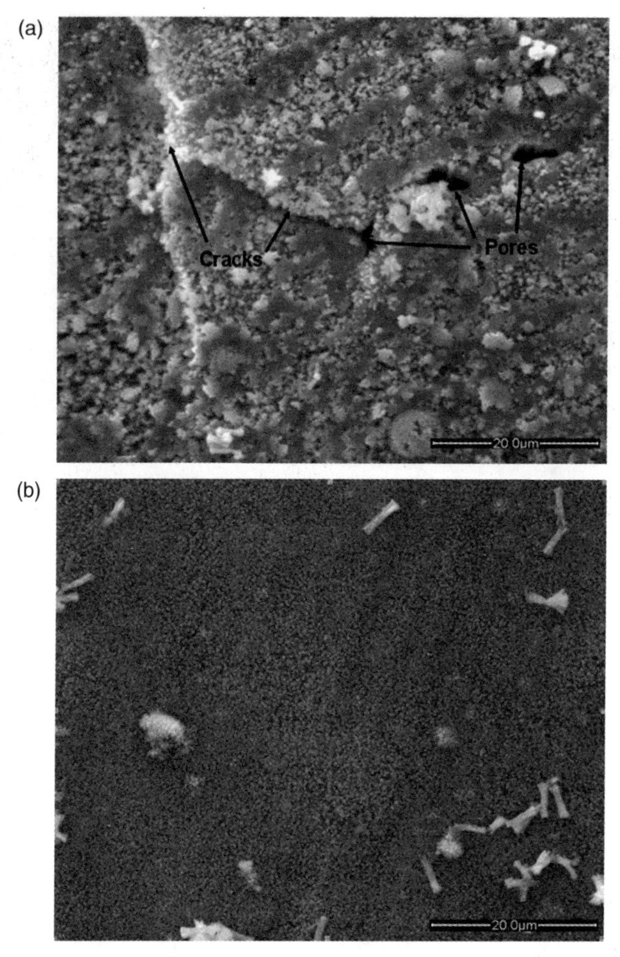

19.8 SEM morphologies of FHA coating prepared by traditional cathodic process (TED)(a) and a pulse reverse current (PRC) process (b) of electrodeposition (Guan *et al.*, 2011c).

alloy were higher than those of the uncoated alloy, meaning the fractured bone could be supported for longer.

In the same study, a rod-like nano-hydroxyapatite (RNHA) was fabricated by electrodeposition on an MAO coating (Guan *et al.*, 2011a). The bonding strength between the HA layer and the MAO coating was increased to 12.3 MPa, almost double that of the direct electrodeposition coating (6.3 MPa). As shown in Fig. 19.11, the distribution of the HA rods was, again, dense and uniform. The diameters of HA rods varied from 95 to 116 nm and the root mean square (RMS) roughness of the composite coating was about 42 nm, rendering it favorable for cellular survival. The composite coating also

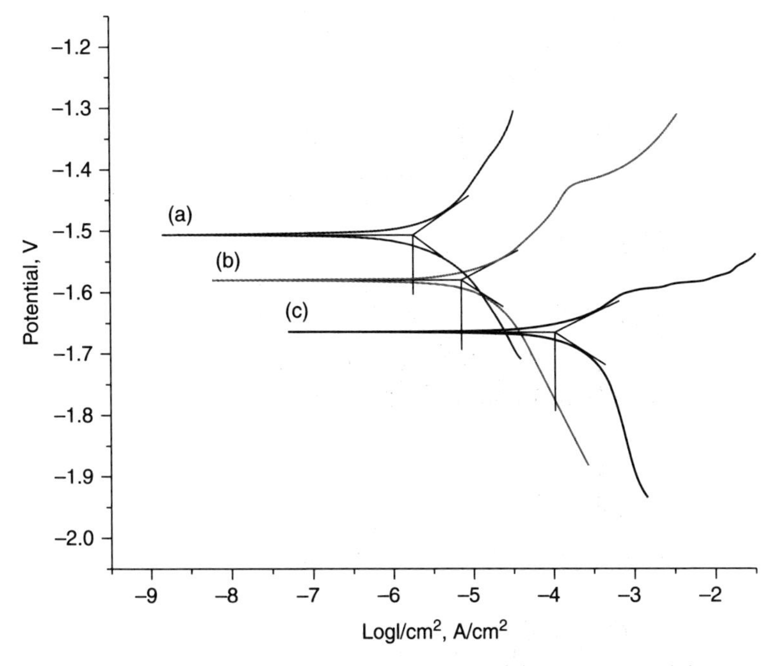

19.9 Polarization curves in SBF of PRC coating (a), TED coating (b) and Mg alloy substrate (c) (Guan *et al.*, 2011c).

enhanced corrosion resistance of the Mg alloy and induced a more rapid precipitation of calcium orthophosphates than was achieved with the conventional HA coating.

19.4.4 Phosphating treatment and chemical deposition

A phosphating treatment was applied on a Mg-Mn-Zn alloy in order to improve its corrosion resistance (Xu *et al.*, 2009b). The reaction layer on the surface of the alloy was mainly composed of brushites ($CaHPO_4{\cdot}2H_2O$) (see Fig. 19.12), which were able to improve the corrosion resistance of the alloy. The brushite layer was then transformed into hydroxyapatite (HA) during an immersion in SBF, suggesting that the brushite layer could provide a good biocompatibility. L-929 cells, moreover, showed significantly good adherence, high growth rate and proliferation characteristics on the coated Mg alloy ($p < 0.05$) in the *in vitro* cells test, as shown in Fig. 19.13, demonstrating that the cytocompatibility of the alloy was greatly improved by the Ca-P coating. The *in vivo* implantation results demonstrated that the Ca-P coating provided magnesium with a better surface bioactivity ($p < 0.05$) and promoted the early bone growth at the implant/bone interface. It was suggested that the brushite coating prepared by phosphating treatment might

19.10 SEM micrographs of the Ca-deficient hydroxyapatite coating obtained by the pulse electrodeposition process deposition process. Two different morphologies of the coating are observed (a): a tabular region composed of an irregular flake-like structure approximately 100 in thickness (b) and a fluffy region composed of a fine acicular structure less than 100 nm in diameter (c), which partly covers the plate-like coating (Guan *et al.*, 2010).

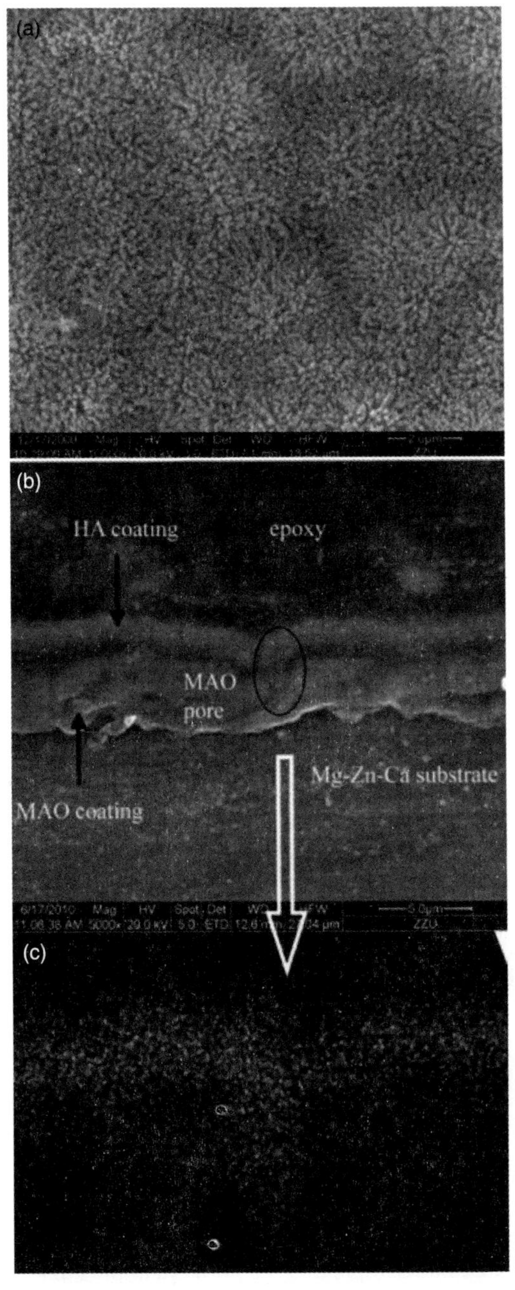

19.11 (a) Surface morphology, (b) cross-section morphology of composite coatings on Mg-Zn-Ca alloy and (c) EDS surficial scanning map of calcium at the position labeled by ellipse in figure (b) (Guan *et al.*, 2011a).

19.12 SEM surface morphology of the Mg-Mn-Zn alloy sample after phosphating treatment for 10 min (Xu *et al.*, 2009b).

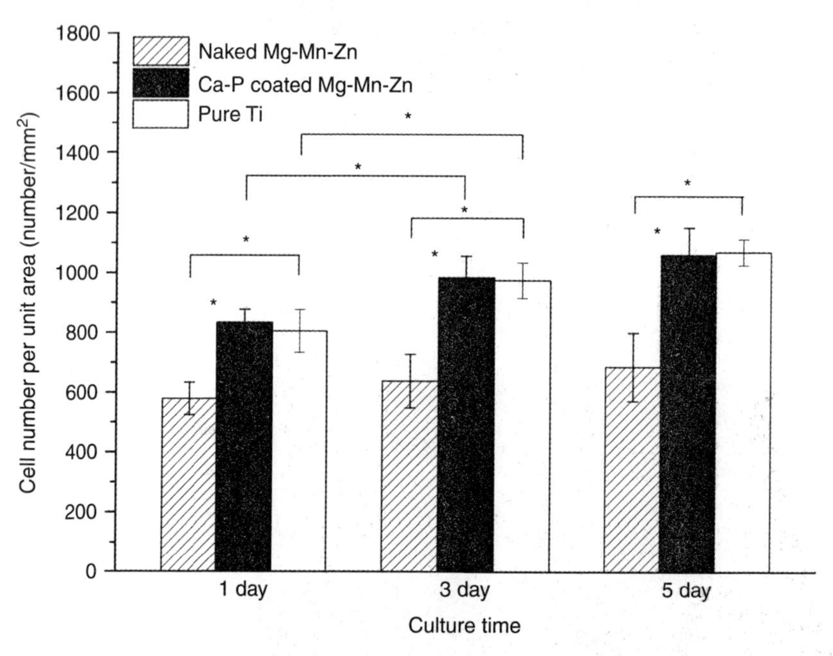

19.13 Growth of L-929 cells versus culturing time on the naked Mg sample, the Ca-P coated sample and the pure Ti sample (*$p < 0.05$) (Xu *et al.*, 2009a).

be an effective method to improve the surface bioactivity of Mg alloys (Xu *et al.*, 2009a).

Biodegradable and bioactive beta-tricalcium phosphate (beta-TCP) coatings could be fabricated on pure Mg by chemical deposition in order to improve its biocompatibility. A cytocompatibility test on the beta-TCP coated Mg showed that MG63 cells could grow well on the sample surface, while its cell viability exceeded levels of 80% during a 10 day co-cultivation with MG63 cells, indicating a good cytocompatibility. It was also found that a bone-like apatite was continually formed on the surface of the sample, after degradation of both the Mg substrate and the beta-TCP coating in the Hank's solution (a SBF) (Yang *et al.*, 2009).

Gray-Munro and Strong (2009) studied the mechanism of Ca-P deposition on the AZ31B alloy, and concluded that the deposition of the coating was related to the anodic dissolution of the substrate.

19.4.5 Other corrosion protection techniques

Besides the wide range of coating based methods mentioned above, some other techniques have also been tried, with good results, for the application of biodegradable magnesium alloys in order to decrease the degradation rate and improve the biocompatibility.

A biomimetic technique was used to prepare a HA coating on magnesium alloy (see Fig. 19.14), and the corrosion rate of the magnesium implants could be closely tailored by adjusting the thickness of apatite coating and thereby monitoring the release of magnesium ions into the body (Wei *et al.*, 2009).

19.14 SEM micrographs of fluoride conversion coating on AZ31B (Tan *et al.*, 2010).

Gu *et al.* (2009a) reported that an alkaline heat treatment could effectively improve the corrosion resistance of Mg substrate, showing no cytotoxicity. The alkaline heat treated coating was mainly composed of MgO and its effectiveness was related to the thickness and defects inside of it.

A fluoride coating was also prepared on the surface of magnesium alloy through chemical conversion (Tan *et al.*, 2010). The surface characterization analysis showed that a dense coating with some irregular pores was formed, as shown in Fig. 19.14. Electrochemical and immersion tests proved that the fluoride conversion coating significantly improved the corrosion resistance of the AZ31B alloy. A three-point bending test, moreover, revealed that the degradation behavior of the fluoride treated AZ31B alloy was able to meet the requirement for a biodegradable material (see Fig. 19.15).

Aerosol deposition (AD) is an alternative coating method for depositing dense ceramic coatings on various substrates such as metals, ceramics and plastics. Dense, adherent HA-chitosan composite coatings were deposited on the AZ31 alloy substrate using AD, in order to improve both the corrosion resistance and the biocompatibility (see Fig. 19.16) (Hahn *et al.*, 2011). All the coatings exhibited high adhesion strengths – ranging from 24.6 to 27.7 MPa – and showed higher corrosion resistances than the bare AZ31 substrate, implying that the corrosion resistance of the AZ31 alloy was enhanced by the aerosol deposited, HA-chitosan composite coating. In

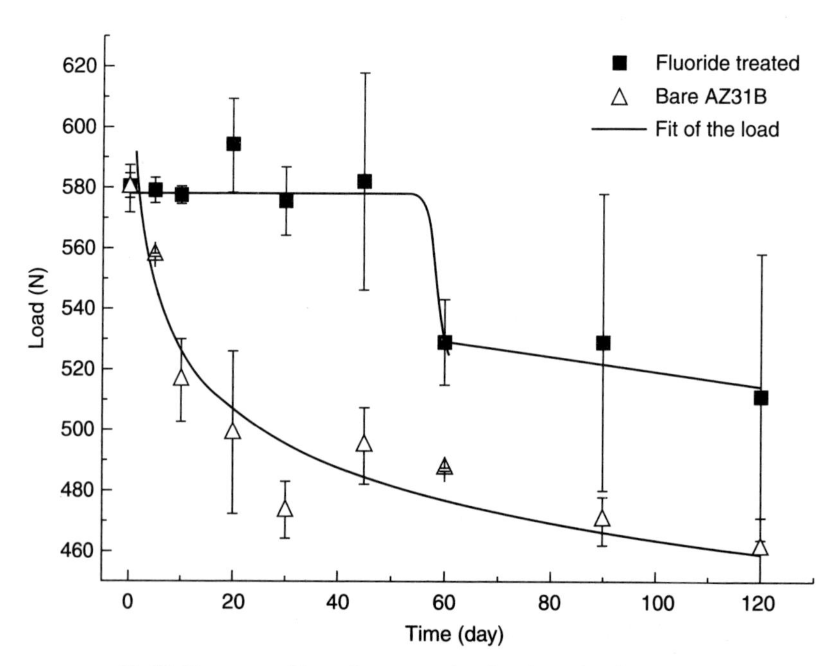

19.15 Changes of bending pressing load on the fluoride treated and bare AZ31B samples with immersion time (Tan *et al.*, 2010).

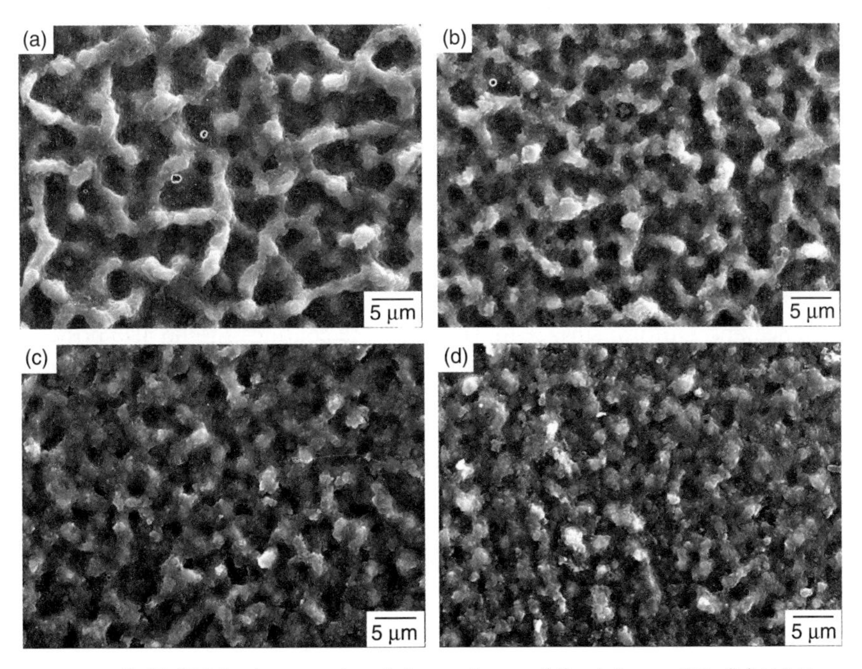

19.16 SEM micrographs of the surfaces of the (a) pure HA, (b) HA-5 wt% chitosan, (c) HA-10 wt% chitosan, and (d) HA-20 wt% chitosan composite coatings deposited on the AZ31 Mg alloy substrates (Hahn *et al.*, 2011).

addition, the biocompatibility of the alloy was remarkably improved by the HA and the incorporated chitosan contained in the coating.

All of the following coatings and methods were also investigated for their potential use in biodegradable materials: the diamond-like-carbon (DLC) coating, as prepared by radio frequency plasma assisted chemical vapor deposition (RFPACVD) (Cui *et al.*, 2010); the TiO_2 coating prepared by liquid phase deposition (LPD) (Junhua *et al.*, 2009); the ZrN/Zr bilayered coating deposited by filtered cathodic arc deposition system (Tang *et al.*, 2009b), and the α-Si:H coating deposited on AZ91 alloy using magnetron sputtering deposition (Tang *et al.*, 2009a). Degradation levels in these coatings still require further study. Low plastic burnishing (LPB) has also emerged as an enabling manufacturing technique to produce superior surface integrity of orthopedic implants to increase the corrosion resistance of Mg-Ca implants (Salahshoor and Guo, 2011).

19.5 Future trends

Biodegradable magnesium alloys and the implants made of them have shown, or been predicted, to induce an improved regeneration ability of the bone tissue and also have other advantages for different medical applications. However, before the clinical application of biodegradable magnesium alloys, some techniques must be developed to satisfy the requirements of medical application. The design and fabrication of the alloys, most notably, must fulfill the requirements of the medical industry with regard to biocompatibility and mechanical integrity. As one of the most important techniques, the corrosion control technique for magnesium alloys in medical applications should be developed for the specific implants.

Aside from corrosion resistance, biocompatibility, good cell adhesion, wear resistance and uniform surface morphology, the coatings should also possess some other specific properties for the applied implants. Fundamental research is largely still required to answer some basic scientific questions about the coatings themselves, such as the degradation mechanism of the magnesium alloys, the biological properties and bio-absorbable routes of the trace elements or the bio-safety of the magnesium alloy implants after breakdown.

19.6 References

Aghion, E. and Levy, G. (2010). The effect of Ca on the in vitro corrosion performance of biodegradable Mg-Nd-Y-Zr alloy. *Journal of Materials Science*, **45**, 3096–3101.

Atrens, A., Liu, M. and Abidin, N. I. Z. (2011). Corrosion mechanism applicable to biodegradable magnesium implants. *Materials Science and Engineering B-Advanced Functional Solid-State Materials*, **176**, 1609–1636.

Cui, F. Z., Zhang, Y., Yang, J. X., Lee, I. S. and Lee, G. H. (2010). Characterization and degradation comparison of DLC film on different magnesium alloys. *Surface and Coatings Technology*, **205**, S15–S20.

Erdmann, N., Angrisani, N., Reifenrath, J., Lucas, A., Thorey, F., Bormann, D. and Meyer-Lindenberg, A. (2011). Biomechanical testing and degradation analysis of MgCa0.8 alloy screws: A comparative in vivo study in rabbits. *Acta Biomaterialia*, **7**, 1421–1428.

Erdmann, N., Bondarenko, A., Hewicker-Trautwein, M., Angrisani, N., Reifenrath, J., Lucas, A. and Meyer-Lindenberg, A. (2010). Evaluation of the soft tissue biocompatibility of MgCa0.8 and surgical steel 316L in vivo: a comparative study in rabbits. *Biomedical Engineering Online*, **9**.

Gray-Munro, J. E. and Strong, M. (2009). The mechanism of deposition of calcium phosphate coatings from solution onto magnesium alloy AZ31. *Journal of Biomedical Materials Research Part A*, **90A,** 339–350.

Gu, X. N., Li, N., Zhou, W. R., Zheng, Y. F., Zhao, X., Cai, Q. Z. and Ruan, L. Q. (2011). Corrosion resistance and surface biocompatibility of a microarc oxidation coating on a Mg-Ca alloy. *Acta Biomaterialia*, **7**, 1880–1889.

Gu, X. N., Zheng, W., Cheng, Y. and Zheng, Y. F. (2009a). A study on alkaline heat treated Mg–Ca alloy for the control of the biocorrosion rate. *Acta Biomaterialia*, **5**, 2790–2799.

Gu, X. N., Zheng, Y. F., Cheng, Y., Zhong, S. P. and Xi, T. F. (2009b). In vitro corrosion and biocompatibility of binary magnesium alloys. *Biomaterials*, **30**, 484–498.

Guan, S. K., Gao, J. H., Chen, J., Wang, L. G., Zhu, S. J., Hu, J. H. and Ren, Z. W. (2011a). Fabrication and characterization of rod-like nano-hydroxyapatite on MAO coating supported on Mg-Zn-Ca alloy. *Applied Surface Science*, **257**, 2231–2237.

Guan, S. K., Gao, J. H., Ren, Z. W., Sun, Y. F., Zhu, S. J. and Wang, B. (2011b). Homogeneous corrosion of high pressure torsion treated Mg-Zn-Ca alloy in simulated body fluid. *Materials Letters*, **65**, 691–693.

Guan, S. K., Meng, E. C., Wang, H. X., Wang, L. G., Zhu, S. J., Hu, J. H., Ren, C. X., Gao, J. H. and Feng, Y. S. (2011c). Effect of electrodeposition modes on surface characteristics and corrosion properties of fluorine-doped hydroxyapatite coatings on Mg-Zn-Ca alloy. *Applied Surface Science*, **257**, 4811–4816.

Guan, S. K., Wang, H. X., Wang, X., Ren, C. X. and Wang, L. G. (2010). In vitro degradation and mechanical integrity of Mg-Zn-Ca alloy coated with Ca-deficient hydroxyapatite by the pulse electrodeposition process. *Acta Biomaterialia*, **6**, 1743–1748.

Guan, S. K., Wen, C. L., Peng, L., Ren, C. X., Wang, X. and Hu, Z. H. (2009). Characterization and degradation behavior of AZ31 alloy surface modified by bone-like hydroxyapatite for implant applications. *Applied Surface Science*, **255**, 6433–6438.

Guo, Y. B. and Salahshoor, M. (2011). Machining Characteristics of High Speed Dry Milling of Biodegradable Magnesium-Calcium Alloy. *Proceedings of the Asme International Manufacturing Science and Engineering Conference 2010*, **1**, 279–286.

Hahn, B. D., Park, D. S., Choi, J. J., Ryu, J. H., Yoon, W. H., Choi, J. H., Kim, H. E. and Kim, S. G. (2011). Aerosol deposition of hydroxyapatite-chitosan composite coatings on biodegradable magnesium alloy. *Surface and Coatings Technology*, **205**, 3112–3118.

Hänzi, A. C., Gerber, I., Schinhammer, M., Löffler, J. F. and Uggowitzer, P. J. (2010). On the in vitro and in vivo degradation performance and biological response of new biodegradable Mg–Y–Zn alloys. *Acta Biomaterialia*, **6**, 1824–1833.

Harandi, S. E., Idris, M. H. and Jafari, H. (2011). Effect of forging process on microstructure, mechanical and corrosion properties of biodegradable Mg-1Ca alloy. *Materials and Design*, **32**, 2596–2603.

Heublein, B., Rohde, R., Kaese, V., Niemeyer, M., Hartung, W. and Haverich, A. (2003). Biocorrosion of magnesium alloys: a new principle in cardiovascular implant technology? *Heart*, **89**, 651–656.

Hort, N., Huang, Y., Fechner, D., Störmer, M., Blawert, C., Witte, F., Vogt, C., Drücker, H., Willumeit, R., Kainer, K. U. and Feyerabend, F. (2010). Magnesium alloys as implant materials – Principles of property design for Mg–RE alloys. *Acta Biomaterialia*, **6**, 1714–1725.

Junhua, H., Guan, S., Caili, Z., Chenxing, R., Cuilian, W., Zhaoqin, Z. and Li, P. (2009). Corrosion protection of AZ31 magnesium alloy by a TiO_2 coating prepared by LPD method. *Surface and Coatings Technology*, **203**, 2017–2020.

Kannan, M. B., Dietzel, W., Raman, R. K. S. and Lyon, P. (2007). Hydrogen-induced-cracking in magnesium alloy under cathodic polarization. *Scripta Materialia*, **57**, 579–581.

Kannan, M. B., Raman, R. K. S., Witte, F., Blawert, C. and Dietzel, W. (2011). Influence of circumferential notch and fatigue crack on the mechanical integrity of biodegradable magnesium-based alloy in simulated body fluid. *Journal of Biomedical Materials Research Part B-Applied Biomaterials*, **96B,** 303–309.

Kim, Y. K., Lee, M. H., Prasad, M. N., Park, I. S., Lee, M. H., Seol, K. W. and Bae, T. S. (2008). Surface characteristics of magnesium alloys treated by anodic oxidation using pulse power. *Multi-Functional Materials and Structures, Pts 1 and 2*, 47–50, 1290–1293.

Kuwahara, H., Al-Abdullat, Y., Mazaki, N., Tsutsumi, S. and Aizawa, T. (2001). Precipitation of magnesium apatite on pure magnesium surface during immersing in Hank's solution. *Materials Transactions*, **42**, 1317–1321.

Loffler, J. F., Zberg, B. and Uggowitzer, P. J. (2009). MgZnCa glasses without clinically observable hydrogen evolution for biodegradable implants. *Nature Materials*, **8**, 887–891.

Lu, P., Fan, H., Liu, Y., Cao, L., Wu, X. and Xu, X. (2011). Controllable biodegradability, drug release behavior and hemocompatibility of PTX-eluting magnesium stents. *Colloids and Surfaces B: Biointerfaces*, **83**, 23–28.

Mantovani, D., Levesque, J., Hermawan, H. and Dube, D. (2008). Design of a pseudo-physiological test bench specific to the development of biodegradable metallic biomaterials. *Acta Biomaterialia*, **4**, 284–295.

McMahon, C. J., Oslizlok, P. and Walsh, K. P. (2007). Early restenosis following biodegradable stent implantation in an aortopulmonary collateral of a patient with pulmonary atresia and hypoplastic pulmonary arteries. *Catheterization and Cardiovascular Interventions*, **69**, 735–738.

Peng, Q., Huang, Y., Zhou, L., Hort, N. and Kainer, K. U. (2010a). Preparation and properties of high purity Mg–Y biomaterials. *Biomaterials*, **31**, 398–403.

Ren, Y. B., Huang, J. J., Yang, K., Zhang, B. C., Yao, Z. M. and Wang, H. (2005). Study of bio-corrosion of pure magnesium. *Acta Metallurgica Sinica*, **41**, 1228–1232.

Rosalbino, F., De Negri, S., Saccone, A., Angelini, E. and Delfino, S. (2010). Bio-corrosion characterization of Mg-Zn-X (X = Ca, Mn, Si) alloys for biomedical applications. *Journal of Materials Science-Materials in Medicine*, **21**, 1091–1098.

Salahshoor, M. and Guo, Y. B. (2011). Contact mechanics in low plasticity burnishing of biomedical magnesium-calcium alloy. *Proceedings of the Stle/Asme International Joint Tribology Conference*, **2010**, 349–351.

Seal, C. K., Vince, K. and Hodgson, M. A. (2009). Biodegradable surgical implants based on magnesium alloys – a review of current research. *Processing, Microstructure and Performance of Materials*, 012011.

Song, G. L. (2007). Control of biodegradation of biocompatable magnesium alloys. *Corrosion Science*, **49**, 1696–1701.

Song, Y. W., Shan, D. Y. and Han, E. H. (2008). Electrodeposition of hydroxyapatite coating on AZ91D magnesium alloy for biomaterial application. *Materials Letters*, **62**, 3276–3279.

Staiger, M. P., Pietak, A. M., Huadmai, J. and Dias, G. (2006). Magnesium and its alloys as orthopedic biomaterials: A review. *Biomaterials*, **27**, 1728–1734.

Tan, L. L., Yan, T. T., Xiong, D. S., Liu, X. J., Zhang, B. C. and Yang, K. (2010). Fluoride treatment and in vitro corrosion behavior of an AZ31B magnesium alloy. *Materials Science and Engineering C-Materials for Biological Applications*, **30**, 740–748.

Tang, G. Y., Xin, Y. C., Jiang, J., Huo, K. F., Tian, X. B. and Chu, P. K. (2009a). Corrosion resistance and cytocompatibility of biodegradable surgical magnesium alloy coated with hydrogenated amorphous silicon. *Journal of Biomedical Materials Research Part A*, **89A,** 717–726.

Tang, G. Y., Xin, Y. C., Liu, C. L., Huo, K. F., Tian, X. B. and Chu, P. K. (2009b). Corrosion behavior of ZrN/Zr coated biomedical AZ91 magnesium alloy. *Surface and Coatings Technology*, **203**, 2554–2557.

Uggowitzer, P. J., Hanzi, A. C. and Sologubenko, A. S. (2009). Design strategy for new biodegradable Mg-Y-Zn alloys for medical applications. *International Journal of Materials Research*, **100**, 1127–1136.

Vaupel, P. (1976). Effect of percentual water-content in tissues and liquids on diffusion-coefficients of O_2, CO_2, N_2, and H_2. *Pflugers Archiv-European Journal of Physiology*, **361**, 201–204.

Wang, Y. M., Wang, F. H., Xu, M. J., Zhao, B., Guo, L. X. and Ouyang, J. H. (2009). Microstructure and corrosion behavior of coated AZ91 alloy by microarc oxidation for biomedical application. *Applied Surface Science*, **255**, 9124–9131.

Wei, M., Zhang, Y. J. and Zhang, G. Z. (2009). Controlling the biodegradation rate of magnesium using biomimetic apatite coating. *Journal of Biomedical Materials Research Part B-Applied Biomaterials*, **89B,** 408–414.

Winzer, N., Atrens, A., Dietzel, W., Song, G. and Kainer, K. U. (2007). Magnesium stress corrosion cracking. *Transactions of Nonferrous Metals Society of China*, **17**, S150–S155.

Winzer, N., Atrens, A., Song, G. L., Ghali, E., Dietzel, W., Kainer, K. U., Hort, N. and Blawert, C. (2005). A critical review of the stress corrosion cracking (SCC) of magnesium alloys. *Advanced Engineering Materials*, **7**, 659–693.

Witte, F., Abeln, I., Switzer, E., Kaese, V., Meyer-Lindenberg, A. and Windhagen, H. (2008a). Evaluation of the skin sensitizing potential of biodegradable magnesium alloys. *Journal of Biomedical Materials Research Part A*, **86A,** 1041–1047.

Witte, F., Fischer, J., Nellesen, J., Crostack, H. -A., Kaese, V., Pisch, A., Beckmann, F. and Windhagen, H. (2006). In vitro and in vivo corrosion measurements of magnesium alloys. *Biomaterials*, **27**, 1013–1018.

Witte, F., Hort, N., Vogt, C., Cohen, S., Kainer, K. U., Willumeit, R. and Feyerabend, F. (2008b). Degradable biomaterials based on magnesium corrosion. *Current Opinion in Solid State and Materials Science*, **12**, 63–72.

Witte, F., Kaese, V., Haferkamp, H., Switzer, E., Meyer-Lindenberg, A., Wirth, C. J. and Windhagen, H. (2005). In vivo corrosion of four magnesium alloys and the associated bone response. *Biomaterials*, **26**, 3557–3563.

Witte, F., Ulrich, H., Palm, C. and Willbold, E. (2007a). Biodegradable magnesium scaffolds: Part II: Peri-implant bone remodeling. *Journal of Biomedical Materials Research Part A*, **81A,** 757–765.

Witte, F., Ulrich, H., Rudert, M. and Willbold, E. (2007b). Biodegradable magnesium scaffolds: Part I: Appropriate inflammatory response. *Journal of Biomedical Materials Research Part A*, **81A,** 748–756.

Witte, F., (2010). The history of biodegradable magnesium implants: A review. *Acta Biomateriallia*, **6**, 1680–1692.

Wu, W., Gastaldi, D., Yang, K., Tan, L., Petrini, L. and Migliavacca, F. (2011). Finite element analyses for design evaluation of biodegradable magnesium alloy stents in arterial vessels. *Materials Science and Engineering B-Advanced Functional Solid-State Materials*, **176**, 1733–1740.

Xin, Y. C., Huo, K. F., Tao, H., Tang, G. Y. and Chu, P. K. (2008). Influence of aggressive ions on the degradation behavior of biomedical magnesium alloy in physiological environment. *Acta Biomaterialia*, **4**, 2008–2015.

Xu, K., Li, H. W., Zhong, H. S., Yang, K., Liu, J., Zhang, B. C., Zheng, F., Xia, Y. H., Tan, L. L. and Hong, D. O. (2011). Enhanced Efficacy of Sirolimus-Eluting Bioabsorbable Magnesium Alloy Stents in the Prevention of Restenosis. *Journal of Endovascular Therapy*, **18**, 407–415.

Xu, L., Yu, G., Zhang, E., Pan, F. and Yang, K. (2007). In vivo corrosion behavior of Mg-Mn-Zn alloy for bone implant application. *Journal of Biomedical Materials Research Part A*, **83A,** 703–711.

Xu, L., Zhang, E., Yin, D., Zeng, S. and Yang, K. (2008). In vitro corrosion behaviour of Mg alloys in a phosphate buffered solution for bone implant application. *Journal of Materials Science: Materials in Medicine*, **19**, 1017–1025.

Xu, L. P., Pan, F., Yu, G. N., Yang, L., Zhang, E. L. and Yang, K. (2009a). In vitro and in vivo evaluation of the surface bioactivity of a calcium phosphate coated magnesium alloy. *Biomaterials*, **30**, 1512–1523.

Xu, L. P., Zhang, E. L. and Yang, K. (2009b). Phosphating treatment and corrosion properties of Mg-Mn-Zn alloy for biomedical application. *Journal of Materials Science-Materials in Medicine*, **20**, 859–867.

Yamamoto, A. and Hiromoto, S. (2009). Effect of inorganic salts, amino acids and proteins on the degradation of pure magnesium in vitro. *Materials Science and Engineering C-Biomimetic and Supramolecular Systems*, **29**, 1559–1568.

Yang, K., Geng, F., Tan, L. L., Jin, X. X. and Yang, J. Y. (2009a). The preparation, cytocompatibility, and in vitro biodegradation study of pure beta-TCP on magnesium. *Journal of Materials Science-Materials in Medicine*, **20**, 1149–1157.

Yang, K., Tan, L., Ren, Y., Zhang, B, Zhang, G. and Ai, H. J. (2009b). Study on biodegradation behavior of AZ31 magnesium alloy. *Rare Metals Letters*, **28**, 26–30.

Yang, K., Tan, L. L., Wang, Q., Geng, F., Xi, X. S. and Qiu, J. H. (2010). Preparation and characterization of Ca-P coating on AZ31 magnesium alloy. *Transactions of Nonferrous Metals Society of China*, **20**, S648–S654.

Yun, Y., Dong, Z. Y., Yang, D. E., Schulz, M. J., Shanov, V. N., Yarmolenko, S., Xu, Z. The in vitro immersion, cytotoxicity assessment and in vivo bone formation study showed that pure magnesium may be more suitable to be used as a biodegradable implant material.

Yun, Y., Dong, Z., Yang, D., Schulz, M. J., Shanov, V. N., Yarmolenko, S., Xu, Z., Kumta, P. and Sfeir, C. (2009). Biodegradable Mg corrosion and osteoblast cell culture studies. *Materials Science and Engineering C-Materials for Biological Applications*, **29**, 1814–1821.

Yun, Y. H., Xue, D. C., Schulz, M. J. and Shanov, V. (2011). Corrosion protection of biodegradable magnesium implants using anodization. *Materials Science and Engineering C-Materials for Biological Applications*, **31**, 215–223.

Zartner, P., Cesnjevar, R., Singer, H. and Weyand, M. (2005). First successful implantation of a biodegradable metal stent into the left pulmonary artery of a preterm baby. *Catheterization and Cardiovascular Interventions*, **66**, 590–594.

Zhang, E., Yang, L., Xu, J. and Chen, H. (2010a). Microstructure, mechanical properties and bio-corrosion properties of Mg-Si(-Ca, Zn) alloy for biomedical application. *Acta Biomaterialia*, **6**, 1756–62.

Zhang, E. L., Xu, L. P., Yu, G. N., Pan, F. and Yang, K. (2009a). In vivo evaluation of biodegradable magnesium alloy bone implant in the first 6 months implantation. *Journal of Biomedical Materials Research Part A*, **90A,** 882–893.

Zhang, X. N., Li, J. N., Song, Y., Zhang, S. X., Zhao, C. L., Zhang, F., Cao, L., Fan, Q. M. and Tang, T. T. (2010b). In vitro responses of human bone marrow stromal cells to a fluoridated hydroxyapatite coated biodegradable Mg-Zn alloy. *Biomaterials*, **31**, 5782–5788.

Zhang, X. N., Song, Y., Zhang, S. X., Li, J. A. and Zhao, C. L. (2010c). Electrodeposition of Ca-P coatings on biodegradable Mg alloy: in vitro biomineralization behavior. *Acta Biomaterialia*, **6**, 1736–1742.

Zhang, X. N., Zhang, S. X., Li, J. A., Song, Y., Zhao, C. L., Xie, C. Y., Zhang, Y., Tao, H. R., He, Y. H., Jiang, Y. and Bian, Y. J. (2009b). In vitro degradation, hemolysis and MC3T3-E1 cell adhesion of biodegradable Mg-Zn alloy. *Materials Science and Engineering C-Materials for Biological Applications*, **29**, 1907–1912.

Zhang, X. N., Zhang, S. X., Zhao, C. L., Li, J. A., Song, Y., Xie, C. Y., Tao, H. R., Zhang, Y., He, Y. H., Jiang, Y. and Bian, Y. J. (2010d). Research on an Mg-Zn alloy as a degradable biomaterial. *Acta Biomaterialia*, **6**, 626–640.

Zheng, Y. F. and Gu, X. N. (2011). Research Activities of Biomedical Magnesium Alloys in China. *Jom*, **63**, 105–108.

Zheng, Y. F., Gu, X. N., Li, X. L., Zhou, W. R. and Cheng, Y. (2010a). Microstructure, biocorrosion and cytotoxicity evaluations of rapid solidified Mg-3Ca alloy ribbons as a biodegradable material. *Biomedical Materials*, **5**, 035013

Zheng, Y. F., Gu, X. N., Zhong, S. P., Xi, T. F., Wang, J. Q. and Wang, W. H. (2010b). Corrosion of, and cellular responses to Mg-Zn-Ca bulk metallic glasses. *Biomaterials*, **31**, 1093–1103.

Zheng, Y. F., Gu, X. N., Zhou, W. R., Cheng, Y., Wei, S. C., Zhong, S. P., Xi, T. F. and Chen, L. J. (2010c). Corrosion fatigue behaviors of two biomedical Mg alloys-AZ91D and WE43-In simulated body fluid. *Acta Biomaterialia*, **6**, 4605–4613.

Zheng, Y. F., Li, Z. J., Gu, X. N. and Lou, S. Q. (2008). The development of binary Mg-Ca alloys for use as biodegradable materials within bone. *Biomaterials*, **29**, 1329–1344.

Index